Nelson MindTap + You = Learning amplified

"I love that everything is interconnected, relevant and that there is a clear learning sequence. I have the tools to create a learning experience that meets the needs of all my students and can easily see how they're progressing."

— Sarah, Secondary School Teacher

NELSON

WAmaths

UNITS 3 + 4

Judith Cumpsty
Dion Alfonsi
Greg Neal

Contributing authors
Sue Garner
George Dimitriadis
Toudi Kouris
Stephen Swift
Neale Woods

mathematics methods 12

Nelson WAmaths Mathematics Methods 12
1st Edition
Judith Cumpsty
Dion Alfonsi
Greg Neal
ISBN 9780170477536

Publisher: Dirk Strasser
Associate product manager: Cindy Huang
Project editor: Tanya Smith
Editor: Karen Chin
Series text design: Alba Design (Rina Gargano)
Series cover design: Nikita Bansal
Series designer: Nikita Bansal
Permissions researcher: Corrina Gilbert
Content developer: Rachael Pictor, Katrina Stavridis
Content manager: Alice Kane
Typeset by: Nikki M Group Pty Ltd

Any URLs contained in this publication were checked for currency during the production process. Note, however, that the publisher cannot vouch for the ongoing currency of URLs.

Acknowledgements
TI-Nspire: Images used with permission by Texas Instruments, Inc.

Casio ClassPad: Images used with permission by Shriro Australia Pty. Ltd.

School Curriculum and Standards Authority: Adapted use of 2016–2021 Mathematics Applications and Mathematics Methods examinations, marking keys and summary examination reports, and ATAR 11 and 12 Mathematics Applications and Mathematics Methods syllabuses. The School Curriculum and Standards Authority does not endorse this publication or product.

Selected VCE Examination questions are copyright Victorian Curriculum and Assessment Authority (VCAA), reproduced by permission. VCE ® is a registered trademark of the VCAA. The VCAA does not endorse this product and makes no warranties regarding the correctness or accuracy of this study resource. To the extent permitted by law, the VCAA excludes all liability for any loss or damage suffered or incurred as a result of accessing, using or relying on the content. Current VCE Study Designs, past VCE exams and related content can be accessed directly at www.vcaa.edu.au

For product information and technology assistance,
in Australia call **1300 790 853**;
in New Zealand call **0800 449 725**

For permission to use material from this text or product, please email
aust.permissions@cengage.com

National Library of Australia Cataloguing-in-Publication Data
A catalogue record for this book is available from the National Library of Australia.

Cengage Learning Australia
Level 5, 80 Dorcas Street
Southbank VIC 3006 Australia

For learning solutions, visit **cengage.com.au**

Printed in China by 1010 Printing International Limited.
1 2 3 4 5 6 7 27 26 25 24 23

Contents

To the teacher

Now there's a better way to WACE maths mastery.

Nelson WAmaths 11–12 is a new WACE mathematics series that is backed by research into the science of learning. The design and structure of the series have been informed by teacher advice and evidence-based pedagogy, with the focus on preparing WACE students for their exams and maximising their learning achievement.

- Using **backwards learning design**, this series has been built by analysing past WACE exam questions and ensuring that all theory and examples are precisely mapped to the SCSA syllabus.
- To reduce the **cognitive load** for learners, explanations are clear and concise, using the technique of **chunking** text with accompanying diagrams and infographics.
- The student book has been designed for **mastery** of the learning content.
- The exercise structure of **Recap**, **Mastery** and **Calculator-free** and **Calculator-assumed** leads students from procedural fluency to **higher-order thinking** using the learning technique of **interleaving**.
- **Calculator-free** and **Calculator-assumed** sections include exam-style questions and past SCSA exam questions.
- The cumulative structure of exercise **Recaps** and chapter-based **Cumulative examinations** is built on the learning and memory techniques of **spacing** and **retrieval**.

About the authors

Judith Cumpsty has taught Mathematics in various Western Australian schools for over thirty years, teaching at many levels, as well as having been Head of Department. She has been involved with projects such as Have Sum Fun and prepared students for OLNA. Judith has been a Team Leader for SCSA WACE marking and a proofreader for examinations published by MAWA.

Dion Alfonsi is Head of Mathematics and a Secondary Mathematics Teacher at Shenton College. In the past, he has had the roles of Years 9 & 10 Mathematics Curriculum Leader and Gifted & Talented/Academic Programs Coordinator. Dion has been a Board Member of MAWA, is a frequent presenter at the MAWA Secondary Conference and a teacher of the MAWA Problem Solving Program.

Greg Neal has taught in regional schools for over 40 years and has co-written several senior textbooks for Cengage Nelson. He has been an examination assessor, presents at conferences and has expertise with CAS technology.

Syllabus grid

Topic	*Nelson WAmaths Mathematics Methods 12* chapter	
Topic 3.1: Further differentiation and applications (20 hours)		
Exponential functions	4	Applying exponential and trigonometric functions
Trigonometric functions	4	Applying exponential and trigonometric functions
Differentiation rules	1	Differentiation
	4	Applying exponential and trigonometric functions
The second derivative and applications of differentiation	2	Applications of differentiation
	7	Calculus of the natural logarithmic function
Topic 3.2: Integrals (20 hours)		
Anti-differentiation	3	Integrals
	4	Applying exponential and trigonometric functions
Definite integrals	3	Integrals
Fundamental theorem	3	Integrals
Applications of integration	3	Integrals
	4	Applying exponential and trigonometric functions
	7	Calculus of the natural logarithmic function
Topic 3.3: Discrete random variables (15 hours)		
General discrete random variables	5	Discrete random variables
Bernoulli distributions	5	Discrete random variables
Binomial distributions	5	Discrete random variables
Topic 4.1: The logarithmic function (18 hours)		
Logarithmic functions	6	Logarithmic functions
Calculus of the natural logarithmic function	6	Logarithmic functions
	7	Calculus of the natural logarithmic function
Topic 4.2: Continuous random variables and the normal distribution (15 hours)		
General continuous random variables	8	Continuous random variables and the normal distribution
Normal distributions	8	Continuous random variables and the normal distribution
Topic 4.3: Interval estimates for proportions (22 hours)		
Random sampling	9	Interval estimates for proportions
Sample proportions	9	Interval estimates for proportions
Confidence intervals for proportions	9	Interval estimates for proportions

About this book

In each chapter

Syllabus coverage and extracts are shown at the start of the chapter along with a listing of **Nelson MindTap chapter resources**.

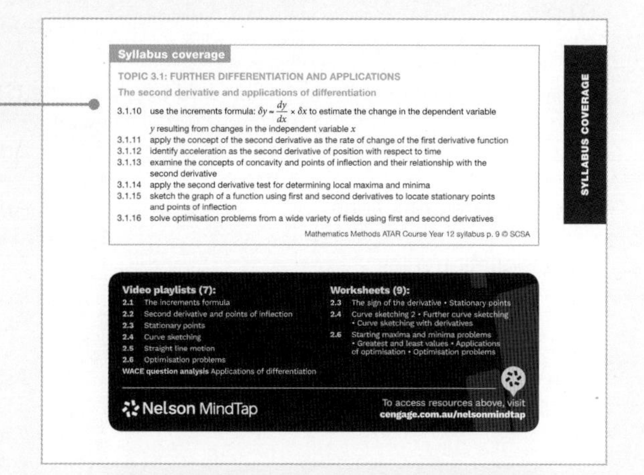

Syllabus coverage

TOPIC 3.1: FURTHER DIFFERENTIATION AND APPLICATIONS

The second derivative and applications of differentiation

3.1.10 use the increments formula: $\delta y \approx \frac{dy}{dx} \times \delta x$ to estimate the change in the dependent variable y resulting from changes in the independent variable x

3.1.11 apply the concept of the second derivative as the rate of change of the first derivative function

3.1.12 identify acceleration as the second derivative of position with respect to time

3.1.13 examine the concepts of concavity and points of inflection and their relationship with the second derivative

3.1.14 apply the second derivative test for determining local maxima and minima

3.1.15 sketch the graph of a function using first and second derivatives to locate stationary points and points of inflection

3.1.16 solve optimisation problems from a wide variety of fields using first and second derivatives

Mathematics Methods ATAR Course Year 12 syllabus p. 9 © SCSA

SYLLABUS COVERAGE

Video playlists (7):
2.1 The increments formula
2.2 Second derivative and points of inflection
2.3 Stationary points
2.4 Curve sketching
2.5 Straight line motion
2.6 Optimisation problems
WACE question analysis Applications of differentiation

Worksheets (9):
2.3 The sign of the derivative • Stationary points
2.4 Curve sketching 2 • Further curve sketching • Curve sketching with derivatives
2.6 Starting maxima and minima problems • Greatest and least values • Applications of optimisation • Optimisation problems

Nelson MindTap

To access resources above, visit cengage.com.au/nelsonmindtap

Important words and phrases are printed in blue and listed in the **Glossary and index** at the back of the book.

Important facts and formulas are highlighted in a shaded box.

Worked examples are explained clearly step-by-step, with the mathematical working shown on the right-hand side.

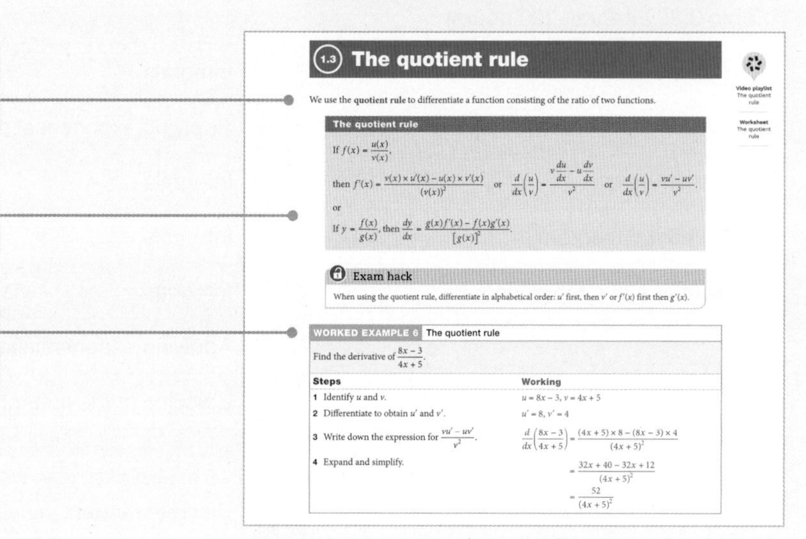

1.3 The quotient rule

We use the quotient rule to differentiate a function consisting of the ratio of two functions.

The quotient rule

If $f(x) = \frac{u(x)}{v(x)}$,

then $f'(x) = \frac{v(x) \times u'(x) - u(x) \times v'(x)}{(v(x))^2}$ or $\frac{d}{dx}\left(\frac{u}{v}\right) = \frac{v\frac{du}{dx} - u\frac{dv}{dx}}{v^2}$ or $\frac{d}{dx}\left(\frac{u}{v}\right) = \frac{vu' - uv'}{v^2}$.

or

If $y = \frac{f(x)}{g(x)}$, then $\frac{dy}{dx} = \frac{g(x)f'(x) - f(x)g'(x)}{[g(x)]^2}$.

Exam hack

When using the quotient rule, differentiate in alphabetical order: u' first, then v' or $f'(x)$ first then $g'(x)$.

Video playlist The quotient rule

Worksheet The quotient rule

WORKED EXAMPLE 6 The quotient rule

Find the derivative of $\frac{8x-3}{4x+5}$.

Steps	Working
1 Identify u and v.	$u = 8x - 3$, $v = 4x + 5$
2 Differentiate to obtain u' and v'.	$u' = 8$, $v' = 4$
3 Write down the expression for $\frac{vu' - uv'}{v^2}$.	$\frac{d}{dx}\left(\frac{8x-3}{4x+5}\right) = \frac{(4x+5) \times 8 - (8x-3) \times 4}{(4x+5)^2}$
4 Expand and simplify.	$= \frac{32x + 40 - 32x + 12}{(4x+5)^2}$
	$= \frac{52}{(4x+5)^2}$

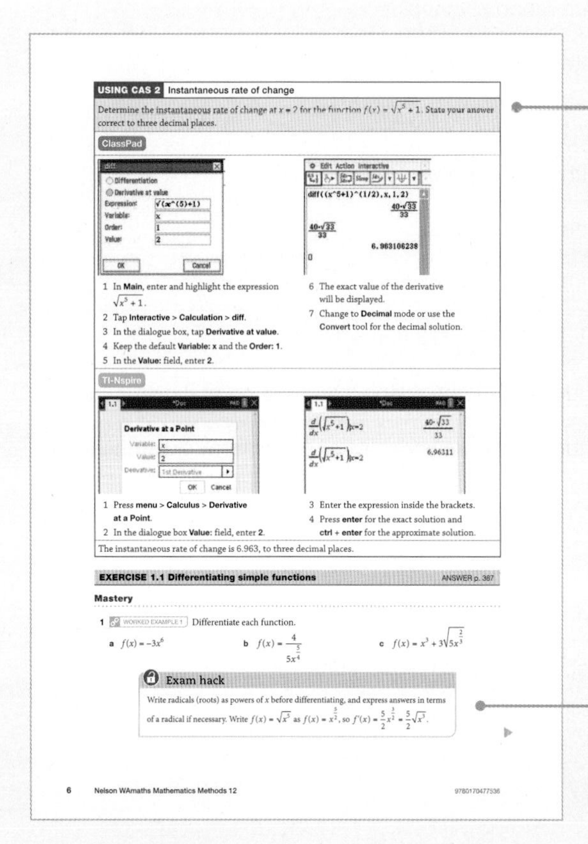

USING CAS 2 Instantaneous rate of change

Determine the instantaneous rate of change at $x = 2$ for the function $f(x) = \sqrt{x^5 + 1}$. State your answer correct to three decimal places.

ClassPad

1 In **Main**, enter and highlight the expression $\sqrt{x^5 + 1}$.
2 Tap **Interactive > Calculation > diff**.
3 In the dialogue box, tap **Derivative at value**.
4 Keep the default **Variable:** x and the **Order: 1**.
5 In the **Value:** field, enter **2**.
6 The exact value of the derivative will be displayed.
7 Change to **Decimal** mode or use the **Convert** tool for the decimal solution.

TI-Nspire

1 Press **menu > Calculus > Derivative at a Point**.
2 In the dialogue box **Value:** field, enter **2**.
3 Enter the expression inside the brackets.
4 Press **enter** for the exact solution and **ctrl + enter** for the approximate solution.

The instantaneous rate of change is 6.963, to three decimal places.

EXERCISE 1.1 Differentiating simple functions ANSWER p. 387

Mastery

1 WORKED EXAMPLE 1 Differentiate each function.

a $f(x) = -3x^6$ b $f(x) = \frac{4}{5x^{\frac{3}{4}}}$ c $f(x) = x^3 + 3\sqrt{5x^{\frac{1}{3}}}$

Exam hack

Write radicals (roots) as powers of x before differentiating, and express answers in terms of a radical if necessary. Write $f(x) = \sqrt{x^5}$ as $f(x) = x^{\frac{5}{2}}$, so $f'(x) = \frac{5}{2}x^{\frac{3}{2}} = \frac{5}{2}\sqrt{x^3}$.

6 Nelson WAmaths Mathematics Methods 12 9780170477536

Using CAS provides clear instructions for Casio ClassPad and TI-Nspire calculators.

Exam hacks highlight valuable exam hints and common student errors.

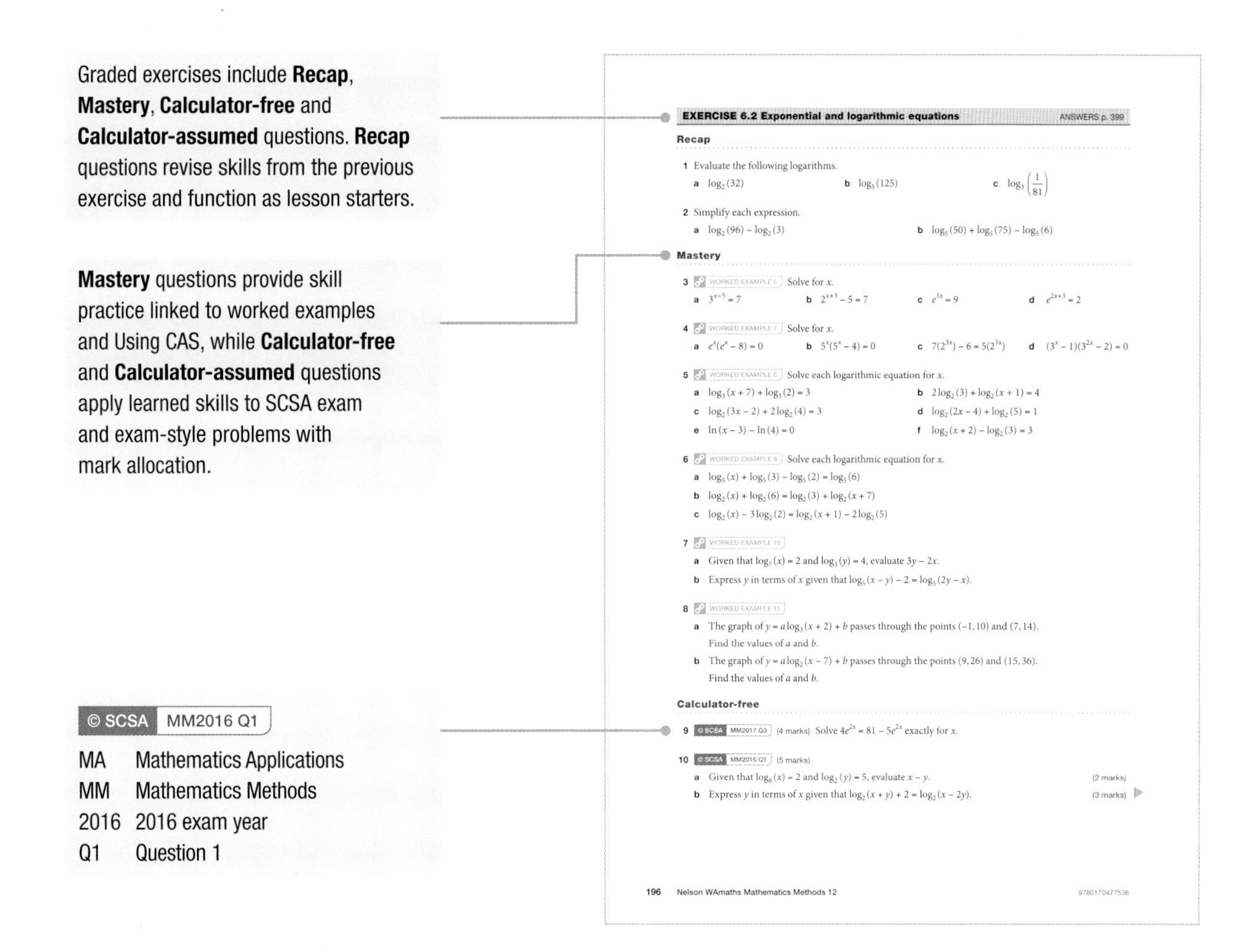

Graded exercises include **Recap**, **Mastery**, **Calculator-free** and **Calculator-assumed** questions. **Recap** questions revise skills from the previous exercise and function as lesson starters.

Mastery questions provide skill practice linked to worked examples and Using CAS, while **Calculator-free** and **Calculator-assumed** questions apply learned skills to SCSA exam and exam-style problems with mark allocation.

© SCSA MM2016 Q1

MA Mathematics Applications
MM Mathematics Methods
2016 2016 exam year
Q1 Question 1

EXERCISE 6.2 Exponential and logarithmic equations ANSWERS p. 399

Recap

1 Evaluate the following logarithms.
a $\log_2(32)$ b $\log_5(125)$ c $\log_3\left(\frac{1}{81}\right)$

2 Simplify each expression.
a $\log_2(96) - \log_2(3)$ b $\log_5(50) + \log_5(75) - \log_5(6)$

Mastery

3 WORKED EXAMPLE 6 Solve for x.
a $3^{x-5} = 7$ b $2^{x+3} - 5 = 7$ c $e^{3x} = 9$ d $e^{2x+3} = 2$

4 WORKED EXAMPLE 7 Solve for x.
a $e^x(e^x - 8) = 0$ b $5^x(5^x - 4) = 0$ c $7(2^{3x}) - 6 = 5(2^{3x})$ d $(3^x - 1)(3^{2x} - 2) = 0$

5 WORKED EXAMPLE 8 Solve each logarithmic equation for x.
a $\log_3(x + 7) + \log_3(2) = 3$ b $2\log_2(3) + \log_2(x + 1) = 4$
c $\log_2(3x - 2) + 2\log_2(4) = 3$ d $\log_2(2x - 4) + \log_2(5) = 1$
e $\ln(x - 3) - \ln(4) = 0$ f $\log_2(x + 2) - \log_2(3) = 3$

6 WORKED EXAMPLE 9 Solve each logarithmic equation for x.
a $\log_5(x) + \log_5(3) - \log_5(2) = \log_5(6)$
b $\log_2(x) + \log_2(6) = \log_2(3) + \log_2(x + 7)$
c $\log_2(x) - 3\log_2(2) = \log_2(x + 1) - 2\log_2(5)$

7 WORKED EXAMPLE 10
a Given that $\log_7(x) = 2$ and $\log_3(y) = 4$, evaluate $3y - 2x$.
b Express y in terms of x given that $\log_5(x - y) - 2 = \log_5(2y - x)$.

8 WORKED EXAMPLE 11
a The graph of $y = a\log_3(x + 2) + b$ passes through the points $(-1, 10)$ and $(7, 14)$. Find the values of a and b.
b The graph of $y = a\log_2(x - 7) + b$ passes through the points $(9, 26)$ and $(15, 36)$. Find the values of a and b.

Calculator-free

9 © SCSA MM2017 Q3 (4 marks) Solve $4e^{2x} = 81 - 5e^{2x}$ exactly for x.

10 © SCSA MM2016 Q1 (5 marks)
a Given that $\log_8(x) = 2$ and $\log_2(y) = 5$, evaluate $x - y$. (2 marks)
b Express y in terms of x given that $\log_2(x + y) + 2 = \log_2(x - 2y)$. (3 marks)

196 Nelson WAmaths Mathematics Methods 12 9780170477536

At the end of each chapter

WACE question analysis leads students through a past WACE exam question that exemplifies the chapter, discussing how to approach the question, providing advice on interpreting the question, common student errors, and a full worked solution with a marking key.

WACE QUESTION ANALYSIS

© SCSA MM2019 Q17 Calculator-assumed (15 marks)

A microbiologist is studying the effect of temperature on the growth of a certain type of bacteria under controlled laboratory conditions. A population of bacteria is incubated at a temperature of 30°C and the size of the population measured at hourly intervals for six hours. The logarithm of the population size appears to lie on a straight line when plotted against time (measured in hours) and the line of best fit shown on the axes below.

$\log_{10}(P)$
6
5
4
3
2
1
0 1 2 3 4 5 6 t (hours)

a i On the basis of the graph above, what is the size of the bacteria population after two hours? (2 marks)
ii The equation of the line can be written in the form $\log_{10}(P) = At + B$. Use the graph to determine the values of A and B. (2 marks)

Another population of the same bacteria is cultured at 40°C. The size of the population, P, after t hours satisfies the equation

$$\log_{10}(P) = \frac{1}{3}t + 2.$$

b i Express the above equation in the form $P = A(10)^{Bt}$. (3 marks)
ii Determine the size of the population after exactly four hours to the nearest whole number. (1 mark)
iii Express the above equation in the form $t = C\log_{10}\left(\frac{P}{D}\right)$. (3 marks)
iv How many minutes does it take for the population to reach a size of 5000? Give your answer to the nearest minute. (2 marks)

c With reference to parts **a** and **b**, describe the effect of temperature on the population growth of this type of bacteria. (2 marks)

Reading the question

- The graph of the population is a log scale, where the log to base 10 of the population is on the vertical axis.
- Highlight the type, accuracy and units of the answer required in each part.
- Take note of the number of marks allocated to each part of the question. This will give an indication of the amount of working required.

Thinking about the question

- This question requires a knowledge of graphs using logarithmic scales.
- You will also need to be able to transform an equation from logarithmic form to exponential form.
- You will need to be able to find the equation of a straight line using a gradient and y-intercept. Remember, in this case, the subject of the linear equation is $\log_{10}(P)$.

6.4

Video WACE question analysis: Logarithmic functions

9780170477536 Chapter 6 Logarithmic functions 207

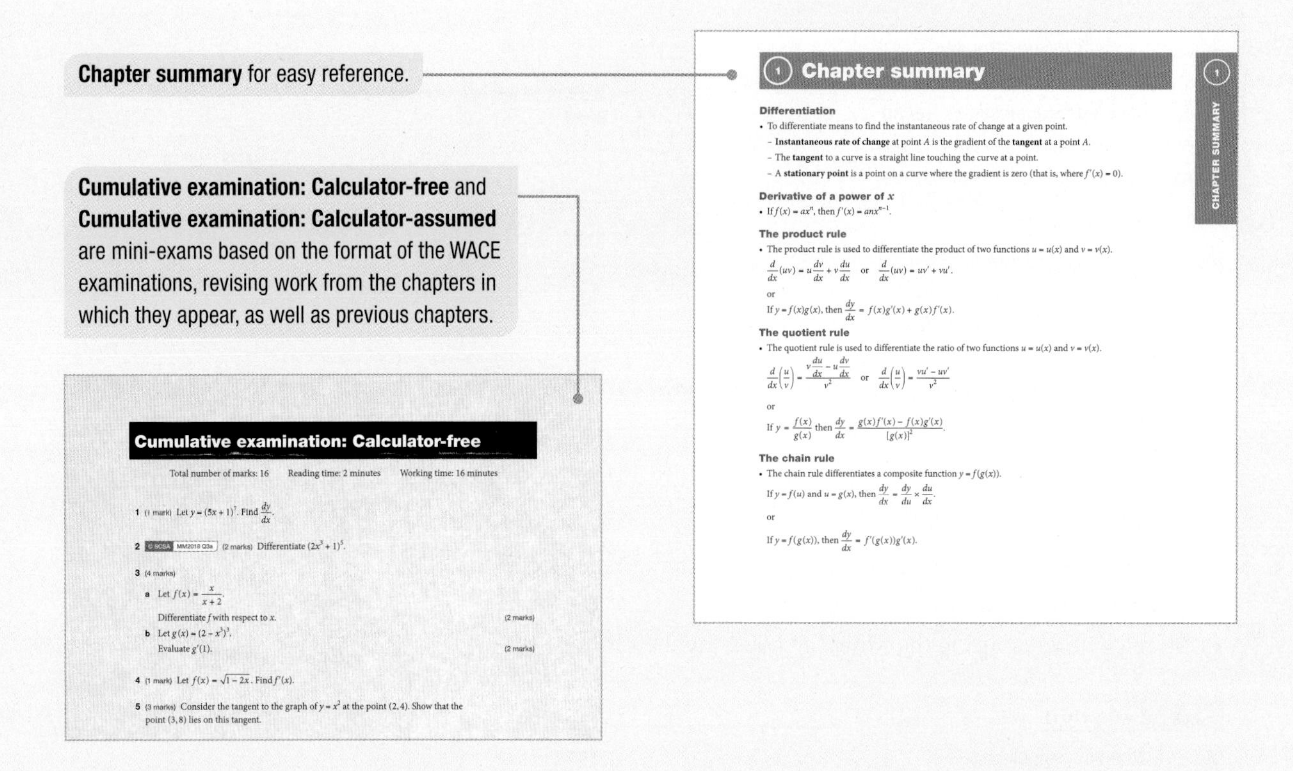

Chapter summary for easy reference.

Cumulative examination: Calculator-free and **Cumulative examination: Calculator-assumed** are mini-exams based on the format of the WACE examinations, revising work from the chapters in which they appear, as well as previous chapters.

1 Chapter summary

Differentiation
- To differentiate means to find the instantaneous rate of change at a given point.
  - **Instantaneous rate of change** at point A is the gradient of the **tangent** at a point A.
  - The **tangent** to a curve is a straight line touching the curve at a point.
  - A **stationary point** is a point on a curve where the gradient is zero (that is, where $f'(x) = 0$).

Derivative of a power of x
- If $f(x) = ax^n$, then $f'(x) = anx^{n-1}$.

The product rule
- The product rule is used to differentiate the product of two functions $u = u(x)$ and $v = v(x)$.

$\frac{d}{dx}(uv) = u\frac{dv}{dx} + v\frac{du}{dx}$ or $\frac{d}{dx}(uv) = uv' + vu'$.

or

If $y = f(x)g(x)$, then $\frac{dy}{dx} = f(x)g'(x) + g(x)f'(x)$.

The quotient rule
- The quotient rule is used to differentiate the ratio of two functions $u = u(x)$ and $v = v(x)$.

$\frac{d}{dx}\left(\frac{u}{v}\right) = \frac{v\frac{du}{dx} - u\frac{dv}{dx}}{v^2}$ or $\frac{d}{dx}\left(\frac{u}{v}\right) = \frac{vu' - uv'}{v^2}$

or

If $y = \frac{f(x)}{g(x)}$ then $\frac{dy}{dx} = \frac{g(x)f'(x) - f(x)g'(x)}{[g(x)]^2}$.

The chain rule
- The chain rule differentiates a composite function $y = f(g(x))$.

If $y = f(u)$ and $u = g(x)$, then $\frac{dy}{dx} = \frac{dy}{du} \times \frac{du}{dx}$.

or

If $y = f(g(x))$, then $\frac{dy}{dx} = f'(g(x))g'(x)$.

CHAPTER SUMMARY

Cumulative examination: Calculator-free

Total number of marks: 16 Reading time: 2 minutes Working time: 16 minutes

1 (1 mark) Let $y = (5x + 1)^7$. Find $\frac{dy}{dx}$.

2 © SCSA MM2018 Q3a (2 marks) Differentiate $(2x^3 + 1)^5$.

3 (4 marks)
a Let $f(x) = \frac{x}{x+2}$.
Differentiate f with respect to x. (2 marks)
b Let $g(x) = (2 - x^3)^3$.
Evaluate $g'(1)$. (2 marks)

4 (1 mark) Let $f(x) = \sqrt{1 - 2x}$. Find $f'(x)$.

5 (3 marks) Consider the tangent to the graph of $y = x^2$ at the point $(2, 4)$. Show that the point $(3, 8)$ lies on this tangent.

At the end of the book

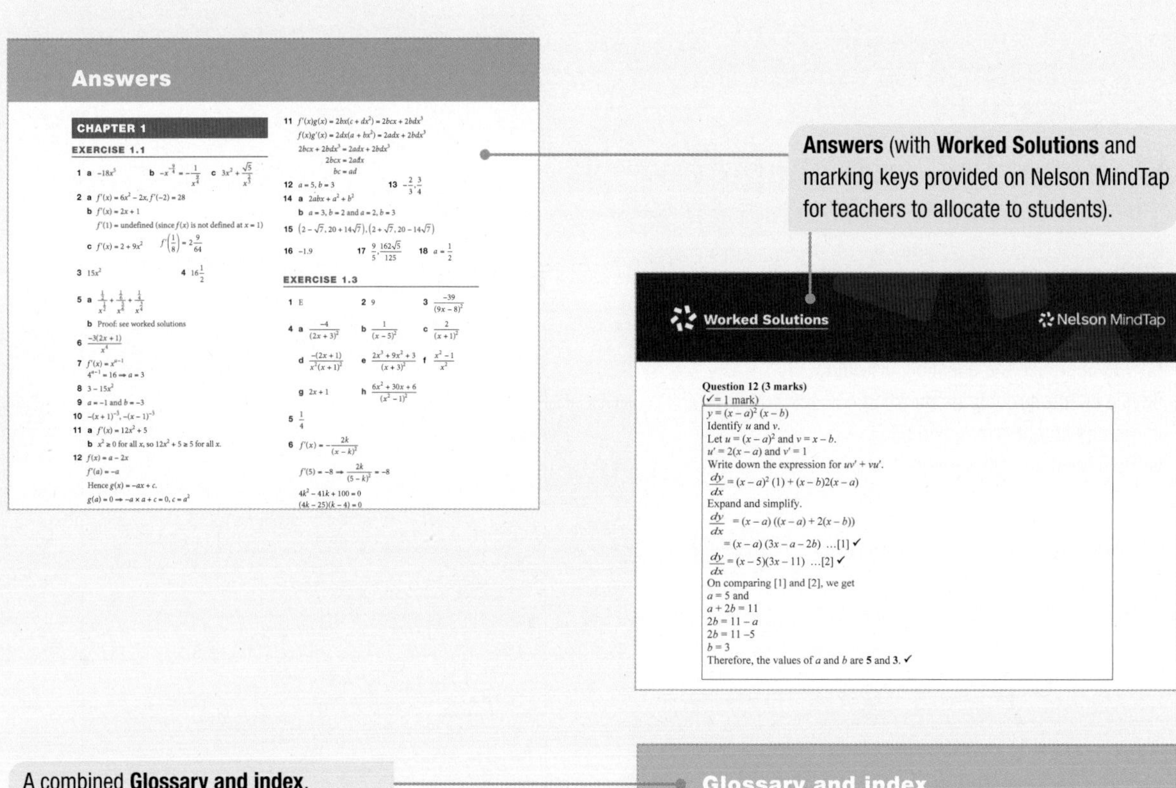

Answers

CHAPTER 1

EXERCISE 1.1

Answers (with **Worked Solutions** and marking keys provided on Nelson MindTap for teachers to allocate to students).

Worked Solutions Nelson MindTap

Question 12 (3 marks)
(✓= 1 mark)
$y = (x - a)^2(x - b)$
Identify u and v.
Let $u = (x - a)^2$ and $v = x - b$.
$u' = 2(x - a)$ and $v' = 1$
Write down the expression for $uv' + vu'$.
$\frac{dy}{dx} = (x - a)^2(1) + (x - b)2(x - a)$
Expand and simplify.
$\frac{dy}{dx} = (x - a)((x - a) + 2(x - b))$
$= (x - a)(3x - a - 2b)$ ...[1] ✓
$\frac{dy}{dx} = (x - 5)(3x - 11)$ ...[2] ✓
On comparing [1] and [2], we get
$a = 5$ and
$a + 2b = 11$
$2b = 11 - a$
$2b = 11 - 5$
$b = 3$
Therefore, the values of a and b are **5** and **3**. ✓

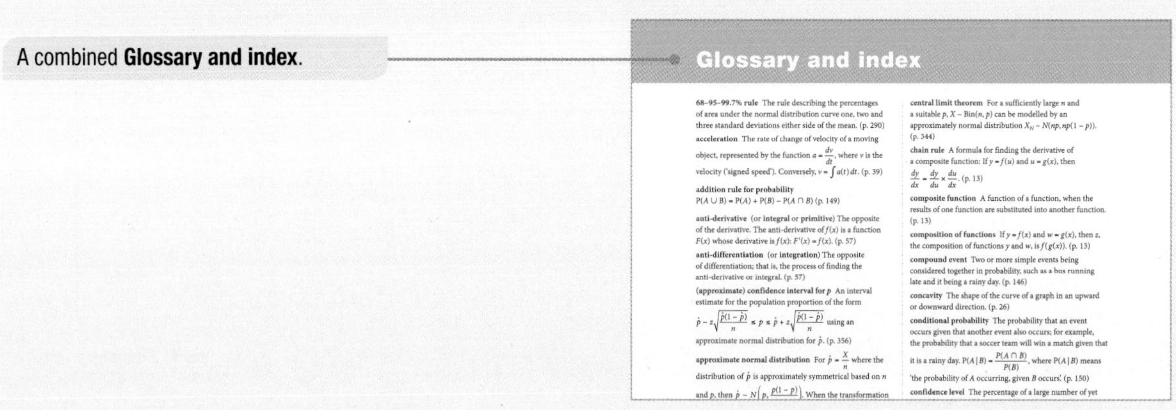

A combined **Glossary and index**.

Glossary and index

68–95–99.7% rule The rule describing the percentages of area under the normal distribution curve one, two and three standard deviations either side of the mean. (p. 290)

acceleration The rate of change of velocity of a moving object, represented by the function $a = \frac{dv}{dt}$, where v is the velocity ('signed speed'). Conversely, $v = \int a(t)\,dt$. (p. 39)

addition rule for probability $P(A \cup B) = P(A) + P(B) - P(A \cap B)$ (p. 149)

anti-derivative (or **integral** or **primitive**) The opposite of the derivative. The anti-derivative of $f(x)$ is a function $F(x)$ whose derivative is $f(x)$: $F'(x) = f(x)$. (p. 57)

anti-differentiation (or **integration**) The opposite of differentiation; that is, the process of finding the anti-derivative or integral. (p. 57)

central limit theorem For a sufficiently large n and a suitable p, $X \sim \text{Bin}(n, p)$ can be modelled by an approximately normal distribution $X_N \sim N(np, np(1 - p))$. (p. 344)

chain rule A formula for finding the derivative of a composite function: If $y = f(u)$ and $u = g(x)$, then $\frac{dy}{dx} = \frac{dy}{du} \times \frac{du}{dx}$. (p. 13)

composite function A function of a function, when the results of one function are substituted into another function. (p. 13)

composition of functions If $y = f(x)$ and $w = g(x)$, then z, the composition of functions y and w, is $f(g(x))$. (p. 13)

compound event Two or more simple events being considered together in probability, such as a bus running late and it being a rainy day. (p. 146)

concavity The shape of the curve of a graph in an upward or downward direction. (p. 26)

conditional probability The probability that an event occurs given that another event also occurs; for example, the probability that a soccer team will win a match given that it is a rainy day. $P(A \mid B) = \frac{P(A \cap B)}{P(B)}$, where $P(A \mid B)$ means 'the probability of A occurring, given B occurs'. (p. 150)

Nelson MindTap

Nelson MindTap is an online learning space that provides students with tailored learning experiences. Access tools and content that make learning simpler yet smarter to help you achieve WACE maths mastery.

Nelson MindTap includes an eText with integrated interactives and online assessment.

Margin links in the student book signpost multimedia student resources found on MindTap.

Nelson MindTap for students:

- **Watch** video tutorials featuring expert teacher advice to unpack new concepts and develop your understanding.
- **Revise** using learning checks, worksheets and skillsheets to practise your skills and build your confidence.
- **Navigate** your own path, accessing the content, analytics and support as you need it.

Video playlists

Skillsheets

Worksheets

Nelson MindTap for teachers*:

- Tailor content to different learning needs – assign directly to the student, or the whole class.
- Monitor progress using the MindTap assessment tools.
- Integrate content and assessment directly within your school's LMS for ease of access.
- Access topic tests, teaching plans and worked solutions to each exercise set.

*Complimentary access to these resources is only available to teachers who use this book as part of a class set, book hire or booklist. Contact your Cengage Education Consultant for information about access codes and conditions.

Nelson WAmaths 11–12 series

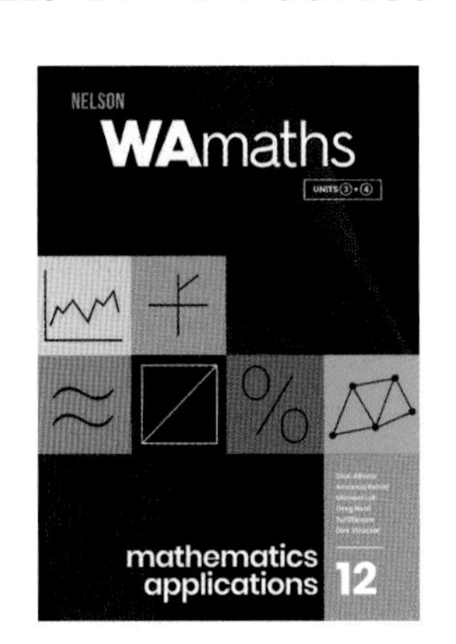

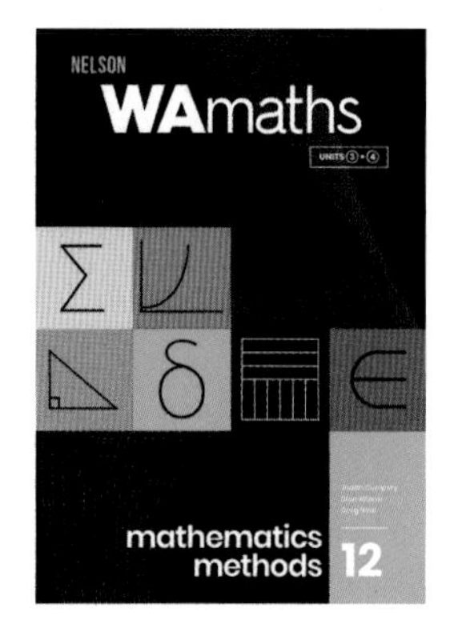

Additional credits

*modified

Chapter 1
Exercise 1.1
Question 11 ©VCAA MM2011 2BQ3a
Exercise 1.3
Question 7 © VCAA MM2012 2BQ2bi
Question 8 © VCAA MM2019N 1Q2
Exercise 1.4
Question 6 © VCAA MM2008 1Q1a
Question 7 © VCAA MM2012 1Q1a
Question 8 © VCAA MM2011 1Q1a
Question 9 © VCAA MM2018 1Q1a
Question 12 © VCAA MM2009 2AQ15*
Question 13 © VCAA MM2007 non-CAS 2AQ16*
Exercise 1.5
Question 4 © VCAA MM2004 non-CAS 2Q1a
Question 5 © VCAA MM2005 1IQ16
Cumulative examination: Calculator-free
Question 1 © VCAA MM2015 1Q1a
Question 3 © VCAA MM2017 1Q1
Question 4 © VCAA MM2016 1Q2a
Question 5 © VCAA MM2015 2AQ4*
Question 6 © VCAA MM2015 2BQ1a

Chapter 2
Exercise 2.3
Question 8 © VCAA MM2009 2AQ2
Question 9 © VCAA MM2007 2AQ12
Question 10 © VCAA MM2009 2AQ21
Question 11 © VCAA MM2014 1Q5ab
Question 12 © VCAA MM2006S 2AQ15
Question 16 © VCAA MM2013 2AQ21
Question 17 © VCAA MM2011 2BQ3
Exercise 2.4
Question 1 © VCAA MM2010 2AQ17
Exercise 2.5
Question 1 © VCAA MM2018 2AQ5*
Exercise 2.6
Question 9 © VCAA MM2008 1Q9
Question 10 © VCAA MM2014 2AQ15
Question 11 © VCAA MM2019 2AQ6
Question 12 © VCAA MM2010 1Q11
Question 13 © VCAA MM2006 1Q9
Question 14 © VCAA MM2007 1Q12
Question 15 © VCAA MM2020 2AQ16
Question 16 © VCAA MM2013 2BQ3
Cumulative examination: Calculator-free
Question 1 © VCAA MM2018 1Q9bc*

Chapter 3
Exercise 3.1
Question 10 © VCAA MM2019 2AQ5
Question 14 © VCAA MM2019N 1Q2
Question 15 © VCAA MM2012 1Q2
Question 16 © VCAA MM2020 2AQ3
Question 17 © VCAA MM2006 2AQ16
Exercise 3.3
Question 11 © VCAA MM2009 1Q2b
Question 14 © VCAA MM2017N 1Q2b
Question 15 © VCAA MM2018 2AQ12*
Question 16 © VCAA MM2008 2AQ4
Question 17 © VCAA MM2018 2AQ8
Question 18 © VCAA MM2018N 2AQ17
Question 19 © VCAA MM2011 2AQ19
Exercise 3.4
Question 2 © VCAA MM2015 2AQ15
Question 9 © VCAA MM2014 1Q5
Question 11 © VCAA MM2017 1Q9
Question 13 © VCAA MM2006 2AQ15
Question 14 © VCAA MM2011 2AQ20
Exercise 3.5
Question 13 © VCAA MM2013 2AQ16
Question 14 © VCAA MM2011 1Q9
Cumulative examination: Calculator-free
Question 1 © VCAA 2015 2AQ19
Question 2 © VCAA 2017N 1Q2b*
Question 3 © VCAA 2017N 1Q2c*
Cumulative examination: Calculator-assumed
Question 1 © VCAA 2018 2AQ5
Question 2 © VCAA 2018 2AQ16
Question 6 © VCAA 2018 2BQ5

Chapter 4
Exercise 4.2
Question 18a © VCAA MM2010 1Q1a
Question 18b © VCAA MM2013 1Q1b
Question 18c © VCAA MM2020 1Q1b
Question 20 © VCAA MM2009 1Q8
Exercise 4.3
Question 13 © VCAA MM2007 1Q9
Question 14 © VCAA MM2013 1Q10
Exercise 4.4
Question 12a © VCAA MM2012 1Q9a
Question 12b © VCAA MM2012 1Q1b
Question 13 © VCAA MM2011 1Q1b
Question 19 © VCAA MM2013 2BQ1e
Exercise 4.5
Question 9a © VCAA MM2010 1Q2a
Question 9b © VCAA MM2014 1Q7
Question 10 © VCAA MM2007 1Q7
Cumulative examination: Calculator-assumed
Question 1 © VCAA MM2019 1Q5

Chapter 5
Exercise 5.1
Question 7 © VCAA MM2018 1Q6
Question 8 © VCAA MM2019 1Q3
Question 9 © VCAA MM2014 1Q9
Question 10 © VCAA MM2016 1Q7
Question 11 © VCAA MM2021 1Q6
Question 13 © VCAA MM2017 2AQ3
Question 14 © VCAA MM2012 2AQ12
Exercise 5.2
Question 9 © VCAA MM2013 1Q7a
Question 10 © VCAA MM2010 1Q8
Question 11 © VCAA MM2008 1Q7b
Question 12 © VCAA MM2012 1Q4bc
Exercise 5.3
Question 9 © VCAA MM2012 1Q4a
Question 10 © VCAA MM2013 1Q7b
Question 11 © VCAA MM2017N 1Q5
Question 14 © VCAA MM2010 2BQ2abd
Exercise 5.4
Question 10 © VCAA MM2016 1Q4
Question 11 © VCAA MM2011 1Q7
Question 12 © VCAA MM2007 1Q5
Question 14 © VCAA MM2013 2BQ2a
Question 15 © VCAA MM2016 2BQ3ab
Question 16 © VCAA MM2015 2BQ3d
Question 18 © VCAA MM2010 2BQ2e
Cumulative examination: Calculator-free
Question 2 © VCAA MM2018 1Q7a
Question 4 © VCAA MM2020 1Q2
Question 5 © VCAA MM2020 1Q5
Cumulative examination: Calculator-assumed
Question 4 © VCAA MM2018 2BQ4bcig

Chapter 6
Exercise 6.2
Question 11a © VCAA MM2013 1Q5
Question 11b © VCAA MM2007 1Q2a
Question 11c © VCAA MM2015 1Q7

Chapter 7
Exercise 7.1
Question 9 © VCAA MM2013 1Q1a
Question 11 © VCAA MM2015 1Q1b
Question 12 © VCAA MM2010 1Q1b
Exercise 7.3
Question 14 © VCAA MM2017 1Q2
Question 15a © VCAA MM2014 1Q2
Question 15b © VCAA MM2010 1Q2b
Exercise 7.4
Question 7 © VCAA MM2010 1Q9
Question 14 © VCAA MM2008 2BQ2
Cumulative examination: Calculator-free
Question 6 © VCAA MM2018 1Q2
Question 7 © VCAA MM2018 1Q8
Cumulative examination: Calculator-assumed
Question 1 © VCAA MM2010 2BQ4
Question 2 © VCAA MM2018 2AQ8

Chapter 8
Exercise 8.1
Question 11 © VCAA MM2012 1Q8b*
Question 12 © VCAA MM2005 1Q2*
Question 15 © VCAA MM2006 2BQ2cd
Question 16 © VCAA MM2008 1Q1c i-ii
Question 17 © VCAA MM2016 2AQ18 and © VCAA MM2017 2AQ19*
Question 18 © VCAA MM2011 2BQ2a ii*
Exercise 8.2
Using CAS 3 © VCAA MM2009 2AQ11*
Question 2 © VCAA MM2014 2AQ14
Question 5 © VCAA MM2008 1Q4*
Question 6 © VCAA MM2006 1Q6*
Question 10 © VCAA MM2014 2AQ16*
Question 11 © VCAA MM2021 1Q7
Question 12 © VCAA MM2014 1Q8*
Question 13 © VCAA MM2010 1Q7*
Question 17 © VCAA MM2011 2AQ6*
Question 18 © VCAA MM2013 2BQ2c
Question 19 © VCAA MM2018N 2BQ2gh*
Exercise 8.3
Worked example 16 © VCAA MM2008 2AQ11*
Question 1 © VCAA MM2005 2BQ2bi*
Question 2 © VCAA MM2016 2BQ3h*
Question 6 © VCAA MM2011 1Q5*
Question 8 © VCAA MM2020 2AQ11*
Question 11 © VCAA MM2017 2BQ3a-d*
Question 12 © VCAA MM2007 2BQ5a-c*
Question 14 © VCAA MM2002 1IQ27*
Question 15 © VCAA MM2019 2AQ18*
Exercise 8.4
Question 11 © VCAA MM2010 1Q5
Question 12 © VCAA MM2015 1Q6*
Question 13 © VCAA MM2006 1Q5*
Question 14 © VCAA MM2012 1Q8a
Question 17 © VCAA MM2019 2AQ14*
Question 22 © VCAA MM2012 2BQ3d*
Cumulative examination: Calculator-free
Question 6 © VCAA MM2013 1Q8*
Cumulative examination: Calculator-assumed
Question 1 © VCAA MM2012 2BQ1

Chapter 9
Exercise 9.2
Worked example 7 © VCAA MM2017 1Q4
Question 6 © VCAA MM2018N 2AQ19*
Question 15 © VCAA MM2017 2AQ16*
Exercise 9.3
Question 13 © VCAA MM2019N 1Q6b*
Question 16 © VCAA MM2016 2BQ3a-g*
Question 17 © VCAA MM2016S 2AQ14*
Question 24 © VCAA MM2017N 2BQ3ghi*
Question 25 © VCAA MM2018N 2BQ2b
Question 26 © VCAA MM2019N 2BQ3d
Cumulative examination: Calculator-assumed
Question 2 © VCAA MM2015 2BQ1
Question 6 © VCAA MM2009 2BQ2

DIFFERENTIATION

Syllabus coverage

TOPIC 3.1: FURTHER DIFFERENTIATION AND APPLICATIONS

Differentiation rules

3.1.7 examine and use the product and quotient rules

3.1.8 examine the notion of composition of functions and use the chain rule for determining the derivatives of composite functions

Video playlists (6):

1.1 Differentiating simple functions
1.2 The product rule
1.3 The quotient rule
1.4 The chain rule
1.5 Combining the rules
WACE question analysis Differentiation

Worksheets (8):

1.1 Derivatives of polynomials • Rates of change 2 • Instantaneous rates of change • Slopes of curves
1.2 The product rule
1.3 The quotient rule
1.4 The chain rule • Mixed differentiation problems

Nelson MindTap

To access resources above, visit **cengage.com.au/nelsonmindtap**

1.1 Differentiating simple functions

In Year 11 we learnt that the derivatives of $f(x) = x$, $f(x) = x^2$, $f(x) = x^3$ and $f(x) = x^4$ can be found using **differentiation** by first principles, with the results shown in the table.

$f(x)$	$f'(x)$	$\frac{d}{dx}$
x	$f'(x) = 1$	$\frac{d}{dx}(x) = 1$
x^2	$f'(x) = 2x$	$\frac{d}{dx}(x^2) = 2x$
x^3	$f'(x) = 3x^2$	$\frac{d}{dx}(x^3) = 3x^2$
x^4	$f'(x) = 4x^3$	$\frac{d}{dx}(x^4) = 4x^3$

We also discovered the pattern in finding the **derivative** of $f(x) = x^n$, where $n = 1, 2, 3 \ldots$ A similar pattern can be found for the derivative of $f(x) = ax^n$.

To differentiate this function, multiply the coefficient a by the power and subtract 1 from the power.

This rule applies to all types of **powers of x** (integers, fractions, surds and irrationals).

If $y = f(x)$ then $f'(a)$ is the **instantaneous rate of change** of the function f at a given point $x = a$.

The **tangent** to a curve is a straight line touching the curve at a point.

A **stationary point** is a point on a curve where the gradient is zero (that is, where $f'(x) = 0$).

Video playlist
Differentiating simple functions

Worksheets
Derivatives of polynomials
Rates of change 2
Instantaneous rates of change
Slopes of curves

The derivative of ax^n

If $f(x) = ax^n$, then $f'(x) = anx^{n-1}$.

To differentiate a function with more than one term, use term-by-term differentiation.

For example, to differentiate $f(x) = 2x^2 + 3x - 1$, differentiate each term separately:

$$\frac{d}{dx}(2x^2) + \frac{d}{dx}(3x) + \frac{d}{dx}(-1), \text{ which gives } f'(x) = 4x + 3.$$

y' means $\frac{dy}{dx}$.

Exam hack

To help when differentiating, make sure that each term is of the form ax^n.

For example, write $\frac{9}{x^2}$ as $9x^{-2}$ and, hence, $\frac{d}{dx}(9x^{-2}) = -18x^{-3} = -\frac{18}{x^3}$.

WORKED EXAMPLE 1	The derivative of ax^n

Differentiate each function.

a $y = 5x^7$ **b** $f(x) = \dfrac{9}{10x^{\frac{2}{3}}}$ **c** $f(x) = x^2 + 2\sqrt{3x^{\frac{7}{2}}}$

Steps	Working
a **1** Write in the form $y = ax^n$.	$y = 5x^7$
2 Differentiate using $y' = a \times nx^{n-1}$.	$y' = 5 \times 7x^{7-1}$
3 Simplify.	$y' = 35x^6$
b **1** Write in the form $f(x) = ax^n$.	$f(x) = \dfrac{9}{10x^{\frac{2}{3}}}$ $= \dfrac{9}{10}x^{-\frac{2}{3}}$
2 Differentiate using $f'(x) = a \times nx^{n-1}$.	$f'(x) = \dfrac{9}{10} \times \left(-\dfrac{2}{3}\right)x^{-\frac{2}{3}-1}$ $= -\dfrac{3}{5}x^{-\frac{5}{3}}$ $= -\dfrac{3}{5x^{\frac{5}{3}}}$
c **1** Write in the form $f(x) = ax^n$.	$f(x) = x^2 + 2\sqrt{3x^{\frac{7}{2}}}$ $= x^2 + 2\left(3x^{\frac{7}{2}}\right)^{\frac{1}{2}}$ $= x^2 + 2\sqrt{3}x^{\frac{7}{4}}$
2 Differentiate using $f'(x) = a \times nx^{n-1}$.	$f'(x) = 2x + 2\sqrt{3} \times \dfrac{7}{4}x^{\frac{7}{4}-1}$ $= 2x + \dfrac{7\sqrt{3}}{2}x^{\frac{3}{4}}$

WORKED EXAMPLE 2 The derivative at a given point

For each function $f(x)$, calculate $f'(x)$ using the given value of x.

a $f(x) = 3x^4 - 2x^2 + 1, f'(1)$ **b** $f(x) = \dfrac{x^2 - 4}{x + 2}, f'(3)$ **c** $f(x) = \dfrac{3x^{\frac{7}{5}} + 4x^{\frac{12}{5}}}{x^{\frac{2}{5}}}, f'\left(\dfrac{1}{8}\right)$

Steps	Working
a **1** Differentiate each term.	$f'(x) = 12x^3 - 4x$
2 Evaluate $f'(1)$.	$f'(1) = 12(1)^3 - 4(1)$ $= 8$
b **1** Factorise and simplify $f(x)$.	$f(x) = \dfrac{x^2 - 4}{x + 2}$ $= \dfrac{(x + 2)(x - 2)}{x + 2}$ $= x - 2$
2 Differentiate each term.	$f'(x) = 1$, for $x \neq -2$
3 Evaluate $f'(x)$.	$f'(3) = 1$
c **1** Simplify $f(x)$.	$f(x) = \dfrac{3x^{\frac{7}{5}} + 4x^{\frac{12}{5}}}{x^{\frac{2}{5}}}$ $= 3x^{\frac{7}{5}-\frac{2}{5}} + 4x^{\frac{12}{5}-\frac{2}{5}}$ $= 3x + 4x^2$
2 Differentiate each term.	$f'(x) = 3 + 8x$
3 Evaluate $f'\left(\dfrac{1}{8}\right)$.	$f'\left(\dfrac{1}{8}\right) = 3 + 8\left(\dfrac{1}{8}\right)$ $= 4$

We can differentiate a function using CAS.

USING CAS 1 Finding the derivative

Find the derivative of $f(x) = x^5$.

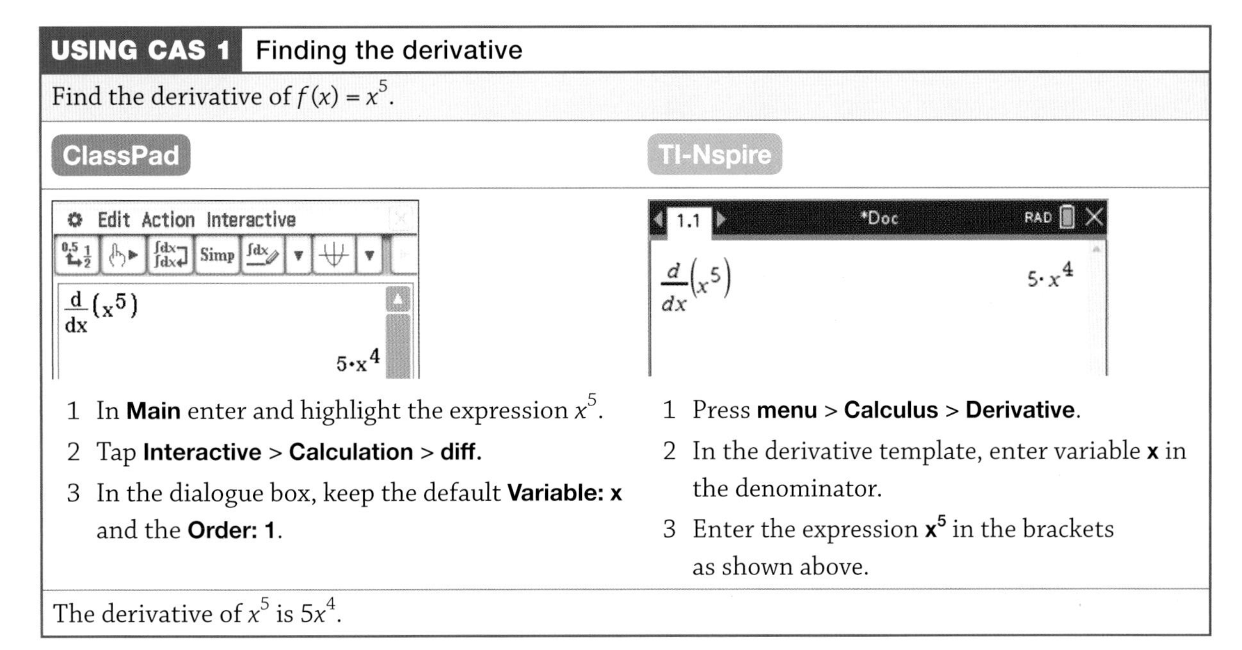

ClassPad

1 In **Main** enter and highlight the expression x^5.
2 Tap **Interactive** > **Calculation** > **diff.**
3 In the dialogue box, keep the default **Variable: x** and the **Order: 1**.

TI-Nspire

1 Press **menu** > **Calculus** > **Derivative**.
2 In the derivative template, enter variable **x** in the denominator.
3 Enter the expression $\mathbf{x^5}$ in the brackets as shown above.

The derivative of x^5 is $5x^4$.

USING CAS 2 Instantaneous rate of change

Determine the instantaneous rate of change at $x = 2$ for the function $f(x) = \sqrt{x^5 + 1}$. State your answer correct to three decimal places.

ClassPad

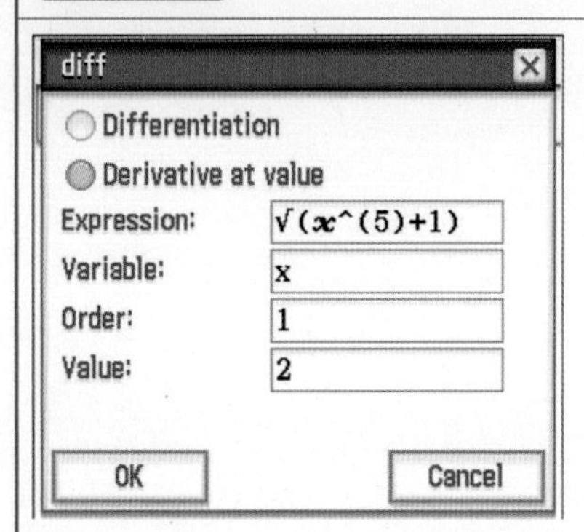

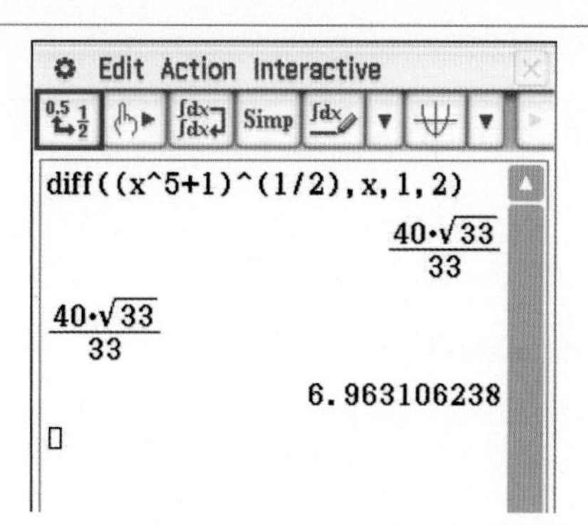

1 In **Main**, enter and highlight the expression $\sqrt{x^5 + 1}$.
2 Tap **Interactive** > **Calculation** > **diff.**
3 In the dialogue box, tap **Derivative at value**.
4 Keep the default **Variable: x** and the **Order: 1**.
5 In the **Value:** field, enter **2**.
6 The exact value of the derivative will be displayed.
7 Change to **Decimal** mode or use the **Convert** tool for the decimal solution.

TI-Nspire

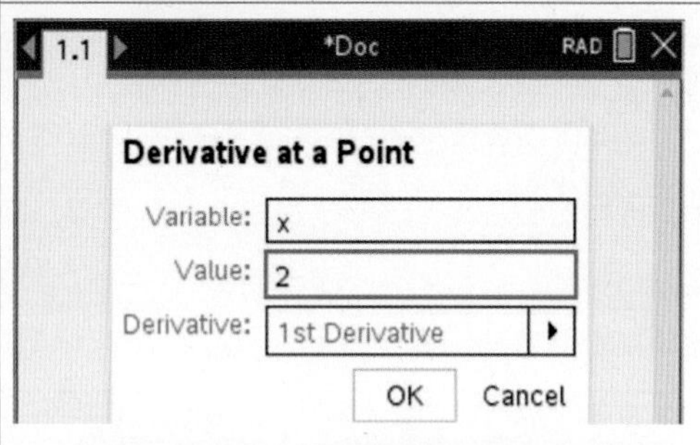

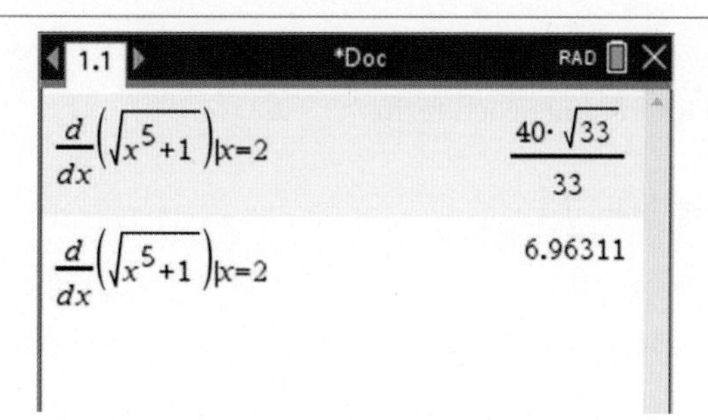

1 Press **menu** > **Calculus** > **Derivative at a Point**.
2 In the dialogue box **Value:** field, enter **2**.
3 Enter the expression inside the brackets.
4 Press **enter** for the exact solution and **ctrl** + **enter** for the approximate solution.

The instantaneous rate of change is 6.963, to three decimal places.

EXERCISE 1.1 Differentiating simple functions

ANSWER p. 387

Mastery

1 WORKED EXAMPLE 1 Differentiate each function.

a $f(x) = -3x^6$

b $f(x) = \dfrac{4}{5x^{\frac{5}{4}}}$

c $f(x) = x^3 + 3\sqrt{5x^{\frac{2}{3}}}$

Exam hack

Write radicals (roots) as powers of x before differentiating, and express answers in terms of a radical if necessary. Write $f(x) = \sqrt{x^5}$ as $f(x) = x^{\frac{5}{2}}$, so $f'(x) = \frac{5}{2}x^{\frac{3}{2}} = \frac{5}{2}\sqrt{x^3}$.

2 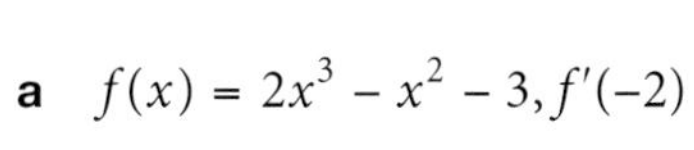 WORKED EXAMPLE 2 For each function, write down $f'(x)$ and calculate its exact value using the given value of x.

a $f(x) = 2x^3 - x^2 - 3, f'(-2)$ **b** $f(x) = \dfrac{x^3 - 1}{x - 1}, f'(1)$ **c** $f(x) = \dfrac{6x^{\frac{5}{3}} + 9x^{\frac{11}{3}}}{3x^{\frac{2}{3}}}, f'\left(\dfrac{1}{8}\right)$

3 Using CAS 1 Find the derivative of $f(x) = 5x^3$.

4 Using CAS 2 Given $f(x) = x^2 - \dfrac{1}{3}x^{\frac{3}{2}}$, determine the instantaneous rate of change at $x = 9$.

Calculator-free

5 (3 marks) Let $f(x) = \dfrac{x + x^{\frac{2}{3}} + x^{\frac{3}{4}}}{x^{\frac{1}{2}}}$.

a Write $f'(x)$ in the form $f'(x) = \dfrac{A}{x^{\frac{1}{2}}} + \dfrac{B}{x^{\frac{5}{6}}} + \dfrac{C}{x^{\frac{3}{4}}}$, where A, B and C are constants to be determined. (2 marks)

b Show that $f'(2^{12}) = \dfrac{13}{3 \times 2^9}$. (1 mark)

6 © SCSA MM2021 Q1a (3 marks) Differentiate $\dfrac{3x + 1}{x^3}$ and simplify your answer.

7 (3 marks) Find the value of a, given that $f(x) = \dfrac{1}{a}x^a + a$ and $f'(4) = 16$.

8 (2 marks) Determine the derivative of $f(x) = \left(\sqrt{3x} + \sqrt{5x^3}\right)\left(\sqrt{3x} - \sqrt{5x^3}\right)$.

9 (2 marks) The function $f(x) = \dfrac{1}{3}x^3 + ax^2 + bx + 1$, where a and b are constants, has a turning point at $x = -1$. Given that $f(1) = -2\dfrac{2}{3}$, find the value of a and b.

10 (2 marks) If $f(x) = \dfrac{1}{2x^2}$, obtain expressions for $f'(x + 1)$ and $f'(x - 1)$.

11 (2 marks) Consider the function $f(x) = 4x^3 + 5x - 9$.

a Find $f'(x)$. (1 mark)

b Explain why $f'(x) \geq 5$ for all x. (1 mark)

12 (3 marks) The functions $f(x) = -x(x - a)$ and $g(x) = mx + c$, where a, m and c are constants, have the same gradient at the non-zero x-intercept of $f(x)$. Show that $c = a^2$.

Calculator-assumed

13 (1 mark) Determine y', given $y = 1 - x + \dfrac{1}{3}x^3$.

14 (1 mark) Determine the gradient function of $f(x) = 2\left(\sqrt{x} - \dfrac{1}{3}\sqrt{x^3}\right)$.

Video playlist
The product rule

Worksheet
The product rule

1.2 The product rule

One way of finding the derivative of $f(x) = (2x + 3)(5x - 1)$ is to first expand brackets to obtain $f(x) = 10x^2 + 13x - 3$ and then differentiate term-by-term to obtain $f'(x) = 20x + 13$.

However, we can also use the **product rule** to differentiate this function because it consists of the product of two functions.

The product rule

If $f(x) = u(x) \times v(x)$, then $f'(x) = u(x) \times v'(x) + v(x) \times u'(x)$.

or

$\frac{d}{dx}(uv)$ is another way of writing the derivative of $f(x) = uv$.

$$\frac{d}{dx}(uv) = u\frac{dv}{dx} + v\frac{du}{dx}$$

This can also be written as $\frac{d}{dx}(uv) = uv' + vu'$.

or

If $y = f(x)g(x)$, then $\frac{dy}{dx} = f(x)g'(x) + g(x)f'(x)$.

To differentiate $f(x) = (2x + 3)(5x - 1)$, let $u = 2x + 3$ and $v = 5x - 1$, and then obtain $u' = 2$ and $v' = 5$.

Now use $\frac{d}{dx}(uv) = uv' + vu'$

$$= (2x + 3) \times 5 + (5x - 1) \times 2$$
$$= 20x + 13$$

or

To differentiate $y = (2x + 3)(5x - 1)$, let $f(x) = 2x + 3$ and $g(x) = 5x - 1$, and then obtain $f'(x) = 2$ and $g'(x) = 5$.

Now use: If $y = f(x)g(x)$, then $\frac{dy}{dx} = f(x)g'(x) + g(x)f'(x)$

$$= (2x + 3) \times 5 + (5x - 1) \times 2$$
$$= 20x + 13$$

Exam hack

The order of the two functions of $f(x)$ does not matter. In the example shown here, we could also have used $u = 5x - 1$ and $v = 2x + 3$.

WORKED EXAMPLE 3 The product rule

Use the product rule to differentiate $y = (5x^3 - 2x)(x^2 + 1)$.

Steps	Working
1 Identify $f(x)$ and $g(x)$.	Let $f(x) = 5x^3 - 2x$ and $g(x) = x^2 + 1$.
2 Differentiate to obtain $f'(x)$ and $g'(x)$.	$f'(x) = 15x^2 - 2$, $g'(x) = 2x$
3 Write down the expression for $f(x)g'(x) + g(x)f'(x)$.	$y' = (5x^3 - 2x) \times 2x + (x^2 + 1) \times (15x^2 - 2)$
4 Expand and simplify.	$= 10x^4 - 4x^2 + 15x^4 - 2x^2 + 15x^2 - 2$ $= 25x^4 + 9x^2 - 2$

WORKED EXAMPLE 4 The product rule with substitution

For the function $f(x) = (2x^3 - 3x + 1)(5x^4 + x^3 - x + 7)$, find $f'(-3)$.

Steps	Working
1 Identify u and v.	Let $u = 2x^3 - 3x + 1$ and $v = 5x^4 + x^3 - x + 7$.
2 Differentiate to obtain u' and v'.	$u' = 6x^2 - 3$, $v' = 20x^3 + 3x^2 - 1$
3 Write down the expression for $u'v + uv'$.	$f'(x) = (6x^2 - 3)(5x^4 + x^3 - x + 7) + (2x^3 - 3x + 1)(20x^3 + 3x^2 - 1)$
4 Substitute the value and simplify.	$f'(-3) = \left(6(-3)^2 - 3\right)\left(5(-3)^4 + (-3)^3 - (-3) + 7\right) + \left((2(-3)^3 - 3(-3) + 1)(20(-3)^3 + 3(-3)^2 - 1)\right)$ $= 51 \times 388 - 44 \times (-514)$ $= 42\,404$

Exam hack

It is usually easier and faster to substitute the value into the long expression than to expand and simplify the long expression and then substitute.

WORKED EXAMPLE 5 Tangents and stationary points

For the function $f(x) = (x - 1)(2x + 3)$, determine

a the equation of the tangent to the curve at $x = 2$

b the coordinates of any stationary points.

Steps	Working
a 1 Differentiate $f(x)$ using the product rule.	$f'(x) = (2x + 3) + 2(x - 1) = 4x + 1$
2 Determine $f(2)$ and $f'(2)$.	$f(2) = 7, f'(2) = 4(2) + 1 = 9$
3 Determine the equation of the tangent.	Using $y = mx + c$, $7 = 9 \times 2 + c \Rightarrow c = -11$ $y = 9x - 11$
b 1 Let $f'(x) = 0$ and solve for x.	$4x + 1 = 0 \therefore x = -\frac{1}{4}$
2 Determine $f\left(-\frac{1}{4}\right)$.	$f\left(-\frac{1}{4}\right) = -\frac{25}{8}$
3 State the coordinates of stationary point.	$\left(-\frac{1}{4}, -\frac{25}{8}\right)$

USING CAS 3 Finding the equations of tangent lines

For the function $y = (x - 3)(x + 2)$, determine the equation of the tangent to the curve at $x = 2$.

ClassPad

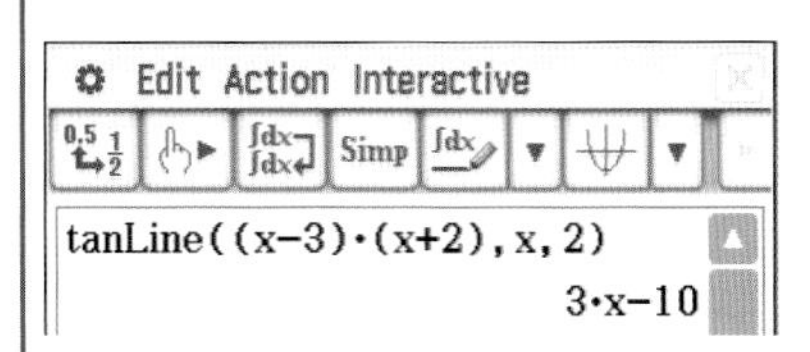

1 Enter and highlight the expression **(x – 3)(x + 2)**.

2 Tap **Interactive > Calculation > line > tanLine**.

3 In the dialogue box, set the **Point** field to 2 and tap **OK**.

4 The expression for the tangent line will be displayed.

TI-Nspire

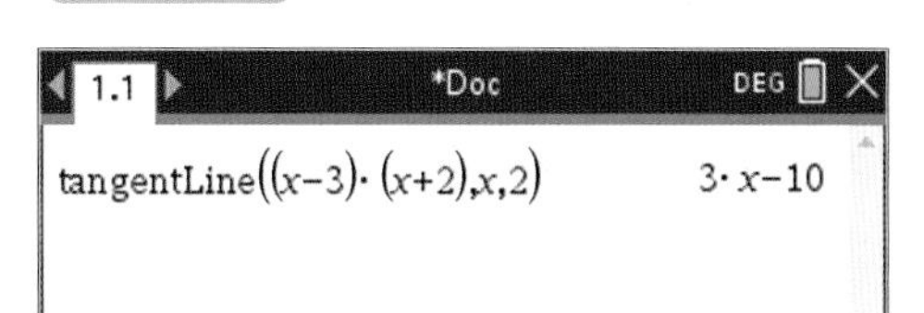

1 Press **menu > Calculus > tangent Line**.

2 Enter **(x – 3)(x + 2),x,2** and press **enter**.

3 The expression for the tangent line will be displayed.

The tangent line is $y = 3x - 10$.

EXERCISE 1.2 The product rule

ANSWERS p. 387

Recap

1 If $f(x) = \frac{1}{3}x^3 - \frac{1}{2}x^2$, then $f'(a)$ is

A a^2 **B** $a^2(a+1)$ **C** $1 - a^2$ **D** $a(a-1)$ **E** $a+1$

2 For the function $f(x) = \sqrt{x} + \frac{1}{\sqrt{x}}$, $f'(x)$ is

A $\frac{\sqrt{x}}{2}\left(1+\frac{1}{x}\right)$ **B** $\frac{1}{2\sqrt{x}}\left(1-\frac{1}{x}\right)$ **C** $\frac{1}{2}\left(1-\frac{1}{\sqrt{x}}\right)$ **D** $\frac{1}{\sqrt{x}}(1+\sqrt{x})$ **E** $\frac{2}{\sqrt{x}}(x-1)$

Mastery

3 WORKED EXAMPLE 3 Use the product rule to differentiate $f(x) = (4x + 3x^2)(7x^2 - 1)$.

4 Differentiate each expression and simplify.

a $x^4(3x+1)$ **b** $(4x+3)(3x-2)$ **c** $7x(8x-5)$

d $-x^5(4-x^2)$ **e** $4x(x^5 - x^2)$ **f** $(5x-7)(5x+7)$

g $(1+3x)(x^2-1)$ **h** $(4x+5)(2x^3-2x+1)$ **i** $(x^2+1)^2$

5 WORKED EXAMPLE 4 Find $f'(-1)$ for the function $f(x) = (1 + 3x - 2x^3)(6 + x^3 - x^5)$.

6 **a** Differentiate $f(x) = (x^4+1)(2x^3+5) + (3x^2-4)(2x^2+5)$.

b Evaluate $f'(-1)$.

7 If $f(x) = (x^2 + x + 2)(x - 3)$, determine $f'(3)$.

8 WORKED EXAMPLE 5 For the function $f(x) = (2x-1)(x-3)$, determine

a the equation of the tangent to the curve at $x = 2$

b the coordinates of any stationary points.

9 Using CAS 3 For the function $y = (x+3)(x-2)$, determine the equation of the tangent to the curve at $x = -2$.

Calculator-free

10 (1 mark) Determine the gradient of the curve $y = \sqrt{x}(x^2+1)$ at $x = 4$.

11 (2 marks) Show that for the functions $f(x) = a + bx^2$ and $g(x) = c + dx^2$, where a, b, c and d are constants, if $f'(x)g(x) = f(x)g'(x)$, then $bc = ad$.

12 (3 marks) Let $y = (x-a)^2(x-b)$.
Determine the values of a and b when $\frac{dy}{dx} = (x-5)(3x-11)$.

13 (3 marks) Given that $f'(6) = 6$ for the function $f(x) = (ax-4)(ax+3)$, where a is a constant, use the product rule to determine the possible values of a.

14 (4 marks) Let $f(x) = (ax+b)(bx+a)$ for positive constants a, b.

a Use the product rule to find $f'(x)$. (2 marks)

b Determine values for a and b when $f'(1) = 25$ and $f'(2) = 37$. (2 marks)

Calculator-assumed

15 (2 marks) Find the coordinates of the turning points on the curve $y = (x - 6)(x^2 - 9)$.

16 (1 mark) Determine the gradient of the tangent to the curve $y = \sqrt[3]{\frac{1}{x^2}}(1 - x^2)$ at $x = 2$, correct to one decimal place.

17 (2 marks) Determine the coordinates of the point on the curve $y = (3 - x)\sqrt{x^3}$ where the gradient is zero.

18 (2 marks) Determine the value of a if $\frac{d}{dx}[x^2(2 - ax^3)] = 4x - \frac{5x^4}{2}$.

1.3

1.3 The quotient rule

Video playlist
The quotient rule

Worksheet
The quotient rule

We use the **quotient rule** to differentiate a function consisting of the ratio of two functions.

The quotient rule

If $f(x) = \frac{u(x)}{v(x)}$,

then $f'(x) = \frac{v(x) \times u'(x) - u(x) \times v'(x)}{(v(x))^2}$ or $\frac{d}{dx}\left(\frac{u}{v}\right) = \frac{v\frac{du}{dx} - u\frac{dv}{dx}}{v^2}$ or $\frac{d}{dx}\left(\frac{u}{v}\right) = \frac{vu' - uv'}{v^2}$.

or

If $y = \frac{f(x)}{g(x)}$, then $\frac{dy}{dx} = \frac{g(x)f'(x) - f(x)g'(x)}{[g(x)]^2}$.

Exam hack

When using the quotient rule, differentiate in alphabetical order: u' first, then v' or $f'(x)$ first then $g'(x)$.

WORKED EXAMPLE 6 The quotient rule

Find the derivative of $\frac{8x - 3}{4x + 5}$.

Steps	Working
1 Identify u and v.	$u = 8x - 3$, $v = 4x + 5$
2 Differentiate to obtain u' and v'.	$u' = 8$, $v' = 4$
3 Write down the expression for $\frac{vu' - uv'}{v^2}$.	$\frac{d}{dx}\left(\frac{8x - 3}{4x + 5}\right) = \frac{(4x + 5) \times 8 - (8x - 3) \times 4}{(4x + 5)^2}$
4 Expand and simplify.	$= \frac{32x + 40 - 32x + 12}{(4x + 5)^2}$ $= \frac{52}{(4x + 5)^2}$

WORKED EXAMPLE 7	The quotient rule with substitution
If $y = \dfrac{x^2 + x}{-2x + 3}$, determine $\dfrac{dy}{dx}$ at $x = 3$.	
Steps	**Working**
1 Identify $f(x)$ and $g(x)$.	$f(x) = x^2 + x, g(x) = -2x + 3$
2 Differentiate to obtain $f'(x)$ and $g'(x)$.	$f'(x) = 2x + 1, g'(x) = -2$
3 Write down the expression for $\dfrac{dy}{dx}$.	$\dfrac{dy}{dx} = \dfrac{(-2x + 3)(2x + 1) - (x^2 + x)(-2)}{(-2x + 3)^2}$
4 Evaluate $\dfrac{dy}{dx}$ at $x = 3$.	$\dfrac{dy}{dx} = \dfrac{(-2(3) + 3)(2(3) + 1) - ((3)^2 + 3)(-2)}{(-2(3) + 3)^2}$ $= \dfrac{1}{3}$

EXERCISE 1.3 The quotient rule

ANSWERS p. 387

Recap

1 The value of $f'(4a^2)$ for the function $f(x) = x(\sqrt{x} - 1)$ is

A $a^2 - 1$ **B** $3a + 3$ **C** $a + 1$ **D** $a^3 + 3$ **E** $3a - 1$

2 Determine the gradient of the function $g(x) = (x + 3)(2x - 5)$ at $x = 2$.

Mastery

3 WORKED EXAMPLE 6 Find the derivative of $\dfrac{6x - 1}{9x - 8}$.

4 Find the derivative of each of the following.

a $\dfrac{2}{2x + 3}$ **b** $\dfrac{4 - x}{x - 5}$ **c** $\dfrac{x - 1}{x + 1}$ **d** $\dfrac{1}{x(x + 1)}$

e $\dfrac{x^3 + x}{x + 3}$ **f** $\dfrac{1 + x + x^2}{x}$ **g** $\dfrac{x^3 - 1}{x - 1}$ **h** $\dfrac{3(2x + 5)}{1 - x^2}$

5 WORKED EXAMPLE 7 If $y = \dfrac{2x - 1}{x + 4}$, determine $\dfrac{dy}{dx}$ at $x = 2$.

Calculator-free

6 (4 marks) Determine the value of the integer constant k in the function $f(x) = \dfrac{x + k}{x - k}$ given that $f'(5) = -8$.

7 (1 mark) Let $f(x) = \dfrac{1}{2x - 4} + 3$. Find $f'(x)$.

8 (2 marks) If $f(x) = \dfrac{2 - x + x^2}{x^2 - 2x + 1}$, determine $f'(2)$.

9 (5 marks) Determine the equation of the tangent to the curve $y = \frac{2+x}{2-x}$ at $x = 4$. Also, determine the coordinates of the x and y intercepts of this tangent.

10 (4 marks) For the function $f(x) = \frac{5}{2x-1}$, find the values of the constants a and b if $(2x-1)f'(x) = -\frac{a}{x+b}$.

11 (3 marks) If $y = \frac{1}{1+x}$, show that $\frac{dy}{dx} = -y^2$.

Calculator-assumed

12 (1 mark) If $y = \frac{x^2}{x+2}$, then determine an expression for $\frac{dy}{dx}$.

13 (2 marks) Given $f(x) = \frac{x^2}{x+1}$ and $g(x) = \frac{3x+2}{x+2}$, find all values of x that satisfy $f'(x) = g'(x)$. Give answers to two decimal places.

14 (3 marks) Determine the exact gradient of the curve $y = \frac{x^2-4}{x^2-3x-4}$ at the points where the curve intersects the coordinate axes.

15 (3 marks) If the derivative of $y = \frac{(x+2)^2}{x^2+a}$ for a unique value of a is $\frac{8-28x-16x^2}{(2x^2+1)^2}$, then determine the value of a.

16 (3 marks) Find the values of a so that the function $f(x) = ax + \frac{x^2+1}{x+1}$ has at least one stationary point.

1.4 The chain rule

Video playlist
The chain rule

Worksheets
The chain rule

Mixed differentiation problems

If $f(x) = x^2 + 2$ and $g(x) = x^3$, then the **composition of functions** f and g, is $f(g(x)) = (x^3)^2 + 2$.

The **chain rule** transforms a **composite function** of the form $y = f(g(x))$ to make differentiation easier or, in some cases, possible.

For example, $y = \sqrt{5x-4}$ is not a standard function, but by letting $u = 5x - 4$, we get $y = \sqrt{u} = u^{\frac{1}{2}}$.

This is differentiated with respect to the variable u to obtain $\frac{dy}{du} = \frac{1}{2}u^{-\frac{1}{2}}$.

Then we differentiate $u = 5x - 4$ and apply the chain rule.

The chain rule

If $y = f(u)$ and $u = g(x)$, then $\frac{dy}{dx} = \frac{dy}{du} \times \frac{du}{dx}$.

or

If $y = f(g(x))$, then $\frac{dy}{dx} = f'(g(x))g'(x)$.

WORKED EXAMPLE 8 The chain rule

Differentiate each function with respect to x using the chain rule.

a $\frac{1}{(3x+1)^4}$ **b** $\sqrt{5x-4}$

Steps	Working
a **1** Write $\frac{1}{(3x+1)^4}$ as a function of a function in index form.	Let $u = 3x + 1$. Then $y = \frac{1}{(3x+1)^4} = (3x+1)^{-4} = u^{-4}$
2 Write the chain rule and find the two derivatives.	$\frac{dy}{dx} = \frac{dy}{du} \times \frac{du}{dx}$ $\frac{dy}{du} = -4u^{-5}, \frac{du}{dx} = 3$
3 Substitute the derivatives.	$\frac{dy}{dx} = -4u^{-5} \times 3$
4 Substitute for u.	$= -12(3x+1)^{-5}$
5 Write the answer.	$\frac{d}{dx}\left[\frac{1}{(3x+1)^4}\right] = -\frac{12}{(3x+1)^5}$
b **1** Determine $f(x)$ and $g(x)$.	$f(x) = \sqrt{x}$, $g(x) = 5x - 4$.
2 Write the chain rule.	$\frac{dy}{dx} = f'(g(x))g'(x)$
3 Differentiate.	$\frac{dy}{dx} = \frac{1}{2}(5x-4)^{-\frac{1}{2}}(5)$
4 Simpify.	$\frac{dy}{dx} = \frac{5}{2\sqrt{5x-4}}$

EXERCISE 1.4 The chain rule

ANSWERS p. 388

Recap

1 Determine $f'(x)$ for the function $f(x) = \frac{x^2}{2x-3}$.

2 The gradient of the tangent to the function $f(x) = \frac{\sqrt[3]{x}}{x+2}$ at $x = 27$ is

A $-\frac{52}{22707}$ **B** $\frac{26}{22707}$ **C** $\frac{52}{7569}$ **D** $\frac{26}{2523}$ **E** $\frac{52}{783}$

Mastery

3 WORKED EXAMPLE 8 Differentiate each expression with respect to x using the chain rule.

a $\frac{2}{(x^3+1)^4}$ **b** $\sqrt{x^2-1}$

4 Differentiate each of the following with respect to x.

a $(x-5)^5$ **b** $(4x-3)^4$ **c** $(2x^3+x)^3$ **d** $(8-2x^2)^6$

e $\left(\frac{1}{2}x-6\right)^9$ **f** $(x^3-2x^2+x+1)^2$ **g** $(4x+6)^{\frac{1}{2}}$ **h** $(2\sqrt{x}-x)^3$

i $\sqrt{5(x+10)}$ **j** $\frac{1}{(2x+7)^2}$ **k** $\frac{1}{\sqrt{4-x}}$ **l** $\frac{5}{\sqrt{(x-8)^3}}$

5 Find the derivative of each expression.

a $(2x-1)^4$ **b** $(3-x^3)^2$

c $(3+4x+2x^2)^7$ **d** $(x^2+6x)^6$

e $(x^3-x^6+1)^5$ **f** $\frac{1}{n+1}(x^{n+1}+1)^{n+1}$ for positive integers, n

Calculator-free

6 (2 marks) Let $y=(3x^2-5x)^5$. Find $\frac{dy}{dx}$.

7 (1 mark) If $y=(x^2-5x)^4$, find $\frac{dy}{dx}$.

8 (1 mark) Differentiate $\sqrt{4-x}$ with respect to x.

9 (1 mark) If $y=(-3x^3+x^2-64)^3$, find $\frac{dy}{dx}$.

10 (2 marks) Given $f(x)=(x^2+ax+1)^3$, state the value of a if $f'(0)=3$.

11 (2 marks) Show that if $x>a$, the gradient at any point on the graph of the function $y=\sqrt{1+(x-a)^2}$ is positive.

12 (2 marks) Show that if $y=\sqrt{1-f(x)}$, $\frac{dy}{dx}$ is equal to $\frac{-f'(x)}{2\sqrt{1-f(x)}}$.

13 (2 marks) Show that if $f(x)=(x-a)^2g(x)$, then the derivative of $f(x)$ is $(x-a)[2g(x)+(x-a)g'(x)]$.

Calculator-assumed

14 (2 marks) Determine the exact value of the instantaneous rate of change in w when $v=3$, given that $w=\frac{3}{\sqrt[3]{9+2v^2}}$.

15 (3 marks) If $f(x)=a(bx+1)^3$, determine the values of a and b given that $f(0)=2$ and $f'(0)=18$.

16 (3 marks) Determine the values of a and b when $f'[g(x)]=\sqrt{a(6x-7)^b}$, $f(x)=\sqrt{x^3}$, and $g(x)=6x-7$.

17 (3 marks) The height, h cm, of a tomato plant is a function of the amount of compost, c grams, it receives. The function is $h(c)=c^2+c+1$, where $c(t)=t^3+t$, with t metres being the thickness of the topsoil. Express the rate of change of the height of the tomato plant with respect to topsoil thickness in the form $\frac{dh}{dt}=f(t)$ and calculate $\frac{dh}{dt}$ when $t=10$ cm, correct to one decimal place.

Video playlist
Combining the rules

1.5 Combining the rules

Sometimes a combination of the product, quotient and chain rules is required. Each question needs to be carefully considered to see which method, or combination of methods, is appropriate.

There may be more than one way to complete a question.

WORKED EXAMPLE 9 Combining the chain and product rules

Differentiate $2x^5(5x + 3)^3$.

Steps	Working
1 Write as a product.	Let $y = uv$, where $u = 2x^5$ and $v = (5x + 3)^3$.
2 Find the derivative of u.	$u' = 10x^4$
3 Write v as a function of a function.	Let $v(x) = p(q(x))$, where $p(q) = q^3$ and $q(x) = 5x + 3$.
4 Write the chain rule.	$v' = \frac{dp}{dx} = \frac{dp}{dq} \times \frac{dq}{dx}$
5 Substitute the derivatives.	$= 3q^2 \times 5$
6 Substitute for q.	$= 15(5x + 3)^2$
7 Write the product rule.	$y = uv' + vu'$
8 Substitute the functions.	$= 2x^5 \times 15(5x + 3)^2 + (5x + 3)^3 10x^4$ $= 30x^5(5x + 3)^2 + 10x^4(5x + 3)^3$
9 Take out the common factor.	$= 10x^4(5x + 3)^2[3x + (5x + 3)]$
10 Simplify and write the answer (optional).	$= 10x^4(5x + 3)^2(8x + 3)$

Video
WACE question analysis: Differentiation

WACE QUESTION ANALYSIS

© SCSA MM2020 Q5 Calculator-free (5 marks)

The graphs of the functions f and g are displayed below.

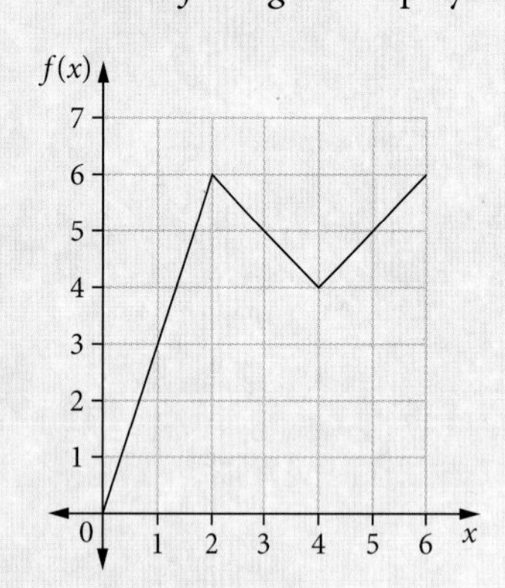

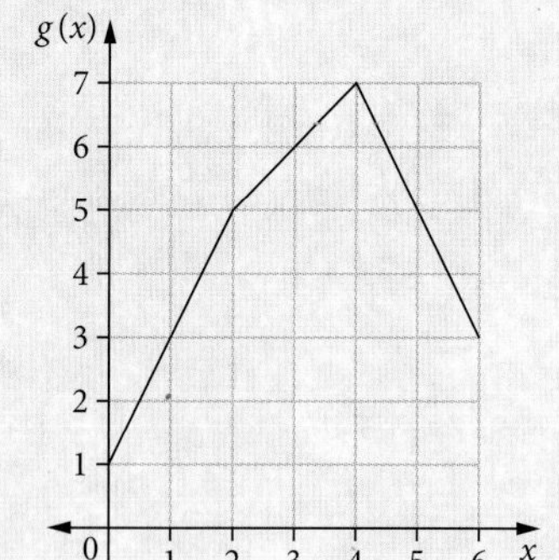

a Evaluate the derivative of $f(x)$ at $x = 3$. (1 mark)

b Evaluate the derivative of $f(x)g(x)$ at $x = 5$. (2 marks)

c Evaluate the derivative of $f(g(x))$ at $x = 1$. (2 marks)

Reading the question

- Understand key words and terms because they provide a good indication of the approach to take. For instance, evaluate means you should provide a numerical value for your answer.
- Note what is asked in each question. For instance, each part is looking for an answer at a particular point. Consider how this fits with the information given in the body of the question and the graphs displayed.
- Take note of the ways the functions are written in parts **b** and **c**. Does this look familiar in terms of the rules you have learnt – product, quotient and/or chain rules?

Thinking about the question

- As parts **b** and **c** are only worth two marks, no working or explanation is required to gain full marks. However, if your final answer is wrong, it may still be possible to obtain part marks for working.
- Note that the graphs relate to functions, but the questions relate to derivatives. Consider how the graphs can give you information on the gradients, and hence the derivatives.
- Remember to answer each part of the question fully.

Worked solution (✓ = 1 mark)

a $f'(3) = -1$

states the correct derivative ✓

b $(fg)'(5) = f'(5)g(5) + g'(5)f(5)$

$= (1)(5) + (-2)(5)$

$= -5$

uses product rule to express derivative ✓

states correct derivative ✓

c $f(g(x)')$ when $x = 1 = f'(g(1))g'(1)$

$= f'(3)2$

$= (-1)2$

$= -2$

uses chain rule to express derivative ✓

states correct derivative ✓

EXERCISE 1.5 Combining the rules

ANSWERS p. 388

Recap

1 The derivative of $(1 - 0.5x^4)^4$ at $x = 1$ is

A -1 **B** $-\frac{1}{2}$ **C** 0 **D** $\frac{1}{4}$ **E** 1

2 $\frac{d}{dx}\left(\sqrt{(x^2+1)^3}\right)$ is

A $\frac{3}{x^2+1}$ **B** $\frac{\sqrt{3x}}{(x^2+1)^2}$ **C** $\frac{3x}{x^2+1}$ **D** $\frac{1}{2\sqrt{x^2+1}}$ **E** $3x\sqrt{x^2+1}$

Mastery

3 WORKED EXAMPLE 9 Differentiate the function $x^2(2x + 1)^3$.

4 Consider the function $f(x) = (x - 1)^2(x - 2) + 1$.

If $f'(x) = (x - 1)(ux + v)$, where u and v are constants, use calculus to find the values of u and v.

5 Determine the derivative of $10p(1 - p)^9$ with respect to p.

Calculator-free

6 (6 marks)

a Differentiate $\frac{x+1}{2x-1}$

i using the quotient rule (2 marks)

ii by expressing the function as a product and using the product rule. (2 marks)

b Show that your answers in part **a** are the same. (2 marks)

7 (4 marks) The gradient of the function $f(x) = \frac{x+a}{x-a}$ (a is an integer) at $x = 1$ is $-\frac{3}{2}$. Find the value of a.

8 (4 marks) If $y = \frac{x}{\sqrt{a^2 - x^2}}$, show that $\frac{dy}{dx} = \frac{a^2}{\sqrt{(a^2 - x^2)^3}}$.

9 (3 marks) If $y = \sqrt{x^2 + 3}$ and $x(t) = 4t^3 + t + 1$, evaluate the rate of change of y with respect to t when $t = 0$.

10 (3 marks) Differentiate $f(x) = \frac{ax^2 + b}{x + b}$ and determine the values of a and b given that they are integers, and $f(1) = 1$ and $f'(-1) = -2$.

11 (3 marks) Let $f(x) = mx$ where $m \neq 0$, and $g(x) = (f(x))^n$ for positive integers n. Show that $g'(x) = \frac{ng(x)}{x}$.

Calculator-assumed

12 (4 marks)

a Differentiate each of the functions $f(x) = (x^2 - x + 1)^2$ and $g(x) = (x + a)^3$. (2 marks)

b Show that at $x = 0$ the tangents to $y = f(x)$ and $y = g(x)$ are perpendicular if $6a^2 = 1$. (2 marks)

13 (3 marks) If $y = \frac{x^2}{1 + x^2}$, show that $(1 + x^2)\frac{dy}{dx} + 2xy = 2x$.

14 (4 marks)

a Give the values of a and b so that the derivative of the function $f(x) = \frac{(6 - x^2)x}{3(2 - x^2)}$ is in the form $\frac{ax^4 + b}{3(x^2 - 2)^2}$. (2 marks)

b Determine the coordinates of the point on the curve of $f(x)$ for $0 \le x \le 5$, where $f(x) = f'(x)$. Give the answer correct to three decimal places. (2 marks)

15 (4 marks) The point $(2, b)$ lies on the graph of the function $y = \frac{a + 4x^2}{2x + 1}$. The gradient of the curve at that point is $\frac{6}{5}$. Determine the value of a and b.

Chapter summary

Differentiation

- To differentiate means to find the instantaneous rate of change at a given point.
  - **Instantaneous rate of change** at point A is the gradient of the **tangent** at a point A.
  - The **tangent** to a curve is a straight line touching the curve at a point.
  - A **stationary point** is a point on a curve where the gradient is zero (that is, where $f'(x) = 0$).

Derivative of a power of x

- If $f(x) = ax^n$, then $f'(x) = anx^{n-1}$.

The product rule

- The product rule is used to differentiate the product of two functions $u = u(x)$ and $v = v(x)$.

  $$\frac{d}{dx}(uv) = u\frac{dv}{dx} + v\frac{du}{dx} \quad \text{or} \quad \frac{d}{dx}(uv) = uv' + vu'.$$

  or

  If $y = f(x)g(x)$, then $\frac{dy}{dx} = f(x)g'(x) + g(x)f'(x)$.

The quotient rule

- The quotient rule is used to differentiate the ratio of two functions $u = u(x)$ and $v = v(x)$.

  $$\frac{d}{dx}\left(\frac{u}{v}\right) = \frac{v\frac{du}{dx} - u\frac{dv}{dx}}{v^2} \quad \text{or} \quad \frac{d}{dx}\left(\frac{u}{v}\right) = \frac{vu' - uv'}{v^2}$$

  or

  If $y = \frac{f(x)}{g(x)}$ then $\frac{dy}{dx} = \frac{g(x)f'(x) - f(x)g'(x)}{[g(x)]^2}$.

The chain rule

- The chain rule differentiates a composite function $y = f(g(x))$.

  If $y = f(u)$ and $u = g(x)$, then $\frac{dy}{dx} = \frac{dy}{du} \times \frac{du}{dx}$.

  or

  If $y = f(g(x))$, then $\frac{dy}{dx} = f'(g(x))g'(x)$.

Cumulative examination: Calculator-free

Total number of marks: 16 Reading time: 2 minutes Working time: 16 minutes

1 (1 mark) Let $y = (5x + 1)^7$. Find $\frac{dy}{dx}$.

2 © SCSA MM2018 Q3a (2 marks) Differentiate $(2x^3 + 1)^5$.

3 (4 marks)

a Let $f(x) = \frac{x}{x + 2}$.

Differentiate f with respect to x. (2 marks)

b Let $g(x) = (2 - x^3)^3$.

Evaluate $g'(1)$. (2 marks)

4 (1 mark) Let $f(x) = \sqrt{1 - 2x}$. Find $f'(x)$.

5 (3 marks) Consider the tangent to the graph of $y = x^2$ at the point $(2, 4)$. Show that the point $(3, 8)$ lies on this tangent.

6 (1 mark) Let $f(x) = \frac{1}{5}(x - 2)^2(5 - x)$.

Write down the derivative $f'(x)$.

7 © SCSA MM2019 Q2ab (4 marks) The values of the functions $g(x)$ and $h(x)$, and their derivatives $g'(x)$ and $h'(x)$ are provided in the table below for $x = 1$, $x = 2$ and $x = 3$.

	$x = 1$	$x = 2$	$x = 3$
$g(x)$	3	5	−3
$h(x)$	2	−2	6
$g'(x)$	−4	1	4
$h'(x)$	0	−6	−5

a Evaluate the derivative of $\frac{g(x)}{h(x)}$ at $x = 3$. (2 marks)

b Evaluate the derivative of $h(g(x))$ at $x = 1$. (2 marks)

Cumulative examination: Calculator-assumed

Total number of marks: 14 Reading time: 2 minutes Working time: 14 minutes

1 (2 marks) Determine the equation of the tangent to the curve $f(x) = (3x + 2)(x^4 + x^3)$ at the point where $x = 1$.

2 (4 marks) Given $y = \dfrac{(5 - x)^3}{\sqrt{2x + 1}}$, show that $\dfrac{dy}{dx} = \dfrac{-(5 - x)^2(5x + 8)}{(2x + 1)^{\frac{3}{2}}}$.

3 (4 marks) The point $(1, -3)$ lies on the curve $y = \dfrac{a + bx}{2x - 5}$ and the gradient at that point is $-\dfrac{11}{3}$.

Determine the value of a and b.

4 (4 marks) Using **calculus techniques** determine the coordinates of all stationary points for the function $y = \dfrac{1}{3}x^3 + 2x^2 + 3x - 2$.

CHAPTER

APPLICATIONS OF DIFFERENTIATION

Syllabus coverage

TOPIC 3.1: FURTHER DIFFERENTIATION AND APPLICATIONS

The second derivative and applications of differentiation

3.1.10 use the increments formula: $\delta y \approx \frac{dy}{dx} \times \delta x$ to estimate the change in the dependent variable y resulting from changes in the independent variable x

3.1.11 apply the concept of the second derivative as the rate of change of the first derivative function

3.1.12 identify acceleration as the second derivative of position with respect to time

3.1.13 examine the concepts of concavity and points of inflection and their relationship with the second derivative

3.1.14 apply the second derivative test for determining local maxima and minima

3.1.15 sketch the graph of a function using first and second derivatives to locate stationary points and points of inflection

3.1.16 solve optimisation problems from a wide variety of fields using first and second derivatives

SYLLABUS COVERAGE

Video playlists (7):

2.1 The increments formula

2.2 Second derivative and points of inflection

2.3 Stationary points

2.4 Curve sketching

2.5 Straight line motion

2.6 Optimisation problems

WACE question analysis Applications of differentiation

Worksheets (9):

2.3 The sign of the derivative • Stationary points

2.4 Curve sketching 2 • Further curve sketching • Curve sketching with derivatives

2.6 Starting maxima and minima problems • Greatest and least values • Applications of optimisation • Optimisation problems

Nelson MindTap

To access resources above, visit **cengage.com.au/nelsonmindtap**

Video playlist
The increments formula

2.1 The increments formula

In Year 11 Methods we learnt that the derivative $\frac{dy}{dx} = \lim_{\delta x \to 0} \frac{\delta y}{\delta x}$, where δy is a small change (small increment) in y and δx is a small change (small increment) in x.

The increments formula

If δx is small, we can say $\frac{dy}{dx} \approx \frac{\delta y}{\delta x}$.

Rearranging gives $\delta y \approx \frac{dy}{dx} \times \delta x$, which is called the **increments formula**.

$\delta y \approx \frac{dy}{dx} \times \delta x$ for small values of δx.

WORKED EXAMPLE 1 Using the increments formula

Consider the function $y = 3x^3 - x + 1$. Using the increments formula, determine the approximate change in y when x changes from 2 to 2.02.

Steps	Working
1 Determine $\frac{dy}{dx}$, x and δx.	$\frac{dy}{dx} = 9x^2 - 1$, $x = 2$ and $\delta x = 0.02$
2 Use the increments formula to determine the approximate change in y.	$\delta y \approx \frac{dy}{dx} \times \delta x$ $= (9x^2 - 1) \times 0.02$ $= (9 \times 2^2 - 1) \times 0.02$ $= 0.7$

WORKED EXAMPLE 2 Working with the increments formula

A spherical balloon has a radius of 10 cm. Using the increments formula, determine the change in the volume of the balloon if the radius changes to 9.97 cm.

Steps	Working
1 Write down a formula for the volume of a sphere.	$V = \frac{4}{3}\pi r^3$
2 Determine $\frac{dV}{dr}$, r and δr.	$\frac{dV}{dr} = 4\pi r^2$, $r = 10$ and $\delta r = -0.03$
3 Use the increments formula to determine the approximate change in volume.	$\delta V \approx \frac{dV}{dr} \times \delta r$ $= (4\pi r^2) \times -0.03$ $= (4\pi \times 10^2) \times -0.03$ $= -37.7\,\text{cm}^3$
4 Comment on change in volume.	The volume of the balloon has decreased by $37.7\,\text{cm}^3$.

EXERCISE 2.1 The increments formula

ANSWERS p. 388

Mastery

1 WORKED EXAMPLE 1 Consider the function $y = \dfrac{3x - 1}{\sqrt{x} + 1}$. Using the increments formula, determine the approximate change in y when x changes from 5 to 5.01.

2 WORKED EXAMPLE 2 A spherical ball has a radius of 4 cm. Using the increments formula, determine the change in the volume of the ball if the radius changes to 3.95 cm.

3 Given the function $f(x) = \sqrt{x + 1} - 4x^2$, use the increments formula to determine the change in y if x changes from 3 to 3.02.

Calculator-free

4 (3 marks) Given that $\sqrt{9} = 3$, use the increments formula to determine an approximation for $\sqrt{9.01}$.

5 (4 marks) Use the increments formula to determine the change in value of the function $y = \dfrac{4}{x^2} - \sqrt{x}$ if the x value changes from 1 to 1.01.

Calculator-assumed

6 (3 marks) The side of a square is 8 cm. How much will the area of the square increase if the length of the side increases by 1 mm?

7 (4 marks) A spherical ball is slowly deflating. Use the increments formula to determine the approximate change in the radius of the ball if its surface area changes from 12.20 cm^2 to 12.15 cm^2.

8 (4 marks) The height of a cylinder is 10 cm and the radius is 5 mm. Find the approximate change in volume of the cylinder if the radius changes to 5.03 mm and the height remains the same.

2.2 Second derivative and points of inflection

Video playlist Second derivative and points of inflection

The **second derivative** is the rate of change of the first derivative function. In other words, the derivative of the derivative is called the second derivative.

> **The second derivative**
>
> For a function $y = f(x)$, the first derivative is written as y' or $f'(x)$ or $\dfrac{dy}{dx}$ and the second derivative as y'' or $f''(x)$ or $\dfrac{d^2y}{dx^2}$.

WORKED EXAMPLE 3 Finding the second derivative

Determine the second derivative of the function $y = (2x^2 - 3)^2$.

Steps	Working
1 Differentiate once for first derivative.	$y' = 2(2x^2 - 3)4x$ $= 16x^3 - 24x$
2 Differentiate again for second derivative.	$y'' = 48x^2 - 24$

USING CAS 1 Finding the second derivative

Determine the value of the second derivative of the function $f(x) = (x - 3)^2(x + 1)$ at the point where $x = 3$.

ClassPad

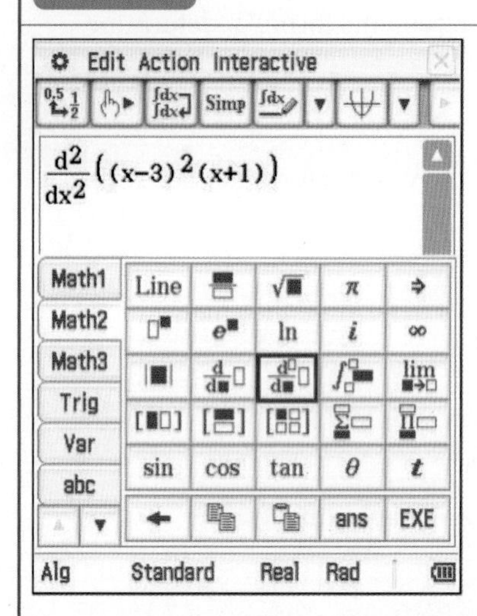

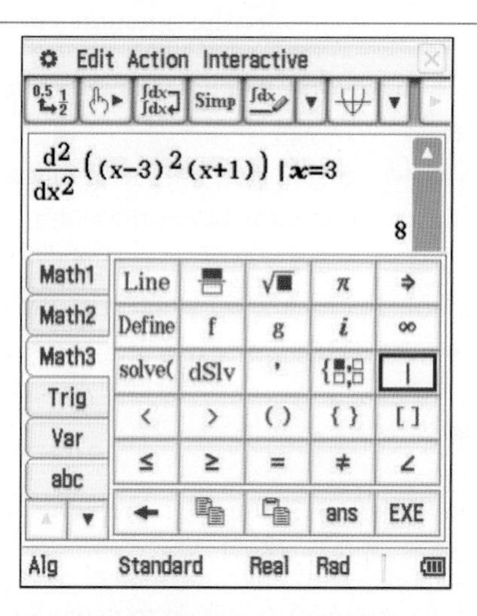

1 Enter the expression **(x – 3)²(x + 1)**.

2 Open the **Keyboard** and tap **Math2**.

3 Tap on the **nth derivative** template.

4 Enter **x** and **2** into the template.

5 Move the cursor to the end of the expression.

6 Tap **Math3**.

7 Tap the | symbol.

8 Enter **x = 3** and press **ENTER**.

Alternatively, tap **Interactive > Calculation diff > Derivative at value**. Complete the fields in the dialogue box and change the **Order** field to **2**.

TI-Nspire

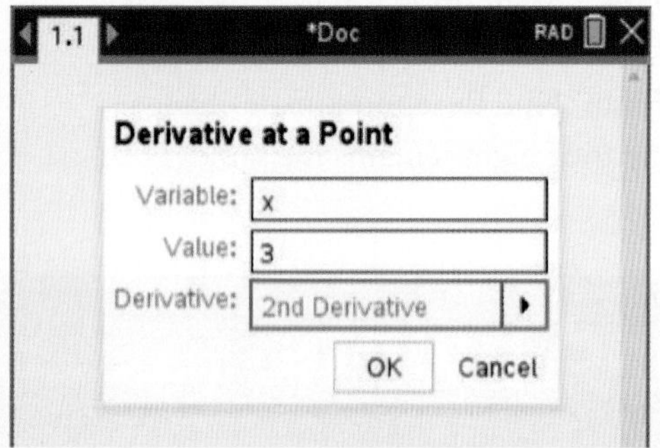

$$\frac{d^2}{dx^2}\left((x-3)^2\cdot(x+1)\right)\Big|_{x=3} \quad 8$$

1 Press **menu > Calculus > Derivative at a Point** to open the dialogue box.

2 For the **Value:** field, enter **3**.

3 Change the **Derivative:** field to **2nd Derivative**.

4 Press **OK**.

5 In the derivative template, enter the expression **(x – 3)²(x + 1)**.

6 Press **enter**.

The value of the second derivative where $x = 3$ is 8.

As with stationary points, a non-stationary **point of inflection** is where a function changes **concavity**. (Stationary points of inflection will be discussed in Section 2.3). For example, for the function $y = (x - 1)(x + 1)(2x + 3)$, as shown on the graph, there is a point of inflection at $(-0.5, -1.5)$. To the left of this point, the graph is concave down and to the right of this point, the graph is concave up.

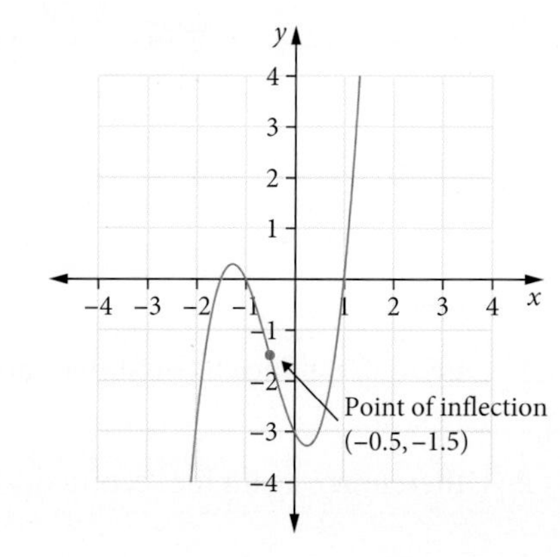

9780170477536

USING CAS 2 Finding points of inflection using graphs

For the function $y = (x - 3)^2(x + 1)$, describe where the function is concave up and concave down and, hence, state the coordinates of any points of inflection (to two decimal places).

ClassPad

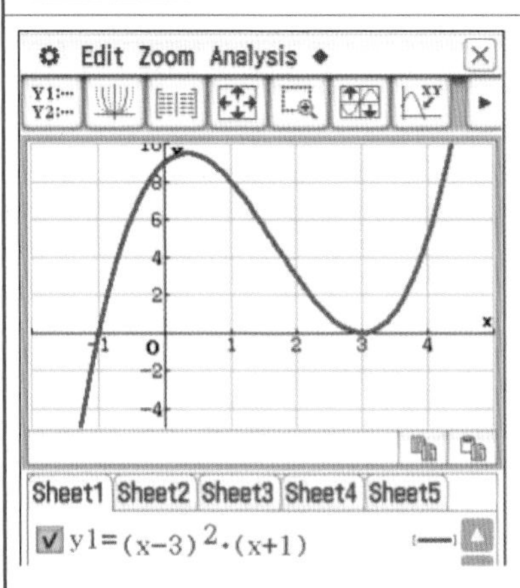

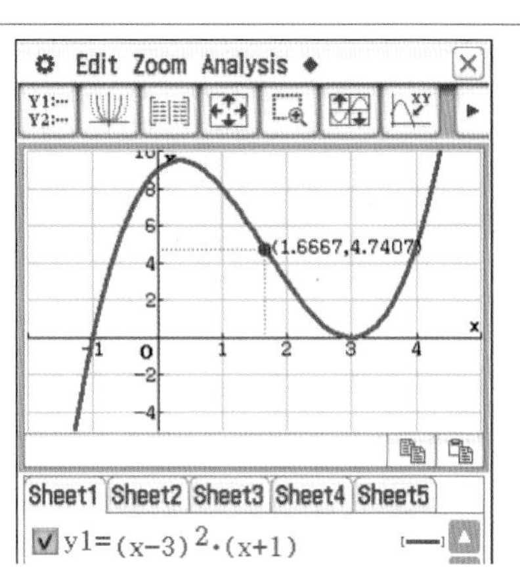

1. Graph the equation $y = (x - 3)^2(x + 1)$.
2. Adjust the window settings to suit.
3. The graph appears to be concave down when x is less than approximately 1.5 and concave up when x is greater than approximately 1.5.
4. Tap **Analysis > G-Solve > Inflection**.
5. The cursor will jump to the point of inflection.
6. Press **EXE** to paste the coordinates of the point of inflection on the graph.

TI-Nspire

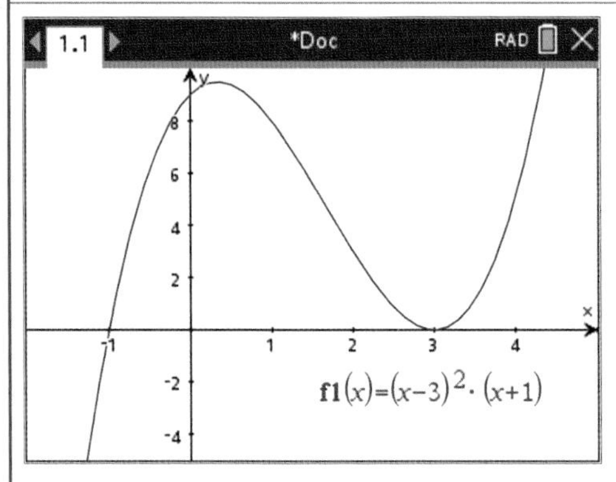

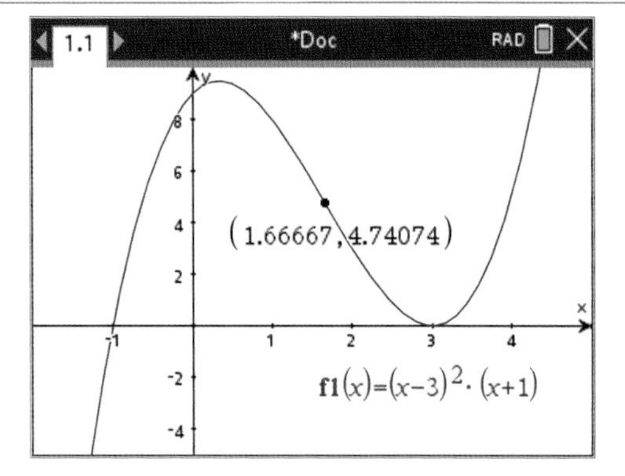

1. Graph the equation $y = (x - 3)^2(x + 1)$.
2. Adjust the window settings to suit.
3. The graph appears to be concave down when x is less than approximately 1.5 and concave up when x is greater than approximately 1.5.
4. Press **menu > Analyze Graph > Inflection**.
5. When prompted for the **lower bound?**, move the cursor to the concave down section and press **enter**.
6. When prompted for the **upper bound?**, move the cursor to the concave up section and press **enter**.
7. The coordinates of the point of inflection will appear on the screen.

Rounding to two decimal places, the graph is concave down for $x < 1.67$ and concave up for $x > 1.67$. The point of inflection is $(1.67, 4.74)$.

EXERCISE 2.2 Second derivative and points of inflection

ANSWERS p. 389

Recap

1 A rectangle is such that the length is three times the width. Determine the change in the perimeter if the width changes from 2 cm to 2.01 cm.

2 If the diameter of a sphere decreases from 25 cm to 24.9 cm, what is the change in the surface area?

Mastery

3 WORKED EXAMPLE 3 Determine the second derivative of each of the functions below.

a $y = (x - 3)^{\frac{1}{2}}$

b $f(x) = \dfrac{x^2 - 1}{2x + 10}$

c $y = 3x - 2$

d $f(x) = \dfrac{3}{2}x^{\frac{3}{2}}$

e $y = (2x + 1)(x^2 + 3)^3$

f $y = 3$

4 Using CAS 1 Determine the value of the second derivative of the function $f(x) = \dfrac{1}{2}(x^2 - 3x + 2)^2$ at the point $x = -2$.

5 Using CAS 2 For the function $f(x) = x(3x - 1)^2$, describe where the function is concave up and concave down and, hence, state the coordinates of any points of inflection. Give your answers to two decimal places.

Calculator-free

6 (2 marks) Give an example of a function where $f'(x) = f''(x)$.

7 (1 mark) How many points of inflection does the function $y = (x - 3)(x + 3)(x - 2)$ have?

Calculator-assumed

8 (4 marks) Determine the value of a and b (where a and b are greater than 0) if $f(x) = (x + a)^2(2x - b)$, $f'(2) = 24$ and $f''(2) = 26$.

9 (3 marks) Given the function $y = \dfrac{x}{\sqrt{x^2 + 2}}$, determine value(s) of x for which $\dfrac{dy}{dx} + \dfrac{d^2y}{dx^2} = 0$.

2.3 Stationary points

Remember that the gradient of the graph of $y = f(x)$ is $f'(x)$ or $\frac{dy}{dx}$.

A positive gradient points up, from left to right. 

A negative gradient points down, from left to right.

A function is **increasing** when it has a positive gradient $\left(\frac{dy}{dx} > 0\right)$ and its graph is pointing up.

A function is **decreasing** when it has a negative gradient $\left(\frac{dy}{dx} < 0\right)$ and its graph is pointing down.

Stationary points

Video playlist Stationary points

Worksheets 'The sign of the derivative

Stationary points

A **stationary point** on the graph of a function is a point where the graph has zero gradient $\left(f'(x) = 0 \text{ or } \frac{dy}{dx} = 0\right)$; that is, the graph is instantaneously flat, neither increasing nor decreasing.

A stationary point is either a **turning point** or a **stationary point of inflection**.

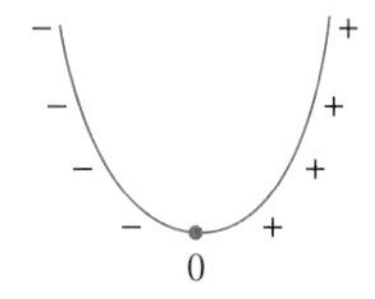

Local minimum turning point

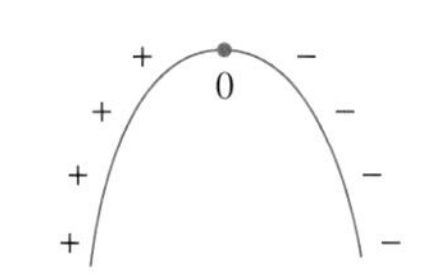

Local maximum turning point

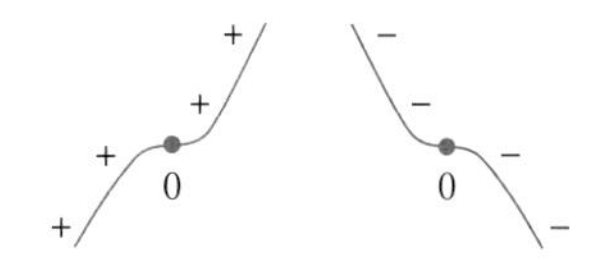

Stationary point of inflection

Turning points

A **turning point** is found where $f'(x) = 0$ and the sign of the gradient changes on either side of the stationary point, from negative to positive (for a minimum point) or from positive to negative (for a maximum point).

Turning points are also called **local maximum** or **local minimum points** because they represent the highest or lowest values of the function in the local vicinity or neighbourhood. They may not actually be the absolute maximum or minimum points for the entire function, which are called the **global maximum** or **global minimum points**.

The second derivative test for finding maximum and minimum points

The second derivative can determine if a turning point is a maximum or minimum.

If $f''(x) > 0$, the turning point is a minimum.

If $f''(x) < 0$, the turning point is a maximum.

The function graphed on the right has stationary points at $(-2, 16)$ and $(2, -16)$. At these points, $f'(x) = 0$.

At the point $(-2, 16)$, $f''(x) < 0$, hence it is a maximum point.

At the point $(2, -16)$, $f''(x) > 0$, hence it is a minimum point.

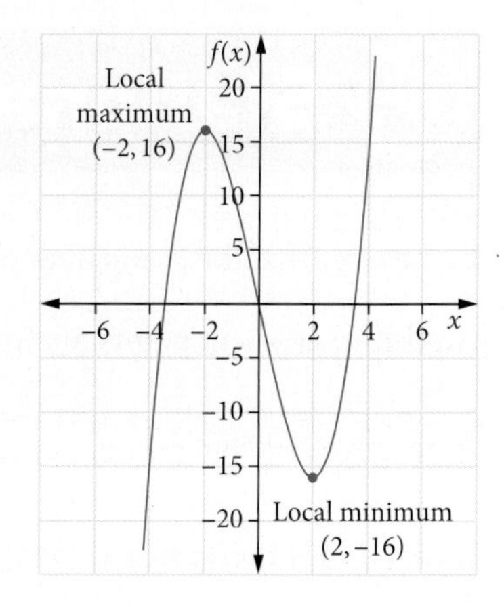

WORKED EXAMPLE 4 Turning points

Find the coordinates and nature of the turning points on the graph of $f(x) = \frac{2}{3}x^3 + \frac{3}{2}x^2 - 2x$.

Steps	Working
1 Solve $f'(x) = 0$ for stationary points.	$f'(x) = 2x^2 + 3x - 2 = 0$ $(x + 2)(2x - 1) = 0$ $x = -2$ or $2x = 1$ $x = -2$ or $x = \frac{1}{2}$
2 Find $f''(x)$ and determine if it's positive or negative to identify the nature of the stationary points.	$f''(x) = 4x + 3$ At $x = -2$, $f''(x) < 0$. At $x = \frac{1}{2}$, $f''(x) > 0$.
Alternative method Check whether the sign of the gradient changes on either side to identify the nature of the stationary points. (Also known as the sign test.)	Around $x = -2$, gradient changes from positive to negative. Around $x = \frac{1}{2}$, gradient changes from negative to positive. Both are turning points.
3 Substitute $x = -2$ and $x = -\frac{1}{2}$ into $f(x)$ to determine each y-coordinate.	$f(-2) = \frac{14}{3}$, $f\left(\frac{1}{2}\right) = -\frac{13}{24}$
4 State the coordinates and nature of the turning points.	The coordinates are $\left(-2, \frac{14}{3}\right)$, which is a local maximum, and $\left(\frac{1}{2}, -\frac{13}{24}\right)$, which is a local minimum.

CAS can be used to find the coordinates and nature of stationary points.

USING CAS 3 Finding stationary points

Find the coordinates and nature of the turning points on the graph of $f(x) = \frac{2}{3}x^3 + \frac{3}{2}x^2 - 2x$.

ClassPad

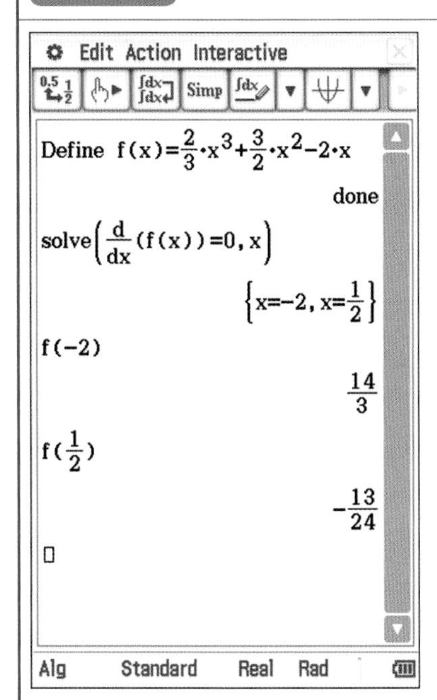

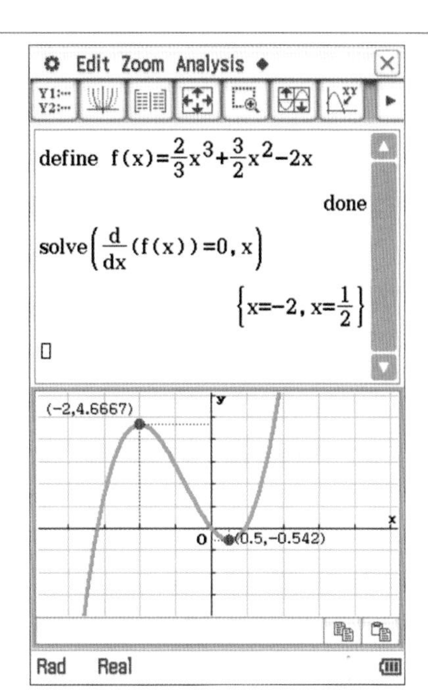

1 Define and highlight **f(x)** as shown above.

2 Derive using **Interactive > Diff**, and equate to zero, then solve using **Equation/Inequality** to solve for x.

3 Substitute the x-coordinates into the defined function to determine the corresponding y-coordinates of the turning points.

4 Graph **f(x)** by dragging into the graph screen.

5 Adjust the window settings to suit.

6 Tap **Analysis > G-Solve > Max** to display the approximate coordinates of the local maximum.

7 Tap **Analysis > G-Solve > Min** to display the approximate coordinates of the local minimum.

The local maximum point is $\left(-2, \frac{14}{3}\right)$.

The local minimum point is $\left(\frac{1}{2}, -\frac{13}{24}\right)$.

Exam hack

Students using the ClassPad often use Zoom Auto to get a better view of their graph.

The more elegant way to find a view that suits is to use .

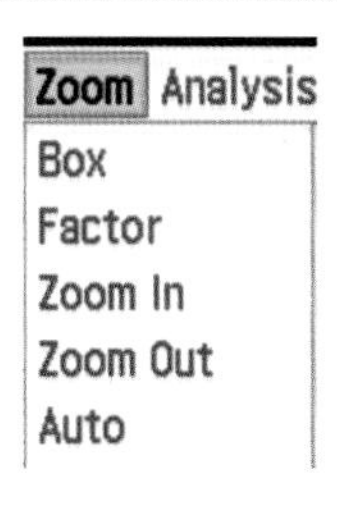

TI-Nspire

Define $f(x)=\frac{2}{3}\cdot x^3+\frac{3}{2}\cdot x^2-2\cdot x$ Done

$\text{solve}\left(\frac{d}{dx}(f(x))=0,x\right)$ $x=-2$ or $x=\frac{1}{2}$

$f(-2)$ $\frac{14}{3}$

$f\left(\frac{1}{2}\right)$ $\frac{-13}{24}$

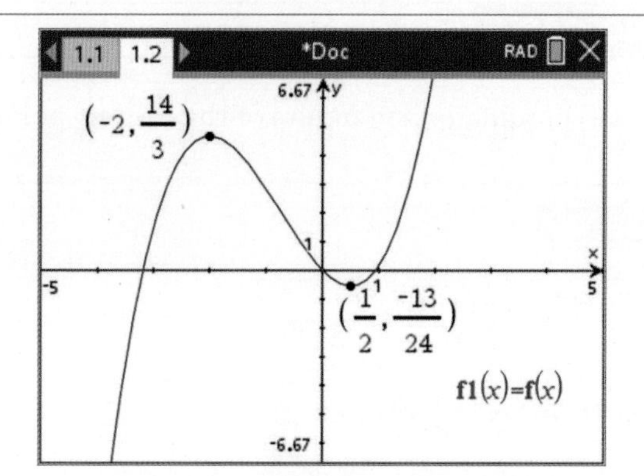

1 Define **f(x)** as shown above.

2 Use the **derivative** template to set the derivative of **f(x) = 0** and solve for **x**.

3 Substitute the x-coordinates into the function to determine the corresponding y-coordinates of the turning points.

4 Graph **f(x)**, which shows the local maximum and minimum.

5 Adjust the window settings to suit.

6 To confirm the turning points, press **menu > Geometry > Points & Lines > Point On**.

7 Click on the graph of the function twice to add two points.

8 Press **esc** to remove the point tool.

9 Click twice on the x-coordinate of the first point and enter **–2**.

10 The exact value of the corresponding y-coordinate will be displayed.

11 Repeat to find the exact coordinates of the second turning point.

The local maximum point is $\left(-2,\frac{14}{3}\right)$.

The local minimum point is $\left(\frac{1}{2},-\frac{13}{24}\right)$.

Stationary points of inflection

A stationary point of inflection is a flat 'bend' in the graph where $f'(x) = 0$; however, the sign of the gradient stays *the same* on either side of the stationary point.

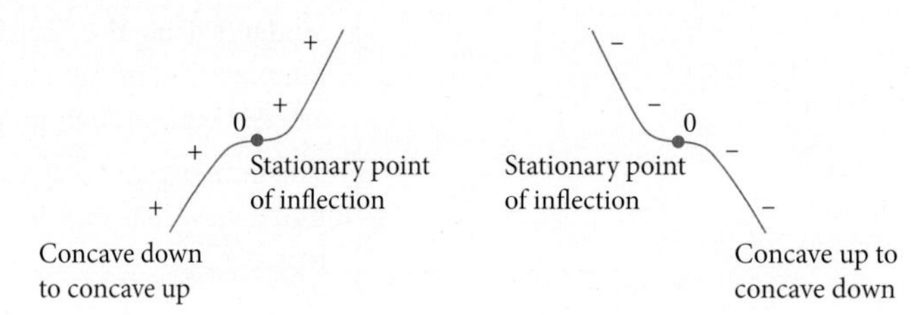

If $f'(x) = 0$ and $f''(x) = 0$ it cannot be assumed that there is a stationary point of inflection. Further investigation of the gradient on either side of the stationary point is needed.

WORKED EXAMPLE 5 Finding stationary points of inflection

Find the coordinates of any stationary points of inflection on the graph $f(x) = (x + 1)^3(x - 2)$.

Steps	Working
1 Find $f'(x)$ and solve for zero.	$f'(x) = 4x^3 + 3x^2 - 6x - 5$ When $f'(x) = 0$, then $x = -1$ and $x = \frac{5}{4}$.
2 Find $f''(x)$ and determine the value to identify the nature of the stationary points.	$f''(x) = 12x^2 + 6x - 6$ $f''(-1) = 0$ (possible stationary point of inflection) $f''\left(\frac{5}{4}\right) > 0$ (not a stationary point of inflection)
3 Check to see if the sign of the gradient changes or stays the same on either side to identify the nature of the stationary points.	On either side of $x = -1$, the gradient remains negative. The stationary point of inflection is at $x = -1$.
4 Find the coordinates of the stationary point of inflection.	$f(-1) = 0$ The stationary point of inflection is at $(-1, 0)$.

Stationary points

$f'(x)$ before the stationary point	$f'(x)$ at the stationary point	$f'(x)$ after the stationary point	$f''(x)$ at the stationary point	Type of stationary point
+	0	−	Less than zero	Local maximum point
−	0	+	Greater than zero	Local minimum point
+	0	+	0	Stationary point of inflection
−	0	−	0	Stationary point of inflection

EXERCISE 2.3 Stationary points

ANSWERS p. 389

Recap

1 Let $f(x) = \dfrac{x^2}{x+1}$. Evaluate $f''(2)$.

2 For the graph shown below, give the approximate coordinates of any inflection points.

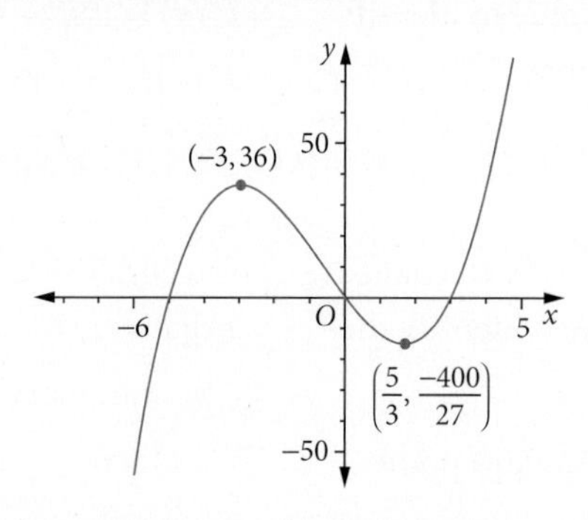

Mastery

3 WORKED EXAMPLE 4 Find the coordinates and nature of the turning point on the graph of $f(x) = x^2 - 4x$.

4 WORKED EXAMPLE 5 Find the coordinates of the stationary point of inflection on the graph of $f(x) = 3(x-1)^3(x+2)$.

5 Using CAS 3 State the coordinates and nature of the turning point on the graph with the rule $f(x) = -\dfrac{1}{2}x^2 - x + 2$.

6 State the local maximum and the stationary point of inflection, respectively, on the graph of $y = -\dfrac{1}{2}x^4 + \dfrac{1}{2}x^3 + \dfrac{3}{2}x^2 - \dfrac{5}{2}x + 1$.

Calculator-free

7 (4 marks) State any stationary point(s) and their nature for the function $y = \dfrac{1}{3}x^3 + x^2 - 3x + 1$.

8 (2 marks) What type of stationary point exists at $(1, 1)$ on the graph of the function with rule $y = (x-1)^3 + 1$?

9 (1 mark) Let $f(x)$ be a function such that $f'(3) = 0$ and $f'(x) < 0$ when $x < 3$ and when $x > 3$. Describe what type of stationary point (if any) exists at $x = 3$.

10 (2 marks) A cubic function has the rule $y = f(x)$. The graph of the derivative function $f'(x)$ crosses the x-axis at $(2, 0)$ and $(-3, 0)$. The maximum value of the derivative function is 10. State the value of x for which the graph of $y = f(x)$ has a local maximum.

11 (6 marks) Consider the function $f(x) = 3x^2 - x^3$.

a Find the coordinates and nature of the stationary points of the function. (4 marks)

b Copy the axes below and on it sketch the graph of $f(x)$. (2 marks)

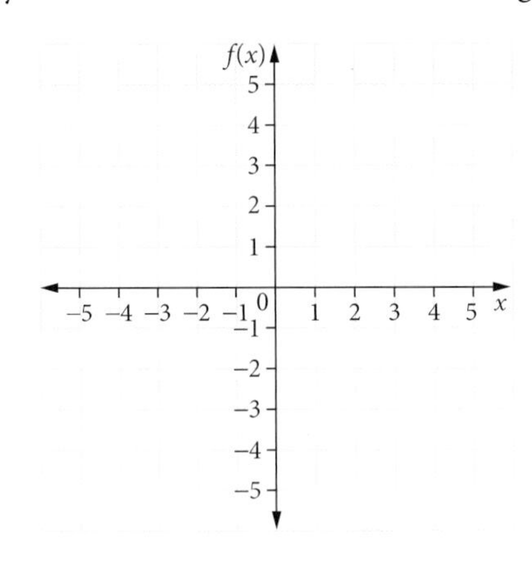

Exam hack

Sketch graphs with smooth curves and not V-shapes at the turning points.

12 (5 marks) A function is such that

- $f'(x) = 0$ at $x = 0$ and $x = 2$
- $f'(x) < 0$ for $0 < x < 2$ and $x > 2$
- $f''(x) < 0$ at $x = 0$.

State whether each of the statements below is true or false.

a The graph of $f(x)$ has a stationary point of inflection at $x = 0$.

b The graph of $f(x)$ has a local maximum point at $x = 2$.

c The graph of $f(x)$ has a stationary point of inflection at $x = 2$.

d The graph of $f(x)$ has a local maximum point at $x = 0$.

e The graph of $f(x)$ has a local minimum point at $x = 2$.

13 (3 marks) The function $f(x) = x^3 + ax^2 + bx$ has a local minimum at $x = 1$ and a local maximum at $x = -3$. Determine the values of a and b.

Calculator-assumed

14 (6 marks)

a Consider the function $y = 2\sqrt{x} - x^2 + x$. Determine the first derivative. (2 marks)

b Use your result from part **a** to show why there is a stationary point at $x = 1$. (2 marks)

c Find the second derivative and use this to describe the nature of the stationary points at $x = 1$. (2 marks)

Exam hack

When a question asks you to 'show' something, you cannot just write the answer from CAS, you need to show some working.

15 (2 marks) Use calculus to show why the function $f(x) = \sqrt{x} + x^2 + 1$ has no stationary points.

16 (2 marks) Explain why the cubic function $f(x) = ax^3 - bx^2 + cx$, where a, b and c are positive constants, has no stationary points when $c > \frac{b^2}{3a}$.

17 (2 marks) Consider the function $f(x) = 4x^3 + 5x - 9$.

a Find $f'(x)$. (1 mark)

b Explain why $f'(x) \geq 5$ for all x. (1 mark)

Exam hack

We need to practise how to answer these one-line 'explain' questions. Referring to the relevant graph is a good strategy. Just saying 'it was translated 5 units up' is not enough detail.

18 (2 marks) The cubic function p is defined by $p(x) = ax^3 + bx^2 + cx + k$, where a, b, c and k are real numbers. If p has m stationary points, what are the possible values of m?

2.4 Curve sketching

Video playlist Curve sketching

Worksheets Curve sketching 2

Further curve sketching

Curve sketching with derivatives

Key features of a graph

When sketching a graph, it is important to identify and show its key features, such as:

- the general shape
- y-intercept
- x-intercept(s)
- stationary points, including their nature.

Exam hack

Don't just rely on CAS for graphing. It is important to have a general idea of the shape of the graph and its significant features for calculator-free questions.

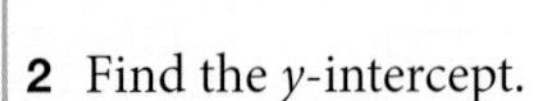

WORKED EXAMPLE 6 Sketching a function

Sketch the graph of $f(x) = -(x + 1)^2(x - 3)$, labelling the key features.

Steps	Working
1 Consider the general shape.	When expanded, the leading coefficient is $-1 < 0$, so the function is an 'upside-down' cubic.
2 Find the y-intercept.	$f(0) = -(0 + 1)^2(0 - 3) = 3$ y-intercept is 3.
3 Find the x-intercept(s).	$f(x) = -(x + 1)^2(x - 3) = 0$ x-intercepts are -1 and 3.
4 Solve $f'(x) = 0$ for stationary points.	$f'(x) = -(x + 1)^2(1) + (x - 3) \times -2(x + 1)$ $= -(x^2 + 2x + 1) + -2(x^2 + x - 3x - 3)$ $= -x^2 - 2x - 1 - 2x^2 + 4x + 6$ $= -3x^2 + 2x + 5$ Solve $-(3x^2 - 2x - 5) = 0$: $-(3x - 5)(x + 1) = 0$ $x = -1, x = \frac{5}{3}$

5 Calculate the y values of the stationary points.	$f(-1) = 0$ and $f\left(\frac{5}{3}\right) = \frac{256}{27}$ Stationary points are $(-1, 0)$ and $\left(\frac{5}{3}, \frac{256}{27}\right)$.
6 Find $f''(x)$ and determine if it's positive or negative to identify the nature of the stationary points.	$f''(x) = -6x + 2$ $f''\left(\frac{5}{3}\right) < 0$, hence a local maximum. $f''(-1) > 0$, hence a local minimum.
7 Sketch the graph. 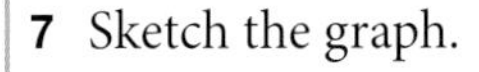	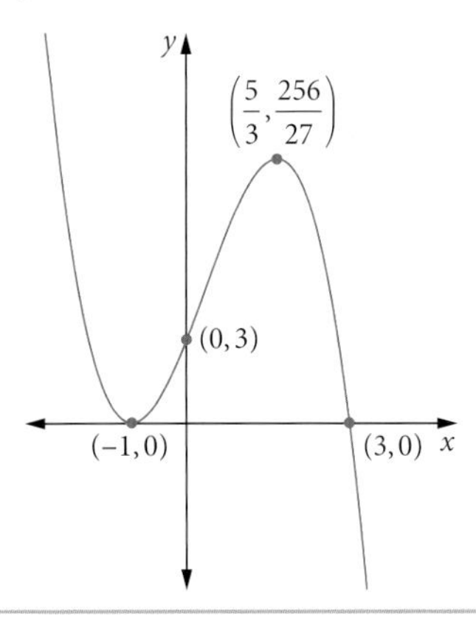

The relationship between a function and its derivative can be shown using a graph. We can explore significant points when comparing both the function graph and the derivative graph on the same axes.

WORKED EXAMPLE 7 Graphing derivative functions

The graph of $y = f(x)$ is shown. Sketch the graph of $y = f'(x)$.

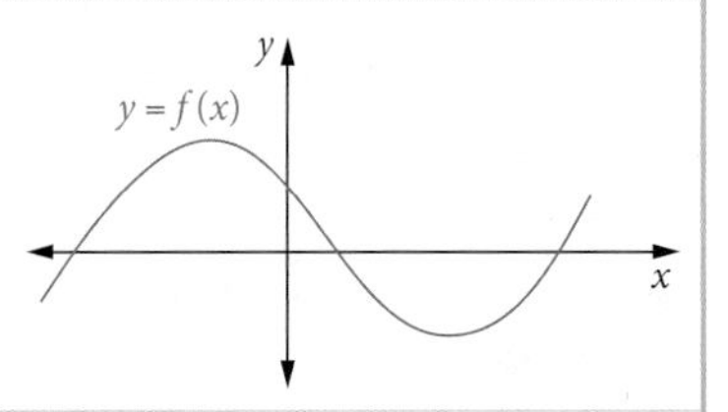

Steps

- The function is increasing at first, so the derivative graph is positive (above the x-axis).
- There is a stationary point just before the y-axis, so the derivative graph is zero (x-intercept).
- Then the function is decreasing, so the derivative graph is negative (below the x-axis).
- At the second stationary point, the derivative graph is zero again.
- Then the function is increasing again, so the derivative graph is positive again.

Graph the derivative function so that the points match the relevant points on the original function.

Working

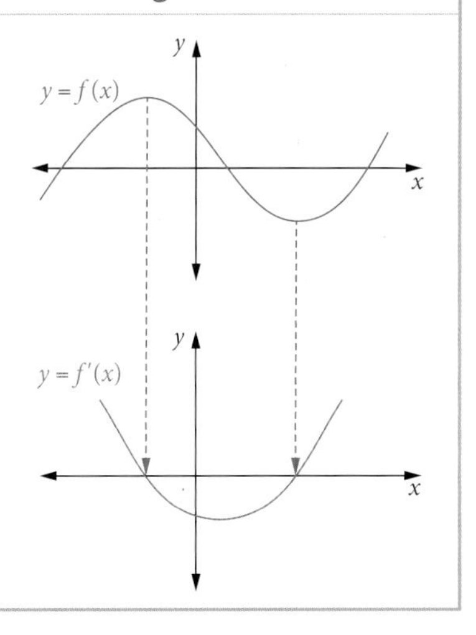

The graphs of $f(x)$ and $f'(x)$

Graph of $f(x)$	increasing	decreasing	stationary	straight line
Graph of $f'(x)$	positive, above x-axis	negative, below x-axis	0 (x-intercept)	horizontal (flat) line

The derivative graph may be required when the rule for the original function is not actually given, as shown on the right. In such cases, technology will be less useful and other methods of noticing patterns and relationships must be used.

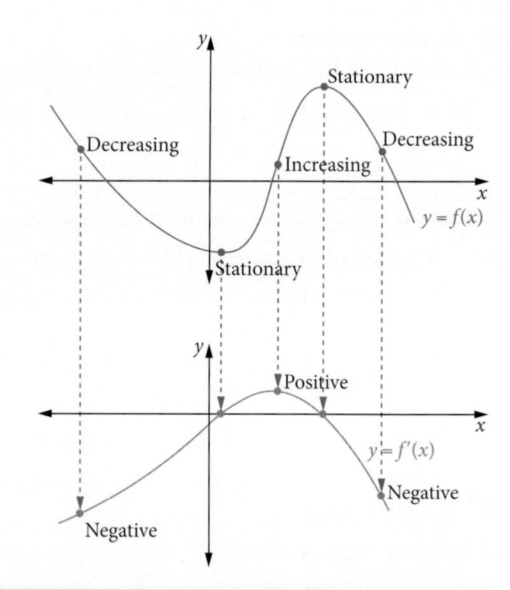

EXERCISE 2.4 Curve sketching

ANSWERS p. 389

Recap

1 The function $f(x)$ satisfies the following conditions.

- $f'(x) < 0$ where $x < 2$
- $f'(x) = 0$ where $x = 2$
- $f'(x) = 0$ where $x = 4$
- $f'(x) > 0$ where $2 < x < 4$
- $f'(x) > 0$ where $x > 4$

Which one of the following is true?

A The graph of $f(x)$ has a local maximum point at $x = 4$.

B The graph of $f(x)$ has a stationary point of inflection at $x = 4$.

C The graph of $f(x)$ has a local maximum point at $x = 2$.

D The graph of $f(x)$ has a local minimum point at $x = 4$.

E The graph of $f(x)$ has a stationary point of inflection at $x = 2$.

2 Find any coordinate(s) where the function $f(x) = x^5 - 1$ has a stationary point of inflection.

Mastery

3 WORKED EXAMPLE 6 Sketch the graph of $y = x^3 + 8$, labelling all key features.

4 WORKED EXAMPLE 7 The graph of $y = f(x)$ is shown below. Sketch the graph of $y = f'(x)$.

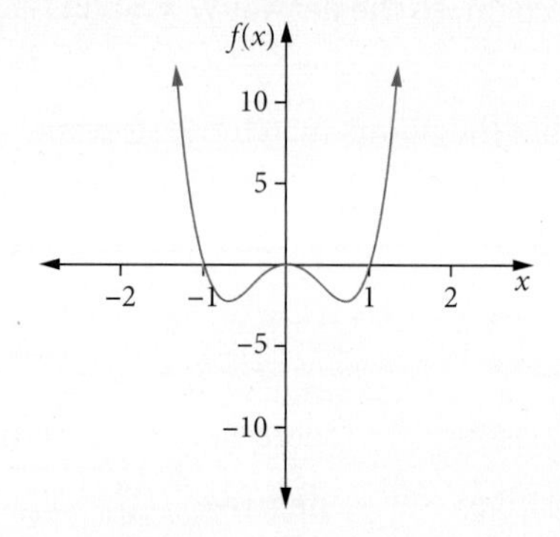

5 **a** Sketch the graph of $f(x) = -x(x + 2)(x - 3)$, labelling all key features.

b Sketch the derivative graph of $f(x) = -x(x + 2)(x - 3)$.

6 (2 marks) Given the graph below, sketch the graph of its derivative function.

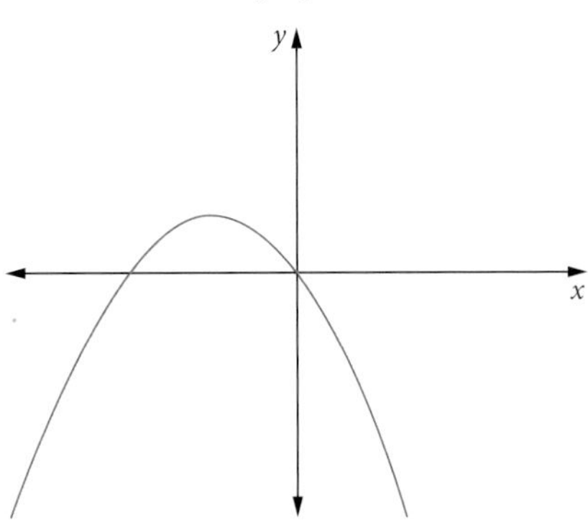

7 (5 marks) A continuous function, f, satisfies the conditions listed below. Using the information, draw a sketch of the function.

- $f(-2) = f'(-2) = f''(-2) = 0$
- $f'\left(\frac{1}{4}\right) = 0, f''\left(\frac{1}{4}\right) = 3$
- Between $-2 < x < \frac{1}{4}$ and $x < -2$, the function has a negative slope.
- $f(x)$ has exactly 2 stationary points.

8 (2 marks) For the graph of $y = f(x)$ shown, state

a the interval(s) when $f'(x)$ is negative (1 mark)

b the coordinate(s) when $f'(x) = 0$ and $f''(x) > 0$. (1 mark)

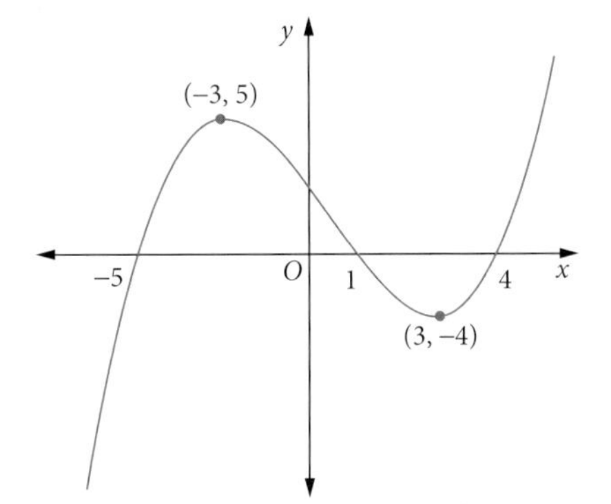

2.5 Straight line motion

Video playlist
Straight line motion

For straight line motion

- **displacement** = $x(t)$ = position of a particle at time t from a chosen origin
- **velocity** = $v(t)$ = velocity at time t, the rate of change of displacement
- **acceleration** = $a(t)$ = acceleration at time t, the rate of change of velocity.

Displacement is a 'signed distance' that can be positive or negative.

Velocity is a 'signed speed' that can be positive or negative.

Hence, velocity is the derivative of displacement, and acceleration is the derivative of velocity (and the second derivative of displacement).

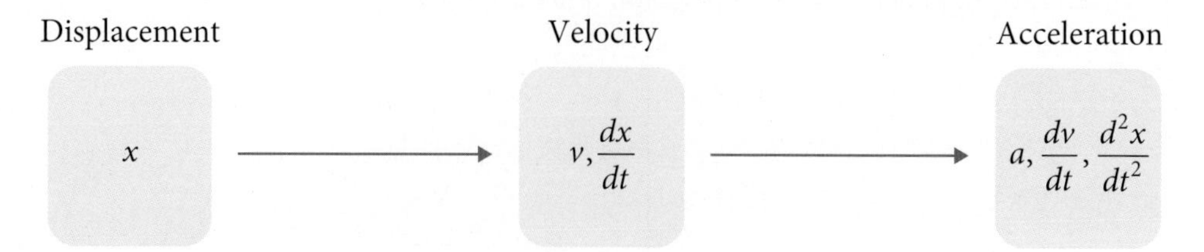

WORKED EXAMPLE 8 Finding straight line motion

The displacement of a particle travelling in a straight line is given by $x(t) = 2t^3 + t$, with x in metres and t in seconds.

a Find an expression for $v(t)$.

b Find the acceleration of the particle at $t = 3$.

Steps	Working
a Write the function for displacement, then use $\frac{dx}{dt}$ to find the velocity.	$x(t) = 2t^3 + t$ $\therefore v(t) = \frac{dx}{dt} = 6t^2 + 1$
b **1** Use $\frac{dv}{dt}$ to find a.	$v(t) = 6t^2 + 1$ $\therefore a(t) = \frac{dv}{dt} = 12t$
2 Find acceleration at $t = 3$.	$a(3) = 36\,\text{m/s}^2$

WORKED EXAMPLE 9 Interpreting straight line motion

A marble dropped into a barrel of water falls such that its height, h metres, after t seconds is given by $h(t) = 1.5 - 0.1t^2$.

a Find the height, velocity and acceleration of the marble at 3 seconds.

b Interpret your answer for the velocity and acceleration in the context of the question.

Steps	Working
a **1** Substitute $t = 3$ into $h(t)$ to find the height.	$h(3) = 1.5 - 0.1(3)^2 = 0.6\,\text{m}$
2 Differentiate $h(t)$ and calculate $v(3)$ to find the velocity.	$\frac{dh}{dt} = v(t) = -0.2t$ $\therefore v(3) = -0.2 \times 3 = -0.6\,\text{m/s}$
3 Differentiate $v(t)$ and calculate $a(3)$ to find the acceleration.	$\frac{d^2h}{dt^2} = \frac{dv}{dt} = a(t) = -0.2$ $\therefore a(3) = -0.2\,\text{m/s}^2$
b Comment.	The marble is moving with a speed of 0.6 m/s in a downward direction. The marble is moving with constant acceleration.

Recap

1 Consider $f(x) = x^2 + \frac{p}{x}$, $x \neq 0$. There is a stationary point on the graph of f when $x = 2$. The value of p is

A -16 **B** -8 **C** 2 **D** 8 **E** 16

2 The graph of the function $y = f(x)$ is shown.

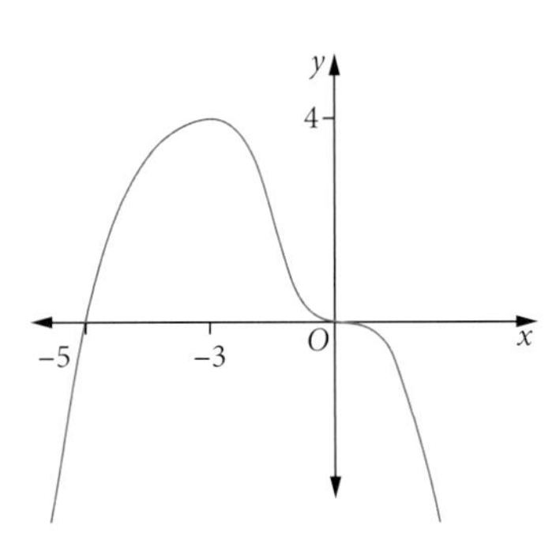

Which of the following could be the graph of the derivative function $y = f'(x)$?

A

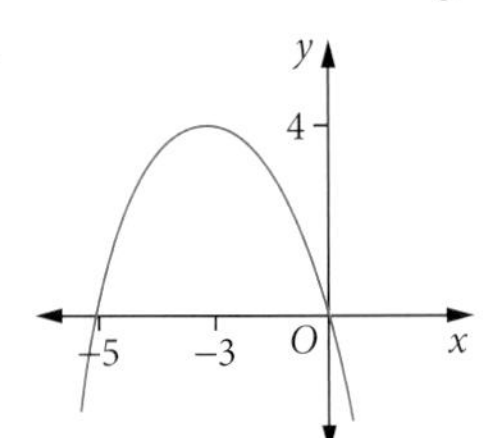

B

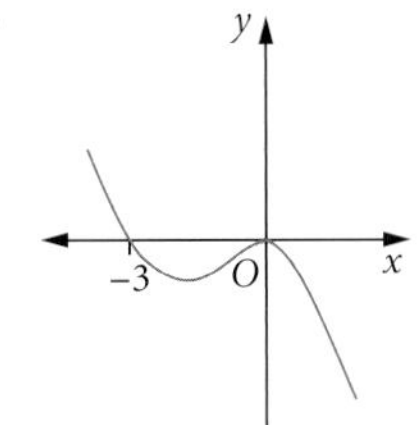

C

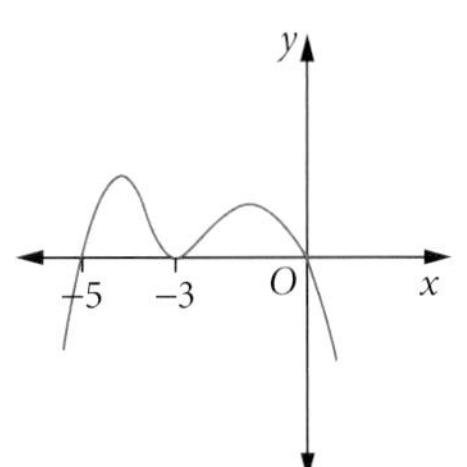

D

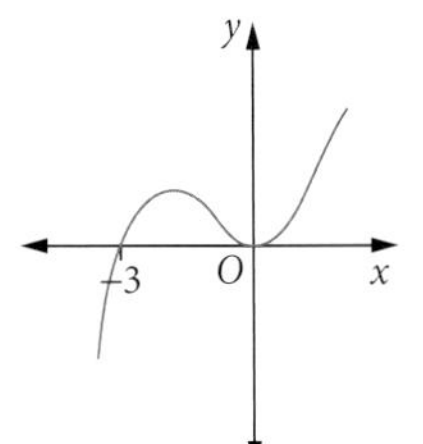

E

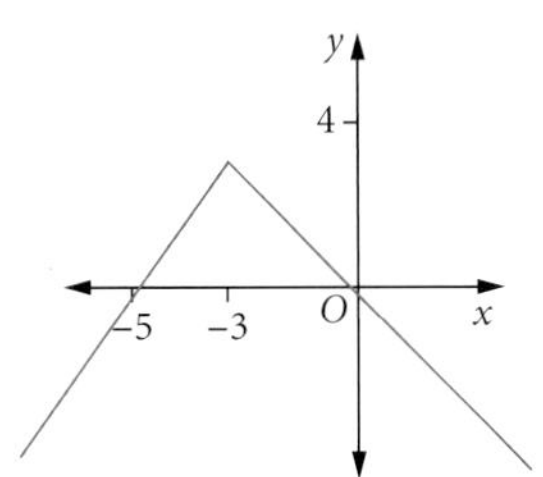

Mastery

3 WORKED EXAMPLE 8 The displacement of a particle travelling in a straight line is given by $x(t) = 2t^2 + 1$, where x is in metres and t is in seconds.

a Find an expression for $v(t)$.

b Find the acceleration of the particle at $t = 2$.

4 WORKED EXAMPLE 9 A ball is thrown into the air such that its height, h metres, after t seconds is given by $h(t) = 1.5t + t^2 - 0.5t^3$ for $0 \leq t \leq 3$.

a Find the height, velocity and acceleration of the ball when $t = 2$ seconds.

b Interpret your answer for the velocity in the context of the question.

5 The displacement of a particle travelling in a straight line is given by $x(t) = 3t^2 + 2t$, where x is in metres and t is in seconds.

a Find an expression for $v(t)$.

b Find the acceleration of the particle at $t = 6$.

Calculator-free

6 (6 marks) An object is travelling along a straight line over time t seconds, with displacement (in metres) according to the equation $x = t^3 + 6t^2 - 2t + 1$.

a Find the equations for its velocity and acceleration. (2 marks)

b What will its displacement be at 5 seconds? (1 mark)

c What will its velocity be at 5 seconds? (1 mark)

d Find the initial acceleration. (1 mark)

e Find its acceleration at 5 seconds. (1 mark)

7 (7 marks) The displacement of an object from a particular point is given by $x(t) = t^3 - 5t^2 + 6t + 10$, where $x(t)$ is in metres and t is in seconds.

a Find the average rate of change in the object's displacement from 2 seconds to 4 seconds. (1 mark)

b Find the velocity at 2 seconds. (1 mark)

c Find the velocity at 4 seconds. (1 mark)

d Find the average of the velocities at 2 seconds and 4 seconds and compare it to the answer for part **a**. (2 marks)

e Find the acceleration at 2 seconds. (1 mark)

f Find the acceleration at 5 seconds. (1 mark)

8 (9 marks) A particle is moving such that its displacement is given by $s = 2t^2 - 8t + 3$, where s is in metres and t is in seconds.

a Find its initial velocity. (1 mark)

b Show that its acceleration is constant and find its value. (2 marks)

c Find the displacement after 5 seconds. (1 mark)

d Determine when the particle will be at rest. (2 marks)

e What will the particle's displacement be at that time? (1 mark)

f Sketch the graph of the displacement against time. (2 marks)

Exam hack

If the particle is at rest, then it is not moving, therefore the velocity is zero.

Calculator-assumed

9 (4 marks) A car is travelling on a day's outing with displacement expressed by $x(t) = 10(t + 1)^2$, where x is the displacement from the driver's home in kilometres and t is the time taken in hours.

a Find its velocity exactly 5 hours from home. (2 marks)

b Find its acceleration exactly 5 hours from home. (2 marks)

10 (2 marks) A toy car is travelling with a velocity expressed by $v(t) = 2t^2 + 1$ m/s. Find its acceleration after travelling for 5 seconds.

11 (2 marks) The velocity of a particle travelling in a straight line is given by $v = (2t - 1)^2$, where v is in m/s and t is in seconds. Determine its acceleration, in m/s^2, at $t = 1$.

12 (4 marks) The position $x(t)$ cm of a particle at time t seconds is given by $x(t) = \frac{t+2}{2t+5}$.

a State the equation for the velocity, $v(t)$, of the particle at time t. (1 mark)

b State the equation for the acceleration, $a(t)$, of the particle at time t. (1 mark)

c Show that the magnitude of the acceleration of the particle is always four times the magnitude of the velocity. (2 marks)

13 (5 marks) A displacement equation is given by $x(t) = pt^2 + qt + r$, where p, q and r are constants and displacement is in metres and time in seconds. Given that $a(3) = -4$, $v(3) = -24$ and $x(3) = -34$, determine the value of p, q and r.

2.6 Optimisation problems

One common application of differentiation is to solve problems involving maximising or minimising a quantity that can be described by a function. These are sometimes called **optimisation problems**. To solve these problems, it is important to follow a planned process.

Video playlist Optimisation problems

Worksheets Starting maxima and minima problems

Greatest and least values

Applications of optimisation

Optimisation problems

Process for solving optimisation problems

1. Read the question, taking note of words such as maximum/minimum, least/highest, largest/smallest.
2. Set up a function to describe what needs to be maximised or minimised.
3. Make sure the quantity to be maximised or minimised is in terms of one variable only. (If not, rearrange variables in terms of each other.)
4. Differentiate the function and solve $f'(x) = 0$.
5. You may need to justify whether your answer is the maximum or minimum. This can be done either by using the second derivative or testing the gradient on either side of the stationary point.
6. Re-read the question to make sure you have answered what is asked.

WORKED EXAMPLE 10 Solving optimisation problems

A rectangular piece of cardboard, measuring 10 cm by 6 cm, has squares of length x cm cut out at each corner. The remaining faces are folded up to make a small packing box for jewellery. Find the maximum possible volume, in cm^3, for these jewellery boxes. Give the answer correct to one decimal place.

Steps	Working
1 Sketch a diagram.	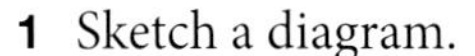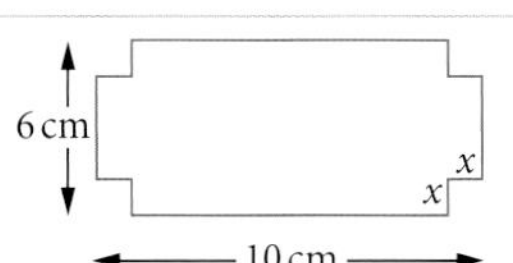
2 Set up an equation to describe the volume of the box.	length = $10 - 2x$, width = $6 - 2x$, height = x $V(x) = x(10 - 2x)(6 - 2x)$ $= 4x^3 - 32x^2 + 60x$
3 Solve $V'(x) = 0$ for local maximum and minimum values.	$V'(x) = 12x^2 - 64x + 60 = 0$ $x = \frac{8 \pm \sqrt{19}}{3}$ ($x = 4.1196$ or $x = 1.2137$) (x cannot be 4.11 cm since the width of the box is only 6 cm.)

4 Use the second derivative to test that the value gives a maximum.	$V''(x) = 24x - 64$ $V''\left(\frac{8-\sqrt{19}}{3}\right) < 0$, maximum

Exam hack

After solving maximum and minimum problems, re-read the question to make sure that you have answered it. One common mistake is to forget to answer the question, even though all the working is correct. Do you need the x or the V value?

5 Substitute this value of x into $V(x)$ to find the maximum volume of the jewellery box.	$V(1.2137) = 32.835$ The maximum possible volume of box is $32.8\,\text{cm}^3$.

USING CAS 4 Solving optimisation problems

Find the minimum distance from a point on the graph of $y = x^2 + 1$ to the point (30, 40), correct to two decimal places.

ClassPad

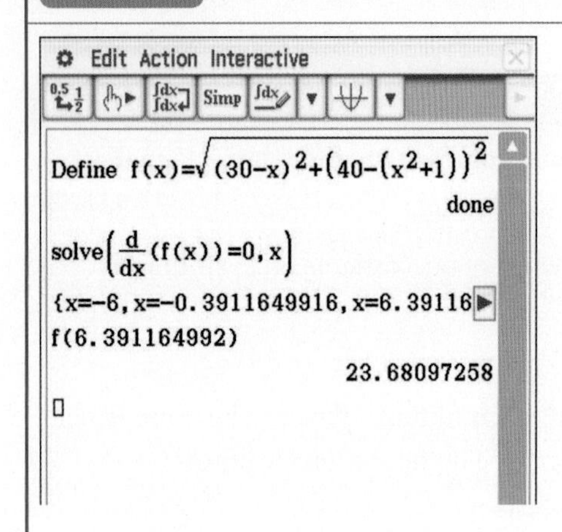

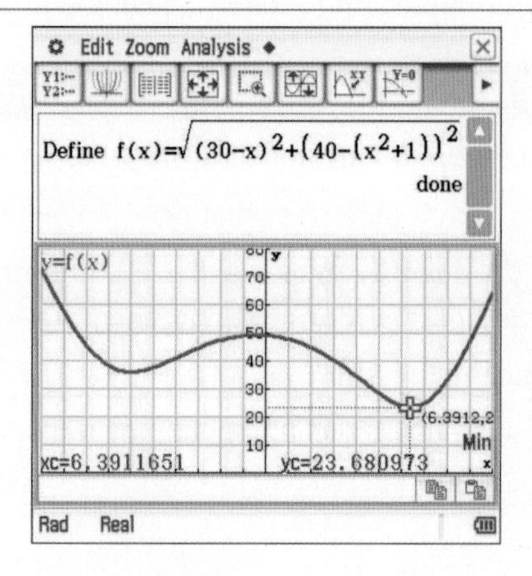

1 Define **f(x)** as shown above.

2 Derive using **Interactive > Calculation > Diff**, then using **Equation/Inequality** solve for **x**.

3 Copy the last solution and calculate **f(6.391164992)**.

4 The minimum distance will be displayed.

5 Graph **f(x)**.

6 Adjust the window settings so that **ymax = 80** to view the local minimums and maximum.

7 Tap **Analysis > G-Solve > Min**.

8 Press the **right arrow** key to move the cursor to the second local minimum.

9 The coordinates of the local minimum will be displayed.

The minimum distance from the function $y = x^2 + 1$ to the point (30, 40) is 23.68.

TI-Nspire

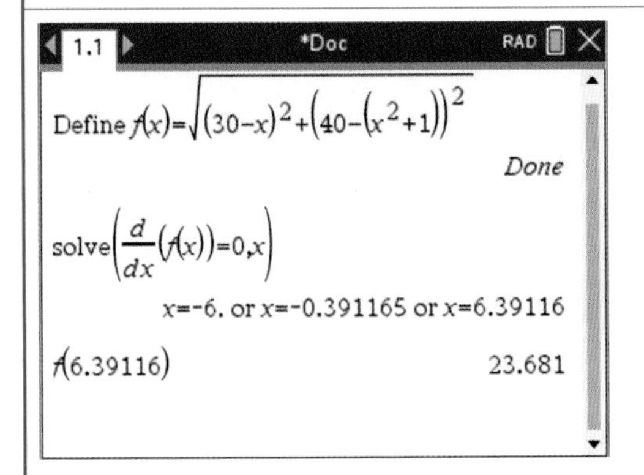

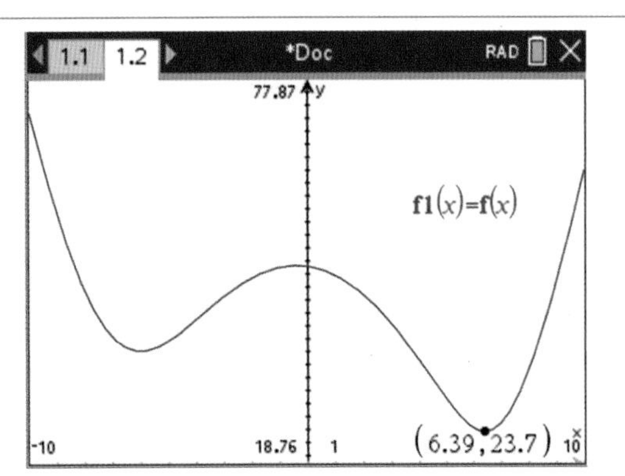

1. Define **f(x)** as shown above.
2. Use the **derivative** template to set the derivative of **f(x) = 0** and solve for **x**.
3. Press **ctrl + enter** for the approximate solutions.
4. Copy the last solution and calculate **f(6.39116)**.
5. The minimum distance will be displayed.
6. Graph **f(x)**.
7. Press **menu > Window / Zoom > Zoom – Fit** to view the local minimums and maximum.
8. Press **menu > Analyze Graph > Minimum**.
9. When prompted for the **lower bound**, click to the left of the second local minimum.
10. When prompted for the **upper bound**, click to the right of the minimum point.
11. The coordinates of the local minimum will be displayed.

The minimum distance from the function $y = x^2 + 1$ to the point $(30, 40)$ is 23.68.

WACE QUESTION ANALYSIS

© SCSA MM2019 Q16 Calculator-assumed (10 marks)

Video
WACE question analysis: Applications of differentiation

A cylindrical glass vase is filled with 20 spherical Christmas decorations as shown below (not all the decorations are visible). All the decorations have a diameter of one-third the internal diameter of the vase and they are completely contained within the vase. For design purposes the sum of the internal diameter of the base of the vase and the vase's internal height is to be 42 cm.

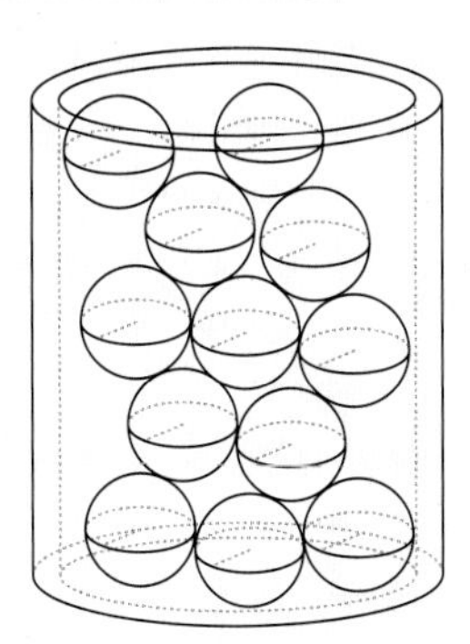

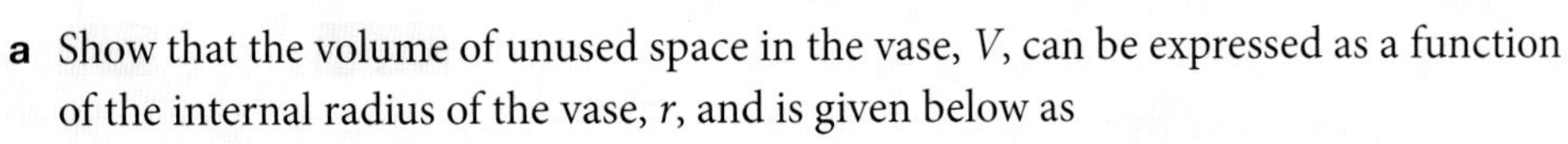

a Show that the volume of unused space in the vase, V, can be expressed as a function of the internal radius of the vase, r, and is given below as (3 marks)

$$V(r) = 2\pi\left(21r^2 - \frac{121}{81}r^3\right).$$

b Use calculus to determine the dimensions of the vase that will maximise the unused space in it. Give your answers rounded to the nearest millimetre. (4 marks)

c Can more than 20 of the spherical decorations fit inside the vase in part **b**? Use calculations to verify your answer. (3 marks)

Reading the question

- You should highlight the fact that the question is about cylinders and spheres, and note any relevant formula on the formula sheet.
- Note that the question gives you extra information about the diameter of the balls, and the relationship between the base and height of the vase. Note that the question uses diameter but the formulas are in terms of radius.
- Part **a** asks you to 'show', which means you need enough working to show how you obtained the answer.
- In part **b** the question is asking you to use calculus, which means you must show some differentiation. Part **b** is also asking for the dimensions of the vase, not the volume. Note that the answer is to be rounded to a particular level of accuracy.
- Part **c** is asking for calculations; just answering 'yes' or 'no' will not suffice.

Thinking about the question

- As each part of the question is worth more than 2 marks, you will need to show some working for each part.
- Even when the question says to 'show' or 'use calculus', you can still use CAS to complete steps such as differentiating. Write down each of these steps as part of your working.
- Note that the question is set up so that even if you cannot complete part **a**, you will be able to use that formula for part **b**.
- For part **b**, you should determine the second derivative to check that your answer is a maximum.

Worked solution (✓ = 1 mark)

a $2r + h = 42$

$h = 42 - 2r$

$$V(r,h) = \pi r^2 h - 20\left(\frac{4}{3}\pi\left(\frac{r}{3}\right)^3\right)$$

$$V(r) = \pi r^2(42 - 2r) - \frac{80\pi}{81}r^3$$

$$= 2\pi\left(21r^2 - r^3 - \frac{40}{81}r^3\right)$$

$$= 2\pi\left(21r^2 - \frac{121}{81}r^3\right)$$

determines an expression for h in terms of r ✓

states an expression for the volume of unused space in terms of r and h ✓

clearly shows that the expression for h in terms of r can substitute into V and simplifies to determine required result ✓

b $V'(r) = 2\pi\left(42r - \dfrac{363r^2}{81}\right)$

$0 = 42r - \dfrac{363r^2}{81}$

$0 = r\left(42 - \dfrac{363r}{81}\right)$

$r = 0$ or $\dfrac{1134}{121}$ $(= 9.372 \text{ (3 d.p.)})$

$V''(r) = 2\pi\left(42 - \dfrac{726r}{81}\right)$

$V''(9.372) = -ve\ (= -84\pi) \Rightarrow \max$

Dimensions are $r = 9.4$ cm and $h = 23.3$ cm.

determines first derivative of $V(r)$ ✓

equates to zero and determines 0 and 9.4 are solutions ✓

clearly shows the use of the second derivative or sign test to show that $r = 9.4$ is a maximum ✓

states the dimensions of the vase that maximise the unused space rounded to the nearest mm ✓

c $V(9.4) = 3863.1\text{ cm}^3$

$V(\text{decoration}) = 127.7\text{ cm}^3$

There is likely space for more decorations, but it is not certain as it would depend on the way the balls were packed into the vase.

states the volume of unused space and the volume of one decoration ✓

infers likely to fit more ✓

states the limitation of packing ✓

EXERCISE 2.6 Optimisation problems

ANSWERS p. 390

Recap

1 If the displacement of an object is given by $x(t) = \dfrac{t+1}{1-2t^2}$, determine the acceleration when $t = 2$.

Give your answer correct to two decimal places.

2 If the displacement of an object is given by $x(t) = 2(t-3)(t+3)(t+3)$ metres in t seconds, determine when the object is at rest.

Mastery

3 WORKED EXAMPLE 10 A right circular cone is made with height h cm and radius r cm, where the height is 2 cm minus the radius.

Find the maximum volume, in cm^3, of this cone. Give your answer correct to two decimal places.

4 Using CAS 4 Find the minimum distance from the origin to a point on the hyperbola with the rule $y = \dfrac{3}{x-1} + 2$, correct to two decimal places.

Calculator-free

5 (4 marks) A rectangular prism has dimensions 4 cm, x cm and $(4 - x)$ cm. Find the maximum possible volume of this rectangular prism.

6 (5 marks) A wire of length P cm is used to construct a square-based rectangular prism of volume V cm^3. Show that the maximum volume occurs when the shape is a cube.

7 (4 marks) A wire frame in the shape of a rectangular box is constructed using P cm of wire. The frame has length x cm, width y cm and height h cm.

a Express the volume, V, of the box in the form $V = axy(P - bx - cy)$, where a, b, c are constants. (2 marks)

b If $y = 2x$, show that the maximum volume is $\dfrac{P^3}{6 \times 18^2}$ cm^3. (2 marks)

8 (4 marks) Let x and y be two non-negative numbers and let S be the sum of their squares.

a Show that the product of the two numbers can be found using $x\sqrt{S - x^2}$. (2 marks)

b Hence show that the maximum product occurs when $x = \sqrt{\dfrac{S}{2}}$. (2 marks)

Calculator-assumed

9 (7 marks) A plastic brick is made in the shape of a right triangular prism. The triangular end is an equilateral triangle with side length x cm and the length of the brick is y cm.

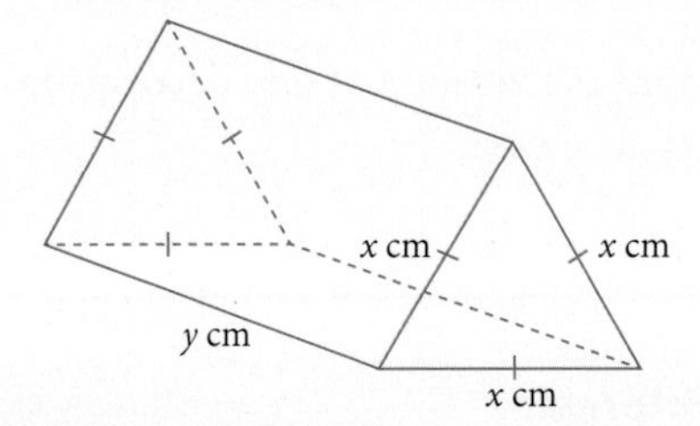

The volume of the brick is 1000 cm^3.

a Find an expression for y in terms of x. (2 marks)

b Show that the total surface area, A cm^2, of the brick is given by

$$A = \frac{4000\sqrt{3}}{x} + \frac{\sqrt{3}x^2}{2}$$ (2 marks)

c Find the value of x for which the brick has the minimum total surface area. (You do not have to find this minimum.) (3 marks)

10 (3 marks) Zoe has a rectangular piece of cardboard that is 8 cm long and 6 cm wide. Zoe cuts squares of side length x centimetres from each of the corners of the cardboard, as shown in the diagram below.

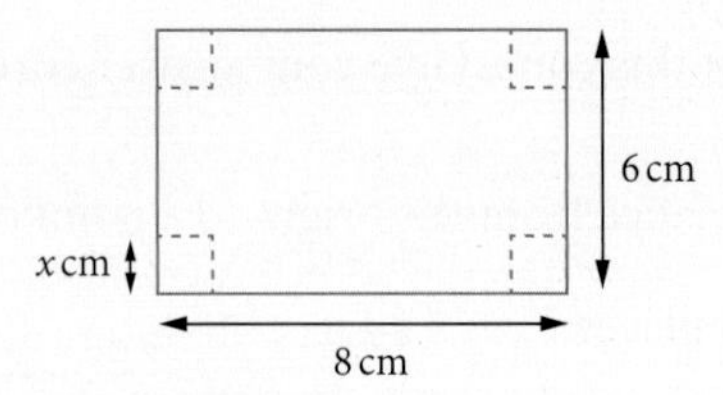

Zoe turns up the sides to form an open box.

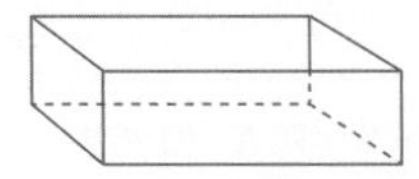

Determine the value of x for which the volume of the box is a maximum.

11 (4 marks) A rectangular sheet of cardboard has a length of 80 cm and a width of 50 cm. Squares, of side length x centimetres, are cut from each of the corners, as shown in the diagram.

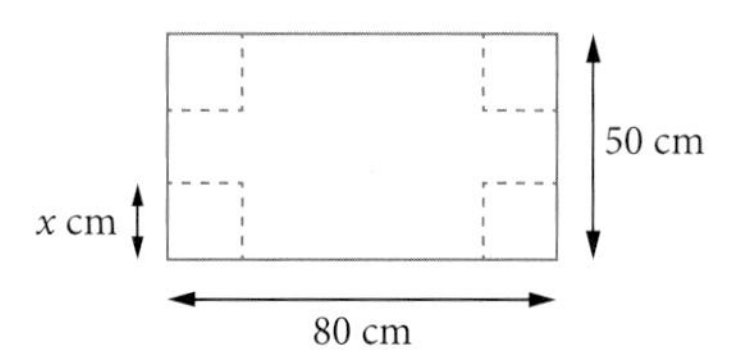

A rectangular box with an open top is then constructed, as shown in the diagram below.

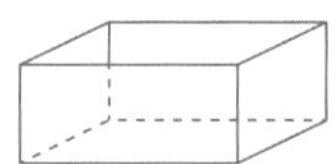

Determine when the volume of the box is a maximum.

12 (5 marks) A cylinder fits exactly in a right circular cone so that the base of the cone and one end of the cylinder are in the same plane, as shown in the diagram. The height of the cone is 5 cm and the radius of the cone is 2 cm.

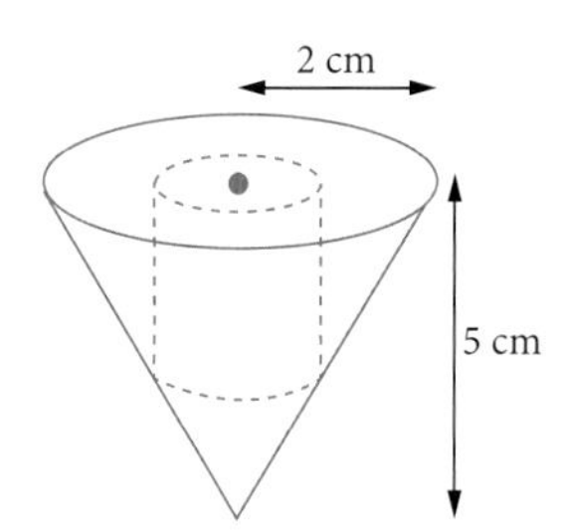

The radius of the cylinder is r cm and the height of the cylinder is h cm, and $h = \dfrac{10 - 5r}{2}$ for the cylinder inscribed in the cone as shown.

a Write a formula for the total surface area (S) of the cylinder in terms of r. (2 marks)

b Find the value of r for which S is a maximum. (3 marks)

13 (4 marks) A rectangle $XYZW$ has two vertices, X and W, on the x-axis and the other two vertices, Y and Z, on the graph of $y = 9 - 3x^2$, as shown in the diagram. The coordinates of Z are (a, b) where a and b are positive real numbers.

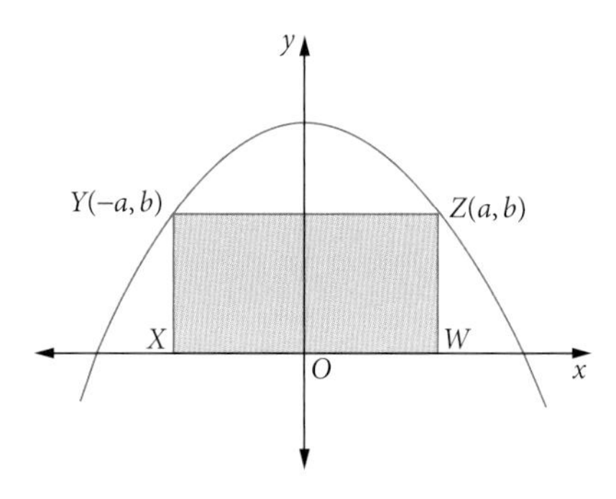

a Find the area, A, of rectangle $XYZW$ in terms of a. (1 mark)

b Find the maximum value of A and the value of a for which this occurs. (3 marks)

14 (4 marks) P is the point on the line $2x + y - 10 = 0$ such that the length of OP, the line segment from the origin O to P, is a minimum. Find the coordinates of P **and** this minimum length.

15 (3 marks) A right-angled triangle, OBC, is formed using the horizontal axis and the point $C(m, 9 - m^2)$, where $0 < m < 3$, on the parabola $y = 9 - x^2$, as shown below.

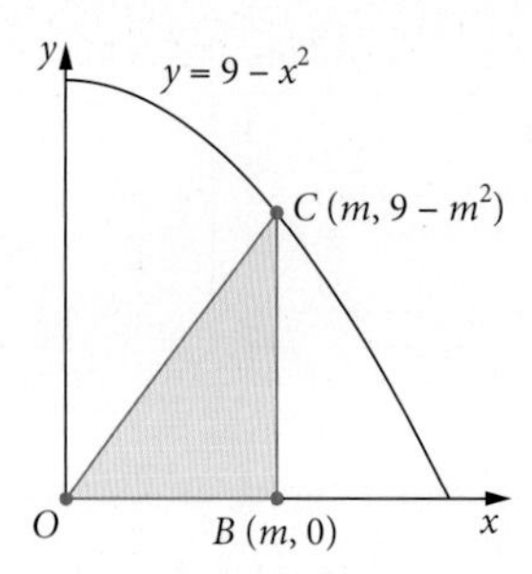

Determine the maximum area of the triangle OBC.

16 (13 marks) Tasmania Jones is in Switzerland. He is working as a construction engineer and he is developing a thrilling train ride in the mountains. He chooses a region of a mountain landscape, the cross-section of which is shown in the diagram.

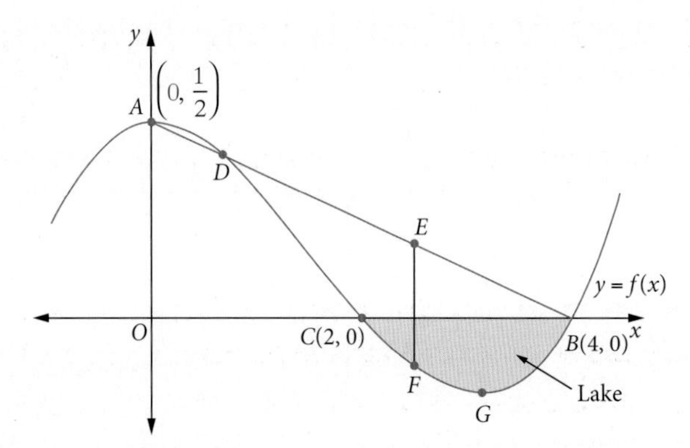

The cross-section of the mountain and the valley shown in the diagram (including a lake bed) is modelled by the function with rule

$$f(x) = \frac{3x^3}{64} - \frac{7x^2}{32} + \frac{1}{2}$$

Tasmania knows that $A\left(0, \frac{1}{2}\right)$ is the highest point on the mountain and that $C(2, 0)$ and $B(4, 0)$ are the points at the edge of the lake, situated in the valley. All distances are measured in kilometres.

a Show use of calculus to find the exact coordinates of G, the deepest point in the lake. (3 marks)

Tasmania's train ride is made by constructing a straight railway line AB from the top of the mountain, A, to the edge of the lake, B. The section of the railway line from A to D passes through a tunnel in the mountain.

b Write down the equation of the line that passes through A and B. (2 marks)

In order to ensure that the section of the railway line from D to B remains stable, Tasmania constructs vertical columns from the lake bed to the railway line. The column EF is the longest of all possible columns.

c Show use of calculus to find the x-coordinate of E. (3 marks)

Tasmania's train travels down the railway line from A to B. The speed, in km/h, of the train as it moves down the railway line is described by the function

$$v(x) = k\sqrt{x} - mx^2 \text{ for } 0 \le x \le 4$$

where x is the x-coordinate of a point on the front of the train as it moves down the railway line, and k and m are positive real constants.

The train begins its journey at $A\left(0, \frac{1}{2}\right)$. It increases its speed as it travels down the railway line. The train then slows to a stop at $B(4, 0)$, that is, $v(4) = 0$.

d Find k in terms of m. (2 marks)

e Show use of calculus to find the value of x for which the speed, v, is a maximum. (3 marks)

2 Chapter summary

The increments formula

If δx is small, we can say $\frac{dy}{dx} \approx \frac{\delta y}{\delta x}$. Rearranging gives $\delta y \approx \frac{dy}{dx} \times \delta x$, which is called the **increments formula**.

Second derivative

The **second derivative** is the rate of change of the first derivative function.

Points of inflection

A non-stationary **point of inflection** is where a function changes **concavity**.

Increasing and decreasing functions

A function is **increasing** when it has a positive gradient $\left(\frac{dy}{dx} > 0\right)$ and its graph is pointing up.

A function is **decreasing** when it has a negative gradient $\left(\frac{dy}{dx} < 0\right)$ and its graph is pointing down.

Stationary points

- A **turning point** (local maximum or minimum point) is found where $f'(x) = 0$ and also where the sign of the gradient changes on either side of the stationary point.
- A **stationary point of inflection** is found where $f'(x) = 0$ and the concavity changes. The sign of the gradient stays *the same* on both sides of the stationary point.
- If $f''(x) > 0$, the turning point is a minimum.
- If $f''(x) < 0$, the turning point is a maximum.
- If $f''(x) = 0$, further investigation into the nature of the stationary point is required.

$f'(x)$ before the stationary point	$f'(x)$ at the stationary point	$f'(x)$ after the stationary point	$f''(x)$ at the stationary point	Type of stationary point
+	0	−	Less than zero	Local maximum point
−	0	+	Greater than zero	Local minimum point
+	0	+	0	Stationary point of inflection
−	0	−	0	Stationary point of inflection

Curve sketching

- Identify **key features** of a graph when sketching:
  - general shape
  - y-intercept
  - x-intercept(s)
  - stationary points, including their nature.

The graph of the derivative function

Graph of $f(x)$	increasing	decreasing	stationary	straight line	discontinuous point, vertical asymptote
Graph of $f'(x)$	positive, above x-axis	negative, below x-axis	0 (x-intercept)	horizontal (flat) line	discontinuous point, asymptote

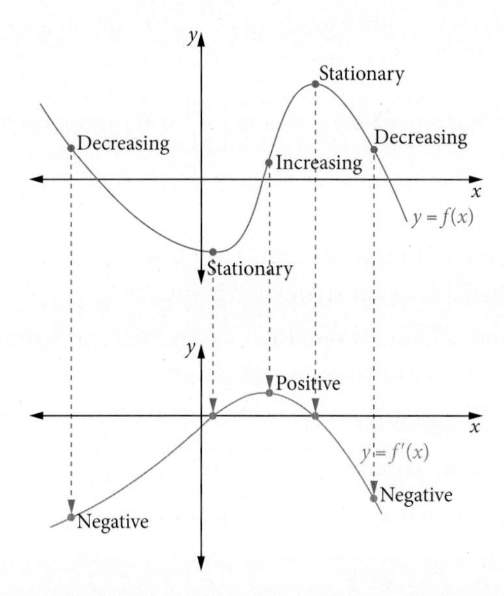

Straight line motion

- **displacement** = $x(t)$ = position of a particle at time t from a chosen origin
- **velocity** = $v(t)$ = velocity at time t, the rate of change of displacement
- **acceleration** = $a(t)$ = acceleration at time t, the rate of change of velocity

Hence, velocity is the derivative of displacement, and acceleration is the derivative of velocity (and the second derivative of displacement).

Optimisation problems

- Techniques for finding maximum and minimum values to help solve problems
  1. Read the question, taking note of words such as maximum/minimum, least/highest, largest/smallest.
  2. Set up a function to describe what needs to be maximised or minimised.
  3. Make sure the quantity to be maximised or minimised is in terms of one variable only. (If not, rearrange variables in terms of each other.)
  4. Differentiate the function and solve $f'(x) = 0$.
  5. You may need to justify whether your answer is the maximum or minimum. This can be done either by using the second derivative or testing the gradient on either side of the stationary point.
  6. Re-read the question to make sure you have answered what is asked.

Cumulative examination: Calculator-free

Total number of marks: 21 Reading time: 2 minutes Working time: 20 minutes

1 (3 marks) Find the equation of the tangent to $y = x^2(x + 1)$ at the point $(-1, 0)$.

2 (6 marks) For the function $f(x) = x^2(2x - 5)$

a determine $f'\left(\frac{5}{3}\right)$ and $f''\left(\frac{5}{3}\right)$ (4 marks)

b in relation to the graph of $f(x)$, explain the meaning of your answers in part **a**. (2 marks)

3 (4 marks) Use the quotient rule to show that the derivative of $\frac{2x^2 + 1}{\sqrt{x}}$ is equal to $\frac{6x^2 - 1}{2x^{\frac{3}{2}}}$.

4 (5 marks) A continuous function, f, satisfies the conditions

- $f(-2) = 0$
- for $x < -\frac{5}{4}, f(x) > 0$
- for $-\frac{5}{4} < x < 1, f(x) < 0$
- f has exactly 2 stationary points
- $f(1) = f'(1) = f''(1) = 0$.

Sketch the function.

5 (3 marks) The graph of $y = f(x)$ is drawn below.

Sketch the graph of $y = f'(x)$.

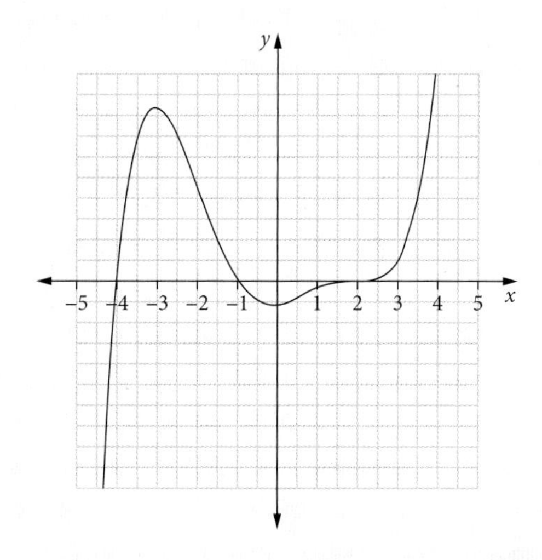

Cumulative examination: Calculator-assumed

Total number of marks: 30 Reading time: 3 minutes Working time: 30 minutes

1 © SCSA MM2016 Q11 (3 marks) The area of a triangle can be found by the formula: Area $= \dfrac{ab\sin C}{2}$.

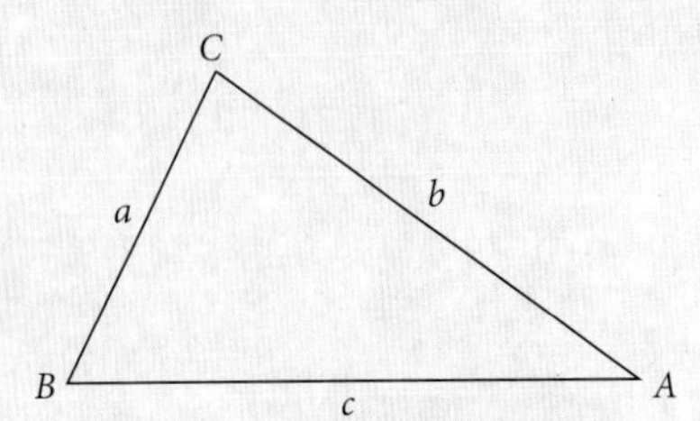

Using the increments formula, determine the approximate change in area of an equilateral triangle, with each side of 10 cm, when each side increases by 0.1 cm.

2 © SCSA MM2017 Q17 (6 marks) A beverage company has decided to release a new product. 'Joosilicious' is to be sold in 375 mL cans that are perfectly cylindrical. {Hint: 1 mL = 1 cm^3}

a If the cans have a base radius of x cm, show that the surface area of the can, S, is given by: $S = 2\pi x^2 + \dfrac{750}{x}$. (2 marks)

b Using calculus methods, and showing full reasoning and justification, find the dimensions of the can that will minimise its surface area. (4 marks)

3 © SCSA MM2018 Q15 (5 marks) The population of mosquitos, P (in thousands), in an artificial lake in a housing estate is measured at the beginning of the year. The population after t months is given by the function, $P(t) = t^3 + at^2 + bt + 2, 0 \le t \le 12$.

The rate of growth of the population is initially increasing. It then slows to be momentarily stationary in mid-winter (at $t = 6$), then continues to increase again in the last half of the year.

Determine the values of a and b.

4 (10 marks) The displacement of a particle s moving along a straight line at time t seconds is given by $s = t^3 - 4t^2 + 4t - 10$ metres.

a Determine the change in displacement in the first 2 seconds. (2 marks)

b Determine the velocity of the particle when $t = 5$ seconds. (2 marks)

c Determine when the particle is instantaneously at rest. (2 marks)

d Determine the initial acceleration of the particle. (2 marks)

e Determine the distance travelled in the first 2 seconds. (2 marks)

5 (6 marks) An **open** (no lid) rectangular tank of volume 8 cm^3 has length x cm and width y cm.

a Show that the surface area of the tank can be written in the form: $A = xy + \dfrac{16}{x} + \dfrac{16}{y}$. (2 marks)

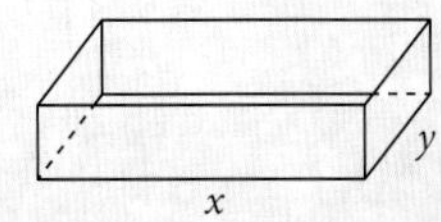

b Due to storage requirements, the length, x, is a fixed constant, $x = k$. The minimum value of A can be expressed in the form: $A = a\sqrt{k} + \dfrac{b}{k}$. Determine the values of the constants a and b. (4 marks)

CHAPTER

3

INTEGRALS

Syllabus coverage

TOPIC 3.2: INTEGRALS

Anti-differentiation

3.2.1 identify anti-differentiation as the reverse of differentiation

3.2.2 use the notation $\int f(x)dx$ for anti-derivatives or indefinite integrals

3.2.3 establish and use the formula $\int x^n dx = \frac{1}{n+1}x^{n+1} + c$ for $n \neq -1$

3.2.6 identify and use linearity of anti-differentiation

3.2.7 determine indefinite integrals of the form $\int f(ax - b)dx$

3.2.8 identify families of curves with the same derivative function

3.2.9 determine $f(x)$, given $f'(x)$ and an initial condition $f(a) = b$

Definite integrals

3.2.10 examine the area problem and use sums of the form $\Sigma_i f(x_i)\,\delta x_i$ to estimate the area under the curve $y = f(x)$

3.2.11 identify the definite integral $\int_a^b f(x)dx$ as a limit of sums of the form $\Sigma_i f(x_i)\,\delta x_i$

3.2.12 interpret the definite integral $\int_a^b f(x)dx$ as area under the curve $y = f(x)$ if $f(x) > 0$

3.2.13 interpret $\int_a^b f(x)dx$ as a sum of signed areas

3.2.14 apply the additivity and linearity of definite integrals

Fundamental theorem

3.2.15 examine the concept of the signed area function $F(x) = \int_a^x f(t)dt$

3.2.16 apply the theorem: $F'(x) = \frac{d}{dx}\left(\int_a^x f(t)\,dt\right) = f(x)$, and illustrate its proof geometrically

3.2.17 develop the formula $\int_a^b f'(x)dx = f(b) - f(a)$ and use it to calculate definite integrals

Applications of integration

3.2.18 calculate total change by integrating instantaneous or marginal rate of change

3.2.19 calculate the area under a curve

3.2.20 calculate the area between curves determined by functions of the form $y = f(x)$

3.2.21 determine displacement given velocity in linear motion problems

3.2.22 determine positions given linear acceleration and initial values of position and velocity

Video playlists (7):

3.1 The anti-derivative

3.2 Approximating areas under curves

3.3 The definite integral and the fundamental theorem of calculus

3.4 Area under a curve

3.5 Areas between curves

3.6 Straight line motion

WACE question analysis Integrals

Worksheets (9):

3.1 Anti-derivatives 1 • The chain rule

3.2 Areas using rectangles

3.3 Definite integrals

3.4 Calculating physical areas

3.5 Calculating areas between curves • Areas between curves 1 • Areas between curves 2

3.6 Displacement, velocity and acceleration

Nelson MindTap

To access resources above, visit **cengage.com.au/nelsonmindtap**

3.1 The anti-derivative

The **anti-derivative** or **integral** or **primitive** of a function $f(x)$ is the function $F(x)$ whose **derivative** is $f(x)$. In other words, if $F(x)$ is the anti-derivative of $f(x)$, then $F'(x) = f(x)$. Finding the anti-derivative is called **anti-differentiation** or **integration** and reverses the process of finding the derivative. When we differentiate a **constant** term (number), we get 0, so when anti-differentiating, we must include '$+ c$' in the answer, where c stands for any number, called the **constant of integration.**

The general anti-derivative of $2x$ is $x^2 + c$, where c is a constant.

Algebraically, this is written as $\int 2x\,dx = x^2 + c$.

This is called the anti-derivative of $2x$, or the integral or primitive of $2x$.

- The symbol $\int$ is read as 'the integral of'.
- 'dx' means 'with respect to x'.

Video playlist The anti-derivative

Worksheet Anti-derivatives 1

The integral of ax^n

$$\int ax^n\,dx = \frac{ax^{n+1}}{n+1} + c \text{ for } n \neq -1$$

where ax^n is called the **integrand**.

For any power function, we say 'add one to the power, divide by the new power'.

WORKED EXAMPLE 1 Finding the anti-derivative

Find the anti-derivative of the expression $3x^2 + 2x - 4$.

Steps	Working
1 Write the function as a derivative.	$\frac{dy}{dx} = 3x^2 + 2x - 4$
2 Integrate each term using the formula $\int ax^n dx = \frac{ax^{n+1}}{n+1} + c$.	$= \int (3x^2 + 2x - 4)\,dx$ $= \frac{3x^3}{3} + \frac{2x^2}{2} - 4x + c$
3 Simplify the expression.	$= x^3 + x^2 - 4x + c$

Exam hack

You can check your answer by differentiating it to see whether you get the original expression.

Worksheet
The chain rule

The integral of $(ax + b)^n$

Consider the derivative of $y = (2x - 1)^3$.

Let $y = u^3$

$\frac{dy}{du} = 3u^2$

Also $u = 2x - 1$

$\frac{du}{dx} = 2$

Using the chain rule

$$\frac{dy}{dx} = \frac{dy}{du} \times \frac{du}{dx}$$
$$= 3u^2 \times 2$$
$$= 6u^2$$
$$= 6(2x - 1)^2$$

So in reverse, $\int 6(2x - 1)^2 dx = (2x - 1)^3 + c$

so $\int (2x - 1)^2 dx = \frac{1}{6}(2x - 1)^3 + c$.

This matches the general statement, 'add one to the power, divide by the new power', but here we also divide by the derivative of the 'inner' expression: $2x - 1$.

When integrating $(2x - 1)^2$, we divide by 3 (the power + 1) and by 2 (the coefficient of x).

Now consider the derivative of any power of $(ax + b)$, such as $y = (ax + b)^{n+1}$.

Let $y = u^{n+1}$ where $u = ax + b$ and use the chain rule.

$$\frac{dy}{dx} = \frac{dy}{du} \times \frac{du}{dx}$$
$$= (n + 1)u^n \times a$$
$$= a(n + 1)u^n$$
$$= a(n + 1)(ax + b)^n$$

So in reverse, $\int a(n + 1)(ax + b)^n dx = (ax + b)^{n+1} + c$,

so $\int (ax + b)^n dx = \frac{1}{a(n + 1)}(ax + b)^{n+1} + c$.

When integrating $(ax + b)^n$, we divide by $n + 1$ (the power + 1) and by a (the coefficient of x).

The integral of $(ax + b)^n$

$$\int (ax + b)^n dx = \frac{(ax + b)^{n+1}}{a(n + 1)} + c$$

Remember this as a **reverse chain rule**.

Exam hack

Many students forget to use this rule in examples

such as $\int \frac{1}{(5x - 3)^2} dx$. This becomes very usefully,

$\int (5x - 3)^{-2} dx$, now of the form $\int (ax + b)^n dx$.

3.1

WORKED EXAMPLE 2 Finding the integral of $(ax + b)^n$

Find $\int (3x + 2)^4 \, dx$.

Steps	Working
1 Integrate using $\int (ax + b)^n \, dx = \dfrac{(ax + b)^{n+1}}{a(n + 1)} + c$.	$\int (3x + 2)^4 \, dx = \dfrac{1}{3(4 + 1)}(3x + 2)^{4+1} + c$
2 Simplify.	$= \dfrac{1}{15}(3x + 2)^5 + c$

We can skip Step 1 and go straight to the answer.

WORKED EXAMPLE 3 Finding y given $\dfrac{dy}{dx}$ and a point

Find y if $\dfrac{dy}{dx} = x^2 - x + 4$ and $x = 1$ when $y = 0$.

For this question, we can use given information to find the value of c.

Steps	Working
1 Write the expression as a derivative.	$\dfrac{dy}{dx} = x^2 - x + 4$
2 Integrate each term and simplify.	$y = \dfrac{x^3}{3} - \dfrac{x^2}{2} + 4x + c$
3 Find the value of c by substituting $x = 1$, $y = 0$.	When $x = 1$, $y = 0$: $0 = \dfrac{1^3}{3} - \dfrac{1^2}{2} + 4(1) + c$ $0 = \dfrac{23}{6} + c$ $c = -\dfrac{23}{6}$
4 State the answer including the value of c.	$y = \dfrac{x^3}{3} - \dfrac{x^2}{2} + 4x - \dfrac{23}{6}$

USING CAS 1 Finding y given $\frac{dy}{dx}$ and a point

Find y if $\frac{dy}{dx} = 3x^2 + x + 3$ and $x = 1$ when $y = 0$.

ClassPad

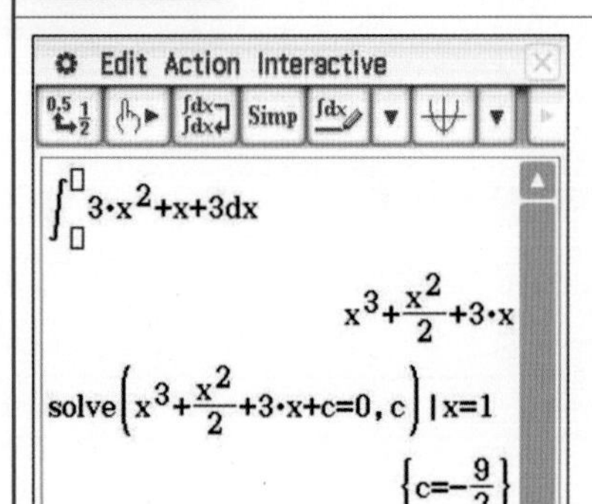

1 Highlight the expression then tap **Interactive > Calculation > ∫**.
2 Add **+c** to the expression as it is not included in the solution.
3 Set the expression including the **+c** equal to **0** and solve by including the condition that **x = 1**.
4 The value of **c** will be displayed.

TI-Nspire

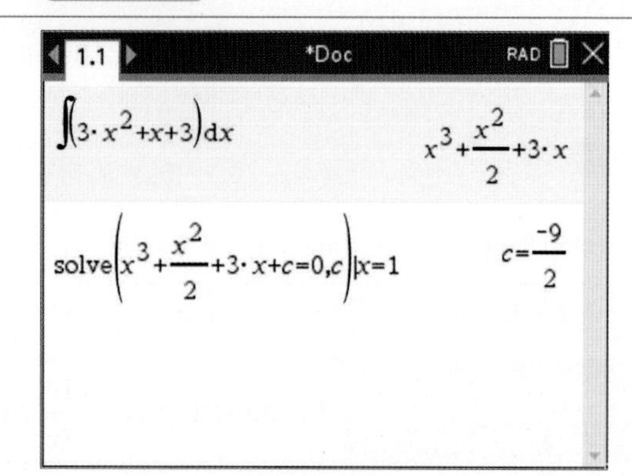

1 Press **menu > Calculus > Integral**.
2 Enter the derivative expression followed by **dx**.
3 Add **+c** to the expression as it is not included in the solution.
4 Set the expression including the **+c** equal to **0** and solve by including the condition that **x = 1**.
5 The value of **c** will be displayed.

The solution is $y = x^3 + \frac{x^2}{2} + 3x - \frac{9}{2}$.

EXERCISE 3.1 The anti-derivative

ANSWERS p. 392

Mastery

1 WORKED EXAMPLE 1 Find the anti-derivative of the expression $x^3 + 3x^2 - 4x$.

2 WORKED EXAMPLE 2 Find $\int (4x - 1)^3 dx$.

3 Determine the anti-derivative for each of the expressions.

a $x^2 - 3x + 2$ **b** $(x - 3)(2x + 4)$ **c** $\frac{x^2 - 2x}{x}$ **d** $\sqrt{x} - \frac{1}{x^2} - 3$

e $\sqrt{(2x - 3)}$ **f** $\sqrt{x}(x^2 - 2x + 3)$ **g** $\frac{1}{(2x - 3)^2}$ **h** $\sqrt[3]{3x - 4}$

> **Exam hack**
>
> There is no reverse of the product or quotient rule. Try simplifying the expression prior to anti-differentiating.

4 WORKED EXAMPLE 3 Determine the function $f(x)$ if $f'(x) = 2x$ and $y = 1$ when $x = 2$.

5 Using CAS 1 Determine the function y if $\dfrac{dy}{dx} = 2x + 4$ and $y = 1$ when $x = 2$.

6 Determine the function $f(x)$ if $f'(x) = \dfrac{4}{\sqrt[3]{4 - 2x}}$ and $f(-2) = 10$.

Calculator-free

7 (2 marks) Determine the anti-derivative of the expression $3x^2 + 4x^3 - 2$.

8 (2 marks) Determine $\displaystyle\int \frac{1}{(3x + 4)^4}\,dx$.

9 (2 marks) Find the anti-derivative of $(4 - 2x)^{-5}$ with respect to x.

10 (2 marks) Let $f'(x) = 3x^2 - 2x$ such that $f(4) = 0$. Determine $f(x)$.

11 (2 marks) If a and b are positive integers and $f'(x) = ax^2 - bx$, draw a possible sketch of $f(x)$.

12 (2 marks) For each graph of a function, sketch a possible graph of its anti-derivative.

a

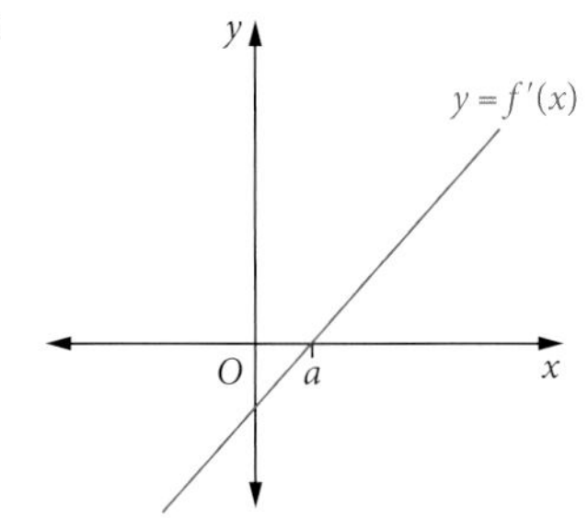

(1 mark)

b

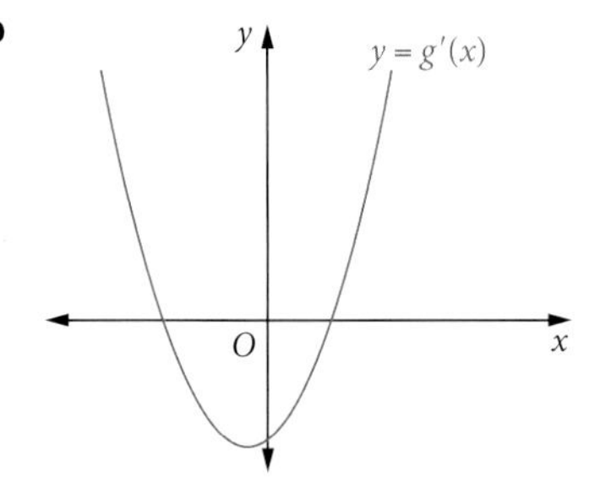

(1 mark)

13 (2 marks) Determine $\displaystyle\int \frac{2x - 3}{\sqrt{x^2 - 3x}}\,dx$.

Calculator-assumed

14 (2 marks) Find $f(x)$ given that $f(1) = -\dfrac{7}{4}$ and $f'(x) = 2x^2 - \dfrac{1}{4}x^{-\frac{2}{3}}$.

15 (2 marks) Find an anti-derivative of $\dfrac{1}{(2x - 1)^3}$ with respect to x.

16 (2 marks) Let $f'(x) = \dfrac{2}{\sqrt{2x - 3}}$. If $f(6) = 4$, determine $f(x)$.

17 (2 marks) If $f'(x) = g'(x) + 3$, $f(0) = 2$ and $g(0) = 1$, show that $f(x) = g(x) + 3x + 1$.

Video playlist
Approximating areas under curves

Worksheet
Areas using rectangles

3.2 Approximating areas under curves

The area under the graph of a function can provide important information in fields such as surveying, physics and the social sciences.

For example, the area under a speed graph shows the distance travelled.

We can estimate the area under a graph using a series of rectangles. The more rectangles we use, the more accurate the area will be. For example, the triangle below has an exact area of 36 units^2. To the right of this, the triangle has been approximated by 2, 3 and 4 rectangles.

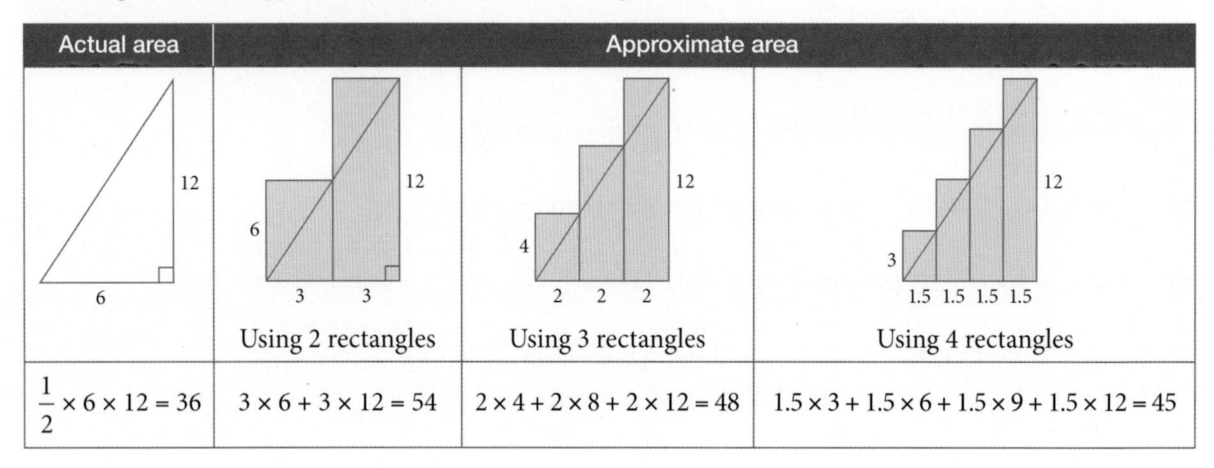

$\frac{1}{2} \times 6 \times 12 = 36$	$3 \times 6 + 3 \times 12 = 54$	$2 \times 4 + 2 \times 8 + 2 \times 12 = 48$	$1.5 \times 3 + 1.5 \times 6 + 1.5 \times 9 + 1.5 \times 12 = 45$

Notice that as the number of rectangles increases, the approximate area gets closer to the actual area of the triangle, in this case, 36 units^2.

We can use a similar idea to approximate the area under curves. When finding the area under any curve, we can draw a series of rectangles or 'vertical slices' to approximate the area.

WORKED EXAMPLE 4 Underestimating the approximate area under the curve

Find an approximation to the area under the curve $y = x^2$ between $x = 1$ and $x = 4$ using the sum of 3 rectangles shown below.

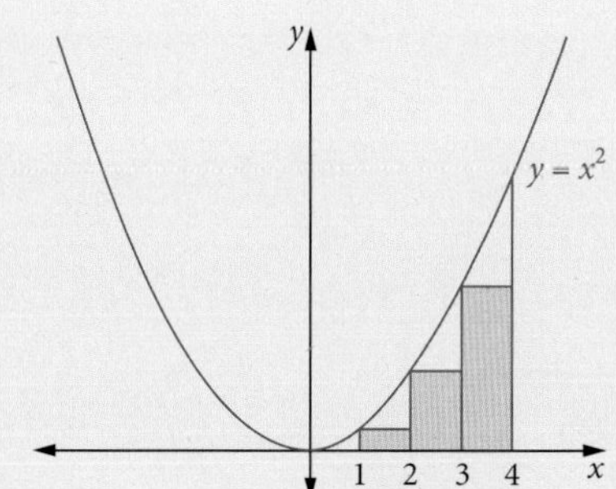

These rectangles are called lower rectangles because their tops touch the curve on their left corner. This is an underestimate of the area, because of the gaps between the tops of the rectangles and the curve.

Steps	Working
1 Find the height of each rectangle.	$f(1) = 1^2 = 1$ $f(2) = 2^2 = 4$ $f(3) = 3^2 = 9$
2 Find the area of each rectangle.	$A_1 = 1 \times 1 = 1$ $A_2 = 1 \times 4 = 4$ $A_3 = 1 \times 9 = 9$
3 Add the areas.	$A = 1 + 4 + 9 = 14$ The area is approximately 14 units^2.

To improve the accuracy of the estimate, an overestimation using rectangles can also be done, and the two values averaged.

The accuracy can be further improved by using more rectangles of a smaller width.

WORKED EXAMPLE 5 Approximation using rectangles

Find an approximation to the area under the curve $y = 12 - x^3$ between $x = 0$ and $x = 2$ with the width of the rectangles being $\frac{1}{2}$ unit, using

a an overestimation of the area **b** an underestimation of the area.

Steps	Working
a 1 Sketch the graph. Draw the rectangles so that each touches the curve at the top left.	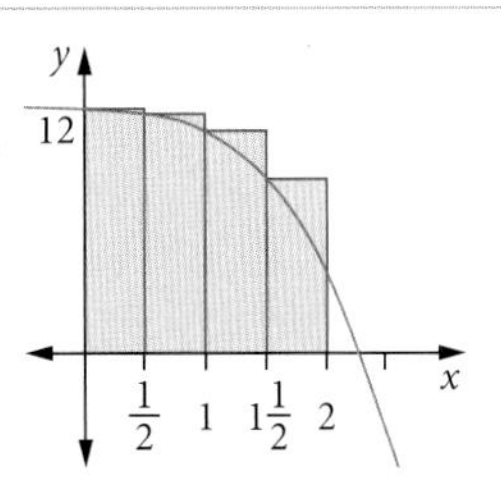
2 State the x value for each rectangle.	The values are at 0, $\frac{1}{2}$, 1 and $\frac{3}{2}$.
3 Find the height of each rectangle.	$f(0) = 12$ $f\left(\frac{1}{2}\right) = \frac{95}{8}$ $f(1) = 11$ $f\left(\frac{3}{2}\right) = \frac{69}{8}$
4 Find the area of each rectangle.	$A_1 = \frac{1}{2} \times 12 = 6$ $A_2 = \frac{1}{2} \times \frac{95}{8} = \frac{95}{16}$ $A_3 = \frac{1}{2} \times 11 = \frac{11}{2}$ $A_4 = \frac{1}{2} \times \frac{69}{8} = \frac{69}{16}$
5 Add the areas.	$A = 6 + \frac{95}{16} + \frac{11}{2} + \frac{69}{16} = \frac{87}{4}$ Total area of left rectangles = $\frac{87}{4}$ units^2
b 1 Sketch the graph. Draw the rectangles so that each touches the curve at the top right.	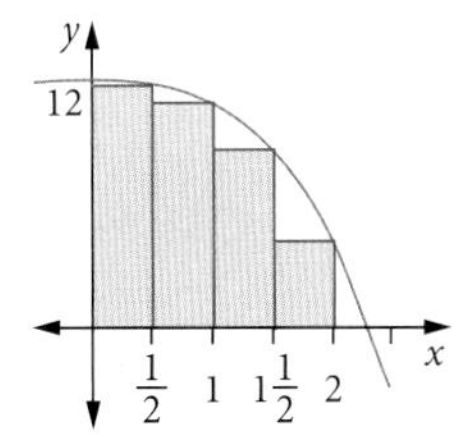
2 State the x value for each rectangle.	The values are at $\frac{1}{2}$, 1, $\frac{3}{2}$ and 2.

3 Find the height of each rectangle.	$f\left(\frac{1}{2}\right) = \frac{95}{8}$ $f(1) = 11$ $f\left(\frac{3}{2}\right) = \frac{69}{8}$ $f(2) = 4$
4 Find the area of each rectangle.	$A_1 = \frac{1}{2} \times \frac{95}{8} = \frac{95}{16}$ $A_2 = \frac{1}{2} \times 11 = \frac{11}{2}$ $A_3 = \frac{1}{2} \times \frac{69}{8} = \frac{69}{16}$ $A_4 = \frac{1}{2} \times 4 = 2$
5 Add the areas.	$A = \frac{95}{16} + \frac{11}{2} + \frac{69}{16} + 2 = \frac{71}{4}$ Total area of right rectangles = $\frac{71}{4}$ units^2

The actual area under the curve must be between $\frac{71}{4}$ and $\frac{87}{4}$ units^2. A more accurate approximation would be to take the mean of the underestimation and overestimation. In the above example this would be

$$\left(\frac{87}{4} + \frac{71}{4}\right) \div 2 = \frac{79}{4} = 19.75 \text{ units}^2$$

We can calculate the above areas efficiently using CAS.

USING CAS 2 Approximating areas under curves

Find an approximation to the area under the curve $y = 12 - x^3$ between $x = 0$ and $x = 2$ with the width of the rectangles being $\frac{1}{2}$ a unit, using an overestimation and underestimation of the area.

ClassPad

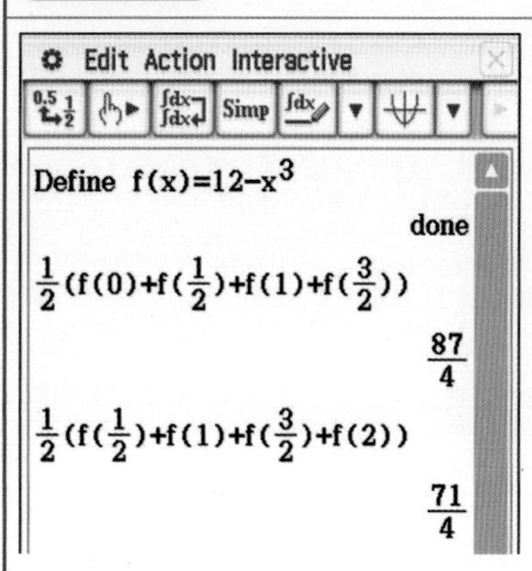

1 Define **f(x)** as shown above.
2 Calculate the sum of the rectangles using the overestimate
3 Calculate the sum of the rectangles using the underestimate.

TI-Nspire

1.1 *Doc RAD

Define $f(x)=12-x^3$ Done

$\frac{1}{2}\cdot\left(f(0)+f\left(\frac{1}{2}\right)+f(1)+f\left(\frac{3}{2}\right)\right)$ $\frac{87}{4}$

$\frac{1}{2}\cdot\left(f\left(\frac{1}{2}\right)+f(1)+f\left(\frac{3}{2}\right)+f(2)\right)$ $\frac{71}{4}$

1 Define **f(x)** as shown above.
2 Calculate the sum of the 4 rectangles using the overestimate.
3 Calculate the sum of the 4 rectangles using the underestimate.

The overestimated area is $\frac{87}{4}$ units^2 and the underestimated area is $\frac{71}{4}$ units^2.

Recap

1 State the anti-derivative of $x^2 + 3x$.

2 Find $f(x)$ if $f'(x) = 2x - x^{\frac{2}{3}}$ and $f(1) = -1$.

Mastery

3 WORKED EXAMPLE 4 Use the rectangles shown to find an approximation to the area under the curve. Give your answer correct to two decimal places.

a $y = x^2 + 2$ between $x = 1$ and $x = 2$, using 2 rectangles.

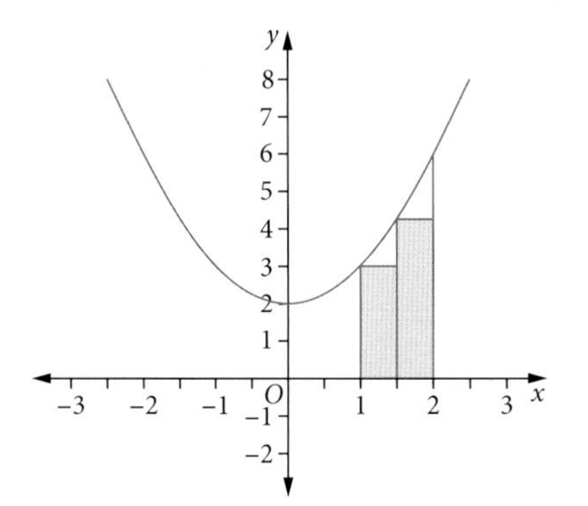

b $y = x^3$ between $x = 1$ and $x = 5$, using 4 rectangles.

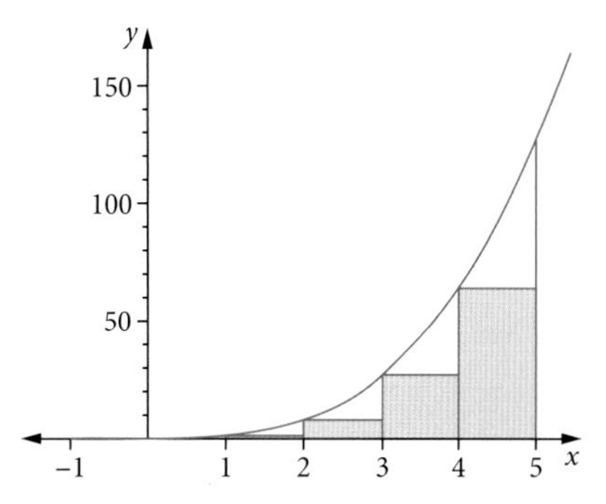

c $y = 4x - x^2$ between $x = 1$ and $x = 4$, using 3 rectangles.

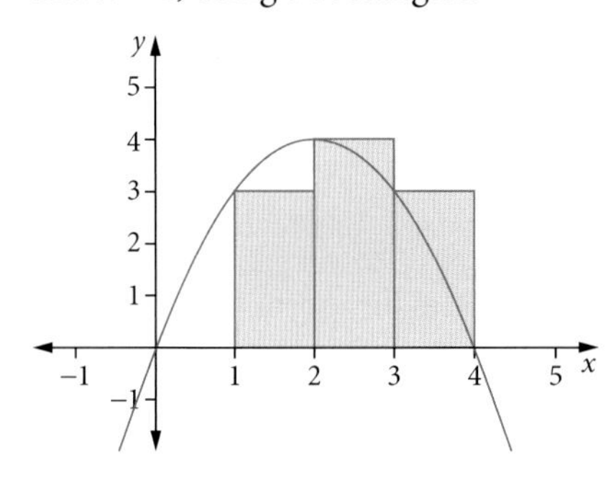

d $y = \sqrt{x+1}$ between $x = 3$ and $x = 7$, using 4 rectangles. Give your answer correct to two decimal places.

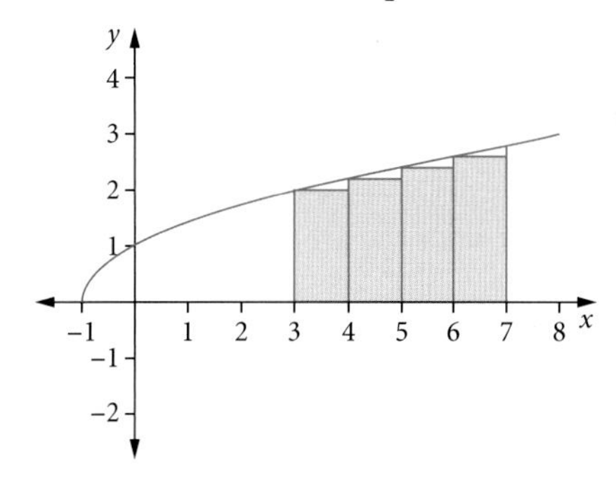

4 WORKED EXAMPLE 5 Use rectangles to find an underestimate and overestimate approximation to each area. Some rectangles have been drawn already.

a $y = x^2$ between $x = 0$ and $x = 1$, using 5 rectangles. (Note: The first rectangle is very low, so doesn't show up well on this diagram.)

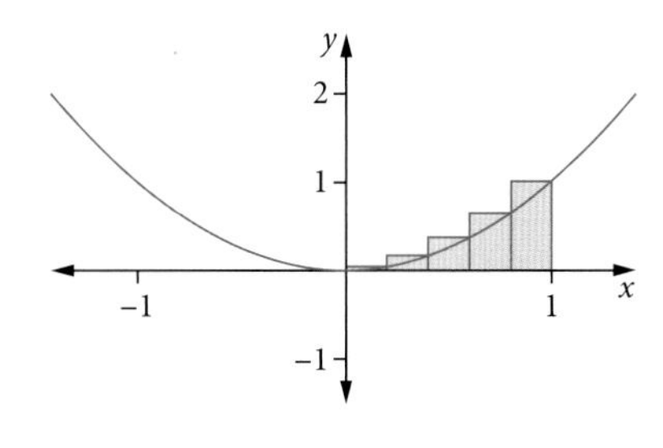

b $f(x) = 2^x + 3$ between $x = 0$ and $x = 3$, using 6 rectangles. Give your answer correct to two decimal places.

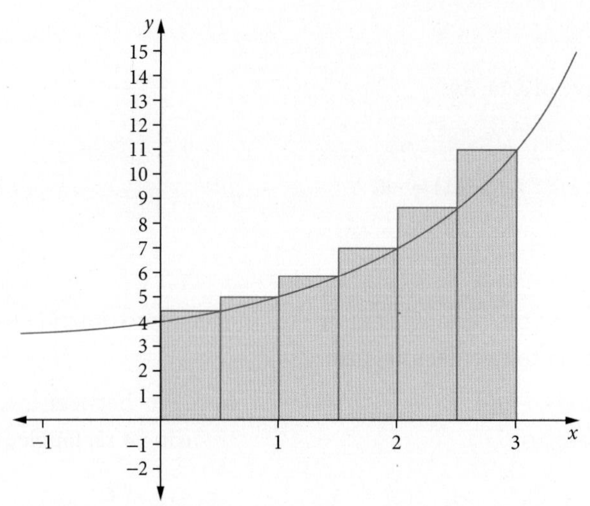

c $f(x) = \frac{2}{x+1}$ between $x = 1$ and $x = 4$, using 3 rectangles. Give your answer correct to two decimal places.

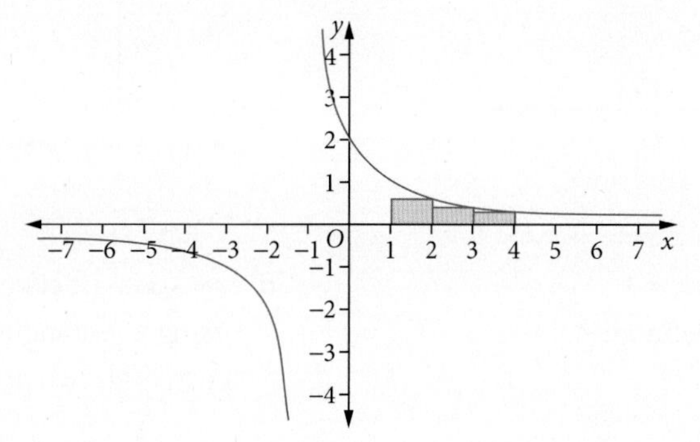

5 Using CAS 2 Consider the function $y = \frac{3}{2(x-1)}$ between $x = 2$ and $x = 4$. Use rectangles to find an approximation for the area described using underestimate and overestimate approximations, with the width of the rectangles being $\frac{1}{2}$ unit.

Calculator-free

6 (3 marks) Use the rectangles to show that the area under the curve $y = \cos(x)$ between $x = 0$ and $x = \frac{\pi}{2}$ is approximately $\frac{\pi}{12}\left[3 + \sqrt{3}\right]$ units^2.

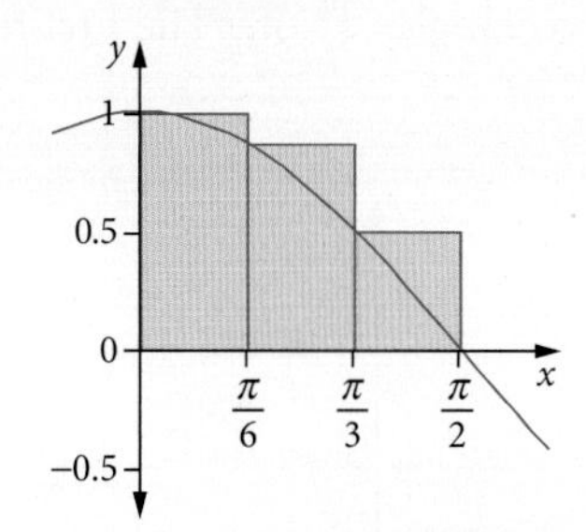

7 (4 marks) Find the approximate area under the curve $f(x) = 9 - x^2$ using rectangles as stated below.

a between $x = 0$ and $x = 3$, with the width of each rectangle being 1 unit (2 marks)

b between $x = 0$ and $x = 2$, with the width of each rectangle being $\frac{1}{2}$ unit (2 marks)

8 (2 marks) An approximation to an area under a curve is to be found by summing the area of n rectangles of width h units that lie under the curve $y = f(x)$ between $x = a$ and $x = b$, as shown below. By considering the values of n and h, under what conditions will the approximation be most accurate?

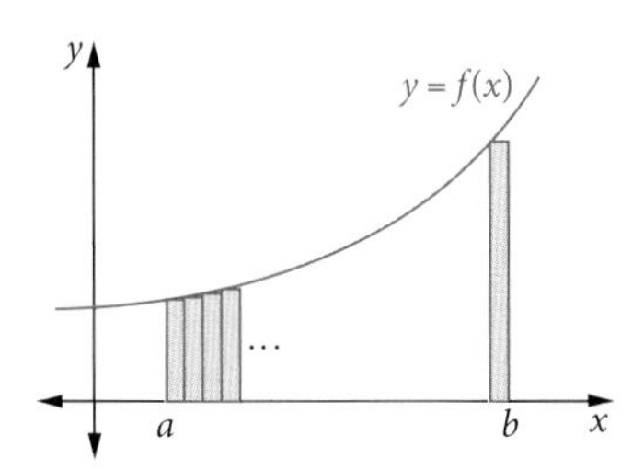

9 © SCSA MM2017 Q9 (8 marks) Consider the function $f(x)$ shown graphed below. The table gives the value of the function at the given x values.

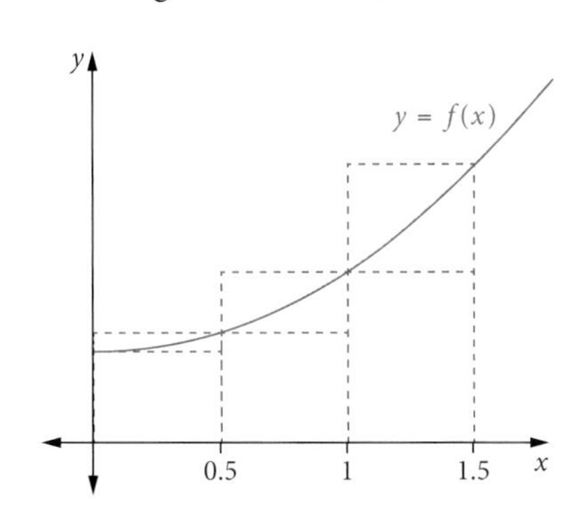

x	0	0.5	1	1.5
$f(x)$	20	21	24	29

a By considering the areas of the rectangles shown, demonstrate and explain why $32.5 < \int_0^{1.5} f(x)\,dx < 37$. (3 marks)

Consider the table of further values of $f(x)$ given below.

x	0	0.5	1	1.5	2	2.5	3
$f(x)$	20	21	24	29	36	45	56

b Use the table values to determine the best estimate possible for $\int_1^3 f(x)\,dx$. (3 marks)

c State **two** ways in which you could determine a more accurate value for $\int_1^3 f(x)\,dx$. (2 marks)

Calculator-assumed

10 (3 marks) Consider the graph of $f(x) = \sqrt{x}$. To find an approximation to the area of the region bounded by the graph of f, the x-axis and the line $x = 4$, four rectangles of equal width are drawn, and their total area calculated. By finding the mean of an underestimation and overestimation, what is this area correct to three decimal places?

11 (2 marks) Using the rectangles shown, would an underestimation of the approximate area under the curve $y = f(x)$ between $x = 1$ and $x = 4$ be found by evaluating $f(1) + f(2) + f(3)$ or $f(2) + f(3) + f(4)$? Explain.

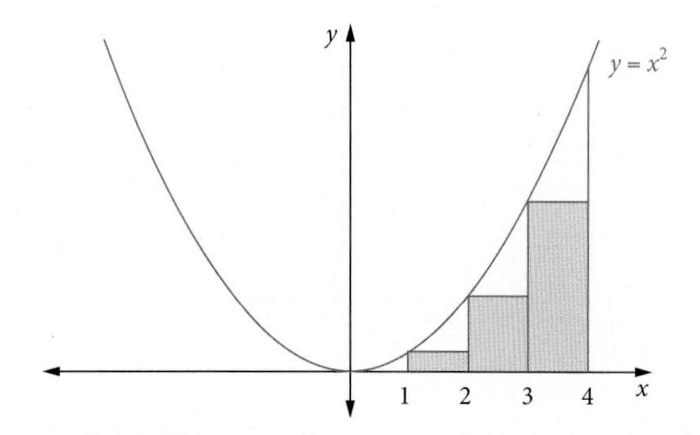

12 (3 marks) Determine an approximation to the area under the curve $y = 10 - x^2$ between $x = 0$ and $x = 2$ using four rectangles as shown.

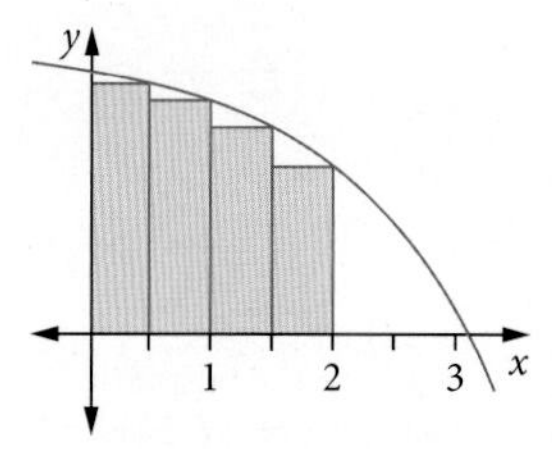

Video playlist
The definite integral and the fundamental theorem of calculus

Worksheet
Definite integrals

3.3 The definite integral and the fundamental theorem of calculus

The definite integral

The area under a curve can be approximated by a sum of the areas of rectangles.

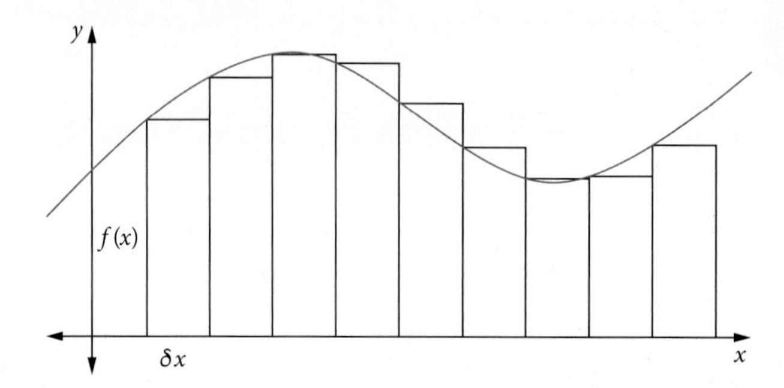

Each rectangle has the same width δx and different height $f(x)$, the y value of the function.

The area A under the curve $y = f(x)$ between $x = a$ and $x = b$ can be approximated by the sum of the areas of rectangles of width δx and height $f(x)$.

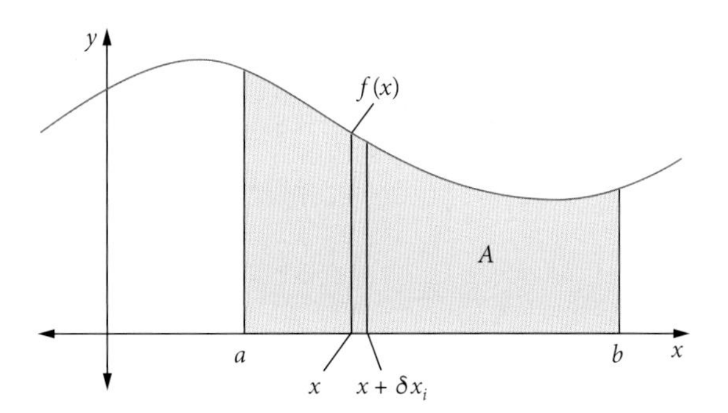

Algebraically, this is written as $A \approx \sum_{x=a}^{b} f(x)\delta x$ where $\sum_{x=a}^{b}$ means 'the sum of the expression for values of x between a and b'.

As more rectangles are used, the sum of the area of the vertical slices that fill the area gets closer and closer to the actual area.

> δx can also be written as Δx.
> δ is the lowercase Greek letter 'delta'.
> Δ is the capital Greek letter 'delta'.
> $\sum$ is the capital Greek letter 'sigma'.

Definite integrals

As more rectangles are used, their widths become smaller, so as δx approaches 0,

$A \approx \sum_{x=a}^{b} f(x)\delta x$ becomes the **definite integral** $A = \int_a^b f(x)\,dx$.

- A definite integral $\int_a^b f(x)\,dx$ with **limits** a and b has a value that is related to the area under a curve and is read as 'the integral of $f(x)$ between a and b with respect to x'.

  Thus, $\int_a^b f'(x)dx = f(b) - f(a)$.

- An **indefinite integral** $\int f(x)\,dx$ is an anti-derivative function and is read as 'the integral of $f(x)$ with respect to x'.

WORKED EXAMPLE 6 Calculating definite integrals

Evaluate each definite integral.

a $\int_0^3 5x^2\,dx$

b $\int_1^2 (x^3 + 4)\,dx$

Steps	Working
a **1** Integrate $5x^2$. **2** Substitute the limits of the integral $x = 3$ and $x = 0$ and subtract: $F(b) - F(a)$.	$\int_0^3 5x^2\,dx = \left[\frac{5x^3}{3}\right]_0^3 = \frac{5(3^3)}{3} - \frac{5(0^3)}{3}$ $= 45 - 0$ $= 45$ The '+ c' is not required when evaluating definite integrals because + c will cancel itself out in the subtraction.

b 1 Integrate $x^3 + 4$.

$$\int_1^2 (x^3 + 4)\,dx = \left[\frac{x^4}{4} + 4x\right]_1^2$$

2 Substitute the limits of the integral $x = 2$ and $x = 1$ and subtract: $F(b) - F(a)$.

$$= \left(\frac{2^4}{4} + 4(2)\right) - \left(\frac{1^4}{4} + 4(1)\right)$$

$$= 12 - \frac{17}{4}$$

$$= \frac{31}{4}$$

USING CAS 3 Definite integrals

Evaluate the definite integral $\int_1^2 (x^3 + 4)dx$.

TI-Nspire

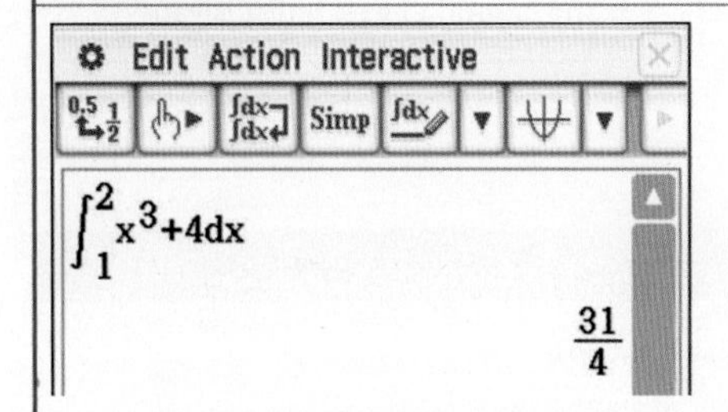

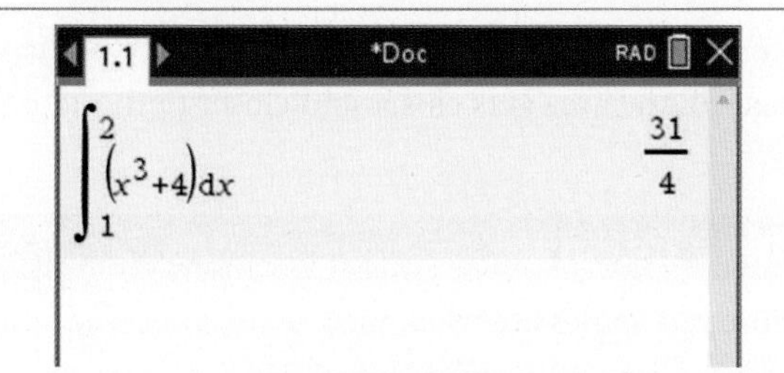

1 Highlight the given expression and tap **Interactive > Calculation > ∫**.
2 In the dialogue box, tap **Definite**.
3 Enter the lower and upper limits.

1 Press **menu > Calculus > Integral**.
2 Enter the lower and upper limits.
3 Enter the expression followed by **dx**.

$\int_1^2 (x^3 + 4)\,dx = \frac{31}{4}$

The fundamental theorem of calculus

The fundamental theorem of calculus is an important concept as it establishes a relationship between differentiation and integration. We begin by defining the **signed area** function $F(x) = \int_a^x f(t)dt$.

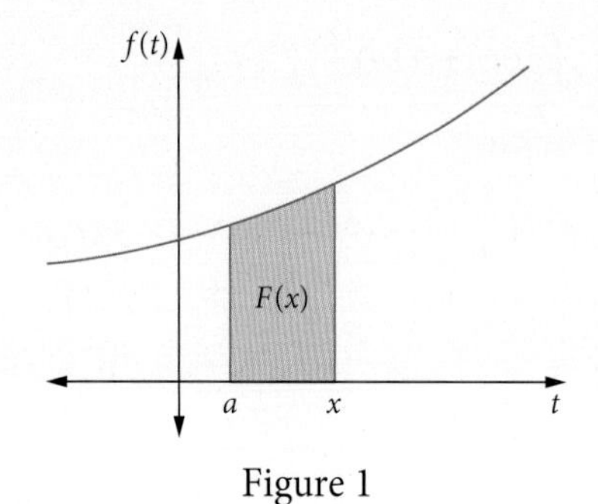

Figure 1

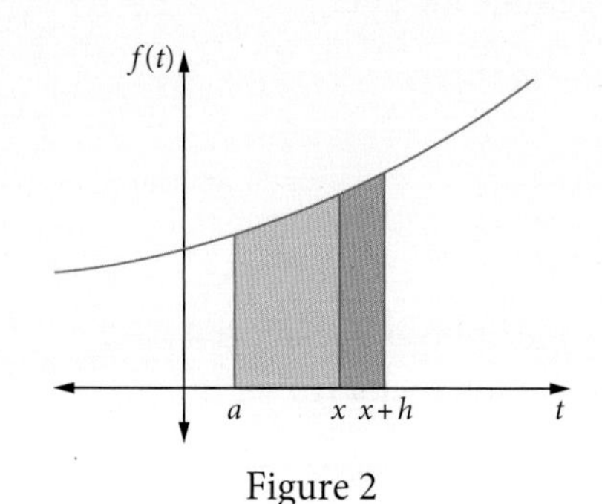

Figure 2

By definition: $F'(x) = \lim\limits_{h\to 0} \dfrac{F(x+h) - F(x)}{h}$ (Figure 1)

Therefore, $F'(x) = \lim\limits_{h\to 0} \dfrac{\int_a^{x+h} f(t)\,dt - \int_a^x f(t)\,dt}{h} = \lim\limits_{h\to 0} \dfrac{\int_x^{x+h} f(t)\,dt}{h} \approx \lim\limits_{h\to 0} \dfrac{hf(x)}{h} = \lim\limits_{h\to 0} f(x) = f(x)$. (Figure 2)

Therefore, $F'(x) = \dfrac{d}{dx}\left(\int_a^x f(t)\,dt\right) = f(x)$.

The fundamental theorem of calculus

The two parts of the fundamental theorem are

$$F'(x) = \frac{d}{dx}\left(\int_a^x f(t)\,dt\right) = f(x) \quad \text{and} \quad \int_a^b f'(x)\,dx = f(b) - f(a)$$

WORKED EXAMPLE 7	Applying the fundamental theorem of calculus
a Determine $\frac{d}{dx}\int_4^x t^2 + 3\,dt$. **b** If $F(x) = \int_{-2}^x \frac{dt}{2t^2+1}$, determine $F'(x)$.	
Steps	**Working**
a Apply the fundamental theorem to replace t with x.	$\frac{d}{dx}\int_4^x t^2 + 3\,dt = x^2 + 3$
b Apply the fundamental theorem to replace t with x.	$F'(x) = \frac{d}{dx}\int_{-2}^x \frac{dt}{2t^2+1} = \frac{1}{2x^2+1}$

Properties of the definite integral

1 $\int_a^b kf(x)\,dx = k\int_a^b f(x)\,dx$	**Constant out:** a constant factor can be taken out of an integral.
2 $\int_a^b f(x) \pm g(x)\,dx = \int_a^b f(x)\,dx \pm \int_a^b g(x)\,dx$	**Split terms:** a sum or difference of terms can be integrated separately.
3 If b is between a and c, then $\int_a^c f(x)\,dx = \int_a^b f(x)\,dx + \int_b^c f(x)\,dx$	**Split limits:** the limits of an integral can be split.
4 $\int_a^b f(x)\,dx = -\int_b^a f(x)\,dx$	**Swap limits:** reversing the order of the limits changes the sign of the definite integral.
5 $\int_a^a f(x)\,dx = 0$	**Same limits:** gives $F(a) - F(a) = 0$.

The next two examples illustrate some exam-type questions where we are not given the equation of the function.

WORKED EXAMPLE 8 Properties of definite integrals

If $\int_0^2 g(x)\,dx = 3$, evaluate $\int_2^0 (g(x) + 2x)\,dx$.

Steps	Working
1 Simplify using Property 2: split terms.	$\int_2^0 (g(x) + 2x)\,dx = \int_2^0 g(x)\,dx + \int_2^0 (2x)\,dx$
2 Simplify using Property 4: swap limits.	$= -\int_0^2 g(x)\,dx - \int_0^2 (2x)\,dx$
3 Substitute $\int_0^2 g(x)\,dx = 3$.	$= -3 - \int_0^2 (2x)\,dx$
4 Evaluate and simplify.	$= -3 - \left[x^2\right]_0^2$
	$= -3 - (2^2 - 0^2)$
	$= -7$

WORKED EXAMPLE 9 Properties of definite integrals

If $\int_1^2 f(x)\,dx = 10$, evaluate $\int_1^2 3(f(x) - 5)\,dx$.

Note carefully where the common factor is placed.

Steps	Working
1 Simplify using Property 1: constant out.	$\int_1^2 3(f(x) - 5)\,dx = 3\int_1^2 (f(x) - 5)\,dx$
2 Simplify using Property 2: split terms.	$= 3\left[\int_1^2 f(x)\,dx - \int_1^2 5\,dx\right]$
3 Substitute $\int_1^2 f(x)\,dx = 10$.	$= 3\left[10 - \int_1^2 5\,dx\right]$
4 Evaluate and simplify.	$= 30 - 3[5x]_1^2$
	$= 30 - 3(10 - 5)$
	$= 15$

EXERCISE 3.3 The definite integral and the fundamental theorem of calculus

ANSWERS p. 392

Recap

1 Use rectangles of width $\frac{1}{2}$ unit to find the area under the curve $y = x^2 + 1$ from $x = 0$ to $x = 2$.

2 When using rectangles to approximate the area under a curve, what effect will reducing the width of the rectangles have on the accuracy of the approximation? Explain.

Mastery

3 WORKED EXAMPLE 6 Evaluate each definite integral.

a $\int_1^3 4x\,dx$ b $\int_0^2 7x^6\,dx$ c $\int_1^2 4x^3\,dx$

d $\int_2^3 (2x-1)^2\,dx$ e $\int_0^4 (x+2)^{-2}\,dx$ f $\int_0^1 (x^3 - 3x^2 + 1)\,dx$

4 Using CAS 3 Evaluate

a $\int_0^2 (2x^2 + 4x)\,dx$ b $\int_2^0 (2x^2 + 4x)\,dx$

5 WORKED EXAMPLE 7

a Determine $\frac{d}{dx}\int_0^x 2t^2 + t - 4\,dt$. b If $F(x) = \int_{-2}^x \frac{3\,dt}{t^2 - 1}$, determine $F'(x)$.

6 WORKED EXAMPLE 8 If $\int_1^3 f(x)\,dx = 3$, evaluate $\int_1^3 -5f(x)\,dx$.

7 WORKED EXAMPLE 9 If $\int_0^2 g(x)\,dx = 5$, evaluate $\int_0^2 4 - 3g(x)\,dx$.

Calculator-free

8 (10 marks) Evaluate each definite integral.

a $\int_0^2 \frac{x^2}{2}\,dx$ b $\int_{-1}^1 (3x^2 + 4x)\,dx$ c $\int_{-1}^2 (x^2 + 1)\,dx$

d $\int_{-2}^3 (4x^3 - 3)\,dx$ e $\int_{-1}^0 (x^2 + 3x + 5)\,dx$

9 (4 marks) Determine

a $\frac{d}{dx}\int_0^x \sqrt{t - \pi}\,dt$ b $\frac{d}{dx}\int_x^0 -2t^2 + t\,dt$

10 (2 marks) If $F(x) = \frac{3x^2}{2} + 2\int_0^x 1 - 2t^2\,dt$, determine $F'(x)$.

11 (3 marks) Evaluate $\int_1^4 (\sqrt{x} + 1)\,dx$.

12 (4 marks) Write each expression as one integral. Do not evaluate.

a $\int_0^1 x^2\,dx + \int_1^5 x^2\,dx$

b $\int_1^4 (x+1)\,dx + \int_4^7 (x+1)\,dx$

c $\int_{-2}^0 (x^3 - x - 1)\,dx + \int_0^2 (x^3 - x - 1)\,dx$

d $\int_0^2 (2x+1)\,dx + \int_2^3 (2x+1)\,dx$

Calculator-assumed

13 (2 marks) Evaluate $\int_0^1 (x^2 - x^3)\,dx$.

14 (2 marks) Evaluate $\int_1^2 \left(3x^2 + \frac{4}{x^2}\right)dx$.

15 (3 marks) The value of the integral of $f(x) = \frac{1}{a-1}(x^2 - 2x)$ over the interval $[1, a]$ is $\frac{13}{3}$. Determine the value of a.

16 (2 marks) If $\int_1^3 f(x)\,dx = 5$, determine the value of $\int_1^3 (2f(x) - 3)\,dx$.

17 (2 marks) If $\int_1^{12} g(x)\,dx = 5$ and $\int_{12}^5 g(x)\,dx = -6$, determine the value of $\int_1^5 g(x)\,dx$.

18 (2 marks) If $F(x)$ is an anti-derivative of $f(x)$ and $F(4) = -6$, then explain why $F(8)$ is equal to $\int_4^8 (-6 + f(x))\,dx$.

19 (2 marks) A part of the graph of $f(x) = x^2$ is shown. Zoe finds the approximate area of the shaded region by drawing rectangles as shown in the second diagram.

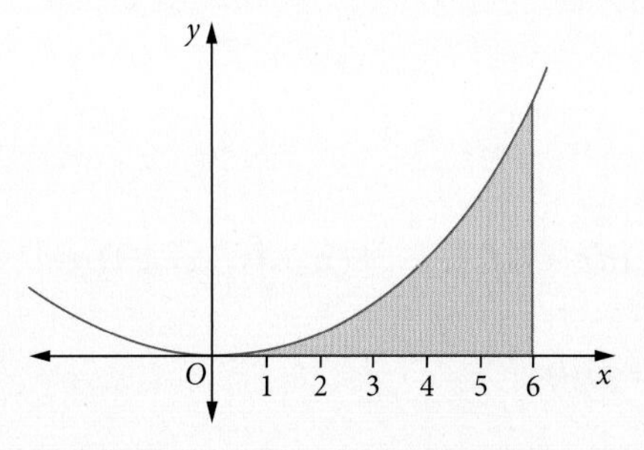

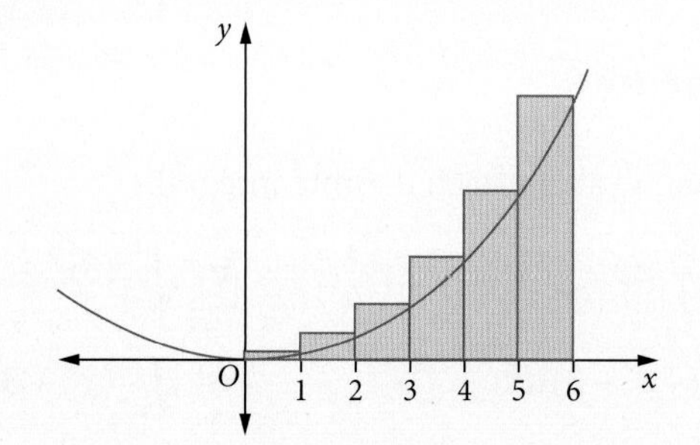

Zoe's approximation is p% more than the exact value of the area. Determine if p is to closer to 20, 25 or 30.

3.4

Area under a curve

The definite integral $\int_a^b f(x)\,dx$ gives the signed area enclosed by the graph of $y = f(x)$ and the x-axis between $x = a$ and $x = b$.

- If the section of a graph is above the x-axis, then the definite integral over that section gives the area under the curve.
- If the section of a graph is below the x-axis, then the definite integral over that section gives the area above the curve and its value is negative.
- If the graph is partly above and partly below the x-axis, then we split the graph and the integral to find the required area.

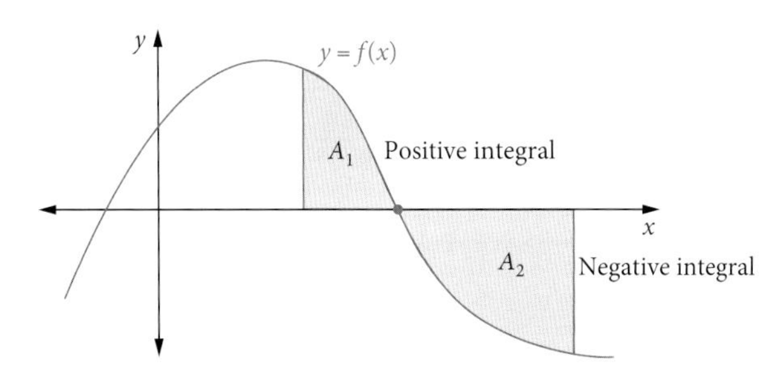

Video playlist
Area under a curve

WORKED EXAMPLE 10 Finding the area under a curve

Find the area enclosed by the graph of $y = x^2 - 1$ and the x-axis between $x = 1$ and $x = 2$.

Steps	Working
1 Sketch the graph showing the positive area required.	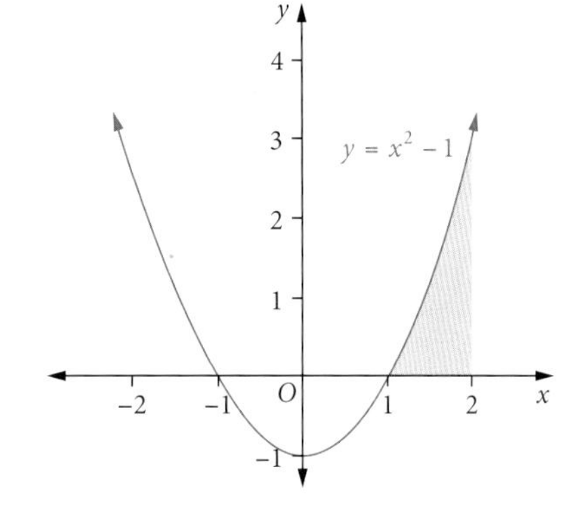
2 Write the integral required to find the area.	$\int_1^2 (x^2 - 1)\,dx$
3 Evaluate the area.	$\int_1^2 (x^2 - 1)\,dx = \left[\frac{x^3}{3} - x\right]_1^2$ $= \left(\frac{2^3}{3} - 2\right) - \left(\frac{1^3}{3} - 1\right)$ $= \frac{4}{3}$ units^2

Exam hack

Always sketch a graph first to see if we are looking for a positive or a negative area or a mixture of both.

USING CAS 4 Area under a curve

Find the area under the curve $y = x(2x - 1)(x + 3)$ enclosed by the graph and the x-axis where the area is positive.

ClassPad

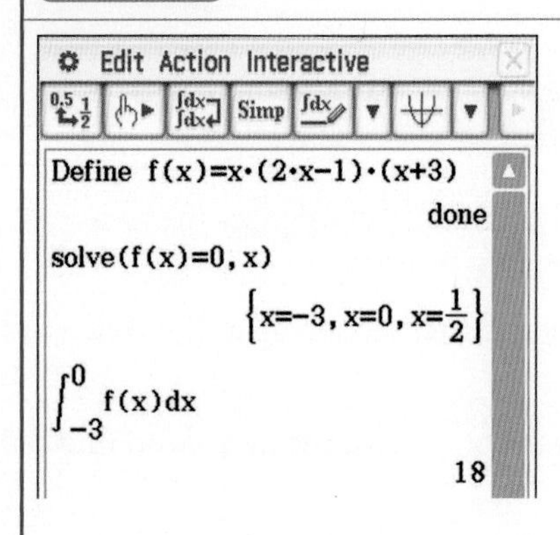

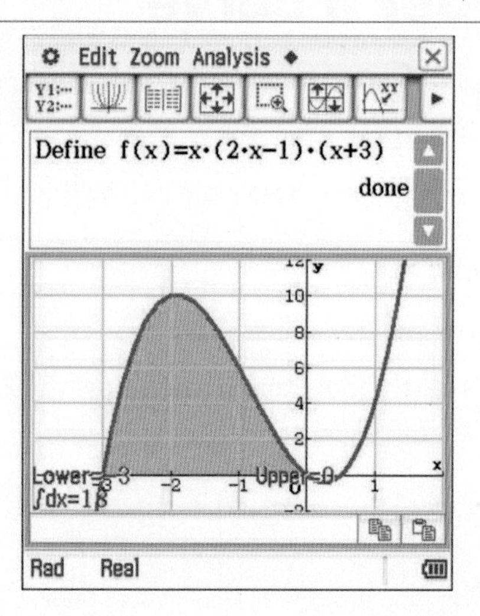

1 Define **f(x)** as shown above.

2 Solve **f(x) = 0** to determine the x-intercepts.

3 Find the definite integral of **f(x)** from **−3** to **0**.

4 The value of the positive area will be displayed.

5 To confirm this result, graph **f(x)**.

6 Adjust the window settings to suit.

7 Tap **Analysis > G-Solve > Integral > ∫dx.**

8 Enter **−3**.

9 A dialogue box will appear with **−3** in the **Lower:** field.

10 Enter **0** in the **Upper:** field.

TI-Nspire

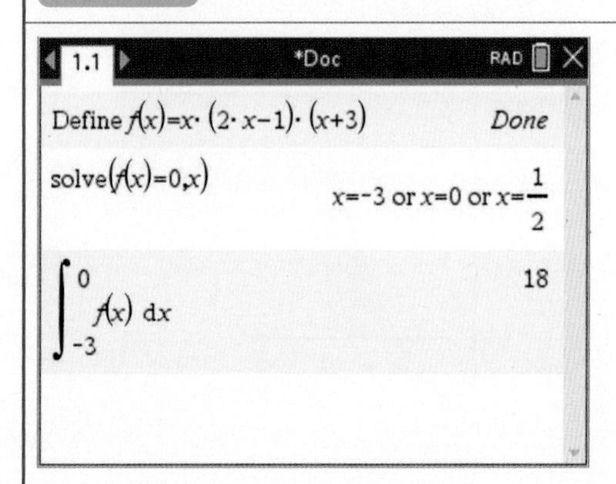

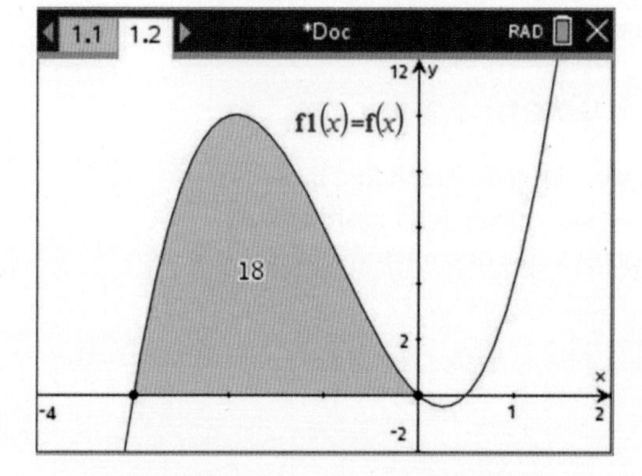

1 Define **f(x)** as shown above.

2 Solve **f(x) = 0** to determine the x-intercepts.

3 Find the definite integral of **f(x)** from **−3** to **0**.

4 The value of the positive area will be displayed.

5 To confirm this result, graph **f(x)**.

6 Adjust the window settings to suit.

7 Press **menu > Analyze Graph > Integral**.

8 When prompted for the **lower bound**, click on **−3** on the x-axis.

9 When prompted for the **upper bound**, click on **0** on the x-axis.

The area under the curve is 18 units^2.

Areas above and below curves

The definite integral is negative for areas below the x-axis. When finding areas using integration, always check if the graph is below the x-axis.

In this diagram, $A_1 = \int_a^b f(x)\,dx$ is positive, whereas $A_2 = \int_b^c f(x)\,dx$ is negative.

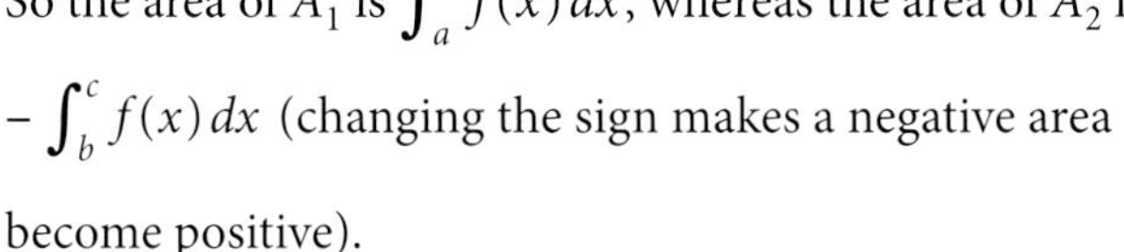

So the area of A_1 is $\int_a^b f(x)\,dx$, whereas the area of A_2 is $-\int_b^c f(x)\,dx$ (changing the sign makes a negative area become positive).

So the sum of the areas of A_1 and A_2 is $\int_a^b f(x)\,dx - \int_b^c f(x)\,dx$.

An alternative method for dealing with the area below the axis is to use the property of definite integrals of **swapping limits**.

In this case, the sum of the areas of A_1 and A_2 is $\int_a^b f(x)\,dx + \int_c^b f(x)\,dx$, where we swap the limits of b and c.

For example, we can calculate the total shaded area in this graph in two different ways.

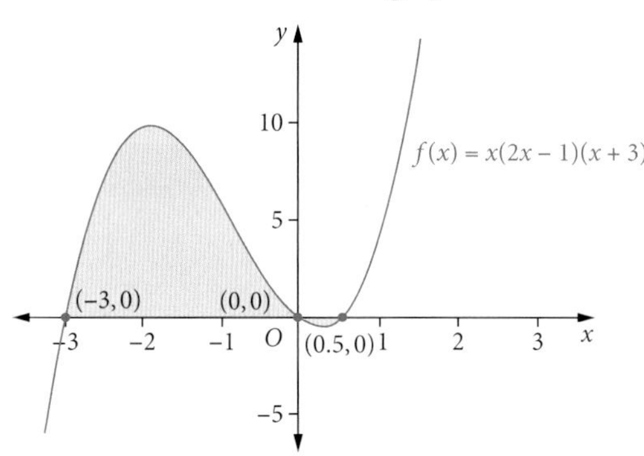

Method 1: Subtract the integral that would give a negative area.

$$\text{Area} = \int_{-3}^{0}(x(2x-1)(x+3))\,dx - \int_{0}^{\frac{1}{2}}(x(2x-1)(x+3))\,dx$$

$$= \frac{1741}{96} \text{ units}^2$$

ClassPad

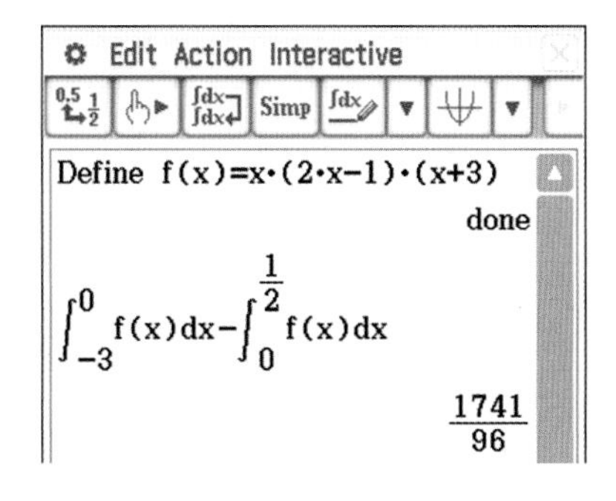

TI-Nspire

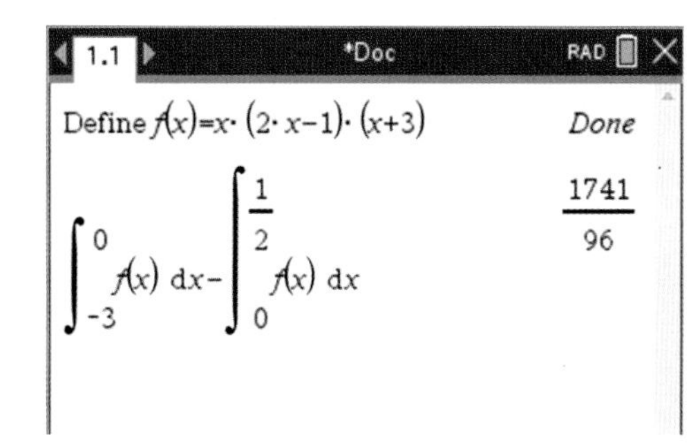

Method 2: Swap the limits of the integral that would give a negative area.

$$\text{Area} = \int_{-3}^{0}(x(2x-1)(x+3))\,dx + \int_{\frac{1}{2}}^{0}(x(2x-1)(x+3))\,dx$$

$$= \frac{1741}{96} \text{ units}^2$$

WORKED EXAMPLE 11 Areas above and below curves

Find the area enclosed by the graph of $y = x^2 - 1$ and the x-axis between $x = 0$ and $x = 2$.

Steps	Working
1 Sketch the graph, showing the area required.	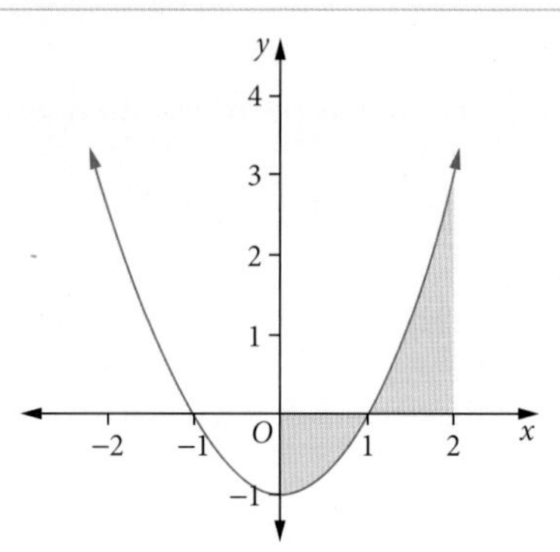
2 Find the x-intercept that splits the area into two regions, above and below the x-axis.	$y = x^2 - 1 = 0$ $x^2 = 1$ $x = \pm 1$ $x = 1$ is where the area is split.
3 Write the integral required to find the split area.	Area $= -\int_0^1 (x^2 - 1)\,dx + \int_1^2 (x^2 - 1)\,dx$
4 Evaluate the integrals.	Area = 2 units^2

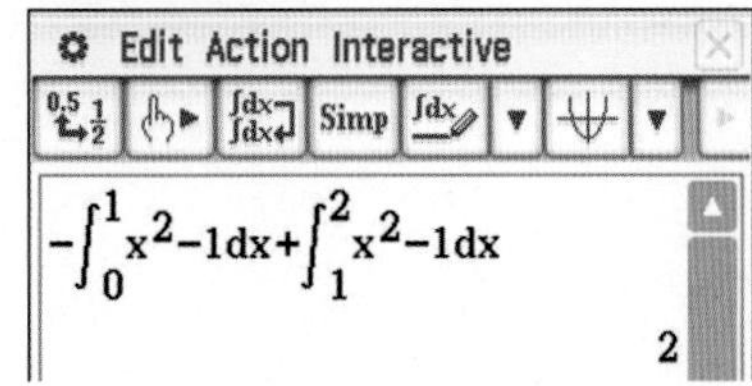

TI-Nspire

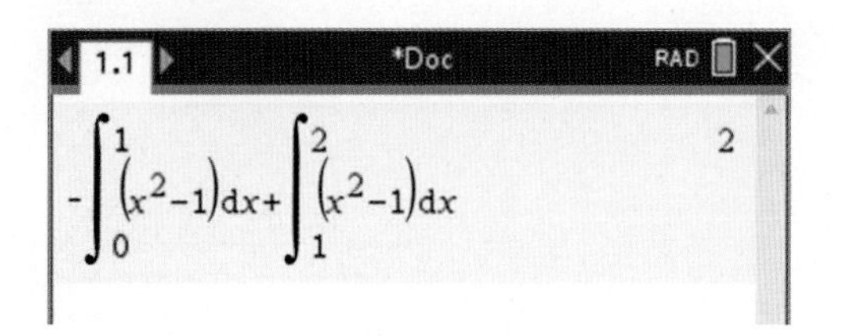

Sometimes we are not given the function but are given information about the areas under the curve. This information can be used to evaluate definite integrals.

WORKED EXAMPLE 12 Using areas under the curve to determine integrals

Given the graph of $f(x)$, determine the exact value of

a $\int_0^2 f(x)\,dx$

b $\int_0^6 f(x)\,dx$

Steps	Working
a Note that the required area is quarter of a circle, with a radius of 2.	$\int_0^2 f(x)\,dx = \frac{\pi(2)^2}{4} = \pi$ As we do not know the function, we use the area formula to get the answer.
b **1** Determine the area of the semicircle.	$\int_0^4 f(x)\,dx = \frac{\pi(2)^2}{2} = 2\pi$ (or double your answer from part **a**)
2 Determine the area of the triangle under the axis (between 4 and 6).	$\int_4^6 f(x)\,dx = \frac{2 \times 2}{2} = 2$ But as the area is under the curve, the answer is −2.
3 Add your answers together.	$\int_0^6 f(x)\,dx = 2\pi - 2$

Exam hack

Although we are using areas to calculate the answers, the question did not mention areas, so do not place units^2 on your answer.

EXERCISE 3.4 Area under a curve

ANSWERS p. 393

Recap

1 Evaluate $\int_1^2 (x^3 - 2x)dx$.

2 If $\int_0^5 g(x)\,dx = 20$ and $\int_0^5 (2g(x) + ax)\,dx = 90$, determine the value of a.

Mastery

3 WORKED EXAMPLE 10 Find the area between the x-axis and the graph of each function.

a $y = x^2 - 5x + 8$ from $x = 1$ to $x = 4$

b $f(x) = 15 + 8x - 6x^2$ between $x = -1$ and $x = 2$

c $y = 4x^3 - 3x^2 + 6x - 2$ between $x = 1$ and $x = 3$

4 Using CAS 4 Determine the area between the graph of $y = x^3 - 2x^2 - 11x + 12$ and the x-axis from $x = 1$ to $x = 4$.

5 WORKED EXAMPLE 11 Determine the area between the graph of $y = x^3 - 2x^2 - 11x + 12$ and the x-axis from $x = -3$ to $x = 4$.

6 WORKED EXAMPLE 12 Given the graph of $f(x)$, determine the exact value of

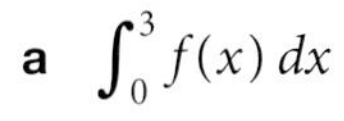

a $\int_0^3 f(x)\,dx$

b $\int_0^7 f(x)\,dx$

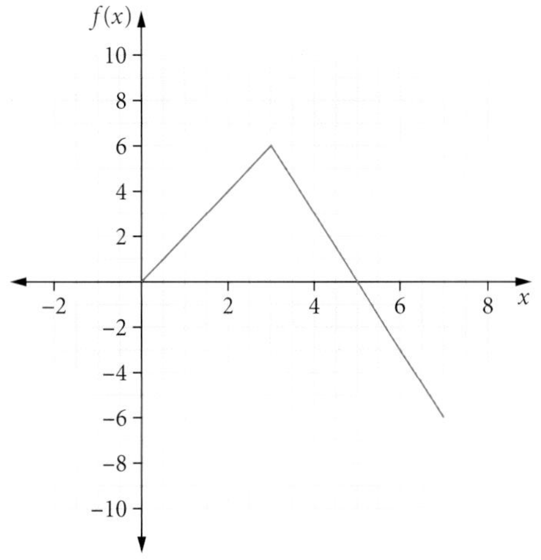

7 (2 marks) Write an integral expression that would determine the area between the graph of $f(x)$, the x-axis and the lines $x = -2$ and $x = -1$.

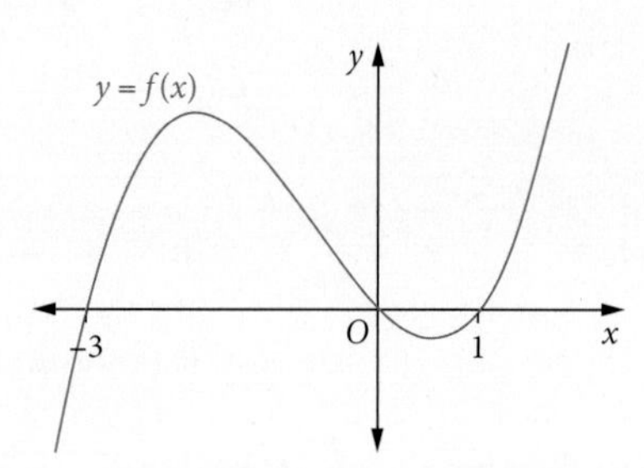

8 (2 marks) Write an integral expression that would determine the area between the graph of $f(x)$, the x-axis and the lines $x = -3$ and $x = 1$.

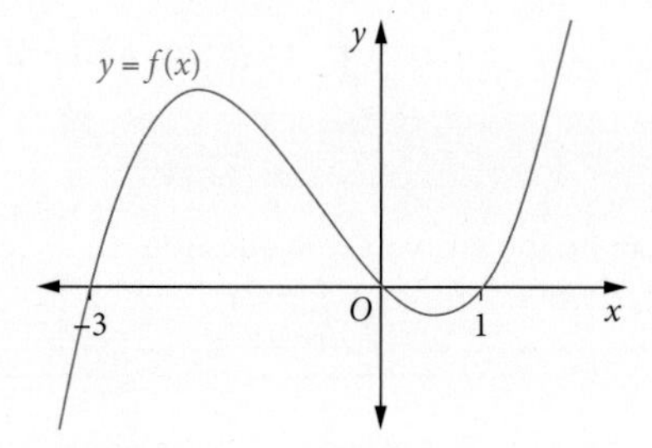

9 (3 marks) Find the area of the region bounded by the function $f(x) = \frac{1}{\sqrt{2}}\sqrt{x}$, the x-axis and the line $x = 2$.

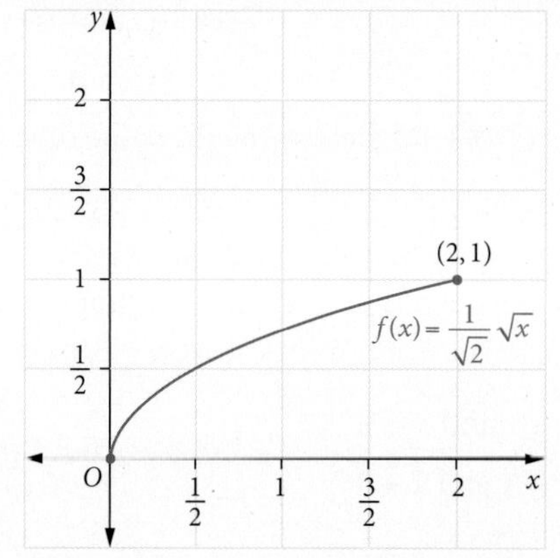

10 (3 marks) Part of the graph $f(x) = \sqrt{x}(1 - x)$ is shown below. Calculate the area between the graph of f and the x-axis.

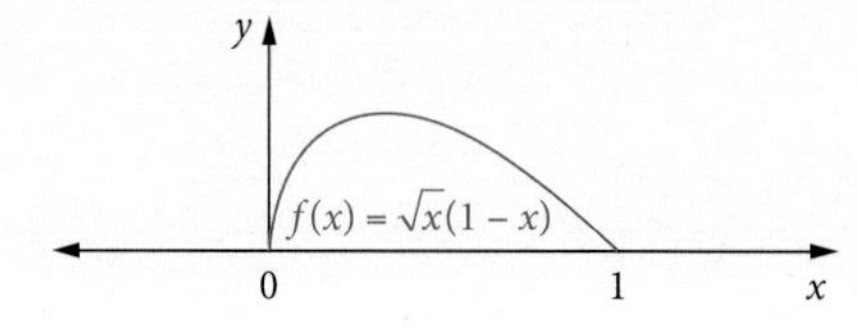

11 (4 marks) Given the graph of $f(x)$, determine the exact value of $\int_2^6 f(x)\,dx$.

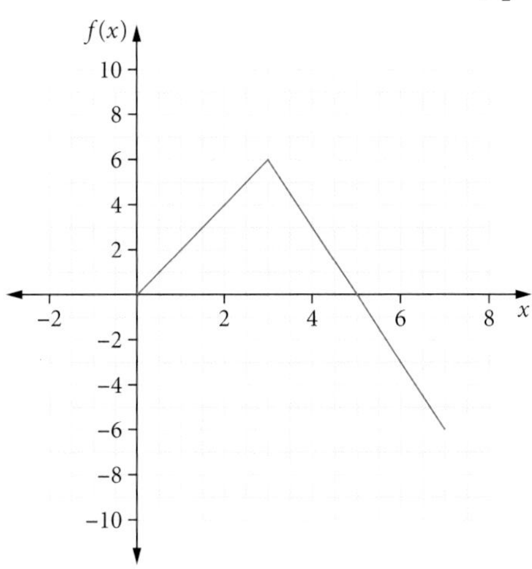

12 (2 marks) Write an integral expression that would determine the total area of the shaded regions in the diagram.

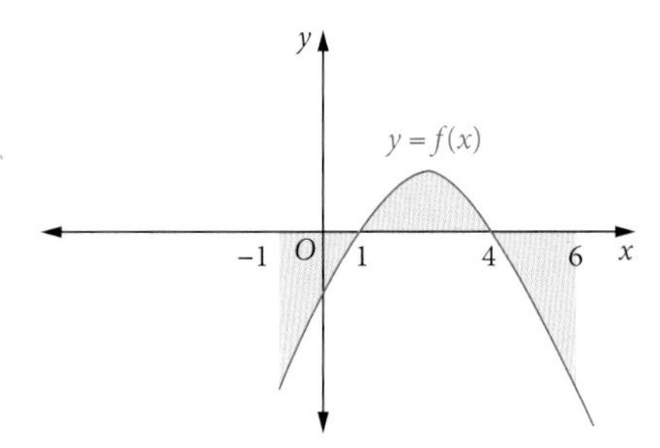

13 (3 marks) A part of the graph of $g(x) = x^2 - 4$ is shown below.

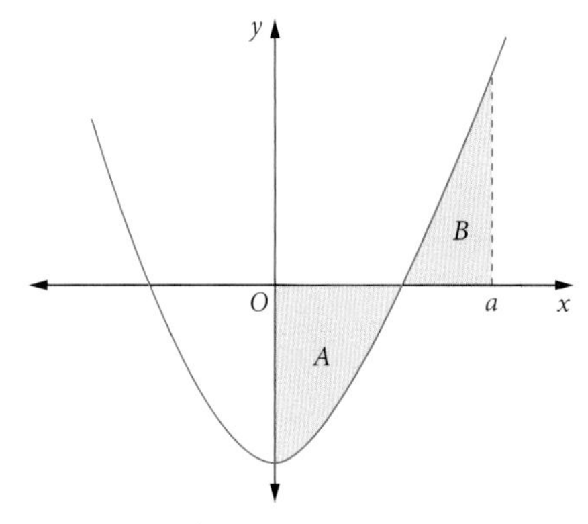

If the area of the region marked A is the same as the area of the region marked B, determine the exact value of a.

14 (3 marks) Determine the area under the curve $y = 2(x + 1)^3$ from $x = -1$ to $x = 2$.

Video playlist
Areas between curves

Worksheets
Calculating areas between curves

Areas between curves 1

Areas between curves 2

3.5 Areas between curves

The area enclosed between two curves can be calculated as the difference between the areas under the two functions. In the diagram, $f(x) > g(x)$, so the shaded area can be found by subtracting areas:

$$A = \int_a^b f(x)\,dx - \int_a^b g(x)\,dx = \int_a^b [f(x) - g(x)]\,dx$$

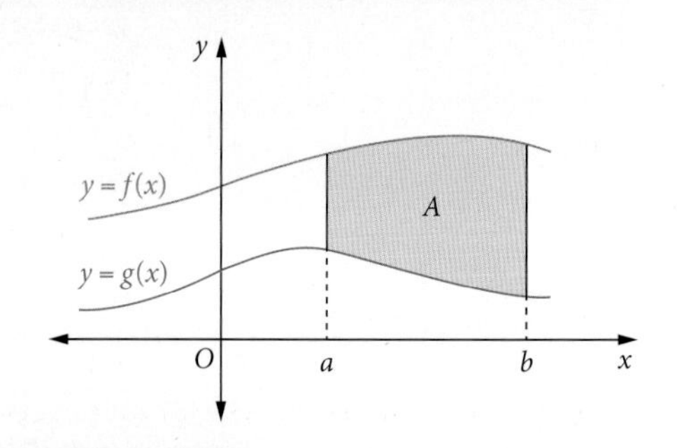

Areas between curves

$$\text{Area} = \int_a^b (\text{upper} - \text{lower})\,dx = \int_a^b [f(x) - g(x)]\,dx$$

For curves that intersect, we need to find their point(s) of intersection and again using properties of definite integrals, split the integral limits, taking note of which function is greater. In the diagram, $g(x)$ is higher on the left, but $f(x)$ is higher on the right.

Note that when finding the area between two curves, the position of the area does not matter. That is, it does not matter if part of the area is below the axes.

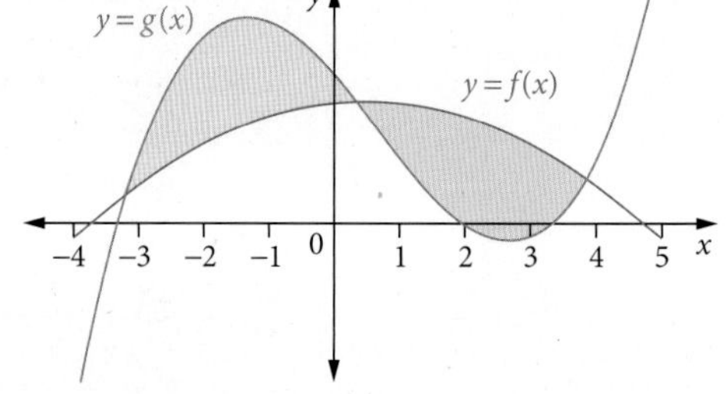
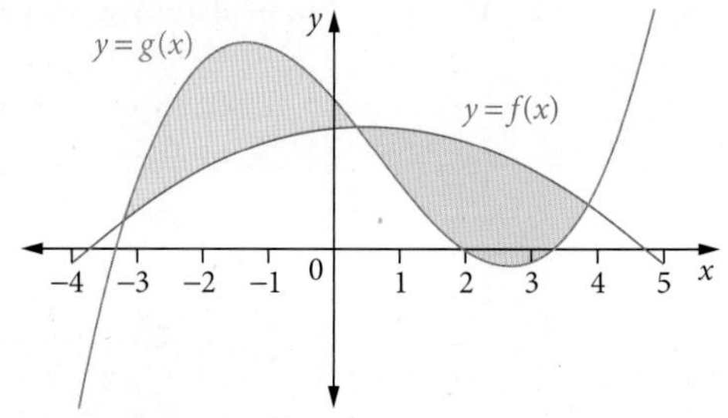

WORKED EXAMPLE 13 Finding the area between curves

Find the area enclosed by the graphs of $y = x^2 - 6x + 13$ and $y = x + 3$.

Steps	Working
1 Sketch the graphs of $y = x^2 - 6x + 13$ and $y = x + 3$. Always draw a sketch first to determine which is the upper function and which is the lower function.	(graph of $y = x^2 - 6x + 13$ and $y = x + 3$)
2 Solve simultaneously to find the points of intersection.	$x^2 - 6x + 13 = x + 3$ $x^2 - 7x + 10 = 0$ $\therefore (x - 2)(x - 5) = 0$ $x = 2, x = 5$
3 Between $x = 2$ and $x = 5$, the line $y = x + 3$ is the upper function. Set up the integral using $\int_a^b (\text{upper} - \text{lower})\,dx$ and simplify the terms.	$\text{Area} = \int_2^5 ((x + 3) - (x^2 - 6x + 13))\,dx$ $= \int_2^5 (-x^2 + 7x - 10)\,dx$

4 Integrate and evaluate.

$$A = \left[-\frac{x^3}{3} + \frac{7x^2}{2} - 10x\right]_2^5$$

$$= \left(-\frac{5^3}{3} + \frac{7(5)^2}{2} - 10(5)\right) - \left(-\frac{2^3}{3} + \frac{7(2)^2}{2} - 10(2)\right)$$

$$= \frac{9}{2}$$

$$\therefore \text{area} = \frac{9}{2} \text{ units}^2$$

Areas of integration calculations are much easier to perform using CAS.

USING CAS 5 Finding the area between curves

Find the area between the curves $f(x) = (x + 1)(x - 1)(x - 2)$ and $g(x) = 4x^2 - 28$.

ClassPad

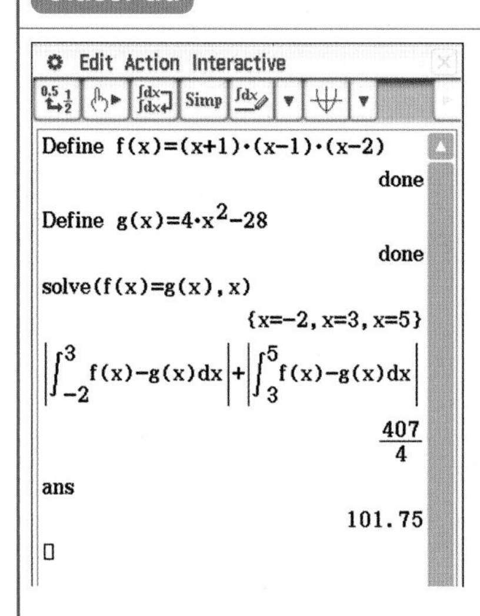

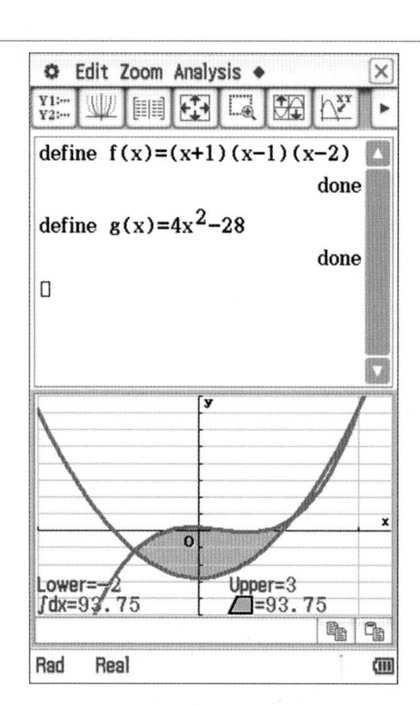

1 Define **f(x)** and **g(x)**, as shown above.
2 Solve **f(x) = g(x)** to determine the x values of the points of intersection.

> The absolute value brackets '||' around an expression make its value always positive whether it is positive or negative.

3 From the **Keyboard**, tap **Math1** or **Math2** to insert the absolute value template **| |**.
4 Find the definite integral of **f(x)−g(x)** from **−2** to **3**.
5 Add the absolute value of the definite integral of **f(x)−g(x)** from **3** to **5**.
6 The sum of the areas between the curves will be displayed.
7 To confirm this result, graph **f(x)** and **g(x)**.
8 Adjust the window settings to suit.
9 Tap **Analysis > G-Solve > Integral > ∫dx Intersection**.
10 With the cursor on the first point of intersection, press **EXE**.
11 Press the right arrow to jump to the second point of intersection, press **EXE**.
12 The area between the first two points of intersection will be displayed.
13 Repeat to find the area between the second and third points of intersection.
14 Add the two values to find the total area.

Total area = 101.75 units^2

TI-Nspire

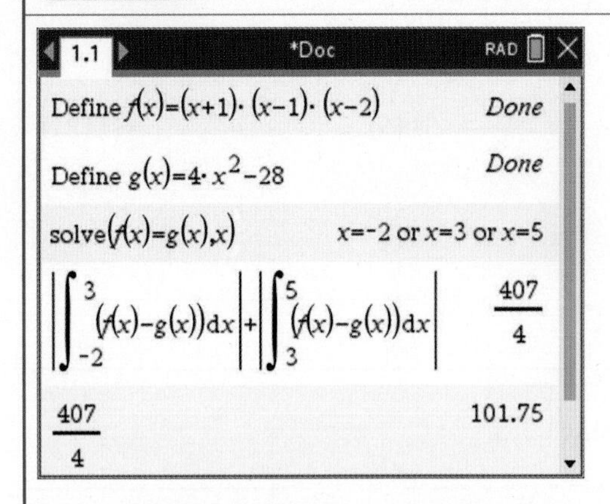

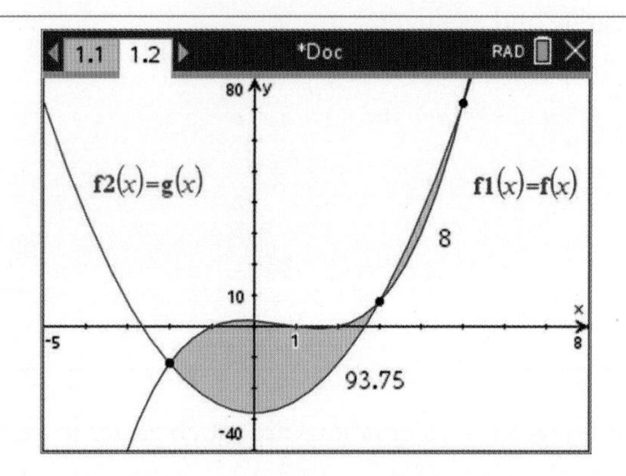

1 Define **f(x)** and **g(x)** as shown above.

2 Solve **f(x) = g(x)** to determine the x values of the points of intersection.

3 Press the **template** key and insert the absolute value template **| |**.

> The absolute value brackets '||' around an expression make its value always positive whether it is positive or negative.

4 Find the definite integral of **f(x)−g(x)** from **−2** to **3**.

5 Add the absolute value of the definite integral of **f(x)−g(x)** from **3** to **5**.

6 The sum of the areas between the curves will be displayed.

7 To confirm this result, graph **f(x)** and **g(x)**.

8 Adjust the window settings to suit.

9 Press **menu > Analyse Graph > Bounded Area**.

10 When prompted for the **lower bound**, click on first point of intersection on the left.

11 When prompted for the **upper bound**, click the second point of intersection.

12 Repeat to find the area between the second and third points of intersection.

13 Add the two values to find the total area.

Total area = 101.75 units^2

EXERCISE 3.5 Areas between curves

ANSWERS p. 393

Recap

1 The area enclosed by the graph of $f(x) = -x^2 + x$ and the x-axis between $x = -1$ to $x = 2$ can be expressed as

A $\int_{-1}^{2} f(x)\,dx$

B $-\int_{-1}^{0} f(x)\,dx + \int_{0}^{2} f(x)\,dx$

C $-\int_{-1}^{0} f(x)\,dx + \int_{0}^{1} f(x)\,dx - \int_{2}^{1} f(x)\,dx$

D $\int_{0}^{1} f(x)\,dx - 2\int_{1}^{2} f(x)\,dx$

E $\int_{0}^{1} f(x)\,dx$

2 Find the area under the curve $y = 2x^3$ from $x = 0$ to $x = 1$.

Mastery

3 WORKED EXAMPLE 13 Find the area enclosed between the curve $y = x^2$ and the line $y = x + 6$.

4 Using CAS 5 Find the area enclosed between the graphs of the functions

a $y = x^2 - 5x$ and $y = -x$ **b** $f(x) = x^2 + 5x$ and $f(x) = x$ **c** $y = x^2 + 5x$ and $y = -x^2$

5 Find the area enclosed between each pair of graphs.

a $y = 2$ and $y = x^2 + 1$

b $y = x^2$ and $y = -6x + 16$

c $y = 2x^2 - 12x + 20$ and $2x + y = 12$

Calculator-free

6 (7 marks) Consider the function $f(x) = 3x^2 - x^3$.

a Find the coordinates of the stationary points of the function. (2 marks)

b Copy the axes below and sketch the graph of $f(x)$. (2 marks)

Exam hack

Make sure your graph has rounded turning points and not V-shaped sharp points.

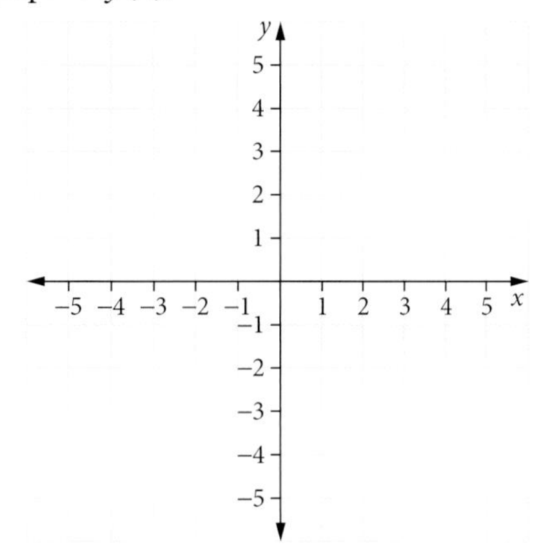

c Find the area enclosed by the graph of the function and the horizontal line given by $y = 4$. (3 marks)

7 (2 marks) Write an integral expression that would determine the shaded area bounded by the curves $y = f(x)$, $y = g(x)$ and the line $x = -3$.

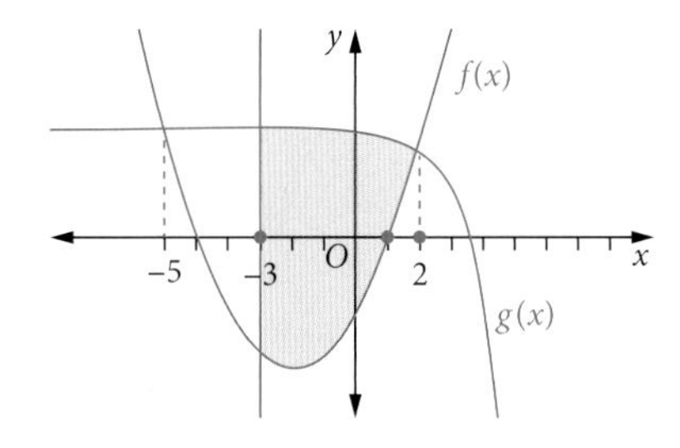

8 (2 marks) Write an integral expression that would determine the shaded area shown.

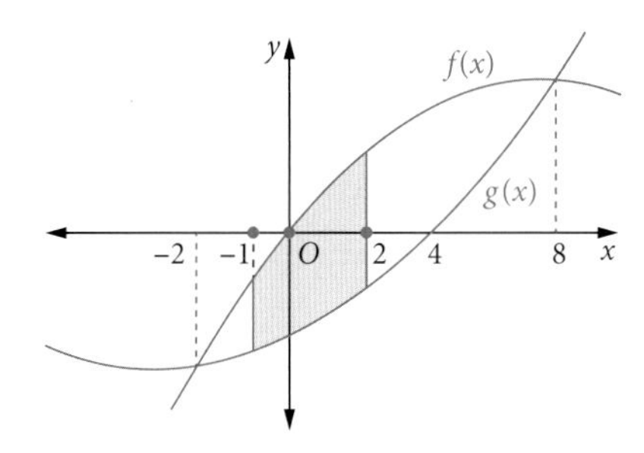

9 (2 marks) Write an integral expression that would determine the total shaded area shown.

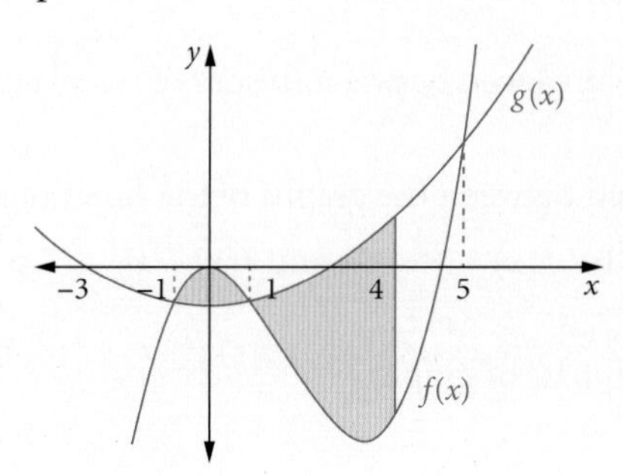

10 (3 marks) Determine the area enclosed between the curves $f(x) = (x + 1)(x - 1)$ and $g(x) = x + 1$.

11 (12 marks) Calculate each shaded area.

a

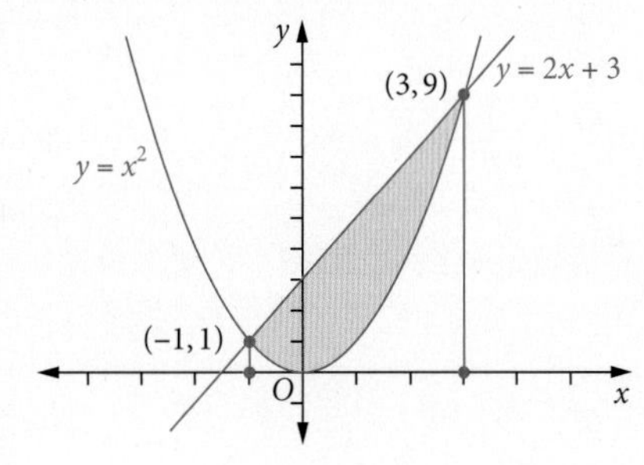

(3 marks)

b

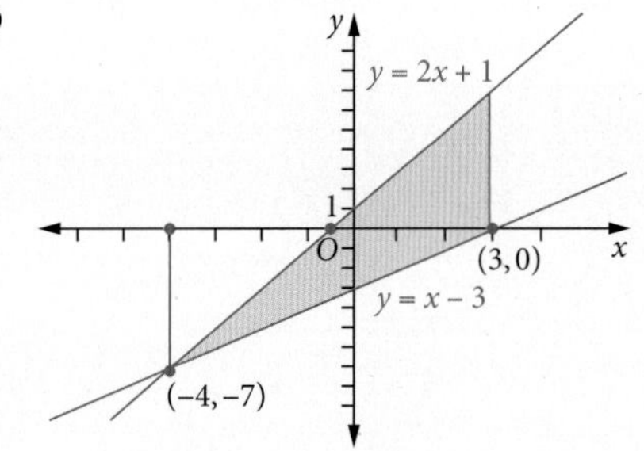

(3 marks)

c

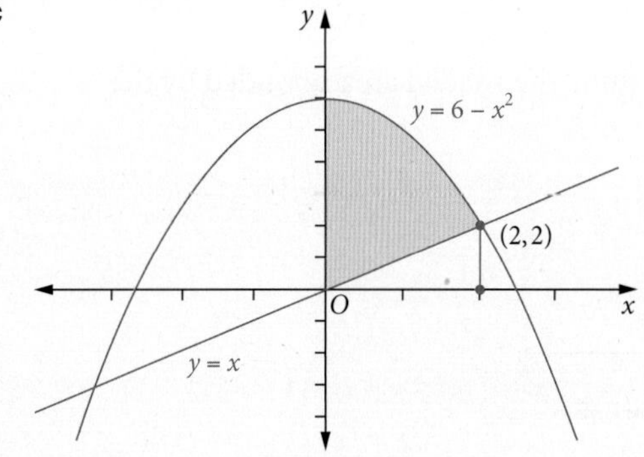

(3 marks)

d

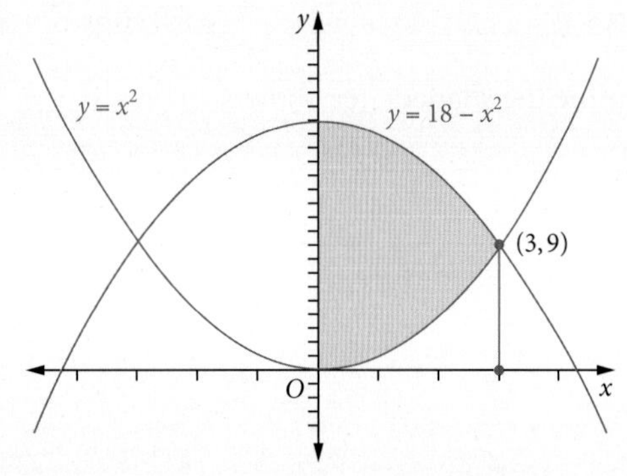

(3 marks)

12 (3 marks) Determine the area enclosed by the two curves $f(x) = x^2 - 3x + 1$ and $y = 2x + 1$.

13 (3 marks) Determine the area between the curves $f(x) = (x + 1)(x^2 - 1)$ and $g(x) = x - 1$.

14 (3 marks) The graph of part of the function of $f(x) = \sqrt{x-1}$ is shown. Determine the area of the shaded region.

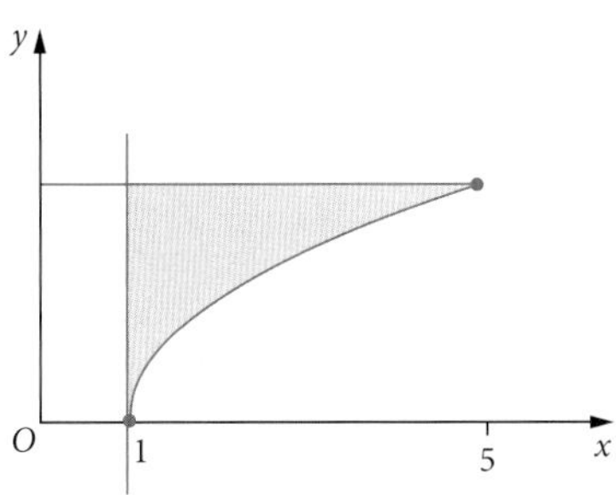

15 (4 marks) Parts of the graphs of the functions

$f(x) = x^3 - ax \qquad a > 0$

$g(x) = ax \qquad a > 0$

are shown in the diagram.

The graphs intersect when $x = 0$ and when $x = m$.

The area of the shaded region is 64.

Find the value of a and the value of m.

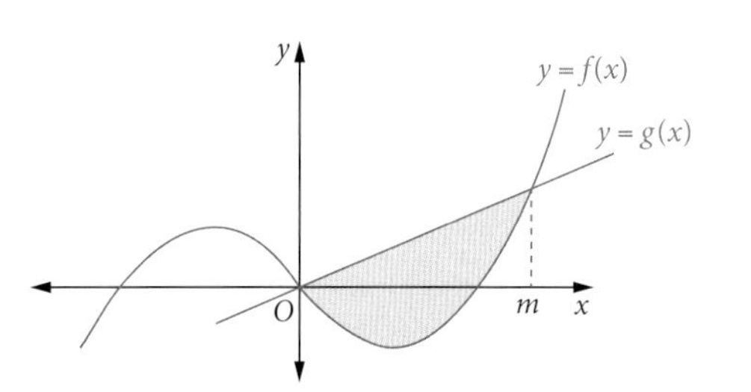

3.6 Straight line motion

Video playlist
Straight line motion

Worksheet
Displacement, velocity and acceleration

Straight line motion

Kinematics is the study of motion using **displacement** (x), **velocity** (v) and **acceleration** (a), all in terms of time t, where $t \geq 0$.

Displacement is a 'signed distance' that can be positive or negative.

We know that $v(t) = \dfrac{dx}{dt}$ and $a(t) = \dfrac{dv}{dt}$.

Velocity is a 'signed speed' that can be positive or negative.

Therefore, displacement, $x = \int v(t)\,dt$ and velocity, $v = \int a(t)\,dt$.

Displacement		Velocity		Acceleration
$x, \int v\,dt$	$\longrightarrow$ / $\longleftarrow$	$v, \dfrac{dx}{dt}, \int a\,dt$	$\longrightarrow$ / $\longleftarrow$	$a, \dfrac{dv}{dt}, \dfrac{d^2x}{dt^2}$

WORKED EXAMPLE 14 Calculating straight line motion

The acceleration of a particle travelling in a straight line is given by $a(t) = 2t$.

a Find an expression for the velocity $v(t)$ if $v = 0$ when $t = 3$.

b Find an expression for the displacement $x(t)$ if the particle started at the origin.

Steps	Working
a 1 State the rule for acceleration.	$a(t) = 2t$
2 Use velocity $= v = \int a(t)\,dt$.	$v = \int (2t)\,dt$ $\therefore v = t^2 + c$
3 Find c using $v = 0$ when $t = 3$.	$0 = 9 + c$ $\therefore c = -9$
4 State the velocity function.	$v(t) = t^2 - 9$
b 1 Use displacement $= x = \int v(t)\,dt$.	$x = \int (t^2 - 9)\,dt$ $\therefore x = \frac{t^3}{3} - 9t + d$ Use d for the constant because we have already used c.
2 Find d using $x = 0$ when $t = 0$.	$x = \frac{0^3}{3} - 9(0) + d$ $\therefore d = 0$
3 State the displacement function.	$x(t) = \frac{t^3}{3} - 9t$

WORKED EXAMPLE 15 Calculating total distance travelled by an object

A particle moves in a straight line such that its acceleration after t seconds is given by $a(t) = 5t - 6\,\text{m/s}^2$. After 2 seconds. the velocity of the particle is 0 m/s.

a Determine an expression for the velocity in terms of t.

b Determine the change in the displacement of the particle between $t = 0$ and $t = 1$. Interpret your answer.

c Determine the total distance travelled by the particle between $t = 0$ and $t = 1$.

Steps	Working
a 1 Integrate acceleration to find velocity.	$\int (5t - 6)\,dt = \frac{5t^2}{2} - 6t + c$
2 Use the given conditions to find the value of c.	$0 = \frac{5(2)^2}{2} - 6(2) + c \Rightarrow c = 2$ Therefore, $v(t) = \frac{5t^2}{2} - 6t + 2$.
b 1 Determine change in displacement.	$\int_0^1 \left(\frac{5t^2}{2} - 6t + 2\right) dt = -\frac{1}{6}\,\text{m}$
2 Interpret your answer.	After one second, the particle has moved $\frac{1}{6}$ m to the left of the origin.

c 1 Determine whether the particle stops and changes direction. That is, does the velocity equal zero between $t = 0$ and $t = 1$?	$0 = \frac{5t^2}{2} - 6t + 2$ $\therefore t = \frac{2}{5}, 2$
2 Integral needs to be split at $t = \frac{2}{5}$.	distance $= \int_0^{\frac{2}{5}} \left(\frac{5t^2}{2} - 6t + 2\right) dt - \int_{\frac{2}{5}}^{1} \left(\frac{5t^2}{2} - 6t + 2\right) dt$ $= \frac{137}{150} \approx 0.913\,\text{m}$

As shown in the above example, there are some important points to note.

- Total distance and displacement will not always give you the same answer. It is important to check if the particle has changed direction (that is, velocity is zero) at any stage.
- The integral $\int_{\frac{2}{5}}^{1} \left(\frac{5t^2}{2} + 6t + 2\right) dt$ was subtracted because its area was below the axis if graphed.
- Distance travelled must always be positive, but displacement can be negative.
- It was not necessary to determine an expression for displacement, as we were integrating velocity using definite integrals.
- Part **c** could be done by using one integral if absolute value signs were used.

  For example, $\int_0^1 \left|\frac{5t^2}{2} - 6t + 2\right| dt = \frac{137}{150}$ m.

WACE QUESTION ANALYSIS

© SCSA MM2016 Q19 Calculator-assumed (8 marks)

The displacement in centimetres of a particle from the point O in a straight line is given by

$x(t) = \frac{1}{3}\left(\frac{t}{2} - 4\right)^2 - 2$ for $0 \leq t \leq 10$, where t is measured in seconds.

Calculate the:

a time(s) that the particle is at rest. (2 marks)

b displacement of the particle during the fifth second. (2 marks)

c maximum speed of the particle and the time when this occurs. (2 marks)

d total distance travelled in the first 10 seconds. (2 marks)

Video WACE question analysis: Integrals

Reading the question

- In part **a**, if the particle is at rest, then the velocity must be equal to zero. The 'times(s)' indicates there may be more than one answer.
- In part **b**, your answer may be negative. Note that it is the 'fifth' second, and not after five seconds.
- In part **c**, care needs to be taken in using the acceleration as this is constant. Therefore, the velocity function (and perhaps a graph) needs to be considered.
- In part **d**, the final answer needs to be positive. Ensure you are finding total distance, and not displacement.

Thinking about the question

- As each question is only worth 2 marks, you do not have to show any working to get full marks. However, it is advised to show some working, so part marks can be attained even if final answer is incorrect.
- As this is a calculator-assumed question, use CAS to calculate integrals and solve equations. The graphing facility may also be useful.

Worked solution (✓ = 1 mark)

a $\frac{dx}{dt} = \frac{1}{3}\left(\frac{t}{2} - 4\right) = 0$

$\frac{t}{2} = 4$

$t = 8$

differentiates to determine velocity ✓

solves for time that velocity equals zero ✓

b ClassPad

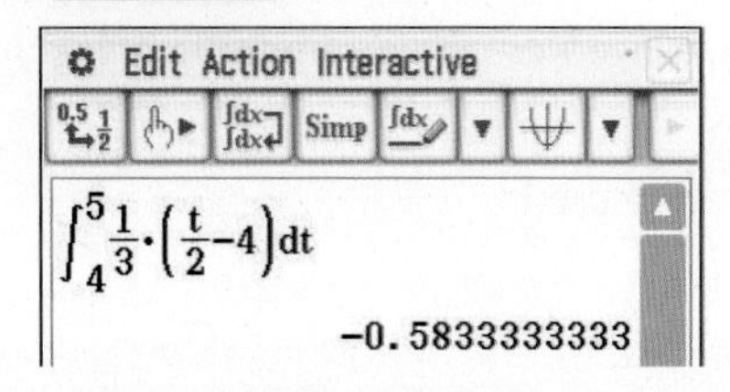

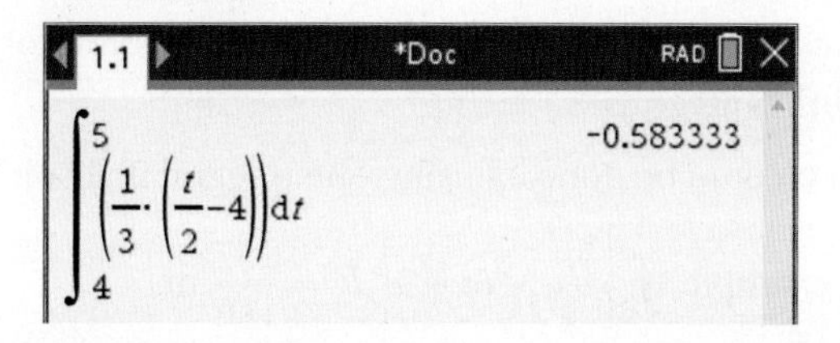

Displacement = −0.5833 cm

examines motion between $t = 4$ and $t = 5$ ✓

determines change in displacement ✓

c $\frac{dx}{dt} = \frac{1}{3}\left(\frac{t}{2} - 4\right), 0 \le t \le 10$

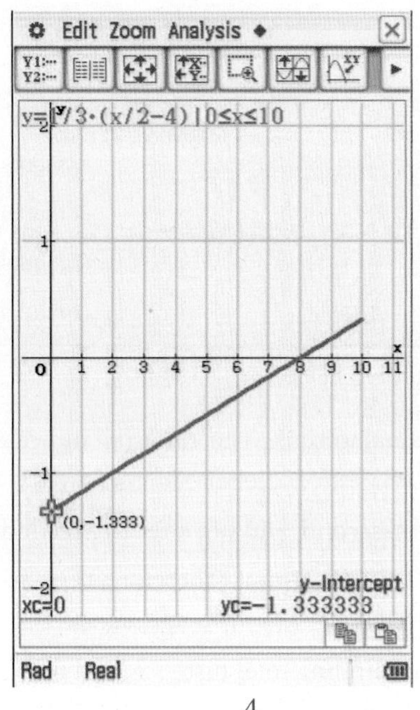

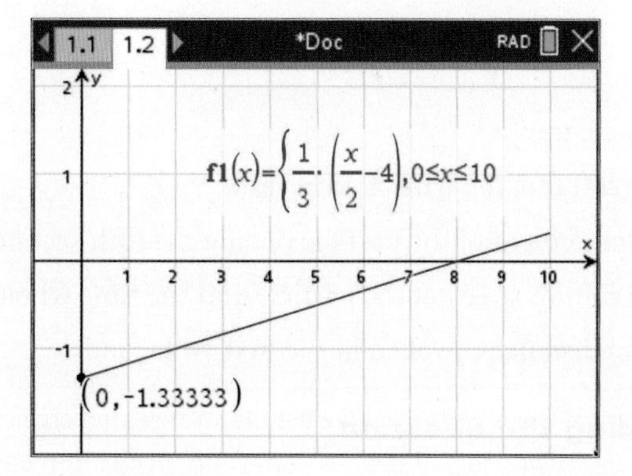

Maximum speed = $\frac{4}{3}$ cm/s at $t = 0$.

examines velocity at endpoints $t = 0$, 10 seconds ✓

determines maximum speed ✓

d ClassPad

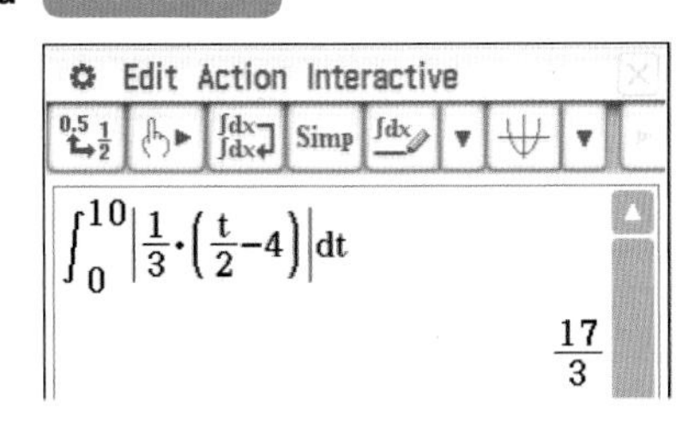

TI-Nspire

1.1 1.2 *Doc RAD

$$\int_0^{10}\left|\frac{1}{3}\cdot\left(\frac{t}{2}-4\right)\right|dt \qquad \frac{17}{3}$$

sets up an integral to determine distance travelled ✓

determines distance travelled ✓

or

$t = 10$, $x = \frac{5}{3}$

$t = 8$, $x = -2$

$t = 0$, $x = \frac{10}{3}$

Distance travelled $= \frac{15}{3} + \frac{1}{3} + \frac{1}{3} = \frac{17}{3}$

sets up a pathway of motion in first 10 seconds ✓

determines distance travelled ✓

EXERCISE 3.6 Straight line motion

ANSWERS p. 393

Recap

1 The area under the curve $y = x^2 + 1$ from $x = 0$ to $x = 3$, in units^2, equals

A 0 **B** 3 **C** 6 **D** 9 **E** 12

2 The area between the curves $f(x) = x^2 + 1$ and $g(x) = 5$, in units^2, is

A $-\frac{32}{3}$ **B** $\frac{32}{3}$ **C** 5 **D** $\frac{88}{3}$ **E** $\frac{28}{3}$

Mastery

3 WORKED EXAMPLE 14 An object's acceleration as a function of time, t, is given by $a(t) = 4t + 1$. Find its velocity $v(t)$ if the object starts at rest.

4 WORKED EXAMPLE 15 A particle moves in a straight line such that its acceleration after t seconds is given by $a(t) = 3 - 2t$ m/s^2. After 3 seconds, the velocity of the particle is 2 m/s.

a Determine an expression for the velocity in terms of t.

b Determine the change in the displacement of the particle between $t = 1$ and $t = 2$. Interpret your answer.

c Determine the total distance travelled by the particle between $t = 1$ and $t = 2$.

5 The velocity of a particle travelling in a straight line is given by $v(t) = -t^2 + t$. Determine an expression for the displacement $x(t)$ if $x = 0$ when $t = 2$.

6 If the velocity of an object in m/s is described by $v(t) = 3t^2 + 4t^3$ and we know that the object starts at the origin, determine its displacement, in metres.

Calculator-free

7 (2 marks) Determine the velocity of an object which has an acceleration of $a(t) = 4t\,\text{m/s}^2$, given the object stops momentarily at $t = 2$.

8 (2 marks) Determine the displacement of an object which has an acceleration of $a(t) = 3 - 4t\,\text{m/s}^2$, and the object starts at rest from the origin.

9 (7 marks) When a train leaves the station, its velocity is 8 m/s and it accelerates constantly at $0.4\,\text{m/s}^2$.

a Find an expression for the train's velocity as a function of time. (2 marks)

b Find an expression for its displacement. (2 marks)

c Find its displacement after 10 seconds. (3 marks)

Calculator-assumed

10 © SCSA MM2021 Q14 (5 marks) The displacement in metres, $x(t)$, of a power boat t seconds after it was launched is given by:

$$x(t) = \frac{5t(t^2 - 15t + 48)}{6}, t \geq 0$$

How far has the power boat travelled before its acceleration is zero?

11 © SCSA MM2021 Q17 (8 marks) A resort in the Swiss Alps features a cable car that travels from the resort station to the mountain station. Engineers are fixing a cable car that unexpectedly stopped shortly before it reached the mountain station. The engineers are ready to test the cable car. For the purposes of the test, the cable car will initially be at rest in its current position, will head up the mountain, stop at the mountain station and immediately return to the resort station where it will stop, and the test will be complete.

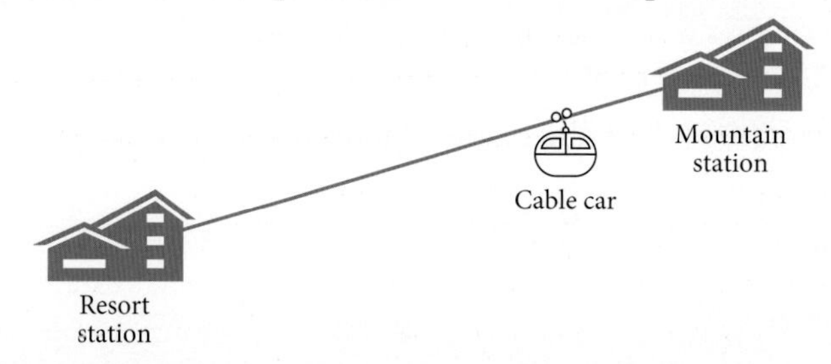

The test begins and engineers believe that the acceleration, $a(t)$, of the cable car during the test will be: $a(t) = kt^2 - 23t + 20k$, measured in m/min^2. The variable t is the number of minutes from the moment the cable car leaves its position and k is a constant. After two minutes, the engineers expect that the cable car will be travelling with velocity 18 m/min and will not yet have reached the mountain station.

a Determine the value of the constant k. (3 marks)

b Once the cable car leaves the mountain station, how long should it take to return to the resort station? (3 marks)

c Unfortunately, 10 minutes into the test, the cable car breaks down again. According to the engineers' model, how far is the cable car from the mountain station at this time? (2 marks)

12 (7 marks) A go kart slows down from an initial velocity of 16 m/s until it is stationary. During this interval, its acceleration (t seconds) after the brakes were applied is given by

$$a(t) = \frac{t}{2} - 4\,\text{m/s}^2$$

a Determine the velocity of the vehicle after four seconds. (3 marks)

b Calculate the distance travelled by the vehicle in the time between the brakes being applied and it becoming stationary. (4 marks)

3 Chapter summary

Integrals

The anti-derivative

- $F(x)$ is the **anti-derivative** or indefinite **integral** or **primitive** of $f(x)$ if $F'(x) = f(x)$.
- $\int ax^n dx = \dfrac{ax^{n+1}}{n+1} + c$, where ax^n, is called the **integrand**.
- $\int (ax+b)^n dx = \dfrac{(ax+b)^{n+1}}{a(n+1)} + c$

Approximating areas under curves

- We can estimate the area under a graph using a series of rectangles.

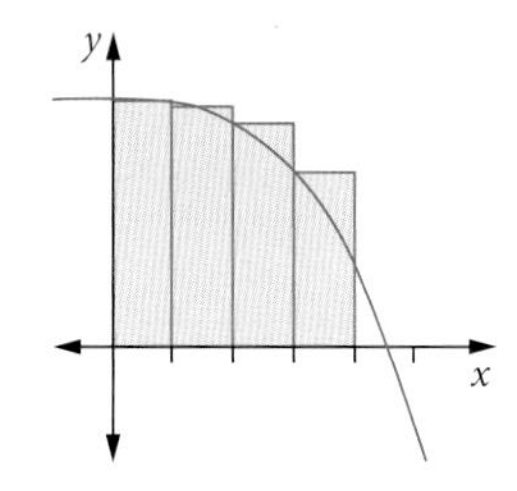

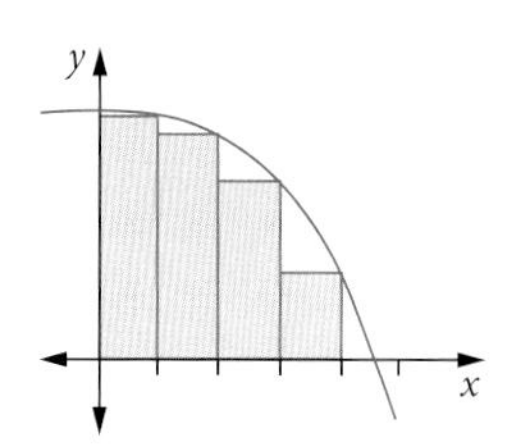

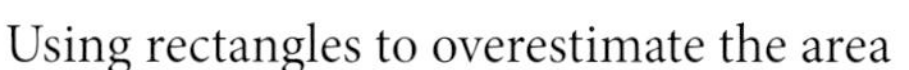

Using rectangles to overestimate the area

Using rectangles to underestimate the area

The definite integral

- A definite integral can be evaluated using $\int_a^b f(x)\,dx = F(b) - F(a)$, where $F(x)$ is an anti-derivative of $f(x)$.

Properties of the definite integral

1 $\int_a^b kf(x)\,dx = k\int_a^b f(x)\,dx$	**Constant out:** a constant factor can be taken out of an integral.
2 $\int_a^b f(x) \pm g(x)\,dx = \int_a^b f(x)\,dx \pm \int_a^b g(x)\,dx$	**Split terms:** a sum or difference of terms can be integrated separately.
3 If b is between a and c, then: $\int_a^c f(x)\,dx = \int_a^b f(x)\,dx + \int_b^c f(x)\,dx$	**Split limits:** the limits of an integral can be split.
4 $\int_a^b f(x)\,dx = -\int_b^a f(x)\,dx$	**Swap limits:** reversing the order of the limits changes the sign of the definite integral.
5 $\int_a^a f(x)\,dx = 0$	**Same limits:** gives $F(a) - F(a) = 0$.

The fundamental theorem of calculus

The two parts of the fundamental theorem are

$$F'(x) = \frac{d}{dx}\left(\int_a^x f(t)\,dt\right) = f(x) \qquad \text{and} \qquad \int_a^b f'(x)dx = f(b) - f(a)$$

Area under a curve

- The definite integral $\int_a^b f(x)\,dx$ gives the **signed area** enclosed by the graph of $y = f(x)$ and the x-axis between $x = a$ and $x = b$.
- If the section of the graph is above the x-axis, then the definite integral gives the area under the curve.
- If the section of the graph is below the x-axis, then the definite integral gives the area above the curve and its value is negative.

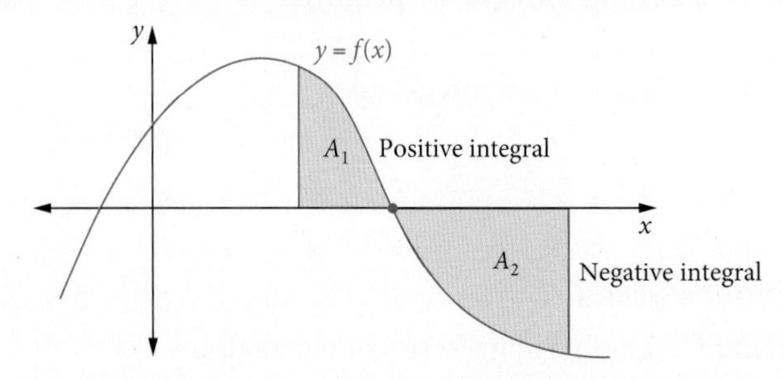

- If we need to find areas using integration, we need to check whether the graph goes below the x-axis and change the sign for areas below the x-axis (to make them positive).

Areas between curves

- The area can be calculated as the difference between the areas under the two functions regardless of the position of the area.

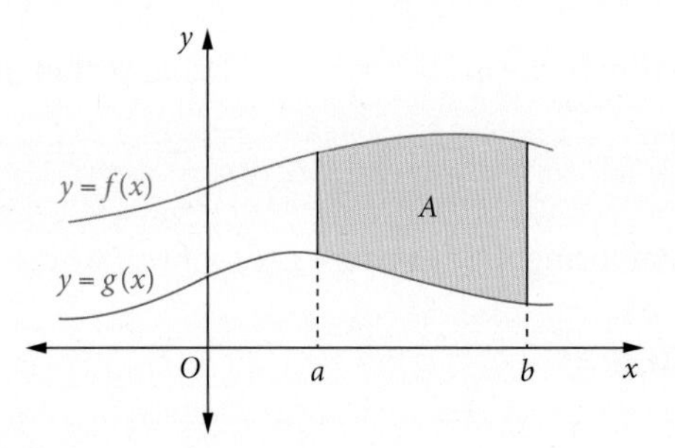

- area $= \int_a^b (\text{upper} - \text{lower})\,dx = \int_a^b [f(x) - g(x)]\,dx$

Straight line motion

- velocity $= v = \int a(t)\,dt$
- displacement $= x = \int v(t)\,dt$

Cumulative examination: Calculator-free

Total number of marks: 26 Reading time: 3 minutes Working time: 26 minutes

1 (2 marks) If $f(x) = \int_0^x (\sqrt{t^2 + 4})\, dt$, determine the value of $f'(-2)$.

2 (2 marks) Evaluate $\int_1^2 \left(3x^2 - \frac{1}{x^2}\right) dx$.

3 (2 marks) Find $f(x)$ given that $f(4) = \frac{64}{3}$ and $f'(x) = x^2 - 10x - x^{-\frac{1}{2}} + 1, x > 0$.

4 © SCSA MM2020 Q3 (7 marks) The graph of the cubic function $f(x) = ax^3 + bx^2 + cx + d$ is shown. A turning point is located at $(1, 0)$ and the shaded region shown on the graph has an area of $\frac{3}{2}$ units^2.

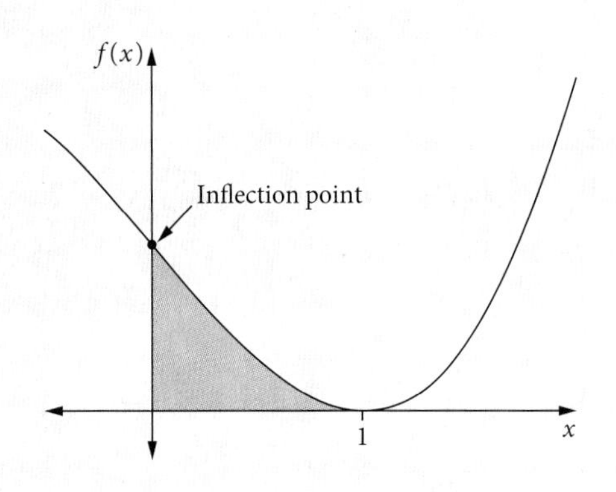

Use the above information to determine the values of a, b, c and d.

5 © SCSA MM2016 Q7 (7 marks) Consider the graph $y = f(x)$. Both arcs have a radius of four units.

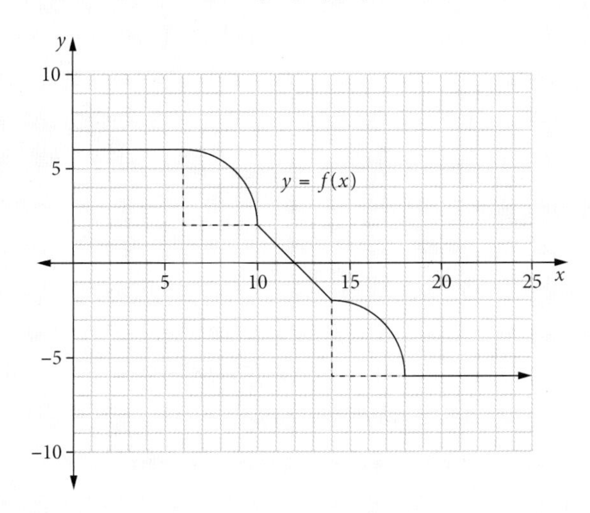

Using the graph of $y = f(x)$, $x \geq 0$, evaluate exactly the following integrals.

a $\int_0^{12} f(x)\, dx$ (3 marks)

b $\int_0^{18} f(x)\, dx$ (2 marks)

c Determine the value of the constant α such that $\int_0^{\alpha} f(x)\, dx = 0$. There is no need to simplify your answer. (2 marks)

6 © SCSA MM2021 Q5 (6 marks)

a Determine the area between the parabola $y = x^2 - x + 3$ and the straight line $y = x + 3$. (4 marks)

b The area between the parabola $y = x^2 - x - 2$ and the straight line $y = x - 2$ is the same as the area determined in part **a**. Explain why this is the case. (2 marks)

Cumulative examination: Calculator-assumed

Total number of marks: 27   Reading time: 3 minutes   Working time: 27 minutes

1 (2 marks) Consider $f(x) = x^2 + \frac{p}{x}$, $x \neq 0$. There is a stationary point on the graph of f when $x = -2$. Determine the value of p.

2 (3 marks) Determine the area enclosed between the graph of $y = -2\sqrt{x-1}$, the x-axis and the line $x = 2$.

3 © SCSA MM2017 Q20 (9 marks) A model train travels on a straight track such that its acceleration after t seconds is given by $a(t) = pt - 13\,\text{cm/s}^2$, $0 \le t \le 10$, where p is a constant.

a Determine the initial acceleration of the model train. (1 mark)

The model train has an initial velocity of 5 cm/s. After 2 seconds it has a displacement of −50 cm. A further 4 seconds later its displacement is 178 cm.

b Determine the value of the constant p. (4 marks)

c When is the model train at rest? (2 marks)

d How far has the model train travelled when its acceleration is $47\,\text{cm/s}^2$? (2 marks)

4 © SCSA MM2018 Q16 (8 marks) Let $f(x)$ be a function such that $f(-2) = 4$, $f(-1) = 0$, $f(0) = -1$, $f(1) = 0$ and $f(3) = 2$. Further, $f'(x) < 0$ for $-2 \le x < 0$, $f'(0) = 0$ and $f'(x) > 0$ for $0 < x \le 3$.

a Evaluate the following definite integrals:

i $\int_0^3 f'(x)\,dx$. (2 marks)

ii $\int_{-2}^3 f'(x)\,dx$. (2 marks)

b What is the area bounded by the graph of $f'(x)$ and the x-axis between $x = -2$ and $x =$ 3? Justify your answer. (4 marks)

5 (5 marks) Consider functions $f(x) = \frac{81x^2(a-x)}{4a^4}$ and $h(x) = \frac{9x}{2a^2}$, where a is a positive real number.

a Find the coordinates of the local maximum of $f(x)$ in terms of a. (2 marks)

b Find the x values of all the points of intersection between the graphs of $f(x)$ and $h(x)$, in terms of a where appropriate. (1 mark)

c Determine the total area of the regions bounded by the graphs of $y = f(x)$ and $y = h(x)$. (2 marks)

APPLYING THE EXPONENTIAL AND TRIGONOMETRIC FUNCTIONS

Syllabus coverage

TOPIC 3.1: FURTHER DIFFERENTIATION AND APPLICATIONS

Exponential functions

3.1.1 estimate the limit of $\frac{a^h - 1}{h}$ as $h \to 0$, using technology, for various values of $a > 0$

3.1.2 identify that e is the unique number a for which the above limit is 1

3.1.3 establish and use the formula $\frac{d}{dx}(e^x) = e^x$

3.1.4 use exponential functions of the form Ae^{kx} and their derivatives to solve practical problems

Trigonometric functions

3.1.5 establish the formulas $\frac{d}{dx}(\sin x) = \cos x$ and $\frac{d}{dx}(\cos x) = -\sin x$ by graphical treatment, numerical estimations of the limits, and informal proofs based on geometric constructions

3.1.6 use trigonometric functions and their derivatives to solve practical problems

Differentiation rules

3.1.9 apply the product, quotient and chain rule to differentiate functions such as xe^x, $\tan x$, $\frac{1}{x^n}$, $x\sin x$, $e^{-x}\sin x$ and $f(ax - b)$

TOPIC 3.2: INTEGRALS

Anti-differentiation

3.2.4 establish and use the formula $\int e^x\,dx = e^x + c$

3.2.5 establish and use the formulas $\int \sin x\,dx = -\cos x + c$ and $\int \cos x\,dx = \sin x + c$

Applications of integration

3.2.18 calculate total change by integrating instantaneous or marginal rate of change

3.2.19 calculate the area under a curve

3.2.20 calculate the area between curves determined by functions of the form $y = f(x)$

3.2.21 determine displacement given velocity in linear motion problems

3.2.22 determine positions given linear acceleration and initial values of position and velocity

Video playlists (6):

4.1 Exponential growth and decay
4.2 Differentiating exponential functions
4.3 Integrating exponential functions
4.4 Differentiating trigonometric functions
4.5 Integrals of trigonometric functions
WACE question analysis Applying the exponential and trigonometric functions

Worksheets (7):

4.1 Exponential functions
4.2 Derivatives of exponential functions
4.4 Further optimisation problems • Trigonometric functions and gradient
4.5 Finding indefinite integrals 1 • Finding indefinite integrals 2 • Finding definite integrals

To access resources above, visit **cengage.com.au/nelsonmindtap**

4.1 Exponential growth and decay

Euler's number, *e*

The function $y = a^x$, where the base a is a positive constant but not equal to 1, is called an **exponential function**. There is a special value of a, called e, discovered by the Swiss mathematician Leonhard Euler in 1731.

$e = 2.71828\ldots$

Euler's number is defined as $\lim_{n\to\infty}\left(1+\frac{1}{n}\right)^n$.

Euler is pronounced 'oiler'.

This number is used in compound interest calculations, where n represents the number of times the interest is compounded in a year. The table shows that as n becomes larger and approaches infinity, and the compounding of interest becomes continuous, the value of $\left(1+\frac{1}{n}\right)^n$ approaches 2.71828…

Video playlist Exponential growth and decay

Worksheet Exponential functions

Frequency of interest compounding	n	$\lim_{n\to\infty}\left(1+\frac{1}{n}\right)^n$
Yearly	1	$\left(1+\frac{1}{1}\right)^1 = 2$
Half-yearly	2	$\left(1+\frac{1}{2}\right)^2 = 2.25$
Quarterly	4	$\left(1+\frac{1}{4}\right)^4 = 2.44140625$
Monthly	12	$\left(1+\frac{1}{12}\right)^{12} \approx 2.6130352902\ldots$
Weekly	52	$\left(1+\frac{1}{52}\right)^{52} \approx 2.69259695444\ldots$
Daily	365	$\left(1+\frac{1}{365}\right)^{365} \approx 2.71456748202\ldots$
Hourly	8760	$\left(1+\frac{1}{8760}\right)^{8760} \approx 2.71812669063\ldots$
Every minute	525 600	$\left(1+\frac{1}{525600}\right)^{525600} \approx 2.718279215\ldots$
Every second	31 536 000	$\left(1+\frac{1}{31536000}\right)^{31536000} \approx 2.71828247254\ldots$

The table shows that $\lim_{n\to\infty}\left(1+\frac{1}{n}\right)^n = e$ and it can also be shown as $\lim_{n\to\infty}\left(1+\frac{x}{n}\right)^n = e^x$.

The exponential function

The function $y = e^x$ is called the **natural exponential function**. Things that grow naturally, such as trees or population size, follow the natural exponential function. e is an irrational number like π.

Since $2 < e < 3$, the graph of the natural exponential function lies between the graphs of $y = 2^x$ and $y = 3^x$, as shown.

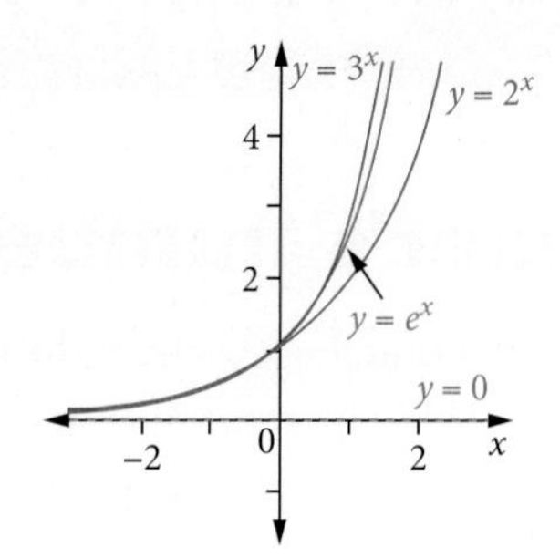

The value of e is stored on the calculator.

This table of values for $y = e^x$ shows y values rounded to three decimal places, and its graph is shown below.

x	−3	−2	−1	0	1	2	3
y	$e^{-3} \approx 0.050$	$e^{-2} \approx 0.135$	$e^{-1} \approx 0.368$	$e^0 = 1$	$e \approx 2.718$	$e^2 \approx 7.389$	$e^3 \approx 20.086$

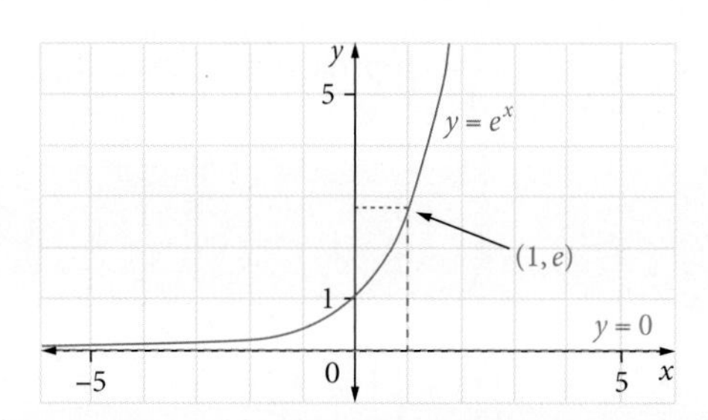

Properties of the natural exponential function, $y = e^x$

- It is a strictly increasing function, increasing slowly at first, then more quickly.
- The gradient of the graph is always increasing.
- The y-intercept is 1 (because $e^0 = 1$).
- The x-axis ($y = 0$) is a horizontal asymptote.

The graph of $y = e^{-x}$ is shown below.

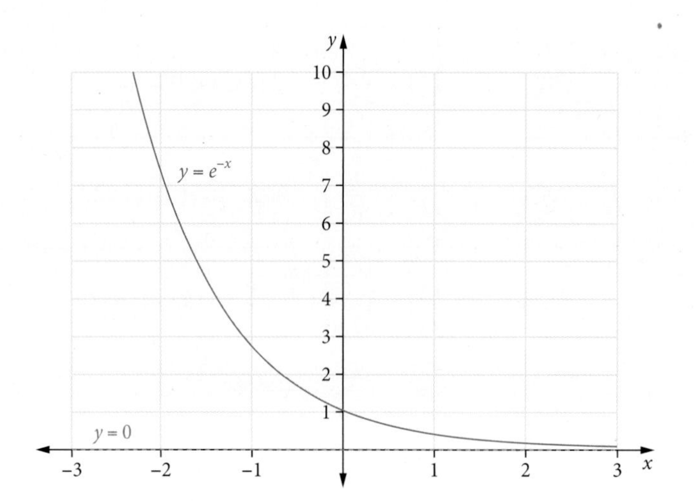

Exponential growth and decay

Any quantity that **increases** according to the exponential function $y = A(e^{kx})$, where $k > 0$, is showing **exponential growth**. Something that grows exponentially increases slowly at first, then more quickly. Examples of exponential growth are population size (people, animals, bacteria), an investment attracting compound interest, a flu or computer virus, and the size of a bushfire.

Any quantity that **decreases** according to the exponential function $y = A(e^{kx})$, where $k < 0$, is showing **exponential decay**. Something that decays exponentially decreases quickly at first, then more slowly. Examples of exponential decay are radioactive decay, the cooling of substances, the intensity of light in water, and the dampening of vibrations.

4.1

Exponential growth and decay

The general form of an exponential growth or decay function is $N(t) = N_0e^{kt}$,

where N_0 is the initial value and the value of k determines the rate of growth or decay.

For a growth function, $k > 0$.

For a decay function, $k < 0$.

WORKED EXAMPLE 1 Modelling an exponential growth problem

The number N of gum trees with pink flowers in a region of Western Australia was studied. The equation $N(t) = 1200e^{0.07t}$ was given as a model for the number of gums, with time t being the number of years since the study began.

a How many trees were there at the beginning of the study?

b How many trees were there after 10 years?

Steps	Working
a 1 This is an example of exponential growth. Substitute $t = 0$ into $N(t) = 1200e^{0.07t}$.	$N(t) = 1200e^{0.07\times0}$ $= 1200e^0$ $= 1200$ N_0 is always the initial value of $N(t) = N_0e^{kt}$.
2 Answer the question.	There were 1200 trees at the beginning.
b 1 Substitute $t = 10$ into $N(t) = 1200e^{0.07t}$.	$N(10) = 1200e^{0.07\times10}$ $= 1200e^{0.7}$ $= 2416.503\ldots$
2 Answer the question.	There were about 2417 trees after 10 years.

WORKED EXAMPLE 2 Finding the parameters in an exponential decay problem

A patient is given 250 mg of an anti-inflammatory drug. Each hour, the amount of drug in the person's system decreases exponentially so that after 1 hour there is 200 mg in the patient's system. The number of mg of the drug D in the patient's system after t hours is given by the rule $D(t) = D_0e^{-kt}$.

a Find the value of D_0.

b Find the value of k correct to four decimal places.

c Find the number of mg of the drug in the patient's system, correct to one decimal place, after 4 hours.

Steps	Working
a Substitute $t = 0$ into $D(t) = 250$.	$D(0) = 250$ $D(t) = D_0e^{-kt}$ $250 = D_0e^0$ $D_0 = 250$
b Substitute $t = 1$ into $D(t) = 200$ and solve using CAS.	$D(t) = 250e^{-kt}$ $200 = 250e^{-k}$ $k = 0.2231\ldots$
c 1 Substitute $t = 4$ into $D(t) = 250e^{-0.2231t}$.	$D(t) = 250e^{-0.2231t}$ $D(t) = 250e^{-0.2231\times4} = 102.4\ldots$
2 Answer the question.	There is 102.4 mg of the drug in the patient's system after 4 hours.

WORKED EXAMPLE 3 Using simultaneous equations to find the parameters in an exponential growth function

The diameter, in centimetres, of a species of gum tree $d(t)$ after t years is given by the rule $d(t) = d_0e^{mt}$. The diameter is 40 cm after 1 year, and 80 cm after 3 years.

a Write two equations that can be used to find the constants d_0 and m.

b Calculate the values of the constants d_0 and m, correct to three decimal places.

Steps	Working
a Substitute $t = 1$ into $d(t) = 40$ and $t = 3$ into $d(t) = 80$ where $d(t) = d_0e^{mt}$.	$d(1) = 40$ $d(t) = d_0e^{mt}$ $40 = d_0e^{m}$ equation 1 $d(3) = 80$ $80 = d_0e^{3m}$ equation 2
b Solve the simultaneous equations using CAS.	$d_0 = 28.283$, $m = 0.347$

ClassPad

```
Edit Action Interactive
{40=de^m
{80=de^3m |d, m
{d=28.28427125, m=0.34657▶
```

TI-Nspire

```
1.1  *Doc  RAD
solve({40=d·e^m, 80=d·e^(3·m)}, {d,m})
d=28.2843 and m=0.346574
```

EXERCISE 4.1 Exponential growth and decay

ANSWERS p. 394

Mastery

1 WORKED EXAMPLE 1 The number of people, N, who have the flu virus at time t months is given by $N(t) = N_0e^{kt}$, where t is the number of months after the outbreak of the virus.

If the number is initially 200 and the number increases to 500 after 1 month, find

- **a** **i** the value of N_0
  - **ii** the value of k, correct to four decimal places
- **b** the number of people infected after 6 months.

2 A biological culture, in a laboratory, contains 100 000 bacteria at 1 pm on Monday. The culture grows exponentially so that the number of bacteria B after t hours after 1 pm Monday is given by the function $B(t) = B_0e^{kt}$. The culture has 105 000 bacteria at 6 pm on Monday. Find

- **a** **i** the value of B_0
  - **ii** the value of k, correct to four decimal places
- **b** the number of bacteria at 1 pm the following Tuesday
- **c** the number of hours, to the nearest hour, for the number of bacteria to double.

3 WORKED EXAMPLE 2 An adult takes 400 mg of ibuprofen. After 1 hour, the amount of ibuprofen in the person's system is 280 mg. The number of mg of the drug D in the patient's system after t hours is given by the rule $D(t) = D_0e^{-kt}$.

- **a** Find the value of D_0.
- **b** Find the value of k correct to four decimal places.
- **c** Find the number of mg of the drug in the patient's system, correct to one decimal place, after 2 hours.

4 The population of Doreen can be modelled by $P(t) = 6191e^{0.04t}$, where t is the number of years since 1990.

- **a** What was the population in 1990?
- **b** What was the population in 1991?
- **c** By what percentage did the population increase in the first year?

5 WORKED EXAMPLE 3 The diameter d, in centimetres, of a species of elm tree after t years is given by the rule $d(t) = d_0e^{mt}$. The diameter is 10 cm after 1 year, and 15 cm after 2 years.

- **a** Write two equations that can be used to find the constants d_0 and m.
- **b** Calculate the values of the constants d_0 and m, correct to three decimal places.

6 In 1985, there were 285 mobile phone subscribers in the small town of Centerville and by 1987 this had grown to 873. The number of subscribers S, t years after 1984 is found to grow exponentially and is given by the rule $S(t) = S_0e^{kt}$.

- **a** Find the value of S_0 to the nearest integer.
- **b** Find the value of k, correct to four decimal places.
- **c** Find the number of mobile phone subscribers in Centerville in 1995.

Calculator-free

7 (3 marks) Cobalt-60 is a radioactive substance whose decay rate can be modelled by the formula $P = P_0e^{kt}$, where P is the mass in grams, t is measured in days, P is the original amount and k is a constant. The time taken to decay to half of the original amount is known as half-life.

The half-life of Cobalt-60 is 5 years and its initial mass is 200 grams.

- **a** Find the value of P_0. (1 mark)
- **b** Show that $e^{-5k} = 2$. (2 marks)

Calculator-assumed

8 © SCSA MM2016 Q9ab (5 marks) Fermium-257 is a radioactive substance whose decay rate can be modelled by the formula $P = P_0e^{kt}$, where P is the mass in grams and t is measured in days and P_0 = original amount and k is a constant. The time taken to decay to half of the original amount is known as half-life. The half-life of Fermium-257 is 100.5 days.

- **a** Determine the value of k to three significant figures. (3 marks)
- **b** How many days will it take for 100 grams of the substance to first decay below five grams? (2 marks)

9 © SCSA MM2018 Q9ab (3 marks) The concentration, C, of a drug in the blood of a patient t hours after the initial dose can be modelled by the equation below.

$C = 4e^{-0.05t}$ mg/L

Patients requiring this drug are said to be in crisis if the concentration of the drug in their blood falls below 2.5 mg/L.

A patient is given a dose of the drug at 9 am.

a What was the concentration in the patient's blood immediately following the initial dose? (1 mark)

b What is the concentration of the drug in the patient's blood at 11:30 am? (2 marks)

10 (5 marks) A radioactive element has a half-life of 10 minutes, meaning it takes 10 minutes for half the remaining atoms to decay. Originally, there are 8.0×10^{20} atoms in a sample of the element. The decay is modelled by $N = N_0e^{-kt}$, where N_0 is the original number of atoms, N is the number of atoms present at time t, and k is the decay rate of the material.

a Find the value of k, correct to four decimal places. (2 marks)

b Find the number of atoms left after 1 hour. (2 marks)

c Is it possible to calculate when there is no sample left? (1 mark)

Video playlist
Differentiating exponential functions

Worksheet
Derivatives of exponential functions

4.2 Differentiating exponential functions

The derivative of $f(x) = 2^x$ by first principles is shown below.

$$\begin{aligned} f'(x) &= \lim_{h \to 0} \frac{f(x+h) - f(x)}{h} \\ &= \lim_{h \to 0} \frac{2^{x+h} - 2^x}{h} \\ &= \lim_{h \to 0} \frac{2^x \times 2^h - 2^x}{h} \\ &= \lim_{h \to 0} \frac{2^x(2^h - 1)}{h} \\ &= 2^x\left(\lim_{h \to 0} \frac{2^h - 1}{h}\right) \end{aligned}$$

We can factorise the 2^x because it does not involve h.

We can use CAS to evaluate the limit $\lim_{h \to 0} \frac{(2^h - 1)}{h}$. Set your calculator to give decimal answers.

(Decimal mode for ClassPad settings and Approximate calculation mode for the TI-Nspire.)

ClassPad

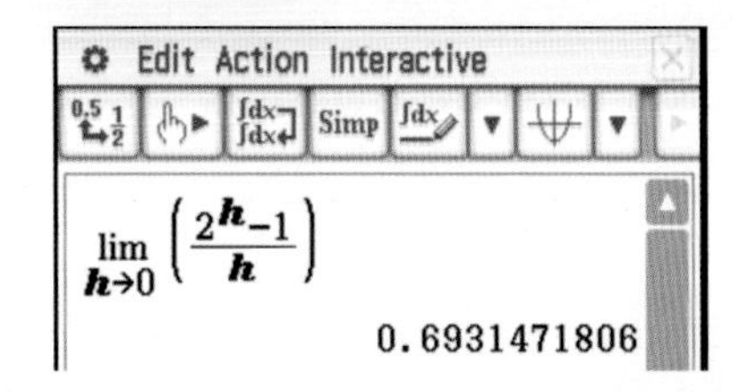

TI-Nspire

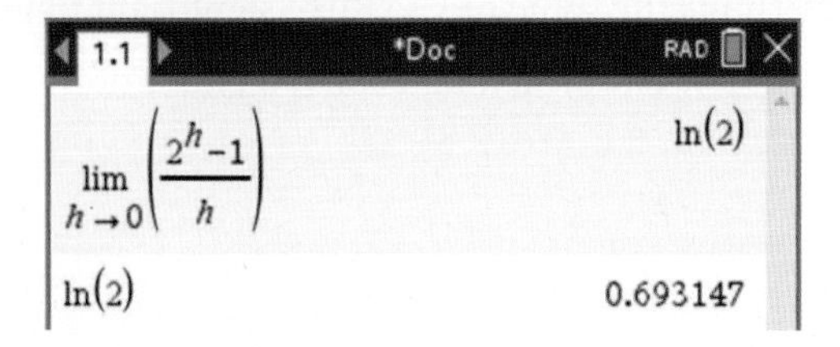

This means that the derivative of $f(x) = 2^x$ is $2^x(0.6931\ldots) = (0.6931\ldots) \times 2^x$.

So, the derivative of $f(x) = 2^x$ is 2^x multiplied by a constant.

Similarly, the derivative of $f(x) = 3^x$ is $3^x \lim_{h\to 0} \frac{(3^h - 1)}{h}$.

When $\lim_{h\to 0} \frac{3^h - 1}{h}$ is evaluated, the answer is 1.0986…

This means that the derivative of $y = 3^x$ is $(1.0986\ldots) \times 3^x$.

Generally, the derivative of $y = a^x$ is a^x multiplied by a constant.

$$\frac{d}{dx}(2^x) = (0.6931\ldots)2^x$$

$$\frac{d}{dx}(3^x) = (1.0986\ldots)3^x$$

It would be convenient if this constant was equal to 1. Then the derivative of $y = a^x$ would be $\frac{dy}{dx} = a^x$ exactly.

WORKED EXAMPLE 4 **Finding the value of a for which $\lim_{h\to 0} \frac{a^h - 1}{h} = 1$**

Using CAS, find $\lim_{h\to 0} \frac{a^h - 1}{h}$ for $a = 2, 2.1, 2.2, 2.3, 2.4, 2.5, 2.6, 2.7, 2.8, 2.9, 3.0$.

Write your answers correct to four decimal places and, hence, determine the best approximation for a for which $\lim_{h\to 0} \frac{a^h - 1}{h} = 1$.

Steps	**Working**
1 Use CAS to find $\lim_{h\to 0} \frac{2^h - 1}{h}$.	$\lim_{h\to 0} \frac{(2^h - 1)}{h} = 0.6931\ldots$
2 Calculate the limit using $a = 2.1$. $\lim_{h\to 0} \frac{2.1^h - 1}{h}$ Repeat this process for the other values of a to complete the table.	(see table below)
3 Find the closest value of a for which $\lim_{h\to 0} \frac{a^h - 1}{h} = 1$.	The value of a for which $\lim_{h\to 0} \frac{a^h - 1}{h} = 1$ is $a = 2.7$.

a	$\lim_{h\to 0} \frac{a^h - 1}{h}$
2	0.6931…
2.1	0.7419…
2.2	0.7885…
2.3	0.8329…
2.4	0.8755…
2.5	0.9163…
2.6	0.9555…
2.7	0.9933…
2.8	1.0296…
2.9	1.0647…
3	1.0986…

From the above, it looks like the base a must lie somewhere between 2.7 and 2.8 and it turns out that this value is $e = 2.718\,28\ldots$, Euler's number.

$$\lim_{h\to 0} \frac{e^h - 1}{h} = 1$$

The derivative and integral of e^x

The natural exponential function is the function $y = e^x$.

The derivative of e^x is e^x: $\frac{d}{dx}(e^x) = e^x$

The derivative of e^{ax-b} is ae^{ax-b}: $\frac{d}{dx}(e^{ax-b}) = ae^{ax-b}$

WORKED EXAMPLE 5 Finding the derivative of an exponential function

Differentiate each exponential function.

a $y = e^{2x}$ **b** $y = e^{x^3+x-2}$

Steps	Working
a Use the rule $\frac{d}{dx}(e^{ax}) = ae^{ax}$.	$\frac{dy}{dx} = 2e^{2x}$
b **1** Use the chain rule: identify u and write y in terms of u.	Let $u = x^3 + x - 2$ so $y = e^u$.
2 Write $\frac{du}{dx}$ and $\frac{dy}{du}$ in terms of x.	$\frac{du}{dx} = 3x^2 + 1$ $\frac{dy}{du} = e^u = e^{x^3+x-2}$
3 Use $\frac{dy}{dx} = \frac{dy}{du} \times \frac{du}{dx}$.	$\frac{dy}{dx} = e^{x^3+x-2} \times (3x^2 + 1)$ $= (3x^2 + 1)e^{x^3+x-2}$

Chain, product and quotient rules

The chain rule: If $y = f(u)$ and $u = g(x)$, then $\frac{dy}{dx} = \frac{dy}{du} \times \frac{du}{dx}$.

or

If $y = f(g(x))$ then $y' = f'(g(x))g'(x)$.

The product rule: $\frac{d}{dx}(f(x)g(x)) = f'(x)g(x) + f(x)g'(x)$ or $\frac{d}{dx}(uv) = u\frac{dv}{dx} + v\frac{du}{dx}$

The quotient rule: $\frac{d}{dx}\left(\frac{f(x)}{g(x)}\right) = \frac{g(x)f'(x) - f(x)g'(x)}{(g(x))^2}$ or $\frac{d}{dx}\left(\frac{u}{v}\right) = \frac{v\frac{du}{dx} - u\frac{dv}{dx}}{v^2}$

We can use the chain rule to create a formula for the derivative of e to the power of a function $f(x)$.

If $y = e^{f(x)}$, then $\frac{dy}{dx} = e^{f(x)} \times f'(x) = f'(x)e^{f(x)}$.

The chain rule for exponential functions

If $y = e^{f(x)}$, then $\frac{dy}{dx} = f'(x)e^{f(x)}$.

For example,

$y = e^{3x^2-7x}$ $\frac{dy}{dx} = (6x - 7)e^{3x^2-7x}$

WORKED EXAMPLE 6	The chain and product rules

Find the derivative of each of the following functions.

a $y = (3e^{5x})^4$ **b** $y = x^2e^{-4x}$

Steps	Working
a **1** Use the chain rule: identify u and write y in terms of u.	Let $u = 3e^{5x}$ so $y = u^4$.
2 Obtain $\frac{du}{dx}$ and $\frac{dy}{du}$ in terms of x.	$\frac{du}{dx} = 3 \times 5e^{5x} = 15e^{5x}$ $\frac{dy}{du} = 4u^3 = 4(3e^{5x})^3$
3 Use $\frac{dy}{dx} = \frac{dy}{du} \times \frac{du}{dx}$.	$\frac{dy}{dx} = 4(3e^{5x})^3 \times (15e^{5x})$ $= 4 \times 27e^{15x} \times 15e^{5x}$ $= 1620e^{20x}$
b **1** Use the product rule: identify $f(x)$ and $g(x)$.	$f(x) = x^2$ and $g(x) = e^{-4x}$
2 Differentiate to obtain $f'(x)$ and $g'(x)$.	$f'(x) = 2x$, $g'(x) = -4e^{-4x}$
3 Write down the expression for $f'(x)g(x) + f(x)g'(x)$.	$\frac{dy}{dx} = f'(x)g(x) + f(x)g'(x)$ $= 2xe^{-4x} - 4x^2e^{-4x}$ $= 2xe^{-4x}(1 - 2x)$

WORKED EXAMPLE 7	The quotient rule

Find the value of the derivative of $f(x) = \frac{e^{3x}}{x^2 - 3x + 4}$ where $x = 2$.

Steps	Working
1 Let $\frac{u}{v} = \frac{e^{3x}}{x^2 - 3x + 4}$	$u = e^{3x}$ $v = x^2 - 3x + 4$ $\frac{du}{dx} = 3e^{3x}$, $\frac{dv}{dx} = 2x - 3$
2 Differentiate using the quotient rule: $\frac{d}{dx}\left(\frac{u}{v}\right) = \frac{v\frac{du}{dx} - u\frac{dv}{dx}}{v^2}$	$f'(x) = \frac{3e^{3x}(x^2 - 3x + 4) - e^{3x}(2x - 3)}{(x^2 - 3x + 4)^2}$
3 Find $f'(2)$ by substituting $x = 2$ into $f'(x)$.	$f'(2) = \frac{3e^{3(2)}((2^2) - 3(2) + 4) - e^{3(2)}(2(2) - 3)}{(2^2 - 3(2) + 4)^2}$ $= \frac{5e^6}{4}$

Exam hack

When finding $f'(x)$ for a particular x value, it's often faster to NOT simplify $f'(x)$ before substituting in the value of x.

4.2

USING CAS 1 Finding the first derivative of functions involving exponentials

Find the first derivative of $f(x) = e^{4x}$ at $x = 1$.

ClassPad

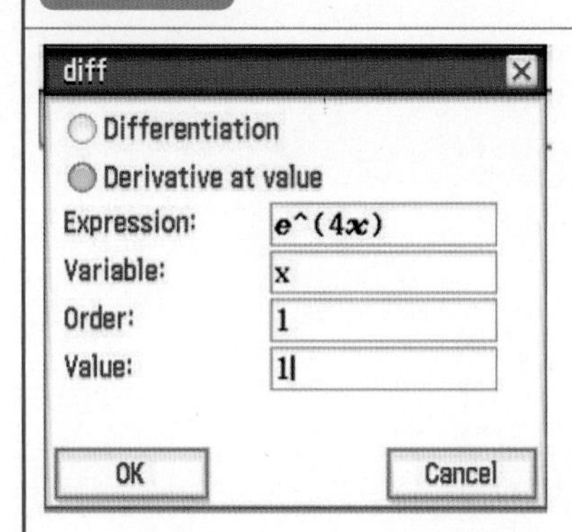

1 In **Main**, enter and highlight the expression.

2 Tap **Interactive > Calculation > diff**.

3 In the dialogue box tap **Derivative at value**.

4 In the **Value:** field, enter **1** and tap **OK**.

TI-Nspire

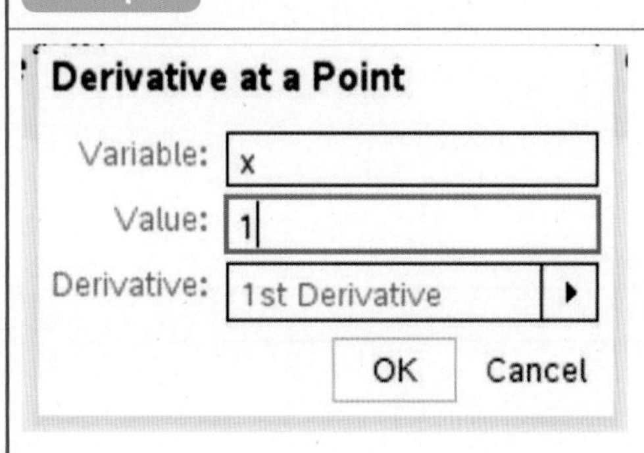

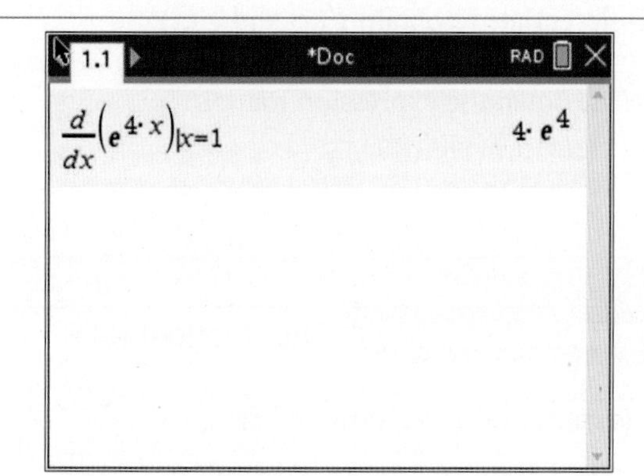

1 Press **menu > Calculus > Derivative at a Point**.

2 In the dialogue box **Value:** field, enter **1**.

3 In the template, enter the expression and press **enter**.

The first derivative at $x = 1$ is $4e^4$.

WORKED EXAMPLE 8 Finding the equation of the tangent to the curve

Find the equation of the tangent to the curve $f(x) = e^{2x+1}$ at $x = 1$.

Steps	Working
1 Find $f'(1)$.	$f'(x) = 2e^{2x+1}$ $f'(1) = 2e^{2(1)+1} = 2e^3$
2 Find $f(1)$.	$f(1) = e^{2(1)+1} = e^3$
3 Write the coordinates of the point on the tangent and the gradient.	Gradient $m = f'(1) = 2e^3$ $y = f(1) = e^3$ The point on the tangent is $(1, e^3)$.
4 Use the formula $y - y_1 = m(x - x_1)$ to find the equation of the tangent.	$y - e^3 = 2e^3(x - 1)$ $y - e^3 = 2e^3x - 2e^3$ $y = 2e^3x - e^3$

USING CAS 2 Finding the equation of the tangent to the curve

Find the approximate equation of the tangent , correct to two decimal places, to the curve $f(x) = \dfrac{e^{\sqrt{2x+5}}}{x}$ at $x = 1$.

ClassPad

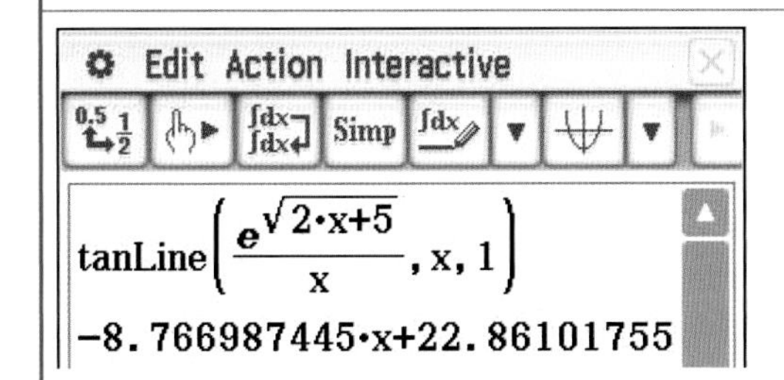

1 In **Main**, enter and highlight the expression $\dfrac{e^{\sqrt{2x+5}}}{x}$.

2 Tap **Interactive > Calculation > line > tanLine**.

3 In the dialogue box, **Point:** field, enter **1**.

4 Tap **OK**.

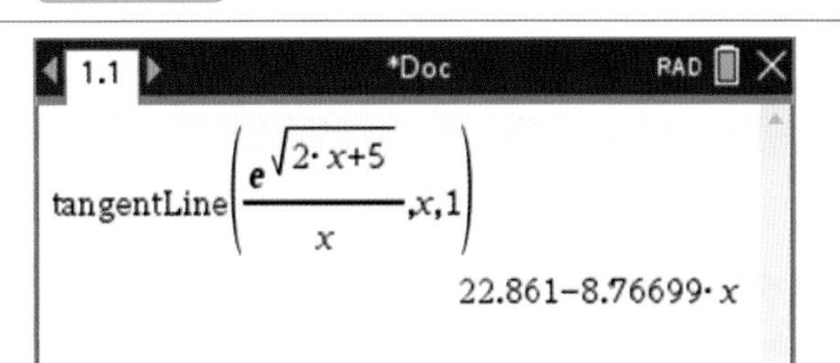

1 Press **menu > Calculus > Tangent Line**.

2 Enter the expression followed by **,x,1**.

3 Press **ctrl + enter** for the approximate solution.

The equation of the tangent is $y = -8.77x + 22.86$, correct to two decimal places.

WORKED EXAMPLE 9 Finding and describing the nature of a stationary point of an exponential function

Consider the function $y = e^{-2x^2}$.

Find

a $\dfrac{dy}{dx}$

b $\dfrac{d^2y}{dx^2}$

c the x value of the stationary point

d the nature of the stationary point.

Steps	Working
a Find the first derivative using the chain rule $\dfrac{dy}{dx} = \dfrac{dy}{du}\dfrac{du}{dx}$.	$u = -2x^2$ $y = e^u$ $\dfrac{du}{dx} = -4x$ $\dfrac{dy}{du} = e^u = e^{-2x^2}$ $\dfrac{dy}{dx} = -4xe^{-2x^2}$
b Find the second derivative using the product rule: $\dfrac{dy}{dx}(uv) = u\dfrac{dv}{dx} + v\dfrac{du}{dx}$	$u = -4x$ $v = e^{-2x^2}$ $\dfrac{du}{dx} = -4$ $\dfrac{dv}{dx} = -4xe^{-2x^2}$ $\dfrac{d^2y}{dx^2} = 16x^2e^{-2x^2} - 4e^{-2x^2}$
c Solve the first derivative equal to zero to find the x value of the stationary points.	$\dfrac{dy}{dx} = 0$ $-4xe^{-2x^2} = 0$ $x = 0$ as $e^{-2x^2} > 0$

d Substitute $x = 0$ into the second derivative to find the nature of the stationary point.	$\frac{d^2y}{dx^2} = 16x^2e^{-2x^2} - 4e^{-2x^2}$ $\frac{d^2y}{dx^2} = 16(0)^2e^{-2(0)^2} - 4e^{-2(0)^2}$ $\frac{d^2y}{dx^2} = -4$ As $\frac{d^2y}{dx^2} < 0$ the stationary point at $x = 0$ is a local maximum.

Exponential modelling and differentiation

There are many natural phenomena that are modelled using simple exponential functions of the form $f(x) = Ae^{kx}$, where A and k are constants. In the case where the variable represents time, we write $f(t) = Ae^{kt}$.

WORKED EXAMPLE 10 **Using the natural exponential function and its derivative**

A metal cools down according to the formula $T = T_0e^{-0.1t}$, where T is the temperature difference between the metal and the surroundings in °C and t is time in minutes. The initial temperature is 228°C and the room is at 20°C.

a Evaluate T_0.

b Find, correct to one decimal place, the temperature difference after 20 minutes.

c What is the temperature after 20 minutes? Answer correct to one decimal place.

d Find, correct to one decimal place, the rate at which the metal is cooling after 20 minutes.

e Use the increments formula at $t = 20$ minutes to estimate the change in temperature for a 30 second change in time.

Steps	Working
a Find the initial temperature difference with the surroundings.	$T_0 = 228 - 20 = 208°C$
b 1 Substitute $t = 20$ into $T = T_0e^{-0.1t}$.	$T = 208e^{-0.1\times20}$ $= 28.149...$
2 State the result.	The temperature difference is about 28.1°C.
c Add the room temperature.	The temperature after 20 minutes is about $28.1 + 20 = 48.1$°C.
d 1 Find the derivative.	$T = 208e^{-0.1t}$ $\frac{dT}{dt} = 208 \times (-0.1e^{-0.1t})$ $= -20.8e^{-0.1t}$
2 Substitute $t = 20$.	rate of change $= -20.8e^{-0.1\times20}$ $= -20.8e^{-2}$ $= -2.814...$ The negative answer indicates cooling.
3 State the result.	After 20 minutes, the metal is cooling at about 2.8°C/min.

e Substitute $t = 20$ and $\delta t = 0.5$ into the formula $\delta T \approx \frac{dT}{dt} \times \delta t$.	$\delta T \approx \frac{dT}{dt} \times \delta t$ At $t = 20$, $\frac{dT}{dt} = -2.814$ and $\delta t = \frac{30}{60} = 0.5$. $\delta T \approx -2.814 \times 0.5 \approx -1.4°C$ The temperature difference decreases by approximately 1.4°C.

WORKED EXAMPLE 11 Using the derivative with exponential decay

Under exponential decay, the amount of radon-222 in milligrams that is present after t days is given by the function $f(t) = ae^{rt}$, where a and r are constants.

a If an initial amount of 100 mg decays to 84.11 mg after one day, show that the value of r is approximately −0.173.

b How much, correct to three decimal places, will 100 mg of radon-222 decay to in three days?

c At what rate is radon-222 decaying after one week? Give your answer to three decimal places.

Steps	Working
a 1 Write the formula.	$f(t) = ae^{rt}$
2 We need to find a and r. Substitute what we know when $t = 0$ (the initial condition).	When $t = 0, f(t) = 100$. $f(0) = ae^0 = 100$ $a = 100$
3 Rewrite the formula with $a = 100$.	$f(t) = 100e^{rt}$
4 Substitute another condition to find r.	When $t = 1, f(t) = 84.11$. $f(1) = 100e^{r(1)} = 84.11$ $e^r = 0.8411$
5 Solve using CAS. r is negative because it is exponential decay.	$r \approx -0.173$
b 1 Write the formula with $r = -0.173$.	$f(t) = 100e^{-0.173t}$
2 Substitute $t = 3$.	$f(3) = 100e^{-0.173 \times 3}$ $= 59.512$
3 State the result.	59.512 mg of radon-222 remains.
c 1 Find $f'(t)$ for the rate of change.	$f(t) = 100e^{-0.173t}$ $f'(t) = 100 \times (-0.173)e^{-0.173t}$ $= -17.3e^{-0.173t}$
2 Substitute $t = 7$ for the rate after one week (7 days).	$f'(7) = -17.3e^{-0.173 \times 7}$ $= -17.3e^{-1.211}$ $= -5.154$
3 State the result.	After one week, radon-222 is decaying at a rate of 5.154 mg/day.

EXERCISE 4.2 Differentiating exponential functions

ANSWERS p. 394

Recap

1 The number of people, N, who have a virus at time t months is given by $N(t) = N_0e^{kt}$, where t is the number of months after the outbreak of the virus.

If the number is initially 50 and the number increases to 150 after one month, find

a the value of N_0

b the value of k, correct to four decimal places.

2 A radioactive element has a half-life of 20 minutes, meaning it takes 20 minutes to reach half its original size. Originally, there are 16.0×10^{20} atoms in the sample of the element. The decay is modelled by $N = N_0e^{-kt}$, where N_0 is the original number of atoms, N is the number of atoms present at time t minutes, and k is a constant. Find the value of k (correct to four decimal places).

Mastery

3 WORKED EXAMPLE 4 Using CAS find $\lim\limits_{h \to 0} \dfrac{a^h - 1}{h}$ for $a = 2.710, 2.711, 2.712, 2.713, 2.714, 2.715, 2.716, 2.717, 2.718, 2.719, 2.720$.

Write your answers correct to six decimal places and hence determine the best approximation for a for which $\lim\limits_{h \to 0} \dfrac{a^h - 1}{h} = 1$.

4 WORKED EXAMPLE 5 Differentiate each function.

a $y = 9e^x$ **b** $y = e^x + x^2$ **c** $y = (2e^x - 3)^6$

d $y = \dfrac{(e^x + e^{-x})^2}{e^x}$ **e** $y = e^{2x-1}$ **f** $y = e^{\sqrt{2x+4}}$

5 WORKED EXAMPLE 6 Find $f'(x)$ for each function.

a $f(x) = xe^x$ **b** $f(x) = (2x + 3)e^x$ **c** $f(x) = 5x^3e^x$

6 WORKED EXAMPLE 7 Find $g'(3)$ if $g(x) = \dfrac{e^x - 4}{\sqrt{e^x + 1}}$.

7 Using CAS 1 Find the derivative of each function.

a $y = \dfrac{e^x}{x^2}$ **b** $y = \dfrac{e^{6x}}{3x}$ **c** $y = \dfrac{2e^{5x}}{5x^3}$ **d** $y = \dfrac{x - 1}{e^x}$ **e** $y = \dfrac{e^x + 1}{e^{2x}}$

8 Given $y = e^{4x}(x^3 - 3x + 5)$, find $\dfrac{dy}{dx}$, for $x = -1$.

9 If $f(x) = \dfrac{xe^{3x} + 5}{x^2 + e}$, find $f'(2)$ correct to one decimal place.

10 $h(x) = 5x^2e^{3x} + e^x$. Find $h'(2)$.

11 Find the value of x such that the rate of change of xe^{2x-1} is $5e^3$.

12 WORKED EXAMPLE 8 Find the equation of the tangent to the curve $f(x) = \sqrt{e^x}$ at the point on the curve where $x = 1$.

13 Using CAS 2 Find the equation of the tangent to the curve $f(x) = \sqrt{e^{3x}}$ at the point where $x = 2$.

14 WORKED EXAMPLE 9 Consider the function $y = e^{2x^2-4x}$.

Find

a $\frac{dy}{dx}$

b $\frac{d^2y}{dx^2}$

c the x value of the stationary point

d the nature of the stationary point.

15 WORKED EXAMPLE 10 A study of swans in an area of Western Australia showed that their numbers were gradually increasing, with the number of swans N over t months given by $N(t) = 1100e^{0.025t}$.

a How many swans were there at the beginning of the study?

b How many swans were there after 5 months?

c At what rate was the number of swans increasing after 5 months?

d Use the increments formula at $t = 5$ months to estimate the change in the number of swans for a change in time of 0.1 month.

16 The area of rainforests is declining in a region of Queensland with the area A hectares over time t years given by $A(t) = 120\,000e^{-0.033t}$.

At what rate is the area of rainforest decreasing in this region after

a 2 years? **b** 15 years? **c** 40 years?

17 WORKED EXAMPLE 11 The amount of $^{226}_{88}Ra$, a common isotope of radium, in milligrams, that is present after t days is given by the function $R(t) = R_0e^{kt}$, where R_0 and k are constants. The initial amount of $^{226}_{88}Ra$ is 200 mg and this decays to 191.6 mg after one day.

Find

a R_0

b k, correct to three decimal places

c the rate at which $^{226}_{88}Ra$ is decaying after seven days.

Calculator-free

18 (7 marks)

a Differentiate x^3e^{2x} with respect to x. (2 marks)

b Let $f(x) = e^{x^2}$. Find $f'(3)$. (3 marks)

c Evaluate $f'(1)$, where $f(x) = e^{x^2-x+3}$. (2 marks)

19 © SCSA MM2016 Q3 (4 marks) Consider the function $f(x) = \dfrac{(x-1)^2}{e^x}$.

a Show that the first derivative is $f'(x) = \dfrac{-x^2 + 4x - 3}{e^x}$. (2 marks)

b Use your result from part **a** to explain why there are stationary points at $x = 1$ and $x = 3$. (2 marks)

20 (3 marks) Let $f(x) = e^x + k$, where k is a real number. The tangent to the graph of f at the point where $x = a$ passes through the point $(0, 0)$. Find the value of k in terms of a.

Calculator-assumed

21 © SCSA MM2020 Q15 (9 marks) A chef needs to use an oven to boil 100 mL of water in five minutes for a new experimental recipe. The temperature of the water must reach 100°C in order to boil. The temperature, T, of 100 mL of water t minutes after being placed in an oven set to T_0°C can be modelled by the equation

$$T(t) = T_0 - 175e^{-0.07t}$$

In a preliminary experiment, the chef placed a 100 mL bowl of water into an oven that had been heated to $T_0 = 200$°C.

a What is the temperature of the water at the moment it is placed into the oven? (1 mark)

b What is the temperature of the water five minutes after being placed in the oven? (1 mark)

c What change could be made to the temperature at which the oven is set in order to achieve the five-minute boiling requirement? (2 marks)

Assume that T_0 is still 200°C.

d Determine the rate of increase in temperature of the water five minutes after being placed in the oven. Give your answer rounded to two decimal places. (2 marks)

e Explain what happens to the rate of change in the temperature of the water as time increases and how this relates to the temperature of the water. (3 marks)

22 © SCSA MM2018 Q14 (5 marks)

a The table below examines the values of $\dfrac{a^h - 1}{h}$ for various values of a as h approaches zero.

Copy and complete the table, rounding your values to five decimal places. (2 marks)

h	$a = 2.60$	$a = 2.70$	$a = 2.72$	$a = 2.80$
0.1	1.002 65		1.052 41	1.084 49
0.001	0.955 97	0.993 75		
0.000 01	0.955 52			1.029 62

It can be shown that $\dfrac{d}{dx}(a^x) = a^h \lim\limits_{h \to 0}\left(\dfrac{a^h - 1}{h}\right)$.

b What is the exact value of a for which $\dfrac{d}{dx}(a^x) = a^x$? Explain how the above definition and the table in part **a** support your answer. (3 marks)

4.3 Integrating exponential functions

We know that:

$$\frac{d}{dx}(e^x) = e^x$$

$$\frac{d}{dx}(e^{ax}) = ae^{ax} \quad \text{(by the chain rule)}$$

We can reverse these rules to integrate exponential functions.

Also consider $y = e^{ax + b}$

$$\frac{dy}{dx} = ae^{ax + b} \quad \text{(by the chain rule)}$$

So, $\int ae^{ax+b}\,dx = e^{ax+b} + c$

and dividing both sides by a gives

$$\int e^{ax+b}\,dx = \frac{1}{a}e^{ax+b} + c$$

Video playlist
Integrating exponential functions

The integral of e^x

The **integral** of e^x is $e^x + c$:

$$\int e^x dx = e^x + c$$

The **integral** of e^{ax} is $\frac{1}{a}e^{ax} + c$:

$$\int e^{ax} dx = \frac{1}{a}e^{ax} + c$$

WORKED EXAMPLE 12 Finding the integral of an exponential function

Find the integral of each of the following.

a $\int e^{3-4x}\,dx$ **b** $\int \frac{e^{5x} + 3 + e^x}{4e^{3x}}\,dx$

Steps	Working
a Use $\int e^{ax+b}dx = \frac{1}{a}e^{ax+b} + c$ with $a = -4$, $b = 3$.	$\int e^{3-4x}dx = -\frac{1}{4}e^{3-4x} + c$
b 1 Separate the terms first.	$\int \frac{e^{5x} + 3 + e^x}{4e^{3x}}dx = \frac{1}{4}\int \frac{e^{5x}}{e^{3x}} + \frac{3}{e^{3x}} + \frac{e^x}{e^{3x}}dx$
2 Simplify by subtracting powers.	$= \frac{1}{4}\int e^{2x} + 3e^{-3x} + e^{-2x}\,dx$
3 Integrate each term using $\int e^{ax}dx = \frac{1}{a}e^{ax} + c$.	$= \frac{1}{4}\left(\frac{1}{2}e^{2x} + \frac{3}{-3}e^{-3x} + \frac{1}{-2}e^{-2x}\right) + c$ $= \frac{1}{8}e^{2x} - \frac{1}{4}e^{-3x} - \frac{1}{8}e^{-2x} + c$

WORKED EXAMPLE 13 Finding $f(x)$ given $f'(x)$ and a point

Find $f(x)$ if its gradient function is $f'(x) = 3e^{2x}$ and $f(0) = 8$.

Steps	Working
1 Integrate $f'(x)$ to find $f(x)$.	$f'(x) = 3e^{2x}$ $f(x) = \frac{3}{2}e^{2x} + c$
2 Substitute $x = 0, f(0) = 8$ to find c.	$8 = \frac{3}{2}e^{0} + c$ $8 - \frac{3}{2} = c$ $c = \frac{13}{2}$
3 Write the equation of the function.	$f(x) = \frac{3}{2}e^{2x} + \frac{13}{2}$

WORKED EXAMPLE 14 Evaluating definite integrals

Evaluate $\int_{-2}^{2} 3e^{-\frac{x}{2}}dx$.

Steps	Working
1 Find the anti-derivative.	$\int_{-2}^{2} 3e^{-\frac{x}{2}}\,dx$ $= \left[\frac{3}{\frac{-1}{2}}e^{\frac{-x}{2}}\right]_{-2}^{2}$ $= \left[-6e^{\frac{-x}{2}}\right]_{-2}^{2}$
2 Evaluate the integral.	$= -6e^{-1} - -6e$ $= 6e - \frac{6}{e}$

USING CAS 3 Integrating exponential functions

Evaluate $\int_{-2}^{3} e^{\frac{2x+1}{3}}\,dx$ correct to three decimal places.

ClassPad

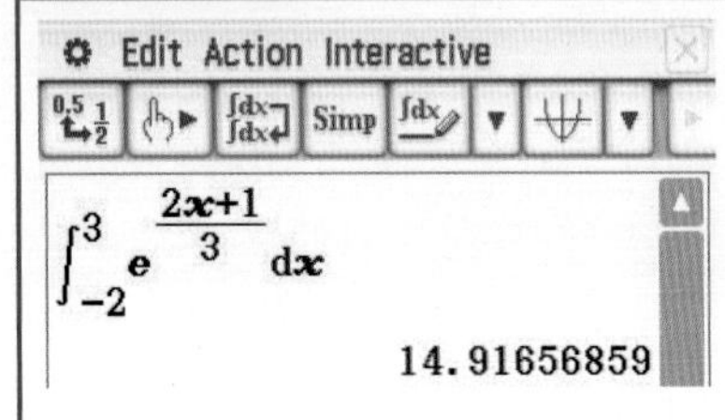

1. In **Main**, enter and highlight the expression $e^{\frac{2x+1}{3}}$.
2. Tap **Interactive > Calculation > ∫**.
3. In the dialogue box, tap **Definite** to enter the lower and upper limits.
4. Tap **OK**.

TI-Nspire

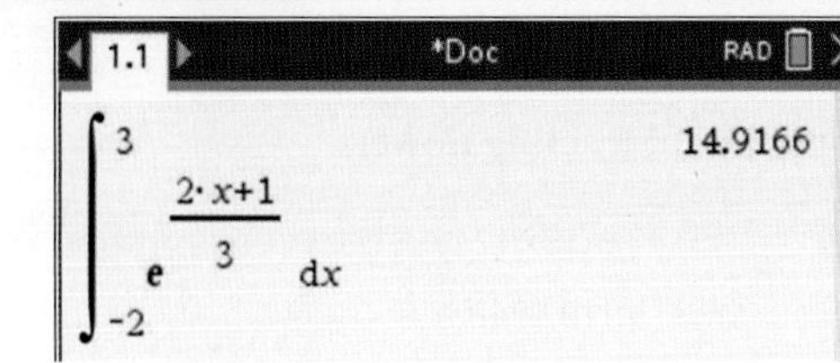

1. Press **menu > Calculus > Integral**.
2. Enter the lower limit, the upper limit and the expression.
3. Press **ctrl + enter** for the approximate solution.

The answer is 14.917, correct to three decimal places.

WORKED EXAMPLE 15 Finding areas using definite integrals

Find the bounded area between $y = e^x$, $y = e$ and the y-axis.

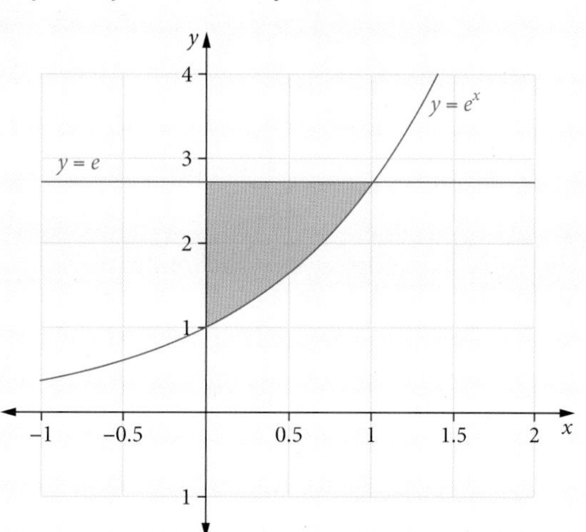

Steps	Working
1 Find the x value of the point of intersection.	$e^x = e$ $e^x = e^1$ $x = 1$
2 Write an integral equation to represent the area bounded by the functions. area = $\int$(upper curve − lower curve) dx	area $= \int_0^1 e - e^x \, dx$
3 Evaluate the integral.	area $= \left[ex - e^x\right]_0^1$ area $= e - e - (0 - e^0) = 1$ unit^2

EXERCISE 4.3 Integrating exponential functions

ANSWERS p. 395

Recap

1 Find $g'(2)$, correct to three decimal places, if $g(x) = e^{x^2} - 7$.

2 Find the gradient of the tangent to the graph of $y = e^{3x}$ at $x = 2$.

Mastery

3 WORKED EXAMPLE 12 Find each integral.

a $\int e^{-2x}\,dx$ **b** $\int 5e^{4x}\,dx$ **c** $\int e^{2x+1}\,dx$

d $\int (3e^{-2x} + e^{4x})\,dx$ **e** $\int \frac{e^{4x} - 1}{e^x}\,dx$ **f** $\int (e^{3x} - e^{-3x})^2\,dx$

4 WORKED EXAMPLE 13 If the gradient function at a point (x, y) on a curve is given by $2e^{4x}$ and the curve passes through $(0, 2)$, find the equation of the curve.

5 If the gradient at a point (x, y) on a curve is given by $\frac{dy}{dx} = \frac{e^{3x} - 1}{e^x}$ and the curve passes through $(0, 11)$, find the equation of the curve.

6 A function $f(x)$ is such that $f'(x) = 5e^{-2x}$ and $f(3) = 2e$. Find $f(x)$.

7 WORKED EXAMPLE 14 Evaluate each definite integral.

a $\int_0^4 e^x\,dx$

b $\int_1^3 5e^x\,dx$

c $\int_2^4 (x^3 - e^x)\,dx$

8 Evaluate each integral, correct to two decimal places.

a $\int_{-\frac{\pi}{6}}^{\frac{\pi}{6}} \left[\sin(3x) + e^{-6x}\right]dx$

b $\int_0^{\pi} \cos\left(\frac{x}{2}\right) + e^{3x}\,dx$

9 Using CAS 3 Evaluate $\int_0^3 5e^x - 2e^{3x}\,dx$ correct to one decimal place.

10 WORKED EXAMPLE 15 Find the bounded area between $y = e^{3x}$, $y = e^3$ and the y-axis.

Calculator-free

11 © SCSA MM2016 Q2 (5 marks)

a Determine $\frac{d}{dx}(2xe^{2x})$. (2 marks)

b Use your answer in part **a** to determine $\int 4xe^{2x}\,dx$. (3 marks)

12 © SCSA MM2016 Q6 (4 marks) The graphs $y = 6 - 2e^{x-4}$ and $y = -\frac{1}{4}x + 5$ intersect at $x = 4$ for $x \geq 0$.

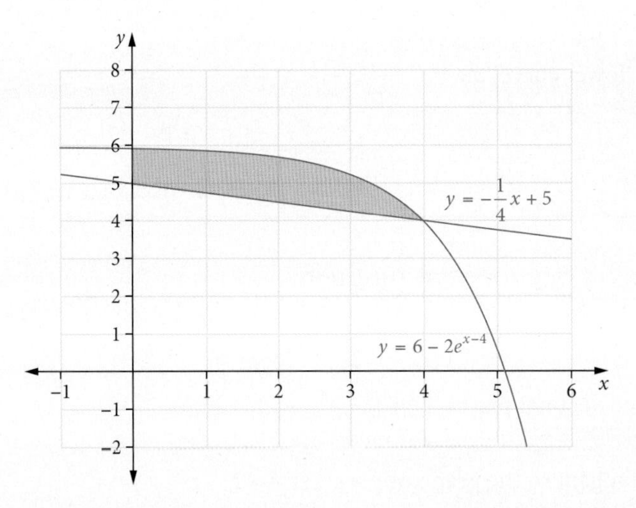

Determine the exact area between $y = 6 - 2e^{x-4}$, $y = -\frac{1}{4}x + 5$ and the y-axis for $x \geq 0$.

13 (5 marks) The graph of $f(x) = e^{\frac{x}{2}} + 1$ is shown.

The normal to the graph of f where it crosses the y-axis is also shown.

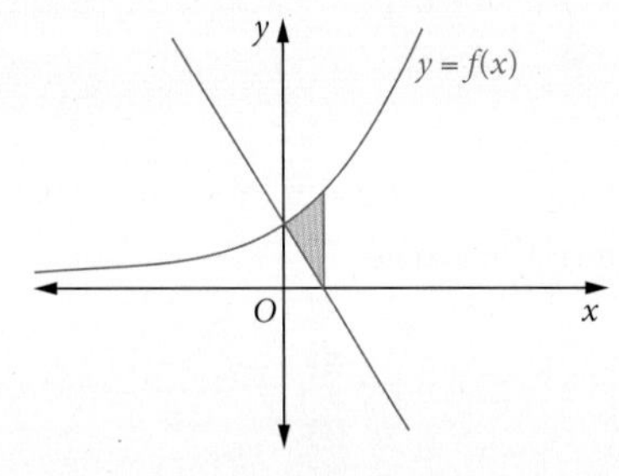

a Find the equation of the normal to the graph of f where it crosses the y-axis. (2 marks)

b Find the exact area of the shaded region. (3 marks)

9780170477536

14 (7 marks) Let $f(x) = 2e^{-\frac{x}{5}}$.

A right-angled triangle OQP has vertex O at the origin, vertex Q on the x-axis and vertex P on the graph of f, as shown. The coordinates of P are $(x, f(x))$.

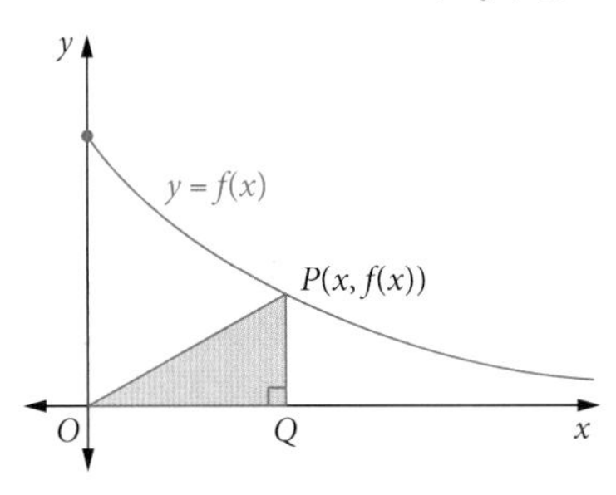

a Find the area, A, of the triangle OQP in terms of x. (1 mark)

b Find the maximum area of triangle OQP and the value of x for which the maximum occurs. (3 marks)

c Let S be the point on the graph of f on the y-axis and let T be the point on the graph of f with the y-coordinate $\frac{1}{2}$. Find the area of the region bounded by the graph of f and the line segment ST. (3 marks)

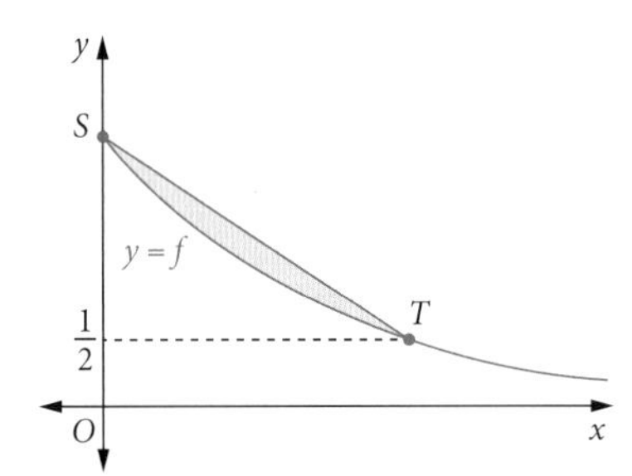

15 © SCSA MM2017 Q15 (10 marks)

The volume $V(h)$ in cubic metres of a liquid in a large vessel depends on the height h (metres) of the liquid in the vessel and is given by

$$V(h) = \int_0^h e^{\left(-\frac{x^2}{100}\right)} dx, 0 \le h \le 15.$$

a Determine $\frac{dV}{dh}$ when the height is 0.5 m. (2 marks)

b What is the meaning of your answer to part **a**? (1 mark)

c The height of the liquid depends on time t (seconds) as follows:

$h(t) = 3t^2 - t + 4, t \ge 0$

i Determine $\frac{dh}{dt}$ when the height is 6 m. (2 marks)

ii Use the chain rule to determine $\frac{dV}{dt}$ when the height is 6 m. (2 marks)

iii Given the volume of the liquid at 2 seconds is 8.439 m^3, use the increments formula to estimate the volume 0.1 second later. (3 marks)

Video playlist
Differentiating trigonometric functions

Worksheets
Further optimisation problems

Trigonometric functions and gradient

4.4 Differentiating trigonometric functions

Derivatives of trigonometric functions

The geometric proof for the derivative of $y = \sin(\theta)$ uses the unit circle definition of $\sin(\theta)$ and the limit definition of the derivative.

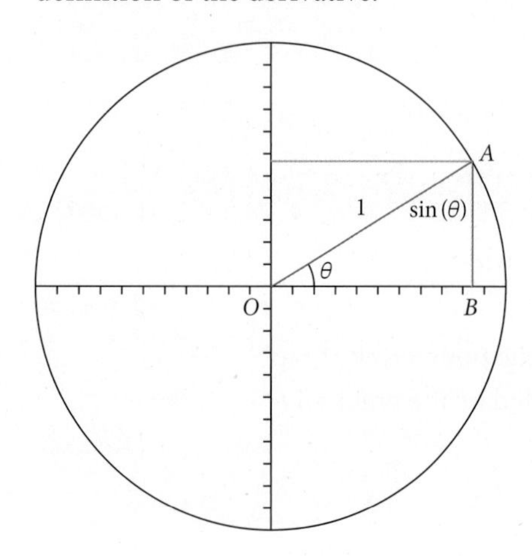

In a unit circle, $\overline{AB} = \sin(\theta)$.

The limit definition for the derivative of $\sin(\theta)$ is $\frac{d(\sin(\theta))}{d\theta} = \lim_{\delta\theta \to 0} \frac{\sin(\theta + \delta\theta) - \sin(\theta)}{\delta\theta}$.

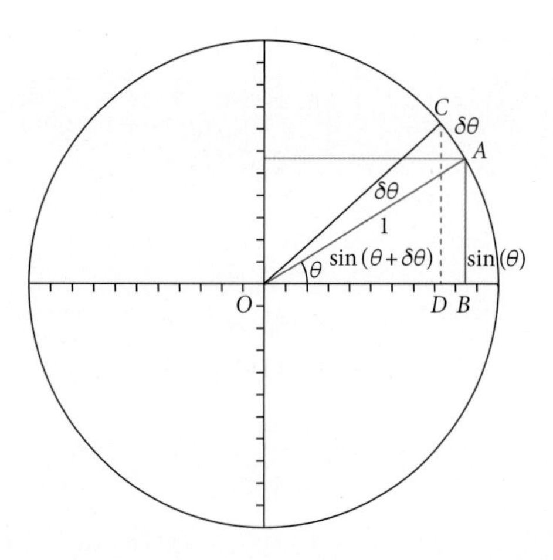

$\delta\theta$ is a small increase in the angle θ.

The arc length in a circle $= r \times \theta$, however, the unit circle has a radius of 1 so the length of arc AC is $\delta\theta$.

$\overline{CD} = \sin(\theta + \delta\theta)$ and $\lim_{\delta\theta \to 0} (\text{arc } AC) = \overline{AC}$.

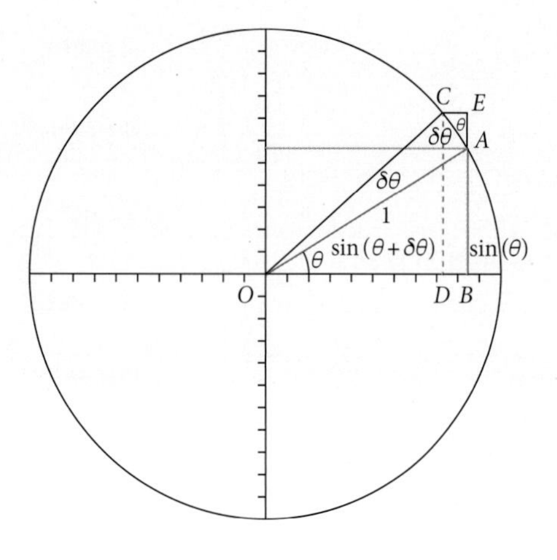

In triangle ACE

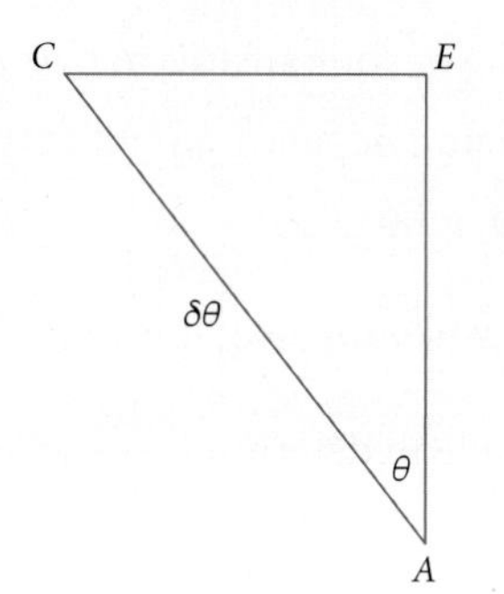

$\cos(\theta) = \frac{AE}{\delta\theta}$ and

$AE = \sin(\theta + \delta\theta) - \sin(\theta)$

9780170477536

$\cos(\theta) = \dfrac{\sin(\theta + \delta\theta) - \sin(\theta)}{\delta\theta}$ and $\dfrac{d(\sin(\theta)}{d\theta} = \lim\limits_{\delta\theta \to 0} \dfrac{\sin(\theta + \delta\theta) - \sin(\theta)}{\delta\theta}$

Therefore, $\dfrac{d(\sin(\theta))}{d\theta} = \lim\limits_{\delta\theta \to 0} \cos(\theta) = \cos(\theta)$

Using the same diagram:

$CE = \cos(\theta) - \cos(\theta + \delta\theta)$ and $\sin(\theta) = \dfrac{CE}{\delta\theta}$

$$\frac{d(\cos(\theta))}{d\theta} = \lim_{\delta\theta \to 0} \frac{\cos(\theta + \delta\theta) - \cos(\theta)}{\delta\theta} = -\sin(\theta)$$

Derivatives of trigonometric functions

Trigonometric function	Derivative	Example
$y = \sin(ax - b)$	$\dfrac{dy}{dx} = a\cos(ax - b)$	$y = 3\sin(3x)$ $\dfrac{dy}{dx} = 3 \times 3\cos(3x)$ $\dfrac{dy}{dx} = 9\cos(3x)$
$y = \cos(ax - b)$	$\dfrac{dy}{dx} = -a\sin(ax - b)$	$y = 4\cos(2x)$ $\dfrac{dy}{dx} = -4 \times 2\sin(2x)$ $= -8\sin(2x)$

WORKED EXAMPLE 16 Using the product rule

Find the first derivative of $y = \sin(x)\cos(x)$.

Steps	Working
1 Use the product rule: identify u and v.	$y = \sin(x)\cos(x)$ $u = \sin(x)$ and $v = \cos(x)$
2 Differentiate to obtain $\dfrac{du}{dx}$ and $\dfrac{dv}{dx}$.	$\dfrac{du}{dx} = \cos(x)$, $\dfrac{dv}{dx} = -\sin(x)$
3 Substitute into the product rule $\dfrac{dy}{dx} = u\dfrac{dv}{dx} + v\dfrac{du}{dx}$ and simplify. For the product rule, either function can be u or v. We can also let $u = \cos(x)$ and $v = \sin(x)$.	$\dfrac{dy}{dx} = u\dfrac{dv}{dx} + v\dfrac{du}{dx}$ $\dfrac{dy}{dx} = \sin(x) \times -\sin(x) + \cos(x) \times \cos(x)$ $\dfrac{dy}{dx} = -\sin^2(x) + \cos^2(x)$

WORKED EXAMPLE 17 Using the quotient rule

Find $\frac{dy}{dx}$ for $y = \frac{2x^2 - x}{\sin(x)}$.

Steps	Working
1 Use the quotient rule: identify u and v where $y = \frac{u}{v}$.	$u = 2x^2 - x$ and $v = \sin(x)$
2 Differentiate to obtain $\frac{du}{dx}$ and $\frac{dv}{dx}$.	$\frac{du}{dx} = 4x - 1$, $\frac{dv}{dx} = \cos(x)$
3 Substitute into the quotient rule $\frac{d}{dx}\left(\frac{u}{v}\right) = \frac{v\frac{du}{dx} - u\frac{dv}{dx}}{v^2}$ and simplify. With the quotient rule, u must be the function in the numerator and v the function in the denominator.	$\frac{d}{dx}\left(\frac{u}{v}\right) = \frac{v\frac{du}{dx} - u\frac{dv}{dx}}{v^2}$ $\frac{dy}{dx} = \frac{\sin(x) \times (4x-1) - (2x^2 - x) \times \cos(x)}{(\sin x)^2}$ $\frac{dy}{dx} = \frac{4x\sin(x) - \sin(x) - 2x^2\cos(x) + x\cos(x)}{\sin^2(x)}$

WORKED EXAMPLE 18 Using the chain rule

Find the first derivative of $y = \sin(x^3 + 2x)$.

Steps	Working
1 Use the chain rule: identify u and write y in terms of u.	Let $u = x^3 + 2x$, so $y = \sin(u)$.
2 Find $\frac{du}{dx}$ and $\frac{dy}{du}$ in terms of x.	$\frac{du}{dx} = 3x^2 + 2$ $\frac{dy}{du} = \cos(u) = \cos(x^3 + 2x)$
3 Use $\frac{dy}{dx} = \frac{dy}{du} \times \frac{du}{dx}$.	$\frac{dy}{dx} = \cos(x^3 + 2x) \times (3x^2 + 2)$ $= (3x^2 + 2)\cos(x^3 + 2x)$

WORKED EXAMPLE 19 Combining the rules

Find $\frac{dy}{dx}$ for $y = x^3\sin(x^2)$.

Steps	Working
1 Identify the differentiation rules to be used.	The product rule and chain rule are needed.
2 Identify u and v in the product rule.	$u = x^3$ $\quad v = \sin(x^2)$
3 Obtain $\frac{du}{dx}$ and $\frac{dv}{dx}$ using the chain rule. The chain rule for $\frac{d(\sin(f(x))}{dx}$ is $f'(x)\cos(f(x))$.	$\frac{du}{dx} = 3x^2$ $\quad \frac{dv}{dx} = 2x\cos(x^2)$
4 Use the product rule.	$\frac{dy}{dx} = \sin(x^2) \times 3x^2 + x^3 \times 2x\cos(x^2)$ $= 3x^2\sin(x^2) + 2x^4\cos(x^2)$

USING CAS 4 Finding the derivative of a trigonometric function at a point

For $f(x) = \dfrac{\cos^2(x)}{\sqrt{\sin(x)}}$, calculate the value of $f'\left(\dfrac{\pi}{4}\right)$ correct to three decimal places.

ClassPad

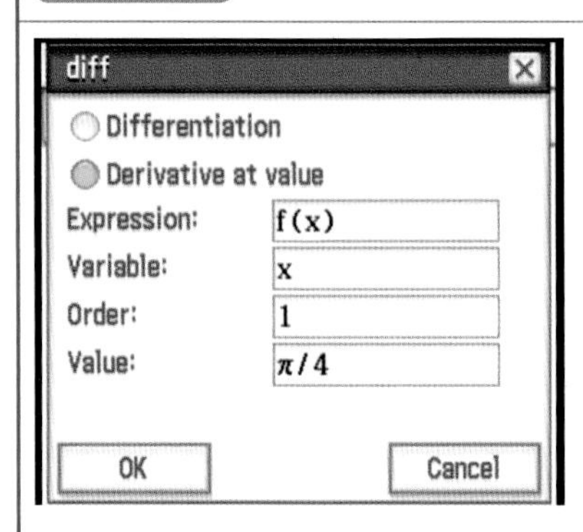

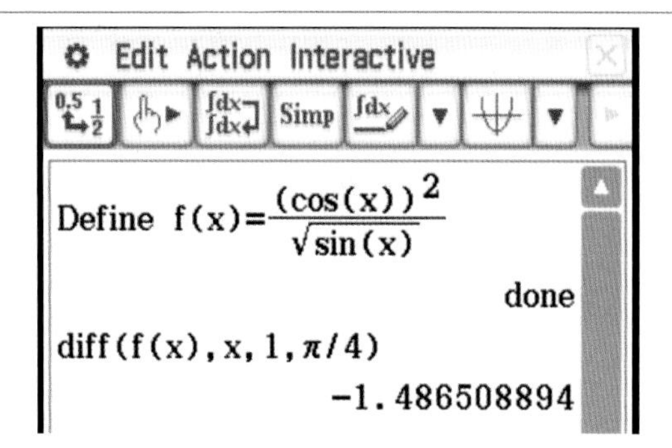

1 Define **f(x)** = $\dfrac{\cos^2(x)}{\sqrt{\sin(x)}}$. Note that $\cos^2(x)$ needs to be entered as $(\cos(x))^2$.

2 Highlight **f(x)** and tap **Interactive > Calculation > diff**.

3 In the dialogue box, tap **Derivative at value**.

4 Enter Value: $\dfrac{\pi}{4}$.

5 Tap **OK** and the answer will be shown.

6 Change to decimals if required using **Convert** or by tapping **Decimal** at the bottom of the screen.

TI-Nspire

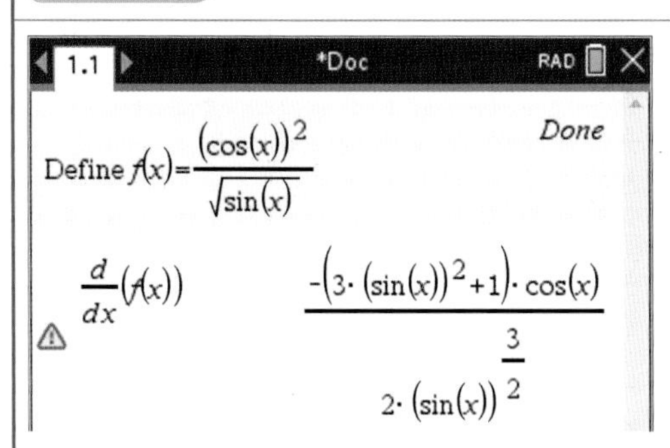

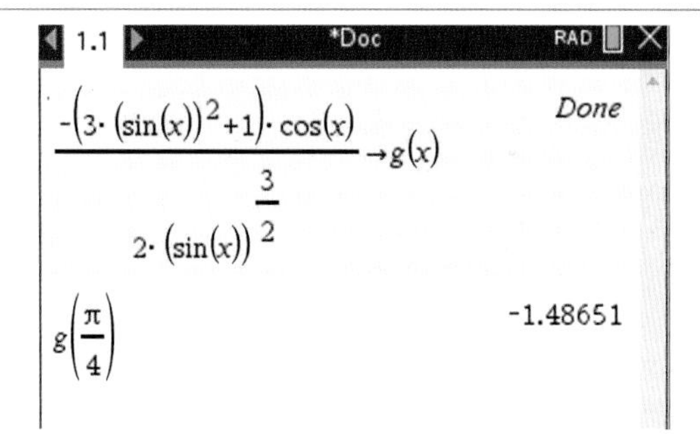

1 Define **f(x)** as shown above. Note that $\cos^2(x)$ needs to be entered as $(\cos(x))^2$.

2 Find the derivative of **f(x)**.

3 Press **ctrl + var** to store the derivative as **g(x)**.

4 Enter $g\left(\dfrac{\pi}{4}\right)$ and press **ctrl + enter** for the approximate answer.

$f'\left(\dfrac{\pi}{4}\right) \approx -1.487$, to three decimal places.

Straight line motion

displacement, x, at time t: $x(t)$

velocity, v, at time t: $v(t) = \dfrac{dx}{dt}$

acceleration, a, at time t: $a(t) = \dfrac{dv}{dt} = \dfrac{d^2x}{dt^2}$

WORKED EXAMPLE 20 Straight line motion

An oscillating spring moves so that its end is x cm from the point P at time t seconds, where $x = 2\sin(4t)$.

a Find an equation for the velocity of the spring.

b What is the initial velocity of the spring?

c What is the maximum acceleration?

Steps	Working
a velocity $= \dfrac{dx}{dt}$	$x = 2\sin(4t)$ $v = \dfrac{dx}{dt} = 2 \times 4\cos(4t)$ $= 8\cos(4t)$
b Initial velocity occurs when $t = 0$.	When $t = 0$, $v = 8\cos(4 \times 0)$ $= 8 \times 1$ $= 8$ cm/s
c Differentiate the velocity equation to obtain acceleration, a.	$v = 8\cos(4t)$ $a = \dfrac{dx}{dt} = -32\sin(4t)$
Find the maximum acceleration.	The maximum acceleration is 32 cm/s^2 when $\sin(4t) = -1$.

WORKED EXAMPLE 21 Finding optimal solutions

The effectiveness (f) of an insect repellent, measured over a 12-hour period, is given by the function $f(t) = e^t \sin\left(\dfrac{\pi t}{12}\right)$, where t is the number of hours after the repellent is applied.

a Show that $f'(t) = e^t\left(\sin\left(\dfrac{\pi t}{12}\right) + \dfrac{\pi}{12}\cos\left(\dfrac{\pi t}{12}\right)\right)$.

b Show that the maximum effectiveness occurs when $\tan\left(\dfrac{\pi t}{12}\right) = -\dfrac{\pi}{12}$.

c Determine the approximate coordinates of the local maximum correct to nearest integer.

Steps	Working
a Use the product rule to find the first derivative.	$u = e^t, \quad v = \sin\left(\dfrac{\pi t}{12}\right)$ $\dfrac{du}{dt} = e^t \quad \dfrac{dv}{dt} = \dfrac{\pi}{12}\cos\left(\dfrac{\pi t}{12}\right)$ $f'(t) = e^t \sin\left(\dfrac{\pi t}{12}\right) + e^t \times \dfrac{\pi}{12}\cos\left(\dfrac{\pi t}{12}\right)$ $f'(t) = e^t\left(\sin\left(\dfrac{\pi t}{12}\right) + \dfrac{\pi}{12}\cos\left(\dfrac{\pi t}{12}\right)\right)$

b Solve $f'(t) = 0$.

Stationary point at $f'(t) = 0$.

$$e^t\left(\sin\left(\frac{\pi t}{12}\right) + \frac{\pi}{12}\cos\left(\frac{\pi t}{12}\right)\right) = 0$$

$$\sin\left(\frac{\pi t}{12}\right) + \frac{\pi}{12}\cos\left(\frac{\pi t}{12}\right) = 0$$

$$\sin\left(\frac{\pi t}{12}\right) = -\frac{\pi}{12}\cos\left(\frac{\pi t}{12}\right)$$

$$\frac{\sin\left(\frac{\pi t}{12}\right)}{\cos\left(\frac{\pi t}{12}\right)} = -\frac{\pi}{12}$$

$$\tan\left(\frac{\pi t}{12}\right) = -\frac{\pi}{12}$$

c Use CAS to find the coordinates of the stationary point.

Coordinates of the local maximum: (11, 15 501)

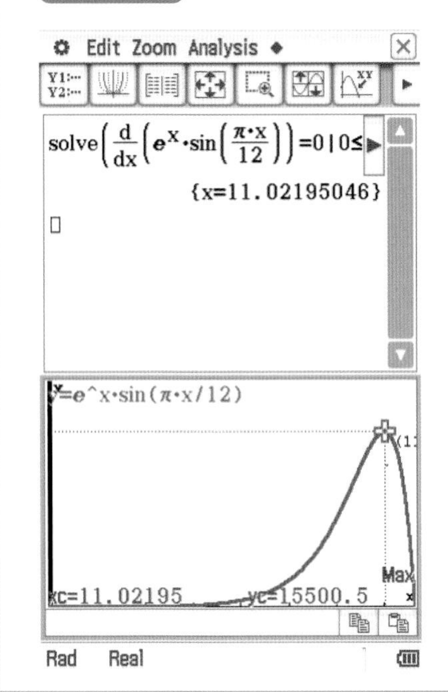

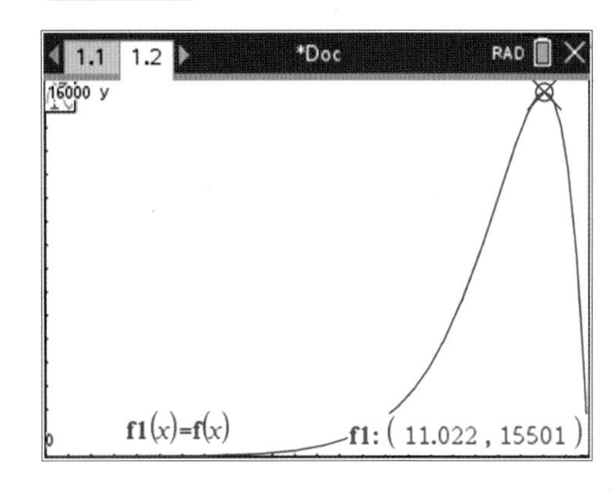

Maximum rate of change

The **maximum rate of increase or decrease** of a function is where the function is increasing or decreasing most rapidly, where the gradient of its graph is steepest in the positive and negative directions. We can identify these points by graphing the derivative function and locating its maximum and minimum points. These show where the function has its maximum rate of increase and decrease, respectively.

Finding the maximum rate of increase or decrease of a function $y = f(x)$

1 Graph $y = f(x)$ and $y = f'(x)$ for the given domain.
2 Find the local maximum and minimum points of $y = f'(x)$.
3 Substitute the x values of these points into $y = f(x)$ to find the points where the function has its maximum rate of increase and decrease, respectively.
4 Substitute the x values into $y = f'(x)$ to find the maximum rate of increase and decrease, respectively.

USING CAS 5 Maximum rate of change

A section of a rollercoaster track is described by the function $f(x) = \frac{1}{5}x^2 \sin\left(\frac{3x}{4}\right)$, where $0 \le x \le \frac{5\pi}{4}$.

Find, correct to two decimal places, the coordinates of the point on the track at which the rollercoaster is experiencing its maximum rate of increase.

ClassPad

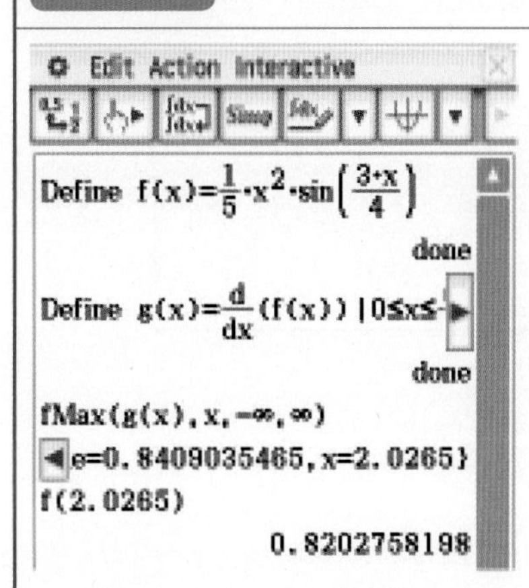

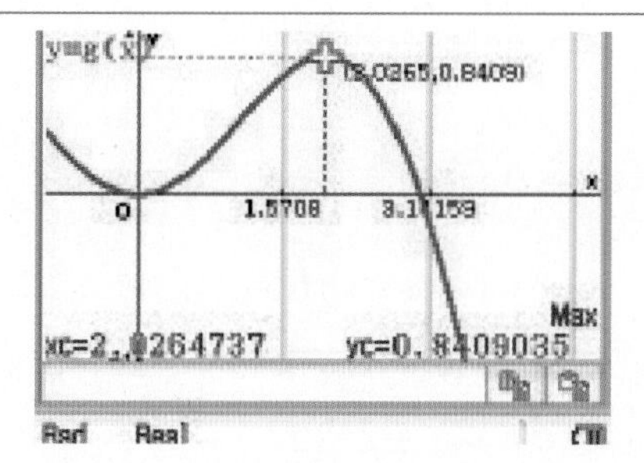

The point (2.0265…,0.8409…) is on the first derivative function. We must substitute $x = 2.0265$ into $f(x)$ to find the actual coordinates of the point of inflection.

1 Define $f(x)$ as shown above.
2 Define $g(x)$ as the derivative of $f(x)$.
3 Use **Math3** to include the domain $0 \le x \le \frac{5\pi}{4}$.
4 Highlight $g(x)$ and tap **Interactive > Calculation > fMin/fMax > fMax.**
5 The maximum value will be displayed.
6 Copy the answer into **f(x)** to find the y coordinate of the maximum point.
7 To confirm the result, graph $g(x)$ in split screen in **Main**. Include the domain $0 \le x \le \frac{5\pi}{4}$.
8 Adjust the window settings to suit.
9 Tap **Analysis > G-Solve > Max.**
10 The coordinates of the maximum point on $g(x)$ will be displayed.

The coordinates of the point of inflection experiencing the maximum positive rate of increase are (2.03, 0.82).

TI-Nspire

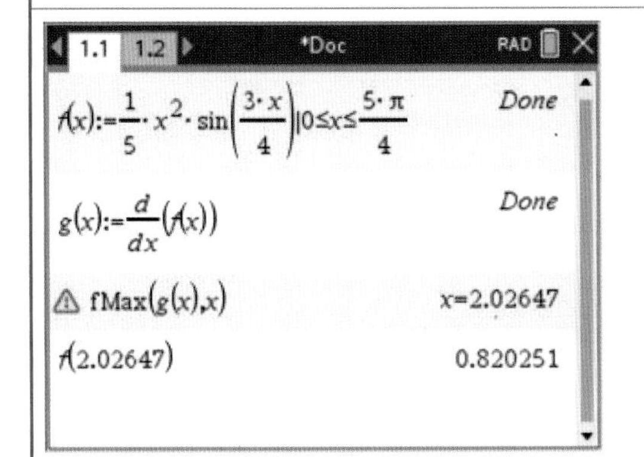

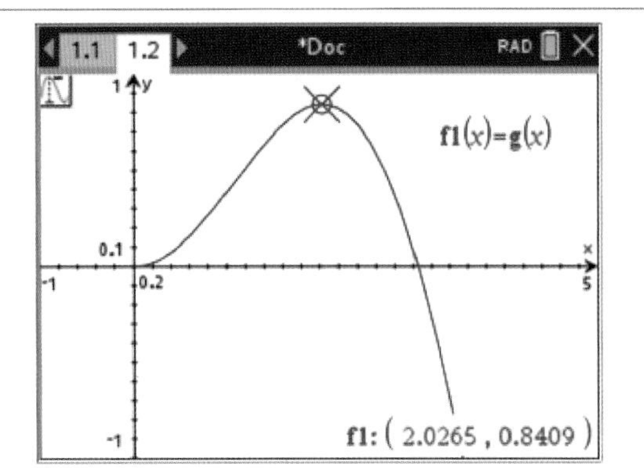

1 Define $f(x)$ as shown above.
2 Define $g(x)$ as the derivative of $f(x)$.
3 Press **catalog** and scroll down to **fMax**.
4 Enter **g(x),x** to find the x coordinate of the maximum point.
5 Copy the answer into **f(x)** to find the y coordinate of the maximum point.
6 To confirm the result, graph $g(x)$.
7 Adjust the window settings to suit.
8 Press **menu > Trace > Graph trace**.
9 Scroll along the graph until maximum appears.

The coordinates of the point of inflection experiencing the maximum positive rate of increase are $(2.03, 0.82)$.

EXERCISE 4.4 Differentiating trigonometric functions

ANSWERS p. 395

Recap

1 Find $\int \left(e^{2x} - e^{-2x}\right)^2 dx$.

2 Evaluate $\int_0^3 4e^{2x} dx$.

Mastery

3 WORKED EXAMPLE 16 Differentiate $y = x^2 \sin(x)$.

4 WORKED EXAMPLE 17 Find $\frac{dy}{dx}$ for $y = \frac{5\cos(3x)}{\sin(2x)}$.

5 WORKED EXAMPLE 18 Find $\frac{dy}{dx}$ if $y = \sin\left(4x^5 - x\right)$.

6 WORKED EXAMPLE 19 Differentiate $y = 4x^2 \cos\left(x^3\right)$.

7 Using CAS 4

a For $f(x) = \sqrt{\cos(x)}$, find $f'\left(\frac{\pi}{6}\right)$ correct to three decimal places.

b Given $f'(c) = -\frac{\sqrt{6}}{4}$, find the value of c, correct to three decimal places if $0 \le c \le \frac{\pi}{2}$.

8 WORKED EXAMPLE 20 A particle moves so that its displacement x is given by the function $x(t) = 3\cos\left(\frac{t}{2}\right)$, $0 \le t \le 2\pi$, where x is in centimetres and t is in seconds.

- **a** Find an equation for the velocity of the particle.
- **b** Find an equation for the acceleration of the particle.
- **c** Find the time(s) when the particle is at $x = 0$.
- **d** Determine the velocity and acceleration when $x = 0$.
- **e** Find the time(s) at which the particle has the greatest acceleration.

9 WORKED EXAMPLE 21 The effectiveness (f) of an insect repellent, measured over a 3-hour period, is given by the function $f(t) = e^{\frac{t}{2}} \sin\left(\frac{\pi t}{3}\right)$, where t is the number of hours after the repellent is applied.

- **a** Show $f'(t) = e^{\frac{t}{2}}\left(\frac{1}{2}\sin\left(\frac{\pi t}{3}\right) + \frac{\pi}{3}\cos\left(\frac{\pi t}{3}\right)\right)$.
- **b** Show the maximum effectiveness occurs when $\tan\left(\frac{\pi t}{3}\right) = -\frac{2\pi}{3}$.
- **c** Determine the approximate coordinates of the local maximum, correct to one decimal place.

10 Using CAS 5

- **a** Consider the function $f(x) = x\sin\left(\frac{x}{4}\right)$, where $0 \le x \le 10$. Find, correct to two decimal places, the coordinates of the point on $f(x)$ where the function has its maximum rate of increase.
- **b** A waterslide track is described by the function $y = \frac{4\sin(x)}{x}$, where $\frac{\pi}{4} \le x \le 3\pi$. Determine the points on the waterslide where the rate of increase is maximum and where the rate of increase is minimum.

Calculator-free

11 © SCSA MM2020 Q2 (4 marks) If $h(x) = \frac{e^{-x}}{\cos x}$, then evaluate $h'(\pi)$.

12 (3 marks)

- **a** Let $f(x) = x\sin(x)$. Find $f'(x)$. (1 mark)
- **b** If $f(x) = \frac{x}{\sin(x)}$, find $f'\left(\frac{\pi}{2}\right)$. (2 marks)

13 (2 marks) If $g(x) = x^2\sin(2x)$, find $g'\left(\frac{\pi}{6}\right)$.

14 (2 marks) P is a point on the curve defined by $y = 4 + 2\cos(3x)$, where $0 \le x \le \frac{\pi}{3}$, and the tangent at P is parallel to the line $y = 1 - 6x$.

- **a** Find the coordinates of P. (1 mark)
- **b** Determine the equation of the tangent line at P. (1 mark)

15 © SCSA MM2016 Q4 (8 marks) The displacement x micrometres at time t seconds of a magnetic particle on a long straight superconductor is given by the rule $x = 5\sin 3t$.

a Determine the velocity of the particle when $t = \frac{\pi}{2}$. (3 marks)

b Determine the rate of change of the velocity when $t = \frac{\pi}{2}$. (3 marks)

Let v = velocity of the particle at t seconds.

c Determine $\int_0^{\frac{\pi}{2}} \frac{dv}{dt}\, dt$. (2 marks)

16 (3 marks) The position of a particle is given by $x = t\cos(2t)$, where x metres is the particle's position from a fixed point after t seconds. Find the acceleration of the particle after $\frac{\pi}{4}$ seconds.

Calculator assumed

17 © SCSA MM2016 Q21 (6 marks) A lighthouse is situated 12 km away from the shoreline, opposite point X as seen in the diagram below. A long brick wall is placed along the shoreline and at night the light from the lighthouse can be seen moving along this wall.

Let y = displacement of light on the wall from point X and θ = angle of the rotating light from the lighthouse.

The light is revolving anticlockwise at a uniform rate of three revolutions per minute $\left(\frac{d\theta}{dt} = 6\pi \text{ radians/minute}\right)$.

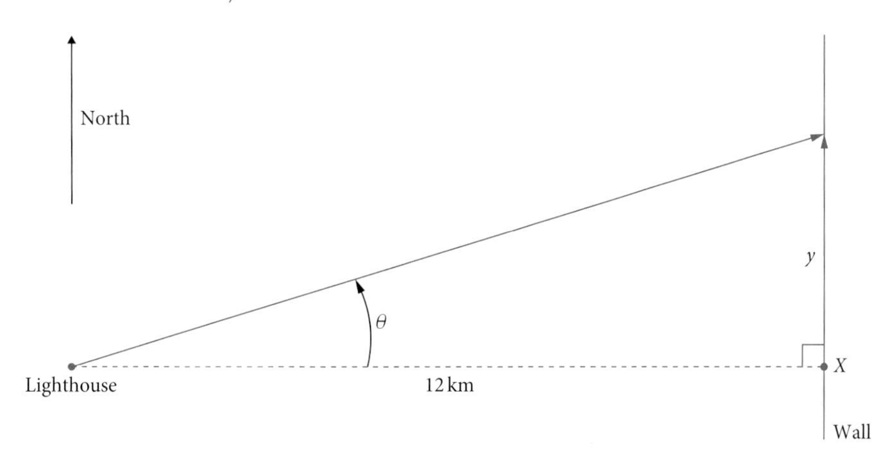

a Show that $\frac{dy}{d\theta} = \frac{12}{\cos^2\theta}$. (3 marks)

b Determine the velocity, in kilometres per minute, of the light on the wall when the light is 5 km north of point X. (3 marks)

$\left(\text{Hint: } \frac{dy}{dt} = \frac{dy}{d\theta} \times \frac{d\theta}{dt}\right)$

18 © SCSA MM2017 Q10 (3 marks) Use the quotient rule to show that $\frac{d}{dx}\tan(x) = \frac{1}{\cos^2(x)}$.

19 (2 marks) Trigg is designing a garden that is to be built on flat ground. In his initial plans, he draws the graph of $y = \sin(x)$ for $0 \le x \le 2\pi$ and decides that the garden beds will have the shape of the shaded regions shown in the diagram below. He includes a garden path, which is shown as line segment PC.

The line through points $P\left(\frac{2\pi}{3}, \frac{\sqrt{3}}{2}\right)$ and $C(c, 0)$ is a tangent to the graph of $y = \sin(x)$ at point P.

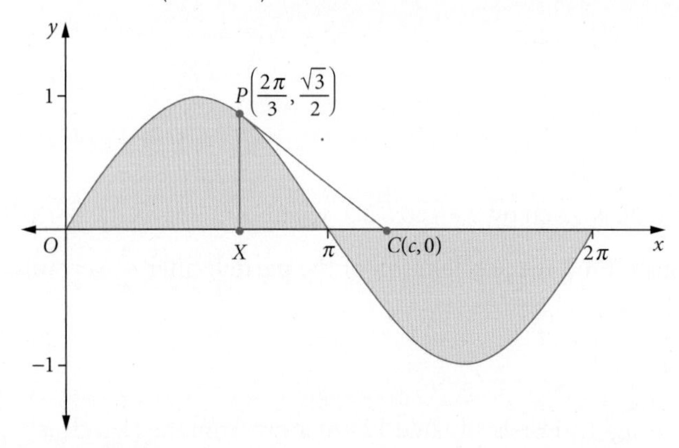

a Find $\frac{dy}{dx}$ when $x = \frac{2\pi}{3}$. (1 mark)

b Show that the value of c is $\sqrt{3} + \frac{2\pi}{3}$. (1 mark)

4.5 Integrals of trigonometric functions

Video playlist Integrals of trigonometric functions

Worksheets Finding indefinite integrals 1

Finding indefinite integrals 2

Finding definite integrals

Recall this table of derivatives of trigonometric functions:

Trigonometric function	Derivative
$y = \sin(ax - b)$	$\frac{dy}{dx} = a\cos(ax - b)$
$y = \cos(ax - b)$	$\frac{dy}{dx} = -a\sin(ax - b)$

We use these derivatives to obtain the anti-derivatives of trigonometric functions:

$$\frac{d}{dx}(\sin(ax - b)) = a\cos(ax - b)$$

$$\int \frac{d}{dx}(\sin(ax - b))\,dx = \int a\cos(ax - b)\,dx$$

$$\int \cos(ax - b)\,dx = \frac{1}{a}\sin(ax - b) + c$$

$$\frac{d}{dx}(\cos(ax - b)) = -a\sin(ax - b)$$

$$\int \frac{d}{dx}(\cos(ax - b))\,dx = \int -a\sin(ax - b)\,dx$$

$$\int \sin(ax - b)\,dx = -\frac{1}{a}\cos(ax - b) + c$$

Integrals of trigonometric functions

$$\int \sin(ax - b)\,dx = -\frac{1}{a}\cos(ax - b) + c$$

$$\int \cos(ax - b)\,dx = \frac{1}{a}\sin(ax - b) + c$$

WORKED EXAMPLE 22 Finding indefinite and definite integrals

a $\int 6\cos(2x) + 12\sin(3x)\,dx$

b $\int_0^{\pi} 4\cos(2x)\,dx$

Steps	Working
a 1 Anti-differentiate each term.	$\int 6\cos(2x)dx = 6 \times \frac{1}{2}\sin(2x) = 3\sin(2x)$ $\int 12\sin(3x)\,dx = -12 \times \frac{1}{3}\cos 3x = -4\cos(3x)$
2 Combine the answers and include the constant of integration.	The anti-derivative of $6\cos(2x) + 12\sin(3x)$ is $F(x) = 3\sin(2x) - 4\cos(3x) + c$.
b Integrate the function and substitute the limits of integration.	$\int_0^{\pi} 4\cos(2x)\,dx = [2\sin(2x)]_0^{\pi}$ $= 2\sin(2\pi) - 2\sin(0)$ $= 0 - 0$ $= 0$

Anti-differentiation by recognition

WORKED EXAMPLE 23 Anti-differentiation by recognition

Given that $\frac{d}{dx}(x\cos(x)) = \cos(x) - x\sin(x)$, write the expression for $\int 6x\sin(x)\,dx$.

Steps	Working
1 Transpose the equation so that $x\sin(x)$ is the subject.	$x\sin(x) = \cos(x) - \frac{d}{dx}(x\cos(x))$
2 Integrate every term in the equation.	$\int x\sin(x)\,dx = \int \cos(x)\,dx - \int \frac{d}{dx}(x\cos(x))\,dx$
3 Simplify the right-hand side of the equation.	$\int x\sin(x)\,dx = \sin(x) - x\cos(x)$
4 Multiple both sides by 6 and then add the constant 'c'.	$\int 6x\sin(x)\,dx = 6\sin(x) - 6x\cos(x) + c$

WORKED EXAMPLE 24 Anti-differentiating to find the constant of integration

The derivative of a function is $f'(x) = -4\sin(3x)$. If $f\left(\frac{\pi}{18}\right) = -\frac{2\sqrt{3}}{3}$, determine the function $f(x)$.

Steps	Working
1 Anti-differentiate $f'(x) = -4\sin(3x)$.	$f(x) = \int -4\sin(3x)\,dx$ $= \frac{4}{3}\cos(3x) + c$

2 Find the value of c using $f\left(\frac{\pi}{18}\right) = -\frac{2\sqrt{3}}{3}$.	When $x = \frac{\pi}{18}$, $f(x) = -\frac{2\sqrt{3}}{3}$. $f(x) = \frac{4}{3}\cos(3x) + c$ $-\frac{2\sqrt{3}}{3} = \frac{4}{3}\cos\left(3 \times \frac{\pi}{18}\right) + c$ $= \frac{4}{3}\cos\left(\frac{\pi}{6}\right) + c$ $= \frac{4}{3} \times \frac{\sqrt{3}}{2} + c$ $= \frac{2\sqrt{3}}{3} + c$ $\therefore c = -\frac{4\sqrt{3}}{3}$
3 State the answer.	$f(x) = \frac{4}{3}\cos(3x) - \frac{4\sqrt{3}}{3}$

WORKED EXAMPLE 25 Finding areas for trigonometric functions by symmetry

a Sketch the graph of the curve $y = \cos(2x)$ in the interval $0 \le x \le \pi$.

b Calculate the area bounded by the curve, the y-axis and the x-axis in the interval $0 \le x \le \frac{3\pi}{4}$.

Steps	Working
a Sketch the graph in the interval $[0, \pi]$.	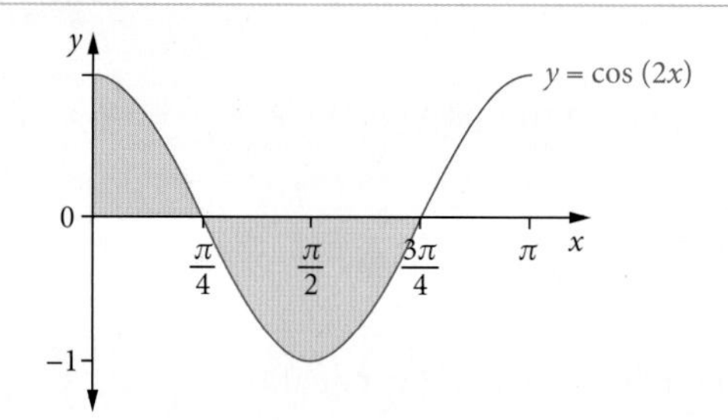
b 1 The total area is three times the area above the x-axis. Write an integral equation for the area and evaluate.	Area $= 3\int_0^{\frac{\pi}{4}} \cos(2x)\, dx$
2 Calculate the total area.	$= 3\left[\frac{1}{2}\sin(2x)\right]_0^{\frac{\pi}{4}}$ $= 3\left[\frac{1}{2}\sin\left(\frac{\pi}{2}\right) - \frac{1}{2}\sin(0)\right]$ $= 3\left[\frac{1}{2} - 0\right]$ $= \frac{3}{2}$ units^2

Straight line motion

WORKED EXAMPLE 26 Straight line motion using integration

It takes an elevator 20 seconds to ascend from the ground floor of a building to the fifth floor. The velocity of the elevator during its ascent is given by

$$v(t) = \frac{7\pi}{20}\sin\left(\frac{\pi t}{20}\right)\text{m/s}.$$

The velocity, v, is measured in metres per second, and the time, t, is measured in seconds.

Find

a the acceleration of the elevator in terms of t

b $\int_0^{\frac{10}{3}} v'(t)\,dt$

c the displacement function of the elevator if the ground floor has displacement zero metres.

Steps	Working
a **1** Graph the function.	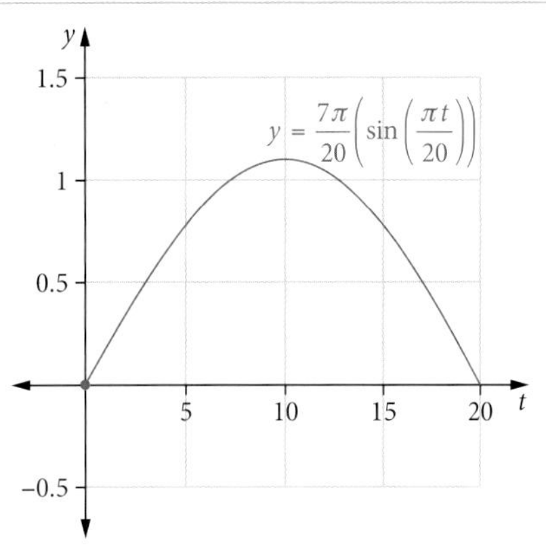
2 Find the derivative $a(t) = \frac{dv}{dt}$.	$a(t) = \frac{dv}{dt}$ $= \frac{7\pi}{20} \times \frac{\pi}{20}\cos\left(\frac{\pi t}{20}\right)$ $= \frac{7\pi^2}{400}\cos\left(\frac{\pi t}{20}\right)\text{m/s}^2$
b Substitute into $\int_a^b f'(x)\,dx = f(b) - f(a)$.	$\int_0^{\frac{10}{3}} v'(t)\,dt = v\left(\frac{10}{3}\right) - v(0)$ $= \frac{7\pi}{20}\sin\left(\frac{\pi}{20} \times \frac{10}{3}\right) - \frac{7\pi}{20}\sin(0)$ $= \frac{7\pi}{20}\sin\left(\frac{\pi}{6}\right)$ $= \frac{7\pi}{40}\text{m/s}$

c 1 Find the integral of $v(t)$: $x(t) = \int v(t)\, dt + c$.	$x(t) = \int \frac{7\pi}{20} \sin\left(\frac{\pi t}{20}\right) dt + c$ $= \frac{7\pi}{20} \int \sin\left(\frac{\pi t}{20}\right) dt + c$ $= \frac{7\pi}{20} \times \frac{20}{\pi} \times -\cos\left(\frac{\pi t}{20}\right) + c$ $= -7\cos\left(\frac{\pi t}{20}\right) + c$
2 Substitute $x = 0$, $t = 0$ to find the constant c.	$x(0) = 0$ $0 = -7\cos(0) + c$ $c = 7$ $x(t) = 7 - 7\cos\left(\frac{\pi t}{20}\right)$

Area between two curves

The area between the curves with equations $y = f(x)$ and $y = g(x)$, where $f(x) > g(x)$, in the interval $a \le x \le b$ is found using area $= \int_a^b [f(x) - g(x)]\, dx$.

USING CAS 6 Area between two curves

Find, correct to three decimal places, the area of the region bounded by the curves $y = \sin(x)$ and $y = \cos(2x)$ in the interval $0 \le x \le \pi$.

ClassPad

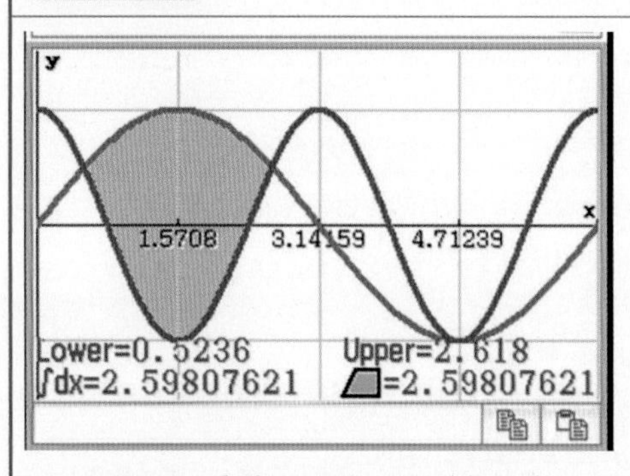

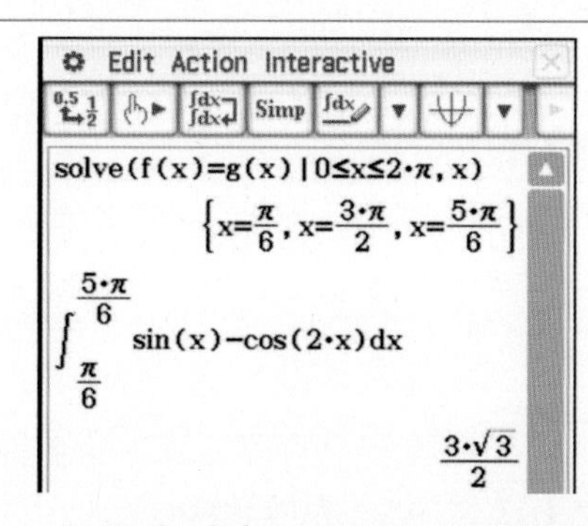

1. In **Main**, define **f(x)** = sin (x) and **g(x)** = cos $(2x)$.
2. Open the graph screen, highlight and drag down $f(x)$ and $g(x)$.
3. Adjust the window settings to suit $0 \le x \le 2\pi$.
4. Tap **Analysis > G-Solve > Integral > ∫dx Intersection**.
5. When the first point of intersection is displayed, press **EXE**, then tap the right arrow.
6. When the second point of intersection is displayed, press **EXE**.
7. The shaded area between the curves and the value will be displayed.
8. To find the exact area, solve the equation **sin (x)=cos (2x)** or **f(x)=g(x),** over the domain $0 \le x \le 2\pi$.
9. Highlight **sin (x)−cos (2x)** or **f(x)−g(x)** and tap **Interactive > Calculation > ∫**.
10. In the dialogue box, tap **Definite**.
11. Enter the two smaller solutions as the lower and upper limits, and enter the expression as shown above.
12. The exact area will be displayed.

The area is 2.598 units^2, correct to three decimal places.

TI-Nspire

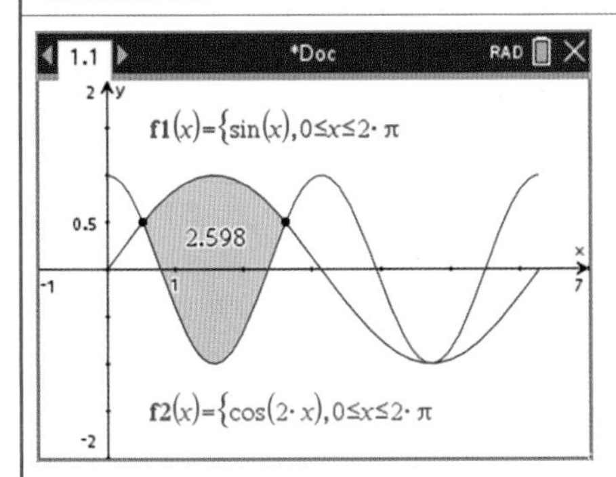

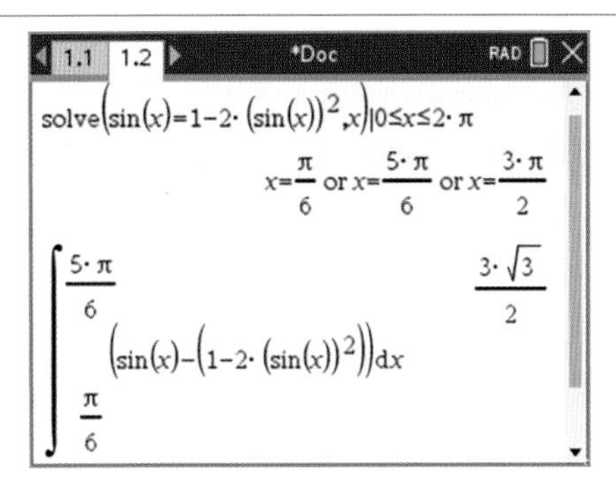

1. Graph **f1(x)** and **f2(x)** as shown above.
2. Adjust the window settings to suit.
3. Press **menu > Analyze Graph > Bounded Area**.
4. When prompted for the **lower bound**, click on the first point of intersection.
5. When prompted for the **upper bound**, click on the second point of intersection.
6. The area will be displayed.
7. To find the exact area, you need to use the identity $\cos(2x) = 1 - 2\sin^2(x)$.
8. Solve the equation as shown above over the domain $0\leq x\leq 2\pi$.
9. Press **menu > Calculus > Integral**.
10. Enter the two smaller solutions as the lower and upper limits, and enter the expression as shown above.
11. The exact area will be displayed.

The area is 2.598 units^2, correct to three decimal places.

WACE QUESTION ANALYSIS

© SCSA MM2021 Q12 Calculator-assumed (15 marks)

Let $f(x) = x^2e^x$.

a Show that $f'(x) = xe^x(x + 2)$. (2 marks)

b Use calculus to determine all the stationary points of $f(x)$ and determine their nature. (7 marks)

c Determine the coordinates of any points of inflection. (2 marks)

d Copy the axes below and hence sketch the graph of $f(x)$, clearly indicating the location of all stationary points and points of inflection. (4 marks)

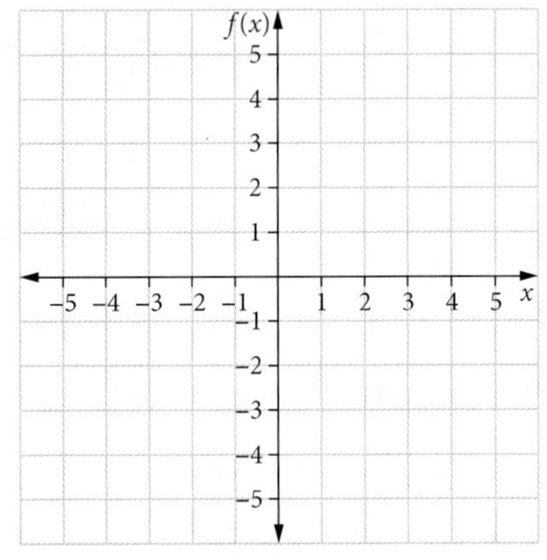

Video WACE question analysis: Applying the exponential and trigonometric functions

Reading the question

- Highlight the type of answer required in each part. This may be coordinates, a first derivative, a second derivative or a graph.
- The 'show that' in part **a** must include all the working for the derivative and cannot be done by CAS.
- Highlight all the details that must be labelled on the graph.

Thinking about the question

- This question requires knowledge of the product and chain rules for differentiation.
- You need to be able to find a second derivative and know how to use it to determine the nature of a stationary point.
- You also need to use a second derivative to determine the coordinates of points of inflection.
- The sketch graph drawn must show the coordinates of all stationary points and points of inflection.

Worked solution (✓ = 1 mark)

a Use the product rule to differentiate.

$u = x^2$ $\quad\quad\quad$ $v = e^x$

$\frac{du}{dx} = 2x$ $\quad\quad\quad$ $\frac{dv}{dx} = e^x$

$f'(x) = 2xe^x + x^2e^x = xe^x(x+2)$

differentiates using product rule ✓

factorises correctly ✓

b $\boldsymbol{f'(x) = 0}$ ✓

$xe^x(x+2) = 0$

stationary points at $x = \mathbf{0}$ and $x = \mathbf{-2}$ ✓

At $x = 0, f(0) = 0$ $\quad\quad$ and at $x = -2$ $\quad f(-2) = (-2)2e^{-2} = \frac{4}{e^2} \approx 0.54.$

coordinates of stationary points $\mathbf{(0,0)}$ ✓ and $\left(\mathbf{-2, \frac{4}{e^2}}\right)$ ✓

Nature of stationary points

$f''(x) = 2xe^x + x^2e^x + 2e^x + 2xe^x$

$\boldsymbol{f''(x) = e^x(x^2 + 4x + 2)}$ ✓

At $x = 0, \quad f''(0) = 2 > 0$

Therefore, $(0,0)$ is a local **minimum.** ✓

At $x = -2, \quad f''(-2) = -2e^{-2} < 0$

Therefore, $\left(-2, \frac{4}{e^2}\right)$ is a local **maximum**. ✓

c **points of inflection** $\boldsymbol{f''(x) = 0}$ ✓

$e^x(x^2 + 4x + 2) = 0$

$$x = \frac{-4 \pm \sqrt{16-8}}{2} = -2 \pm \sqrt{2}$$

Points are $(\mathbf{-3.4, 0.38})$ and $(\mathbf{-0.59, 0.19})$. ✓

d

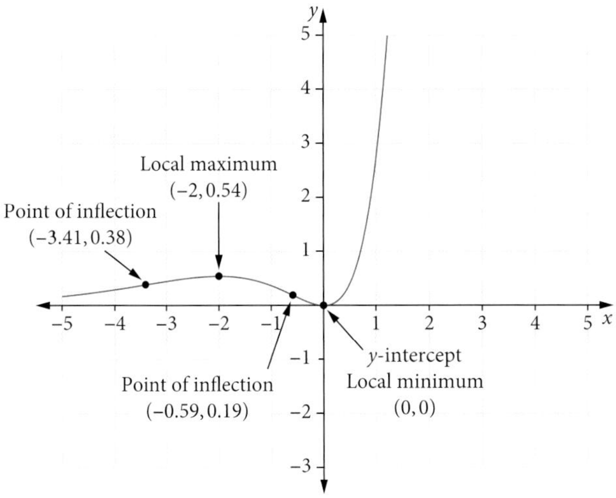

Local maximum $(-2, 0.54)$

Point of inflection $(-3.41, 0.38)$

Point of inflection $(-0.59, 0.19)$

y-intercept
Local minimum $(0, 0)$

indicates local minimum ✓

indicates local maximum ✓

indicates points of inflections ✓

overall shape ✓

EXERCISE 4.5 Integrals of trigonometric functions

ANSWERS p. 395

Recap

1 An object's velocity v m/s at time t s is $v = \frac{3}{4}\tan(8t - \pi)$. Find the object's acceleration in m/s^2 when $t = \frac{5\pi}{32}$.

2 If $\frac{dy}{dx} = \sqrt{2}\cos(\pi - 3x)$, find the value of x when y is a maximum for $0 < x < \pi$.

Mastery

3 WORKED EXAMPLE 22 Find

a $\int 6\sin(2x)\,dx$

b $\int \frac{1}{2}\cos\left(\frac{x}{2}\right)dx$

c $\int 3\sin\left(\frac{1}{2}(5x - 7)\right)dx$

d $\int_0^{\frac{\pi}{12}} 2\cos(2x)\,dx$

e $\int_{\frac{\pi}{12}}^{\frac{\pi}{6}} 4\sin(2x)\,dx$

f $\int_0^{\frac{\pi}{12}} \sin(2x) - \cos(3x)\,dx$

4 WORKED EXAMPLE 23 Given that $\frac{d}{dx}(3x\sin(2x)) = 3\sin(2x) + 6x\cos(2x)$, write the expression for $\int 12x\cos(2x)\,dx$.

5 WORKED EXAMPLE 24 If $f'(x) = \sin(3x) + \cos(x)$ and $f\left(\frac{\pi}{2}\right) = 3$, find the function $f(x)$.

6 WORKED EXAMPLE 25 Find the area bounded by the graph $y = 4\sin(3x)$ between $x = 0$ and $x = \pi$.

7 WORKED EXAMPLE 26 It takes an elevator 40 seconds to ascend from the ground floor of a building to the tenth floor. The velocity of the elevator during its ascent is given by

$$v(t) = \frac{7\pi}{8}\sin\left(\frac{\pi t}{40}\right)\text{m/s}.$$

The velocity, v, is measured in metres per second and the time, t, is measured in seconds.

Find

a the acceleration of the elevator in terms of t

b $\int_0^{\frac{20}{3}} v'(t)\,dt$

c the displacement function of the elevator if the ground floor has displacement zero metres.

8 Using CAS 6

Find, correct to two decimal places, the area of the two regions bounded by the curves $y = \sin(2x)$ and $y = \cos(x)$ in the interval $0 \le x \le \frac{\pi}{2}$.

Calculator-free

9 (4 marks)

a Find an anti-derivative of $\cos(2x+1)$ with respect to x. (1 mark)

b If $f'(x) = 2\cos(x) - \sin(2x)$ and $f\left(\frac{\pi}{2}\right) = \frac{1}{2}$, find $f(x)$. (3 marks)

10 (3 marks) If $f(x) = x\cos(3x)$, then $f'(x) = \cos(3x) - 3x\sin(3x)$.

Use this fact to find the anti-derivative of $x\sin(3x)$.

11 (2 marks) The acceleration $a\,\text{m/s}^2$ of a particle at time t s is given by $a(t) = 2\sin(t) - 18\cos(3t)$.

If the speed of the particle is 0 m/s after π seconds and its position is $-\pi$ m after $\frac{\pi}{2}$ s, determine the equation for the object's position.

12 (2 marks) The motion of an object is described by $a(t) = \cos(t)$, where $a(t)$ is the object's acceleration at time t. Determine the equation for the object's position, $s(t)$, given that $s\left(\frac{\pi}{2}\right) = \frac{\pi}{2} + 2$ and $s(\pi) = \pi + 3$.

13 © SCSA MM2017 Q8 (5 marks)

a Differentiate $2x\sin(3x)$ with respect to x. (2 marks)

b Hence show that $\int x\cos(3x)\,dx = \frac{3x\sin(3x) + \cos(3x)}{9} + c$. (3 marks)

14 © SCSA MM2019 Q9 (8 marks) It takes an elevator 16 seconds to ascend from the ground floor of a building to the sixth floor. The velocity of the elevator during its ascent is given by

$$v(t) = \frac{9\pi}{16}\sin\left(\frac{\pi t}{16}\right)\text{m/s.}$$

The velocity, v, is measured in metres per second, while the time, t, is measured in seconds.

a Determine the acceleration of the elevator during its ascent and provide a sketch of the acceleration function for $0 \le t \le 16$. (2 marks)

b With reference to your answer from part **a**, explain what is happening to the velocity of the elevator in the interval $0 < t < 8$ and in the interval $8 < t < 16$. (3 marks)

c Suppose that the ground floor has displacement $x = 0$ m. Determine the displacement function of the elevator and hence determine the height above the ground floor of the sixth floor. (3 marks)

15 © SCSA MM2020 Q17ab (8 marks) David and Katrina have a small farm and wish to fence off an area of their land so they can raise sheep. The area they have chosen has one border along a road as shown in the diagram below.

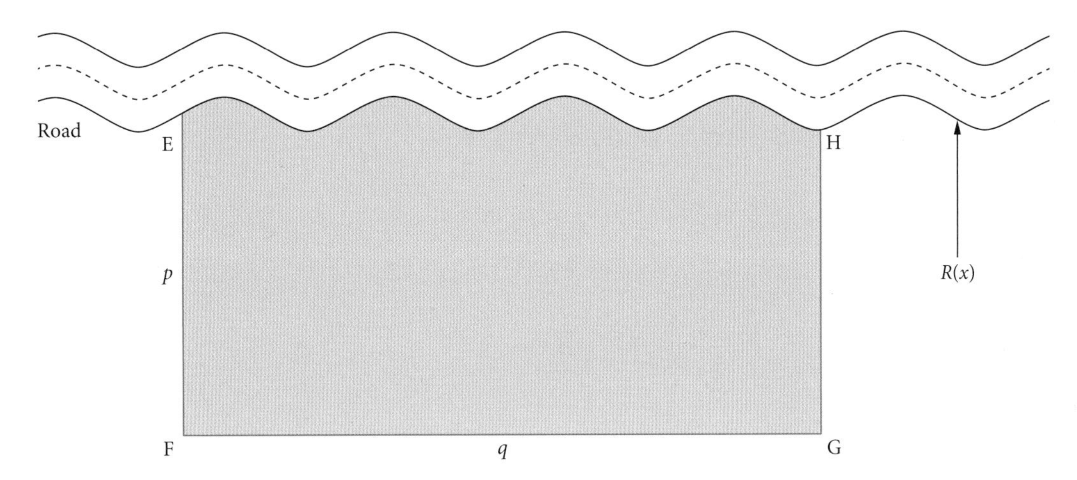

The enclosure is shown as the shaded area above and has right angles at points F and G. David and Katrina want the combined lengths of the fencing from E to F and F to G to equal 500 metres. Let the length of fence EF be equal to p metres and the length of fence FG be equal to q metres. If we locate the origin at point F and the x-axis along the line FG, the equation defining the fence along the road is given by:

$$R(x) = 10\sin\left(\frac{x}{15}\right) + p$$

a Show that the equation defining the area of the enclosure, $A(q)$, can be given in terms of q as follows:

$$A(q) = 500q - 150\cos\left(\frac{q}{15}\right) - q^2 + 150$$

(4 marks)

b Determine, to the nearest metre, the value of q that will allow the sheep to graze over the maximum area and state this maximum area. (4 marks)

16 © SCSA MM2021 Q9abc (5 marks) The Interesting Architecture company has designed a building with a constant cross-section shown in the figure below.

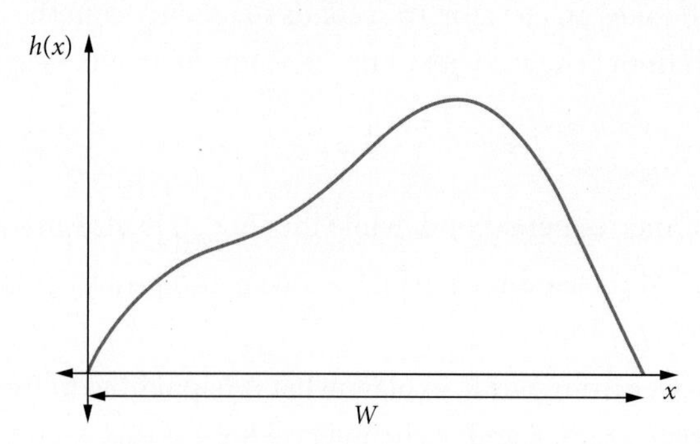

With reference to the figure, the height $h(x)$ of the building at a point x along its width is given by

$h(x) = 4\sin\left(x - \frac{3\pi}{2}\right) - x^2 + 3\pi x - 4$ where h and $0 \le x \le W$ are measured in metres.

a Determine the width W of the building to the nearest centimetre. (2 marks)

b Determine $h'(x)$. (1 mark)

c Determine, to the nearest centimetre, the value of x at which the height of the building is maximum and state this maximum height. (2 marks)

17 (2 marks) Jamie approximates the area between the x-axis and the graph of $y = 2\cos(2x) + 3$, over the interval $0 \le x \le \frac{\pi}{2}$, using the three rectangles shown below.

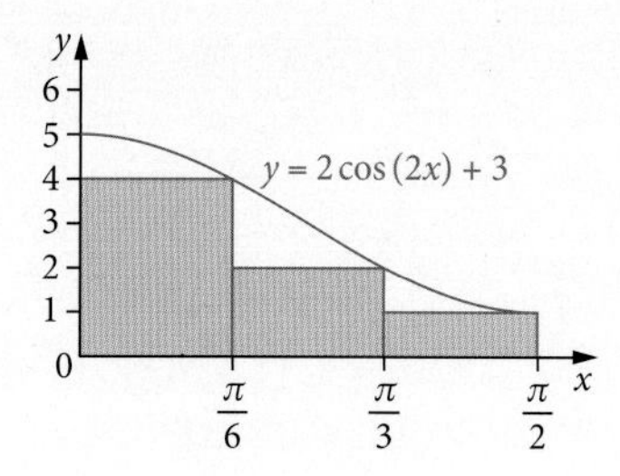

Determine Jamie's approximation as a fraction of the exact area.

4 Chapter summary

The natural exponential function $y = e^x$

- $e = 2.718\,28\ldots$ is Euler's number.
- Euler's number is defined as $\lim_{n\to\infty}\left(1+\frac{1}{n}\right)^n$. As $\lim_{n\to\infty}\left(1+\frac{x}{n}\right)^n = e^x$.
- The gradient of the graph is always increasing.
- The y-intercept is 1 (because $e^0 = 1$).
- The x-axis ($y = 0$) is a horizontal asymptote.

Exponential growth and decay

- The general form of an **exponential growth or decay function** is $N(t) = N_0e^{kt}$, where N_0 is the initial value and the value of k determines the rate of growth or decay.
- For a **growth function**, $k > 0$.
- For a **decay function**, $k < 0$.

Differentiating exponential functions

- $\frac{d}{dx}(e^x) = e^x$
- $\frac{d}{dx}(e^{ax-b}) = ae^{ax-b}$
- If $y = e^{f(x)}$, then $\frac{dy}{dx} = f'(x)e^{f(x)}$ by the chain rule.

Integrating exponential functions

- $\int e^x dx = e^x + c$
- $\int e^{ax} dx = \frac{1}{a}e^{ax} + c$

Derivatives of trigonometric functions

Trigonometric function	Derivative
$y = \sin(ax)$	$\frac{dy}{dx} = a\cos(ax)$
$y = \cos(ax)$	$\frac{dy}{dx} = -a\sin(ax)$

Integrals of trigonometric functions

$$\int \sin(ax - b)\,dx = -\frac{1}{a}\cos(ax - b) + c$$

$$\int \cos(ax - b)\,dx = \frac{1}{a}\sin(ax - b) + c$$

Straight line motion

- displacement x at time t: $x(t) = \int v(t)\,dt$
- velocity v at time t: $v(t) = \frac{dx}{dt}$ or $v(t) = \int a(t)\,dt$
- acceleration a at time t: $a(t) = \frac{dv}{dt} = \frac{d^2x}{dt^2}$

Cumulative examination: Calculator-free

Total number of marks: 20   Reading time: 2 minutes   Working time: 20 minutes

1 (6 marks)

Consider the function $y = (x + 1)(x^2 - 3x + 3)$.

a Determine the gradient of the tangent to the curve at $x = 3$. (3 marks)

b Using calculus techniques, determine the nature of the stationary point at $x = \frac{4}{3}$. (3 marks)

2 © SCSA MM2018 Q3a–ci (7 marks)

a Differentiate $(2x^3 + 1)^5$. (2 marks)

b Given $g'(x) = e^{2x}\sin(3x)$, determine a simplified value for the rate of change of $g'(x)$ when $x = \frac{\pi}{2}$. (3 marks)

c Determine $\int 3\cos(2x)\,dx$. (2 marks)

3 © SCSA MM2016 Q8 (7 marks) An isosceles triangle ΔPQR is inscribed inside a circle of fixed radius r and centre O. Let θ be defined as in the diagram below.

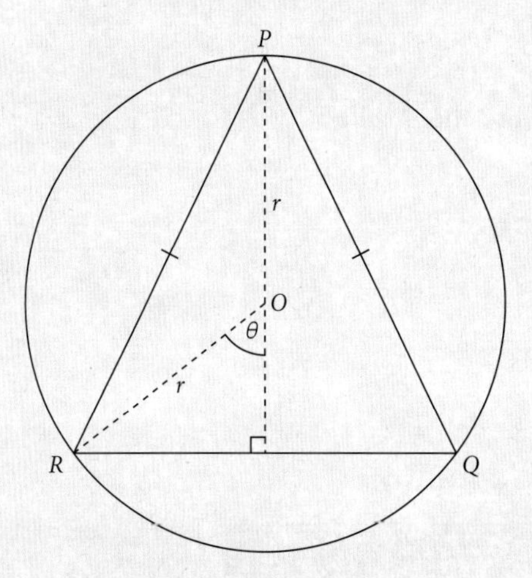

a Show that the area A of the triangle ΔPQR is given by $A = r^2\sin\theta(1 + \cos\theta)$. (2 marks)

b Using calculus, determine the value of θ that maximises the area A of the inscribed triangle. State this area in terms of r exactly. Justify your answer. (Hint: you may need the identity $\sin^2 x = 1 - \cos^2 x$ in your working.) (5 marks)

Cumulative examination: Calculator-assumed

Total number of marks: 21 Reading time: 3 minutes Working time: 21 minutes

1 (5 marks) Let $f(x) = \dfrac{2}{(x-1)^2} + 1$.

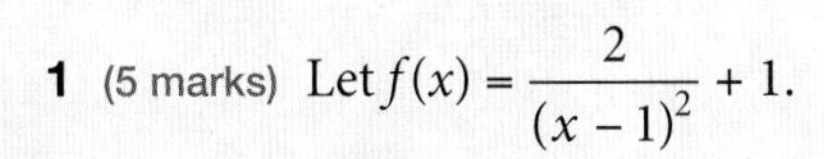

a **i** Evaluate $f(-1)$. (1 mark)

ii Copy the axes below and sketch the graph of f, labelling all asymptotes with their equations. (2 marks)

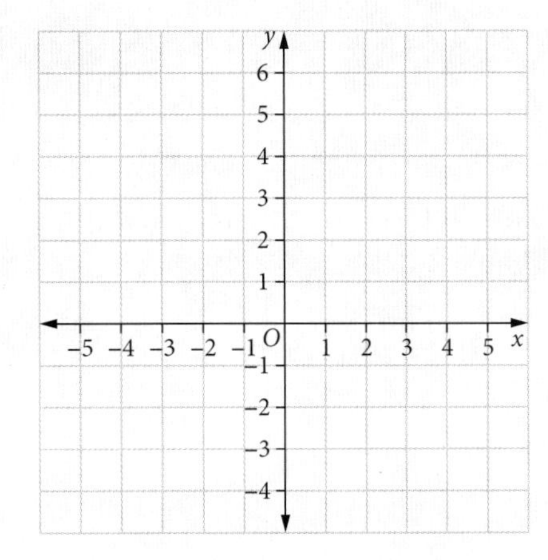

b Find the area bounded by the graph of f, the x-axis, the line $x = -1$ and the line $x = 0$. (2 marks)

2 © SCSA MM2017 Q16 (8 marks) A group of biologists has decided that colonies of a native Australian animal are in danger if their populations are less than 1000. One such colony had a population of 2300 at the start of 2011. The population was growing continuously such that $P = P_0e^{0.065t}$ where P is the number of animals in the colony t years after the start of 2011.

a Determine, to the nearest 10 animals, the population of the colony at the start of 2014. (2 marks)

b Determine the rate of change of the colony's population when $t = 2.5$ years. (2 marks)

c At the beginning of 2017, a disease caused the colony's population to decrease continuously at the rate of 8.25% of the population per year. If this rate continues, when will the colony become 'in danger'? Give your answer to the nearest month. (4 marks)

3 © SCSA MM2018 Q11 (8 marks) Ava is flying a drone in a large open space at a constant height of 5 metres above the ground. She flies the drone due north so that it passes directly over her head and then, sometime later, reverses its direction and flies the drone due south so it passes directly over her again. With $t = 0$ defined as the moment when the drone first flies directly above Ava's head, the velocity of the drone, at time t seconds, is given by:

$$v = 2\sin\left(\frac{t}{3} + \frac{\pi}{6}\right) \text{m/s} \quad 0 \le t \le 16$$

a Determine $x(t)$, the displacement of the drone at t seconds, where $x(0) = 0$. (3 marks)

b Where is the drone in relation to the pilot after 16 seconds? (2 marks)

c At a particular time, the drone is heading due south and it is decelerating at $0.5\,\text{m/s}^2$. How far has the drone travelled from its initial position directly above Ava's head until this particular time? (3 marks)

CHAPTER

DISCRETE RANDOM VARIABLES

SYLLABUS COVERAGE

Syllabus coverage

TOPIC 3.3: DISCRETE RANDOM VARIABLES

General discrete random variables

3.3.1 develop the concepts of a discrete random variable and its associated probability function, and their use in modelling data
3.3.2 use relative frequencies obtained from data to obtain point estimates of probabilities associated with a discrete random variable
3.3.3 identify uniform discrete random variables and use them to model random phenomena with equally likely outcomes
3.3.4 examine simple examples of non-uniform discrete random variables
3.3.5 identify the mean or expected value of a discrete random variable as a measurement of centre, and evaluate it in simple cases
3.3.6 identify the variance and standard deviation of a discrete random variable as measures of spread, and evaluate them using technology
3.3.7 examine the effects of linear changes of scale and origin on the mean and the standard deviation
3.3.8 use discrete random variables and associated probabilities to solve practical problems

Bernoulli distributions

3.3.9 use a Bernoulli random variable as a model for two-outcome situations
3.3.10 identify contexts suitable for modelling by Bernoulli random variables
3.3.11 determine the mean p and variance $p(1 - p)$ of the Bernoulli distribution with parameter p
3.3.12 use Bernoulli random variables and associated probabilities to model data and solve practical problems

Binomial distributions

3.3.13 examine the concept of Bernoulli trials and the concept of a binomial random variable as the number of 'successes' in n independent Bernoulli trials, with the same probability of success p in each trial
3.3.14 identify contexts suitable for modelling by binomial random variables
3.3.15 determine and use the probabilities $P(X = x) = \binom{n}{x} p^x(1 - p)^{n-x}$ associated with the binomial distribution with parameters n and p; note the mean np and variance $np(1 - p)$ of a binomial distribution
3.3.16 use binomial distributions and associated probabilities to solve practical problems

Video playlists (5):

5.1 Review of probability
5.2 Discrete probability distributions
5.3 Measures of centre and spread
5.4 The Bernoulli and binomial distributions
WACE question analysis Discrete random variables

Worksheets (13):

5.2 Random variables • Discrete probability distributions 1 • Discrete probability distributions 2
5.3 Expected values • Expected values using a calculator • Using expected values • Expected value • Variance and standard deviation
5.4 The Bernoulli distribution • Binomial probability experiments • Using the binomial probability distribution • Binomial probability – Mean and standard deviation • Applications of binomial probability

To access resources above, visit **cengage.com.au/nelsonmindtap**

Video playlist
Review of probability

5.1 Review of probability

The probability of an event occurring must have a value between 0 and 1, where 0 represents an impossible event and 1 represents a certain event.

Probability of an event

$$P(A) = \frac{\text{number of ways } A \text{ can occur}}{\text{total number of possible outcomes}}$$

Tree diagrams for compound events

A **compound event** consists of two or more simple events being considered together, such as tossing 2 tails on 3 coins, or winning first or second prize in a raffle. A **tree diagram** is a useful way of representing compound events.

WORKED EXAMPLE 1 Finding probabilities using a tree diagram

A barrel contains 20 balls, of which 8 are multicoloured. Two balls are randomly selected from the barrel, **without replacement**. This means that a ball is selected and the ball is **not** replaced **before** the next ball is selected. Find the probability that one of the two balls is multicoloured.

Steps	Working
1 On each selection, the ball selected can be either multicoloured (C) or *not multicoloured* (C'). The first selection is made from 8 balls that are multicoloured and 12 balls that are not multicoloured. In the second selection, the number of multicoloured and not multicoloured balls is determined by the type of ball selected first.	On the first selection: $n(C) = 8$, $n(C') = 12$; $P(C) = \frac{8}{20}$, $P(C') = \frac{12}{20}$. On the second selection: If the first ball selected is multicoloured: $n(C) = 7$, $n(C') = 12$; $P(C) = \frac{7}{19}$, $P(C') = \frac{12}{19}$. If the first ball selected is not multicoloured: $n(C) = 8$, $n(C') = 11$; $P(C) = \frac{8}{19}$, $P(C') = \frac{11}{19}$
2 Represent with a tree diagram.	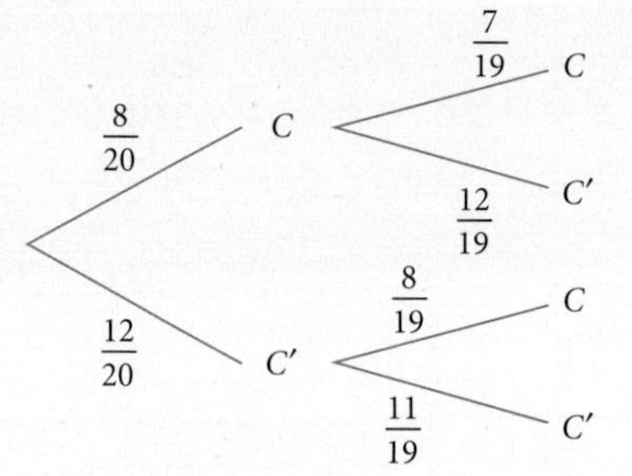
3 Identify the branches where there is one multicoloured ball and one ball that is not multicoloured.	$P(1 \text{ multicoloured}) = P(CC') + P(C'C)$
4 Multiply the probabilities along the branches and add the products.	$P(1 \text{ multicoloured}) = \frac{8}{20} \times \frac{12}{19} + \frac{12}{20} \times \frac{8}{19} = \frac{48}{95}$

CAS can be used to calculate the probabilities when selections are done **without replacement**. This method uses combinations ${}^nC_r$ or $\binom{n}{r}$, which gives the number of ways of choosing r objects from n objects.

5.1

USING CAS 1 Selection without replacement probabilities

A bag contains 8 black discs and 7 white discs. If 3 discs are chosen from the bag, without replacement, find the probability of choosing exactly 2 black discs.

Two black discs must be chosen from 8 black discs in the bag and one white disc must be chosen from 7 white discs in the bag.

Combinations can be used to find the number of ways this can occur.

$$n(2\text{B}, 1\text{W}) = \binom{8}{2} \times \binom{7}{1}$$

The total number of ways of choosing 3 discs from 15 discs is $\binom{15}{3}$.

$$\text{P}(2\text{B},1\text{W}) = \frac{\binom{8}{2} \times \binom{7}{1}}{\binom{15}{3}}$$

ClassPad **TI-Nspire**

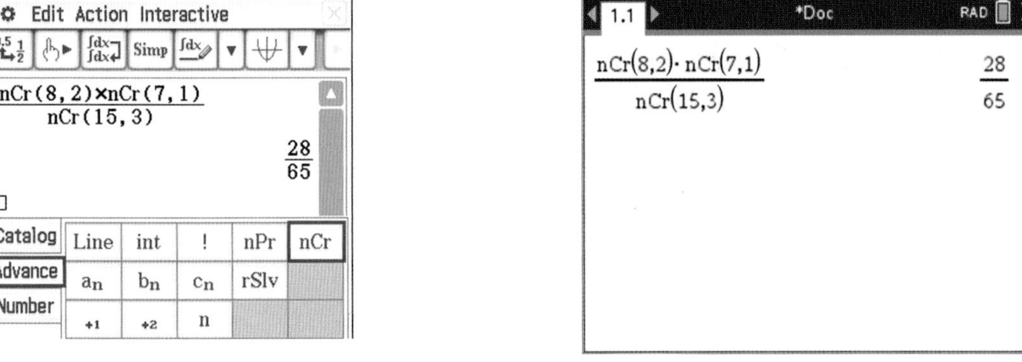

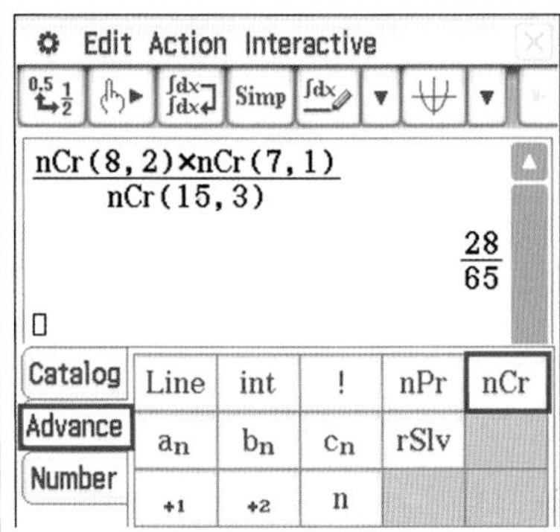

ClassPad

1. Open the soft Keyboard and tap the downward arrow.
2. Tap **Advance** and find the **nCr** symbol.
3. Enter the combinations and values as shown above.

TI-Nspire

1. Press **menu** > **Probability** > **Combinations**.
2. Enter the combinations and values as shown above.

The probability is $\frac{28}{65}$.

Arrays

An **array**, **grid** or lattice is a set of numbers arranged in row and column format. It can be useful to show the outcomes of a two-step experiment. An array is a better choice than a tree diagram to represent situations where two simple events would result in a large number of branches.

WORKED EXAMPLE 2 Finding probabilities using an array

The spinner shown is spun twice.

a Illustrate the sample space using an array.

b Use this diagram to find the probability that

i the first spin is even and the second spin is a 5

ii the sum of the two numbers is 8.

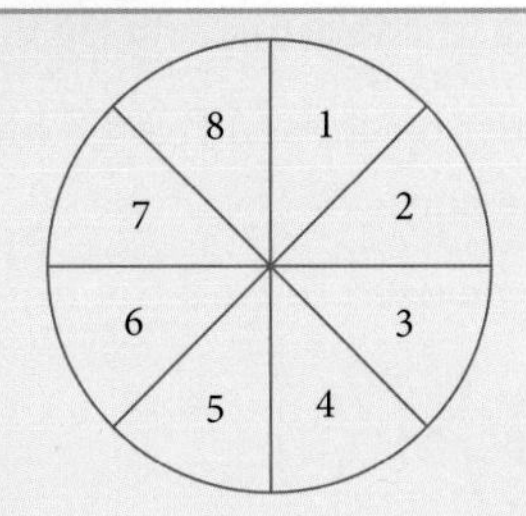

Steps	Working
a Draw an 8 × 8 grid and let the columns represent the first spin and the rows represent the second spin.	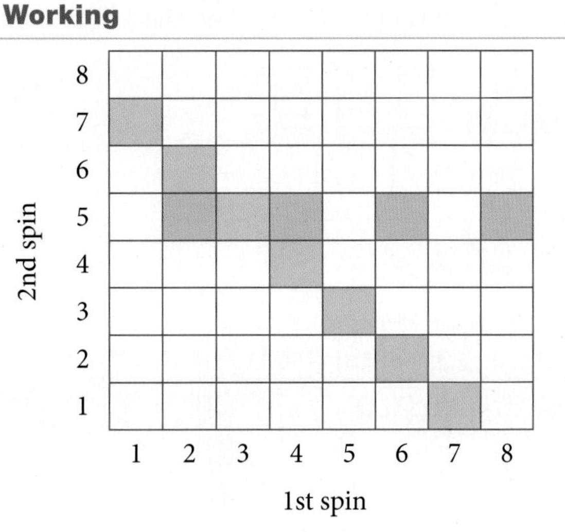
b **i** Identify the four cells in the array where the first number is even and the second number is a 5. Use this to calculate the probability. These cells are shaded blue in the grid.	$\text{P(even, 5)} = \frac{4}{64} = \frac{1}{16}$
ii Identify the seven cells in the array where the first and second numbers sum to 8. The cells in which the sum is 8 are shaded green.	$\text{P(sum of 8)} = \frac{7}{64}$

Independent events

Two events A and B are **independent** if one event does not influence the outcome of the other event. For example, where event A is Fremantle Dockers winning a football match at home and event B is West Coast Eagles winning a different football match in Adelaide.

If $P(A) = 0.6$ and $P(B) = 0.5$ then the probability Fremantle Dockers AND West Coast Eagles both win can be found using the multiplication rule.

$P(A \cap B) = 0.6 \times 0.5 = 0.3.$

Independent events

For two **independent events** A and B, $P(A \cap B) = P(A) \times P(B)$.

WORKED EXAMPLE 3	**Finding probabilities for independent events**
Jordyn is training to improve the accuracy of her tennis serve by attempting to hit a can on the centre line. The probability of hitting the can on any attempt is 0.6 and her success on any serve is independent of the result on the previous serve. Find the probability Jordyn hits the can on two out of three attempts.	
Steps	**Working**
1 Let event H represent Jordyn hits the can and event M represent Jordyn misses the can. List the different ways of getting two hits and one miss.	two hits = $\{H, H, M\}$ or $\{H, M, H\}$ or $\{M, H, H\}$
2 The three serves are independent so multiply the probabilities for each combination of three shots.	$\text{P}(H, H, M) = 0.6 \times 0.6 \times 0.4 = 0.144$ $\text{P}(H, M, H) = 0.6 \times 0.4 \times 0.6 = 0.144$ $\text{P}(M, H, H) = 0.4 \times 0.6 \times 0.6 = 0.144$
3 Add the probabilities for each possibility where there are two hits.	P(2 hits and 1 miss) $= 0.144 + 0.144 + 0.144 = 0.432$

The addition rule for probability

$\text{P}(A \cup B)$ means the probability of events A or B or both occurring.

The addition rule for probability

The **addition rule for probability** is

$$\text{P}(A \cup B) = \text{P}(A) + \text{P}(B) - \text{P}(A \cap B)$$

WORKED EXAMPLE 4	**Finding probabilities using the addition rule**
If $\text{P}(A \cup B) = \frac{2}{5}$, $\text{P}(A \cap B) = \frac{1}{5}$ and $\text{P}(A) = 2 \times \text{P}(B)$, find $\text{P}(A)$.	
Steps	**Working**
1 Let $\text{P}(B) = b$. Write $\text{P}(A)$ in terms of b.	Let $\text{P}(B) = b$. $\text{P}(A) = 2 \times \text{P}(B) = 2b$
2 Substitute into the addition rule.	$\text{P}(A \cup B) = \text{P}(A) + \text{P}(B) - \text{P}(A \cap B)$ $\frac{2}{5} = b + 2b - \frac{1}{5}$ $\frac{3}{5} = 3b$ $b = \frac{1}{5}$
3 Substitute the value of b into $\text{P}(A) = 2b$.	$\text{P}(A) = 2 \times \frac{1}{5} = \frac{2}{5}$

Conditional probability

The probability of an event occurring if another event has also occurred is a **conditional probability**. In these situations, the sample space is reduced because of the condition placed on the probability.

Conditional probability

$$P(A|B) = \frac{P(A \cap B)}{P(B)}$$

where $P(A|B)$ means 'the probability of A occurring, given B occurs'.

If events A and B are independent, then $P(A|B) = P(A)$.

WORKED EXAMPLE 5 Finding conditional probabilities

Farmer Bob owns a tractor, a motorcycle and a utility. Each morning he starts each vehicle but they do not always start on the first attempt. The respective probabilities of each starting on the first attempt are 0.2, 0.7 and 0.6. On Wednesday morning, two vehicles started on the first attempt. Find the probability the two that started were the tractor and the utility.

Steps	Working
1 Write the conditional probability in the form $P(A\|B) = \frac{P(A \cap B)}{P(B)}$.	$P(\text{tractor and utility}\|\text{two starts})$ $= \frac{P(\text{tractor and utility start AND two starts})}{P(\text{two starts})}$
2 Find the probability that the tractor and the utility start but the motorcycle does not start. Let T = tractor starts M = motorcycle starts and U = utility starts.	$P(T) = 0.2$, $P(M) = 0.7$, $P(U) = 0.6$ $P(M') = 1 - 0.7$ $= 0.3$ $P(T, M', U) = 0.2 \times 0.3 \times 0.6$ $= 0.036$
3 Find the probability that two vehicles start, that is, $P(T, M, U') + P(T, M', U) + P(T', M, U)$.	$P(\text{two starts})$ $= P(T, M, U') + P(T, M', U) + P(T', M, U)$ $= 0.2 \times 0.7 \times 0.4 + 0.036 + 0.8 \times 0.7 \times 0.6$ $= 0.428$
4 Substitute into the conditional probability formula.	$P(\text{tractor and utility}\|\text{two starts}) = \frac{0.036}{0.428}$ $= \frac{9}{107}$

EXERCISE 5.1 Review of probability

ANSWERS p. 397

Mastery

1 WORKED EXAMPLE 1 A bag contains 18 tulip bulbs and 12 daffodil bulbs. Three bulbs are selected *without replacement*. Find the probability of selecting three bulbs of the same type.

2 Using CAS 1 A bag contains fifteen red marbles and five green marbles. If four marbles are chosen from the bag, *without replacement*, find the probability of choosing three green marbles.

3 WORKED EXAMPLE 2 A tetrahedral die, with four faces numbered 2, 4, 6 and 8, is rolled twice and the number facedown is noted.

a Illustrate the sample space using an array.

b Use this diagram to find the probability that

i the same number occurs on both rolls

ii the sum of the two numbers is 10.

4 WORKED EXAMPLE 3 Emily exercises each morning before school. She will either go for a run or a swim. The probability that she will go for a run on any morning is 0.3 and the type of exercise she does on any morning is independent of the exercise done on the previous morning. Find the probability that Emily will go for a run on either Monday or Tuesday.

5 WORKED EXAMPLE 4 If $P(A \cup B) = 0.7$, $P(A \cap B) = 0.05$ and $P(B) = 4 \times P(A)$, find $P(A)$.

6 WORKED EXAMPLE 5 Four identical balls are numbered 1, 2, 3 and 4 and put into a box. A ball is randomly drawn from the box, and not returned to the box. A second ball is then randomly drawn from the box. Find the probability that the first number drawn is numbered 2 if it is known that the sum of the numbers is at least 4.

Calculator-free

7 (4 marks) Two boxes each contain four stones that differ only in colour. Box 1 contains four black stones and Box 2 contains two black stones and two white stones. A box is chosen randomly, and one stone is drawn randomly from it. Each box is equally likely to be chosen, as is each stone.

a What is the probability that the randomly drawn stone is black? (2 marks)

b It is not known from which box the stone has been drawn. Given that the stone that is drawn is black, what is the probability that it was drawn from Box 1? (2 marks)

8 (3 marks) The only possible outcomes when a coin is tossed are a head or a tail. When an unbiased coin is tossed, the probability of tossing a head is the same as the probability of tossing a tail. Jo has three coins in her pocket; two are unbiased and one is biased. When the biased coin is tossed, the probability of tossing a head is $\frac{1}{3}$. Jo randomly selects a coin from her pocket and tosses it.

a Find the probability that she tosses a head. (2 marks)

b Find the probability that she selected an unbiased coin, given that she tossed a head. (1 mark)

9 (6 marks) Sally aims to walk her dog, Mack, most mornings. If the weather is pleasant, the probability that she will walk Mack is $\frac{3}{4}$ and if the weather is unpleasant, the probability that she will walk Mack is $\frac{1}{3}$.

Assume that pleasant weather on any morning is independent of pleasant weather on any other morning. In a particular week, the weather was pleasant on Monday morning and unpleasant on Tuesday morning.

a Find the probability that Sally walked Mack on at least one of these two mornings. (2 marks)

b **i** In the month of April, the probability of pleasant weather in the morning was $\frac{5}{8}$. Find the probability that on a particular morning in April, Sally walked Mack. (2 marks)

ii Using your answer from part **b i**, or otherwise, find the probability that on a particular morning in April, the weather was pleasant, given that Sally walked Mack that morning. (2 marks)

10 (3 marks) A company produces motors for refrigerators. There are two assembly lines, Line A and Line B. 5% of the motors assembled on Line A are faulty and 8% of the motors assembled on Line B are faulty. In one hour, 40 motors are produced from Line A and 50 motors are produced from Line B. At the end of an hour, one motor is selected at random from all the motors that have been produced during that hour.

a What is the probability that the selected motor is faulty? Express your answer in the form $\frac{1}{b}$, where b is a positive integer. (2 marks)

b The selected motor is found to be faulty. What is the probability that it was assembled on Line A? Express your answer in the form $\frac{1}{c}$, where c is a positive integer. (1 mark)

11 (3 marks) An online shopping site sells boxes of doughnuts.

A box contains 20 doughnuts. There are only four types of doughnuts in the box. They are:

- glazed, with custard
- glazed, with no custard
- not glazed, with custard
- not glazed, with no custard.

It is known that, in the box:

- $\frac{1}{2}$ of the dougnuts are with custard
- $\frac{7}{10}$ of the doughnuts are not glazed
- $\frac{1}{10}$ of the doughnuts are glazed, with custard.

a A doughnut is chosen at random from the box. Find the probability that it is not glazed, with custard. (1 mark)

b The 20 doughnuts in the box are randomly allocated to two new boxes, Box A and Box B. Each new box contains 10 doughnuts. One of the two new boxes is chosen at random and then a doughnut from that box is chosen at random.

Let g be the number of glazed doughnuts in Box A. Find the probability, in terms of g, that the doughnut comes from Box B given that it is glazed. (2 marks)

Calculator-assumed

12 © SCSA MM2021 Q10b (2 marks) A charity organisation has printed 'Lucky 7' scratchie tickets as a fundraiser for use at two special events. The tickets contain two panels. Each ticket has the same numbers as the sample ticket shown below, arranged randomly and hidden within each panel.

A player scratches one section of each panel to reveal a number. The two numbers revealed are then added together. If the total is 7 or higher, the player wins a prize.

At the first event, 400 tickets are purchased, and a prize is won on 124 occasions.
Let p denote the probability that a prize is won. Show that the probability p of winning a prize is $\frac{7}{24}$.

13 (3 marks) A box contains five red marbles and three yellow marbles. Two marbles are drawn at random from the box without replacement. Find the probability that the marbles are

a different colours (2 marks)

b the same colour. (1 mark)

14 (3 marks) Demelza is a badminton player. If she wins a game, the probability that she will win the next game is 0.7. If she loses a game, the probability that she will lose the next game is 0.6. Demelza has just won a game.

Find the probability that

a she wins her next two games (1 mark)

b she wins exactly one of her next two games. (2 marks)

5.2 Discrete probability distributions

Video playlist
Discrete probability distributions

Worksheets
Random variables

Discrete probability distributions 1

Discrete probability distributions 2

A **random variable,** X, is a variable whose value is determined by the outcome of a random experiment. There are two types of random variables: **discrete** and **continuous**.

A **discrete random variable** can only take a countable number of values, for example, the number of pets in a household.

A **continuous random variable** can take any value in a given interval, for example, the height of students in a class.

If a coin is tossed three times and the number of heads is recorded, then the discrete random variable X represents the number of heads in three tosses. The values that this variable can take is represented by x, where $x = 0, 1, 2, 3$.

The **probability distribution** of a discrete random variable shows all the possible values of the variable and the probabilities associated with those values. This function can be represented numerically as a table, in graphical form or as a formula.

Let $p(x)$ or $P(X = x)$ be the probability of obtaining x heads in three tosses of a coin.

The probability distribution of X is

x	0	1	2	3
$P(X = x)$	$\frac{1}{8}$	$\frac{3}{8}$	$\frac{3}{8}$	$\frac{1}{8}$

Properties of a probability distribution

- All probabilities must be between 0 and 1 inclusive: $0 \le p(x) \le 1$.
- The sum of all probabilities must equal 1: $\Sigma p(x) = 1$.

The probability distribution for a uniform discrete random variable

A discrete uniform distribution is one in which all the outcomes have the same probability of occurring. A simple example of a uniform distribution is rolling a die. The discrete random variable X represents the number rolled.

The probability distribution of X is

x	1	2	3	4	5	6
$P(X = x)$	$\frac{1}{6}$	$\frac{1}{6}$	$\frac{1}{6}$	$\frac{1}{6}$	$\frac{1}{6}$	$\frac{1}{6}$

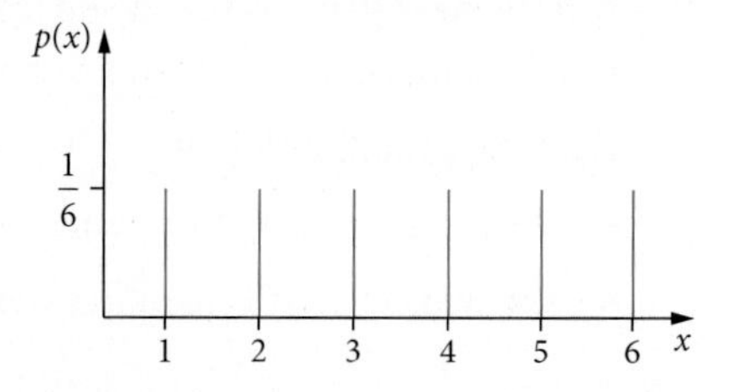

If X is a uniform discrete random variable where $x = 1, 2, 3, \dots N$ then $P(X = x) = \frac{1}{N}$.

WORKED EXAMPLE 6 **Finding the probability distribution of a uniform discrete random variable**

A disc is randomly selected from a bag that contains five discs numbered 1, 2, 3, 4 and 5. The random variable X represents the number selected.

a Identify the probability distribution of X.

b Complete the probability distribution of X.

x						
$P(X = x)$						

Steps	Working
a A discrete uniform distribution is one in which all the outcomes have the same probability of occurring.	This is a discrete uniform distribution.
b There are five outcomes and each has the same probability of $\frac{1}{5}$ of occurring.	see table below

x	1	2	3	4	5
$P(X = x)$	$\frac{1}{5}$	$\frac{1}{5}$	$\frac{1}{5}$	$\frac{1}{5}$	$\frac{1}{5}$

Non-uniform distributions

Many discrete random variables do not have uniform distributions. Consider the situation where we toss a coin twice and the discrete random variable X represents the number of heads that occur.

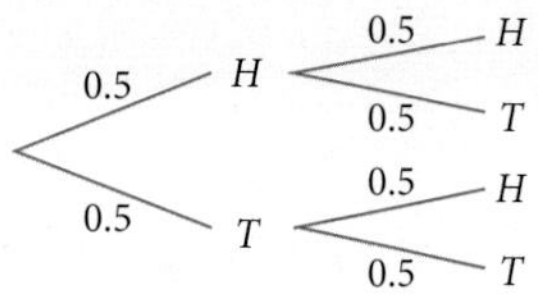

Tree diagram: first toss H (0.5), T (0.5); from H: H (0.5), T (0.5); from T: H (0.5), T (0.5).

x	0	1	2
$P(X = x)$	$\frac{1}{4}$	$\frac{1}{2}$	$\frac{1}{4}$

The probability of rolling one head is not the same as the probability of rolling no heads or the probability of rolling two heads. This discrete random variable does not have a uniform distribution.

WORKED EXAMPLE 7 Finding the probability distribution of a non-uniform discrete random variable

A bag contains ten marbles of which six are red. Two marbles are taken from the bag. If X represents the number of red marbles taken, list the probability distribution of X.

x	0	1	2
$P(X = x)$			

Steps	Working
1 Draw a tree diagram to illustrate the problem.	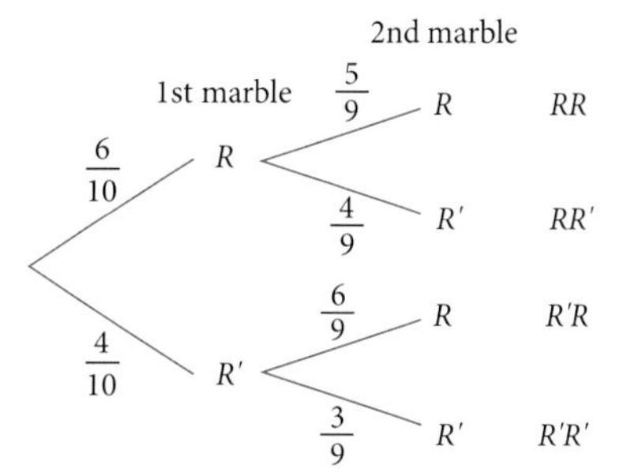

2 Use the tree diagram to find the probabilities.

$$P(X = 0) = \frac{4}{10} \times \frac{3}{9} = \frac{2}{15}$$

$$P(X = 1) = \frac{6}{10} \times \frac{4}{9} + \frac{4}{10} \times \frac{6}{9} = \frac{8}{15}$$

$$P(X = 2) = \frac{6}{10} \times \frac{5}{9} = \frac{1}{3}$$

3 Write the probabilities in the table.

The probability distribution of X is

x	0	1	2
$P(X = x)$	$\frac{2}{15}$	$\frac{8}{15}$	$\frac{1}{3}$

The probabilities could also be calculated using combinations:

$$P(X = 1) = \frac{\binom{4}{1} \times \binom{6}{1}}{\binom{10}{2}} = \frac{8}{15}$$

 Exam hack

Always check that the sum of all probabilities is 1.

WORKED EXAMPLE 8 Finding a probability for a given probability distribution

The probability distribution of a discrete random variable, X, is given by the table below.

Find the value of p.

x	1	2	3	4
$P(X = x)$	$p^2 - 0.1$	p	$p^2 - 0.1$	0.2

Steps	Working
1 As this is a probability distribution, the sum of the probabilities is one.	$\Sigma p(x) = 1$ $p^2 - 0.1 + p + p^2 - 0.1 + 0.2 = 1$ $2p^2 + p - 1 = 0$
2 Factorise and solve for p.	$(2p - 1)(p + 1) = 0$ $p = -1$ or $\frac{1}{2}$
3 As p is a probability, $p \geq 0$.	$p = \frac{1}{2}$

WORKED EXAMPLE 9 Compound event probabilities

The Mercury newspaper publishes three puzzles every day. John attempts these puzzles each day and has calculated the probability of solving 0, 1, 2 or 3 puzzles. The probability distribution of the number of puzzles solved X, is given in the table below.

Find the probability that John solves the same number of puzzles on Monday and Tuesday.

x	0	1	2	3
$P(X = x)$	0.4	0.3	0.2	0.1

Steps	Working
1 List the possible options where John solves the same number of puzzles on two days. P(0, 0) is the probability of 0 puzzles solved on Monday and on Tuesday.	P(same on Mon and Tues) $= P(0,0) + P(1,1) + P(2,2) + P(3,3)$
2 Find the probabilities from the table and multiply the probabilities for each pair.	$= 0.4 \times 0.4 + 0.3 \times 0.3 + 0.2 \times 0.2 + 0.1 \times 0.1$ $= 0.16 + 0.09 + 0.04 + 0.01$ $= 0.3$

EXERCISE 5.2 Discrete probability distributions

ANSWERS p. 397

Recap

1 A box contains four red marbles and two yellow marbles. Two marbles are drawn at random from the box without replacement. The probability that the marbles are the same colour is

A $\frac{7}{30}$ **B** $\frac{2}{5}$ **C** $\frac{7}{15}$ **D** $\frac{8}{15}$ **E** $\frac{5}{9}$

2 Two cubes are randomly selected, with replacement, from a bag that contains seven cubes numbered 1, 2, 3, 4, 5, 6 and 7. The discrete random variable X is the sum of the two selected cubes. Find

a $P(X = 5)$ **b** $P(X \geq 5)$

Mastery

3 WORKED EXAMPLE 6 A disc is randomly selected from a bag that contains eight discs numbered 1, 2, 3, 4, 5, 6, 7 and 8. The random variable X represents the number selected.

a Identify the probability distribution of X.

b List the probability distribution of X.

4 WORKED EXAMPLE 7 A coin is biased so that the probability of tossing a tail is $\frac{2}{5}$. List the probability distribution of the number of tails that are obtained on two tosses of the coin.

5 Julie has a bag that contains three blue discs and five yellow discs. She selects two discs, without replacement. The discrete random variable, X, represents the number of yellow discs selected. List the probability distribution of X.

6 WORKED EXAMPLE 8 The probability distribution of a discrete random variable, X, is given by the table below.

x	0	1	2	3
$\mathrm{P}(X = x)$	p	$0.3p$	$p^2 - 0.2$	$p^2 - 0.3$

Find the value of p.

7 The probability distribution function of a discrete random variable, X, is given by the function

$f(x) = kx, x = 1, 2, 3, 4.$

Find the value of k.

8 WORKED EXAMPLE 9 Emma drives through two intersections with traffic lights on her way to work. The probability distribution of the number of green lights that Emma gets on any trip is given below.

x	0	1	2
$\mathrm{P}(X = x)$	0.5	0.3	0.2

Find the probability that on two successive trips Emma gets a total of two green lights.

Calculator-free

9 (3 marks) The probability distribution of a discrete random variable, X, is given by the table below.

x	0	1	2	3	4
$\mathrm{P}(X = x)$	0.2	$0.6p^2$	0.1	$1 - p$	0.1

Show that $p = \frac{2}{3}$ or $p = 1$.

10 (3 marks) The discrete random variable, X, has the probability distribution given by the table below.

x	−1	0	1	2
$\mathrm{P}(X = x)$	p^2	p^2	$\frac{p}{4}$	$\frac{4p+1}{8}$

Find the value of p.

11 (2 marks) Jane drives to work each morning and passes through three intersections with traffic lights. The number, X, of traffic lights that are red when Jane is driving to work is a random variable with probability distribution given by

x	0	1	2	3
$P(X = x)$	0.1	0.2	0.3	0.4

Jane drives to work on two consecutive days. What is the probability that the number of traffic lights that are red is the same on both days?

12 (4 marks) On any given day, the number, X, of telephone calls that Daniel receives is a random variable with probability distribution given by

x	0	1	2	3
$P(X = x)$	0.2	0.2	0.5	0.1

a What is the probability that Daniel receives only one telephone call on each of three consecutive days? (1 mark)

b Daniel receives telephone calls on both Monday and Tuesday. What is the probability that Daniel receives a total of four calls over these two days? (3 marks)

Calculator-assumed

13 © SCSA MM2016 Q15abii (4 marks) A tetrahedral die has the numbers 1 to 4 on each face. When thrown, each side is equally likely to land facedown. Let X be defined as the sum of the numbers on the facedown side when the die is thrown twice.

a Copy and complete the following table. (1 mark)

		Roll two			
	Sum of two rolls	1	2	3	4
Roll one	1	$1 + 1 = 2$	3		
	2	3			
	3		5		
	4				

b **i** Hence, or otherwise, complete the probability distribution of X, which is given by the following table. (1 mark)

x	2	3	4	5	6	7	8
$P(X = x)$	$\frac{1}{16}$						$\frac{1}{16}$

ii Calculate the probability of obtaining a sum of five or less. (2 marks)

5.3 Measures of centre and spread

The three **measures of centre** of a probability distribution are the **mean**, the **mode** and the **median**. They measure the expected or likely outcome for the distribution. The mean is the only measure of centre covered in this course.

The expected value (mean)

The mean of a discrete random variable is called the **expected value**. For a discrete random variable, X, the expected value is written in mathematical notation as $E(X)$ or μ.

The expected value is found by summing the product of x and $P(X = x)$ for all possible values of x.

Remember that $p(x)$ is a short way of writing $P(X = x)$.

The expected value of a discrete probability distribution

$$\mu = E(X) = \Sigma x \cdot p(x)$$

Video playlist
Measures of centre and spread

Worksheets
Expected values

Expected values using a calculator

Using expected values

Expected value

The effects of linear changes of scale and origin on the expected value of $aX + b$

The discrete random variable X represents the number selected at random from the set $\{1, 2, 3\}$. The probability distribution of X is

x	1	2	3
$p(x)$	$\frac{1}{3}$	$\frac{1}{3}$	$\frac{1}{3}$

The discrete random variable X has a mean $\mu = E(X) = 2$.

Change the scale

If we change the scale and multiply all these values by 2, then the distribution of $2X$ is

$2x$	2	4	6
$p(x)$	$\frac{1}{3}$	$\frac{1}{3}$	$\frac{1}{3}$

The discrete random variable $2X$ has a mean $\mu = E(2X) = 2 \times 2 = 4$.

Change the origin

If we now add 5 to all the $2x$ values then the distribution of $2X + 5$ is

$2x + 5$	7	9	11
$p(x)$	$\frac{1}{3}$	$\frac{1}{3}$	$\frac{1}{3}$

The discrete random variable $2X + 5$ has a mean $\mu = E(2X + 5) = 2 \times 2 + 5 = 9$.

Expected value of $aX + b$

For a linear function $aX + b$ of a discrete random variable X:

$$E(aX + b) = aE(X) + b$$

WORKED EXAMPLE 10 Finding the expected value from a discrete probability distribution

The probability distribution of a discrete random variable X is shown below.

x	0	1	2	3
$P(X = x)$	0.25	0.3	0.2	0.25

Find

a the mean of X

b $E(10X + 20)$

c $P(X \leq \mu)$.

Steps

a **1** Rewrite the table with rows as columns and add a column headed $x \times p(x)$. Calculate the product of the x values and their probabilities.

2 The sum of the $x \times p(x)$ column is the expected value of X or $E(X)$.

Working

x	$p(x)$	$x \times p(x)$
0	0.25	0
1	0.3	0.3
2	0.2	0.4
3	0.25	0.75
Total		1.45

$\mu = E(X) = \Sigma x \cdot p(x)$

The mean or $E(X)$ is 1.45.

b Substitute into the formula:

$E(aX + b) = aE(X) + b$

$$\begin{aligned} E(10X + 20) &= 10E(X) + 20 \\ &= 10 \times 1.45 + 20 \\ &= 34.5 \end{aligned}$$

c Identify the x values in the table that are less than or equal to 1.45 and add their probabilities.

$$\begin{aligned} P(X \leq \mu) &= P(X \leq 1) \\ &= P(X = 0) + P(X = 1) \\ &= 0.25 + 0.3 \\ &= 0.55 \end{aligned}$$

WORKED EXAMPLE 11 Finding probabilities given the expected value of a discrete probability distribution

The probability distribution of a discrete random variable X is shown below.

x	1	2	3	4
$P(X = x)$	a	0.1	0.2	b

The mean of the distribution is 2.4. Find the values of a and b.

Steps

1 Rewrite the table with rows as columns and add a column headed $x \times p(x)$.

Working

x	$p(x)$	$x \times p(x)$
1	a	a
2	0.1	0.2
3	0.2	0.6
4	b	$4b$
Total		

2 Calculate the product of the x values and their probabilities.

The sum of the $p(x)$ column must be 1.

The sum of the $x \times p(x)$ column must be 2.4.

x	$p(x)$	$x \times p(x)$
1	a	a
2	0.1	0.2
3	0.2	0.6
4	b	$4b$
Total	1.0	2.4

The mean E(X) is 2.4.

3 Write the equations for the total of the $p(x)$ and $x \times p(x)$ columns.

$a + 0.1 + 0.2 + b = 1$

$a + b = 0.7 \quad [1]$

$a + 0.2 + 0.6 + 4b = 2.4$

$a + 4b = 1.6 \quad [2]$

4 Solve the simultaneous equations [1] and [2] to find the values of a and b.

[2] – [1] gives

$3b = 0.9$

$b = 0.3$

Substitute in [1]:

$a + 0.3 = 0.7$

$a = 0.4$

So, $a = 0.4$, $b = 0.3$.

The variance and standard deviation

Worksheet Variance and standard deviation

The most useful measures of spread for a probability distribution are the **variance** and the **standard deviation**.

The variance and standard deviation of a discrete random variable, X, measure the spread of the variable about its mean.

$$\text{Var}(X) = \sigma^2 = \Sigma(x - \mu)^2 p(x)$$

$$\text{SD}(X) = \sqrt{\text{Var}(X)}$$

There is another formula, called the computational formula, that can be used to find the variance. The computational formula for variance can be found algebraically.

$$\sum(x - \mu)^2 p(x) = \text{E}((X - \mu)^2)$$
$$= \text{E}(X^2 - 2\mu X + \mu^2)$$

Using $\text{E}(aX + b) = a\text{E}(X) + b$, this expression can be simplified to

$$= \text{E}(X^2) - \text{E}(2\mu X) + \text{E}(\mu^2)$$
$$= \text{E}(X^2) - 2\mu\text{E}(X) + \mu^2$$
$$= \text{E}(X^2) - 2\mu^2 + \mu^2$$
$$= \text{E}(X^2) - \mu^2$$

Another form of the variance formula is $\text{Var}(X) = \text{E}(X^2) - \mu^2$.

The variance formula

A simpler computational formula for the variance is:

$$\text{Var}(X) = \sigma^2 = \text{E}(X^2) - \mu^2$$

where $\text{E}(X^2) = \Sigma x^2 \cdot p(x)$ and $\mu = \text{E}(X)$.

standard deviation $\text{SD}(X) = \sqrt{\text{Var}(X)}$

Standard deviation is the square root of the variance.

The symbols for the variance and standard deviation are $\sigma^2 = \text{Var}(X)$ and $\sigma = \text{SD}(X)$.

Exam hack

The formula sheet has the variance formula given as:

$$\text{Var}(X)\text{: } \sigma^2 = \Sigma(x - \mu)^2 p(x)$$

The computational formula

$$\text{Var}(X)\text{: } \sigma^2 = \text{E}(X^2) - \mu^2$$

is more useful when calculating the variance of a discrete distribution where the mean μ is a fraction or decimal.

WORKED EXAMPLE 12 **Finding the variance and standard deviation of a discrete random variable X, using the computational formula: $\text{Var}(X)\text{: } \sigma^2 = \text{E}(X^2) - \mu^2$**

For the probability distribution of a discrete random variable X, find the

x	0	1	2	3	4
$p(x)$	0.4	0.1	0.1	0.2	0.2

a expected value

b variance

c standard deviation, correct to three decimal places.

Steps	Working
a 1 Add two extra columns headed $x \times p(x)$ and $x^2 \times p(x)$ for calculating $\text{E}(X)$ and $\text{E}(X^2)$ respectively.	
2 Multiply x by $p(x)$, enter the results in the $x \times p(x)$ column and find the total.	
3 Multiply x by $x \times p(x)$, enter the results in the $x^2 \times p(x)$ column and find the total.	
4 The total of the $x \times p(x)$ column is $\text{E}(X)$.	$\text{E}(X) = \Sigma x \cdot p(x) = 1.7$
b 1 The total of the $x^2 \times p(x)$ column is $\text{E}(X^2)$.	$\text{E}(X^2) = 5.5$
2 Use the computational formula to find $\text{Var}(X)$.	$\text{Var}(X) = \text{E}(X^2) - \mu^2$ $\text{Var}(X) = 5.5 - 1.7^2$ $= 2.61$
c Find the standard deviation using the formula: $\text{SD}(X) = \sqrt{\text{Var}(X)}$	$\text{SD}(X) = \sqrt{2.61} \approx 1.616$

x	$p(x)$	$x \times p(x)$	$x^2 \times p(x)$
0	0.4	0	0
1	0.1	0.1	0.1
2	0.1	0.2	0.4
3	0.2	0.6	1.8
4	0.2	0.8	3.2
Total	1.0	$\text{E}(X) = 1.7$	$\text{E}(X^2) = 5.5$

WORKED EXAMPLE 13 Finding the variance and standard deviation of a discrete random variable X, using the formula: $\text{Var}(X)$: $\sigma^2 = \Sigma(X - \mu)^2 p(x)$

Find the variance of the discrete random variable X.

x	0	1	2	3
$p(x)$	0.5	0.2	0.1	0.2

Steps

1. Add two extra columns headed $x \times p(x)$ and $(x - \mu)^2 \times p(x)$ for calculating $E(X)$ and $\text{Var}(X)$ respectively.
2. Multiply x by $p(x)$, enter the results in the $x \times p(x)$ column and find the total. The mean $\mu = 1$, which makes it easier to use the formula $\sigma^2 = \Sigma(x - \mu)^2 p(x)$.
3. Subtract $E(X)$ from x, square, then multiply the result by $p(x)$, enter the results in the $(x - \mu)^2 \times p(x)$ column and find the total.
4. The total of the $x \times p(x)$ column is the mean and the total of $(x - \mu)^2 \times p(x)$ is the variance.

Working

x	$p(x)$	$x \times p(x)$	$(x - \mu)^2 \times p(x)$
0	0.5	0	$(-1)^2 \times 0.5 = 0.5$
1	0.2	0.2	$(0)^2 \times 0.2 = 0$
2	0.1	0.2	$(1)^2 \times 0.1 = 0.1$
3	0.2	0.6	$(2)^2 \times 0.2 = 0.8$
		$\mu = 1$	$\sigma^2 = 1.4$

$\text{Var}(X)$: $\sigma^2 = 1.4$

The effects of linear changes of scale and origin on the variance of $aX + b$

Earlier in this exercise we examined the effects of linear changes of scale and origin on the mean of $aX + b$ for the probability distribution of a discrete random variable X. We will now look at how these changes affect the variance of a distribution.

The probability distribution of the discrete random variable X is

x	1	2	3
$p(x)$	$\frac{1}{3}$	$\frac{1}{3}$	$\frac{1}{3}$

The variance is the average of the squared deviations from the mean.

The discrete random variable X has a variance $\sigma^2 = \frac{2}{3}$.

Change the scale

If we change the scale and multiply all these values by 2, then the distribution of $2X$ is

$2x$	2	4	6
$p(x)$	$\frac{1}{3}$	$\frac{1}{3}$	$\frac{1}{3}$

When we multiply the x values by 2 the deviations from the mean also multiply by 2 and the squared deviations multiply by $2^2 = 4$.

The discrete random variable $2X$ has a variance $\sigma^2 = \frac{2}{3} \times 2^2 = \frac{8}{3}$.

Change the origin

If we now add 5 to all the $2x$ values, then the distribution of $2X + 5$ is

$2x + 5$	7	9	11
$p(x)$	$\frac{1}{3}$	$\frac{1}{3}$	$\frac{1}{3}$

When we add 5 to all the $2x$ values, the spread of the $2x$ values will not change.

The discrete random variable $2X + 5$ also has a variance $\sigma^2 = \frac{2}{3} \times 2^2 = \frac{8}{3}$.

Variance of $aX + b$

For a linear function $aX + b$ of a discrete random variable X:

$$\text{Var}(aX + b) = a^2\,\text{Var}(X)$$

WORKED EXAMPLE 14 **Finding the expected value and variance of $aX + b$**

The scores X, on a fitness test at school, are a discrete random variable. The variable X has an expected value of 12.5 and a variance of 2.5.

Find

a $\text{E}(3X + 4)$

b $\text{Var}(3X + 4)$.

The scores need to be rescaled to be compatible with other classes, so the teacher applies the scaling $Y = aX + b$, where a and b are positive constants, so that the mean is now 80 and the variance 40.

c Find the values of a and b.

Steps	**Working**
a Use the formula: $\text{E}(aX + b) = a\,\text{E}(X) + b$	$\text{E}(3X + 4) = 3\,\text{E}(X) + 4$ $= 3 \times 12.5 + 4$ $= 41.5$
b Use the formula: $\text{Var}(aX + b) = a^2\,\text{Var}(X)$	$\text{Var}(3X + 4) = 3^2 \times \text{Var}(X)$ $= 9 \times 2.5$ $= 22.5$
c **1** Substitute $\text{E}(aX + b) = 80$ into $\text{E}(aX + b) = a\text{E}(X) + b$ and $\text{Var}(aX + b) = 40$ into $\text{Var}(aX + b) = a^2\,\text{Var}(X)$.	$\text{E}(aX + b) = a\text{E}(X) + b$ $80 = 12.5a + b$ [1] $\text{Var}(aX + b) = a^2\,\text{Var}(X)$ $40 = 2.5a^2$ [2]
2 Solve the simultaneous equations.	From [2]: $a^2 = 16$ $a = 4$ Substitute into [1]: $12.5 \times 4 + b = 80$ $50 + b = 80$ $b = 30$

CAS can be used to find the expected value, the variance and the standard deviation of a random variable X, given its probability distribution.

5.3

USING CAS 2 Expected value, variance and standard deviation

Using the results from the following table, find the mean, variance and standard deviation of X.

x	0	1	2	3
$P(X = x)$	0.4	0.25	0.2	0.15

ClassPad

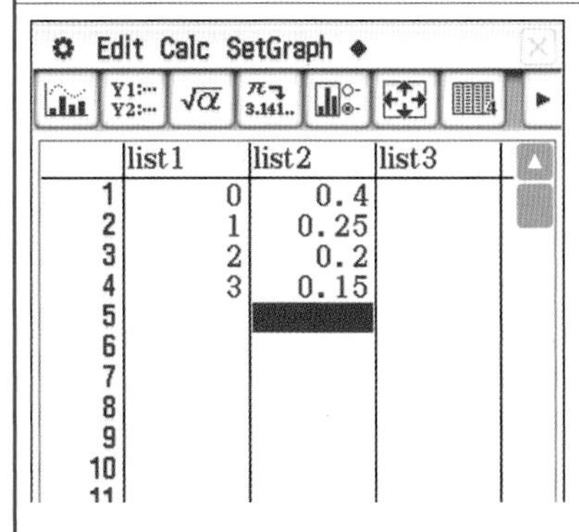

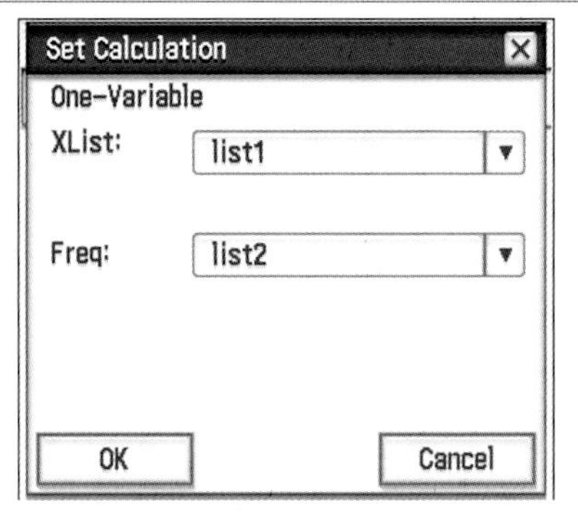

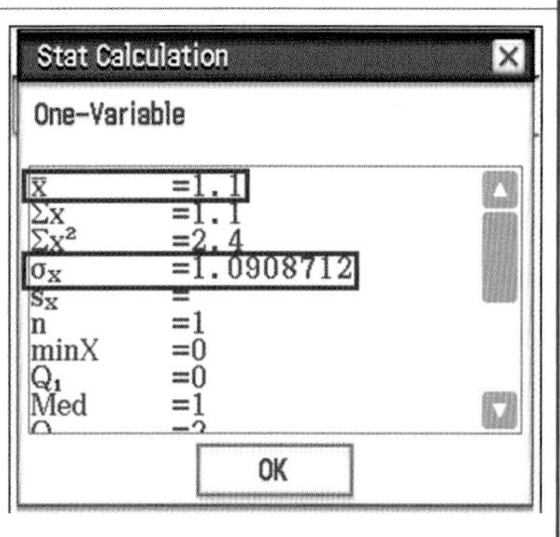

1 Tap **Menu** > **Statistics**.

2 Clear all lists.

3 Enter the values from the table into **list1** and **list2** as shown above.

4 Tap **Calc** > **One-Variable**.

5 Keep the **XList:** field set to **list1**.

6 Change the **Freq:** field to **list2**.

7 The labels and corresponding values will be displayed.

8 The **mean and standard deviation** values are highlighted in blue.

9 To calculate the variance, square the standard deviation.

TI-Nspire

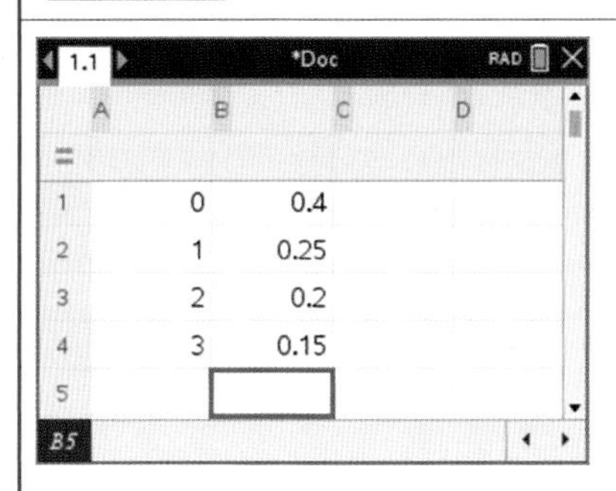

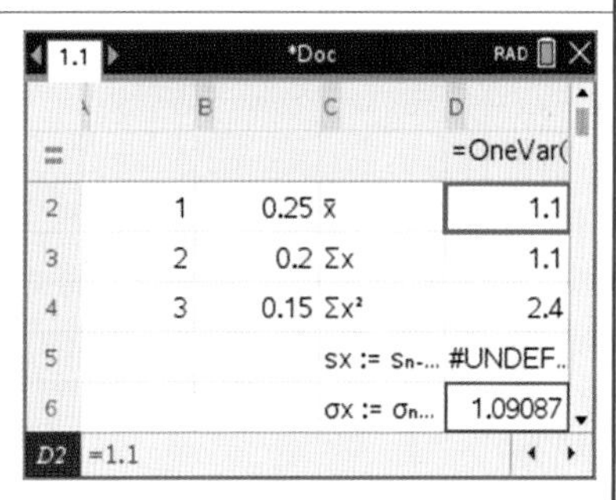

1 Add a **List & Spreadsheet** page.

2 Enter the data from the table in columns **A** and **B** as shown above.

3 Press **menu** > **Statistics** > **Stat Calculations** > **One-Variable Statistics**.

4 On the next screen, keep the **Num of Lists**: default setting of 1.

5 On the next screen shown above, in the **X1 List**: field, enter **a[ ]**.

6 In the **Frequency List**: field, enter **b[ ]**.

7 The labels and corresponding values will be displayed in columns **C** and **D**.

8 The **mean and standard deviation** values are highlighted in blue.

9 To calculate the variance, square the standard deviation.

mean = 1.1, variance = 1.19, standard deviation = 1.09

EXERCISE 5.3 Measures of centre and spread

ANSWERS p. 397

Recap

1 On any given hour of the day, the number, X, of text messages that Danni receives is a random variable with probability distribution given by

x	0	1	2	3
$P(X = x)$	0.1	0.3	0.4	0.2

The probability that Danni receives five text messages over two consecutive hours is

A 0.08 **B** 0.09 **C** 0.16 **D** 0.3 **E** 0.6

2 The discrete random variable X has a probability distribution given by

x	1	2	3	4
$P(X = x)$	0.15	$4p$	0.25	$2p$

The value of p is

A $\frac{1}{15}$ **B** 0.1 **C** $\frac{1}{6}$ **D** 0.4 **E** 0.6

Mastery

3 WORKED EXAMPLE 10 Find the expected value of the discrete random variable X with the following probability distribution.

x	0	1	2	3	4	5
$P(X = x)$	0.02	0.13	0.38	0.17	0.05	0.25

4 WORKED EXAMPLE 11 The probability distribution of a discrete random variable X is given by the table below. Find the values of a and b if the mean of the distribution is 6.4.

x	5	6	7	8	9
$P(X = x)$	a	0.35	b	0.15	0.1

5 WORKED EXAMPLE 12 For the given probability distribution of a discrete random variable X, find the

x	1	2	3	4
$p(x)$	0.1	0.4	0.1	0.4

a expected value

b variance

c standard deviation, correct to three decimal places.

6 WORKED EXAMPLE 13 Find the variance of the discrete random variable X.

x	0	1	2	3
$p(x)$	0.1	0.2	0.3	0.4

9780170477536

7 WORKED EXAMPLE 14 At the Mt Magnet archery competition, competitors are scored for accuracy and technique. The scores at this competition are a discrete random variable X which has an expected value of 15 and a variance of 12. Find

a $E(2X - 3)$

b $Var(2X - 3)$

The scores need to be rescaled to be compatible with other archery competitions in Western Australia. The scaling $Y = aX + b$, where a and b are positive constants, is applied so that the mean is 100 and the variance is 48.

c Find the values of a and b.

5.3

8 Using CAS 2 The discrete random variable X has a probability distribution given by the following table.

x	5	10	15	20
$P(X = x)$	0.12	0.25	0.28	0.35

Find the mean, the variance and the standard deviation of X, correct to two decimal places.

Calculator-free

9 (2 marks) On any given day, the number X of telephone calls that Daniel receives is a random variable with probability distribution given by

x	0	1	2	3
$P(X = x)$	0.2	0.2	0.5	0.1

Find the mean of X.

10 (3 marks) The probability distribution of a discrete random variable X is given by the table below.

x	0	1	2	3	4
$P(X = x)$	0.2	$0.6p^2$	0.1	$1 - p$	0.1

Let $p = \frac{2}{3}$.

a Calculate $E(X)$. (2 marks)

b Find $P(X \geq E(X))$. (1 mark)

11 © SCSA MM2018 Q4 (4 marks) Ten shop owners in a coastal resort were asked how many extra staff they intended to hire for the next holiday season. Their responses are shown below.

3, 0, 2, 1, 2, 1, 1, 0, 2, 1

If N = number of additional staff,

a copy and complete the probability distribution of N below. (2 marks)

n	0	1	2	3
$P(N = n)$				

b what is the mean number of staff the shop owners intend to hire? (2 marks)

12 © SCSA MM2016 Q17 (7 marks) A school has analysed the examination scores for all its Year 12 students taking Methods as a subject. Let X = the examination percentage scores of all the Methods Year 12 students at the school. The school found that the mean was 75 with a standard deviation of 22.

Determine

a $E(X + 5)$ (1 mark)

b $\text{Var}(25 - 2X)$. (2 marks)

The school has decided to scale the results using the transformation $Y = aX + b$ where a and b are constants and Y = the scaled percentage scores. The aim is to change the mean to 60 and the standard deviation to 15.

c Determine the values of a and b. (4 marks)

13 (10 marks) Victoria Jones runs a small business making and selling statues.

The statues are made in a mould, then finished (smoothed and then hand-painted using a special gold paint) by Victoria herself. Victoria sends the statues **in order of completion** to an inspector, who classifies them as either 'Superior' or 'Regular', depending on the quality of their finish.

If a statue is Superior, then the probability that the next statue completed is Superior is p.

If a statue is Regular, then the probability that the next statue completed is Superior is $p - 0.2$.
On a particular day, Victoria knows that $p = 0.9$.

On that day

a if the **first statue inspected is Superior**, find the probability that the third statue is Regular. (2 marks)

b if the **first statue inspected is Superior**, find the probability that the next three statues are Superior. (1 mark)

On another day, Victoria finds that if the **first statue inspected is Superior** then the probability that the third statue is Superior is 0.7.

c **i** Show that the value of p on this day is 0.75. (3 marks)

On this day, a group of three consecutive statues is inspected. Victoria knows that the **first** statue of the three statues is **Regular**.

ii Find the expected number of these three statues that will be Superior. (4 marks)

5.4 The Bernoulli and binomial distributions

The Bernoulli distribution

The **Bernoulli distribution** is a discrete distribution that has two possible outcomes, $x = 1$ and $x = 0$. The outcome $x = 1$, described as success, has probability p and the outcome $x = 0$, described as failure, has probability $1 - p$. The notation for a Bernoulli distribution is $X \sim \text{Bern}(p)$.

An example of a Bernoulli distribution is the single toss of a coin where we observe whether the outcome of a head occurs. The discrete random variable X can be thought of as the number of heads obtained. We will either get 1 head or 0 heads. The outcome of a head is success ($x = 1$) and the outcome of a tail is failure ($x = 0$).

However, a Bernoulli random variable will not necessarily always be about the number of times something occurs. For example, a light switch being in the ON position could be considered a success ($x = 1$) and the OFF position could be considered a failure ($x = 0$).

A **Bernoulli trial** is a trial, or result, of a Bernoulli distribution.

The probability distribution function for a Bernoulli distribution is

$P(X = x) = p^x(1 - p)^{1-x}$ where $x = 0, 1$.

mean: $\mu = p$ variance: $\sigma^2 = p(1 - p)$

Video playlist The Bernoulli and binomial distributions

Worksheets The Bernoulli distribution

Binomial probability experiments

Using the binomial probability distribution

WORKED EXAMPLE 15 Applying the Bernoulli distribution

One disc is removed from a box that contains 1 green and 5 blue discs. The discrete random variable X is the number of blue discs selected.

a Copy and complete the probability distribution for X shown below.

x	0	1
$P(X = x)$		

b State the distribution of X.

c Determine the mean and standard deviation of the distribution.

Steps	Working
a The probability of selecting a blue disc ($x = 1$) is $\frac{5}{6}$ and the probability of a green disc ($x = 0$) is $\frac{1}{6}$.	see table below
b There are two possible outcomes. Either $x = 1$ or $x = 0$ so this satisfies the conditions for a Bernoulli distribution.	This is a Bernoulli distribution.

x	0	1
$P(X = x)$	$\frac{1}{6}$	$\frac{5}{6}$

c Calculate the mean and variance.

> The mean and variance can also be found using the formula
>
> $E(X) = p = \frac{5}{6}$
>
> $\text{Var}(X) = p(1-p) = \frac{1}{6} \times \frac{5}{6} = \frac{5}{36}$

x	$p(x)$	$x \times p(x)$	$x^2 \times p(x)$
0	$\frac{1}{6}$	0	0
1	$\frac{5}{6}$	$\frac{5}{6}$	$\frac{5}{6}$
Total		$\frac{5}{6}$	$\frac{5}{6}$

$E(X) = \frac{5}{6}$

$E(X^2) = \frac{5}{6}$

$\text{Var}(X) = \sigma^2 = E(X^2) - \mu^2 = \frac{5}{6} - \left(\frac{5}{6}\right)^2 = \frac{5}{36}$

The **binomial distribution** is a special discrete probability distribution of the results of a series of Bernoulli trials. The Bernoulli distribution is a special case of the binomial distribution where the number of trials (n) is 1.

Properties of binomial distributions

- Every outcome has two possibilities, which are categorised as 'success' or 'failure'.
- There is a series of n independent trials.
- The probability of success, denoted as p, remains constant on each trial.
- Selections with replacement produce a binomial distribution.
- The probability of x successes is given by $P(X = x) = \binom{n}{x} p^x (1-p)^{n-x}$.
- The notation for a discrete random variable X with a binomial distribution is $X \sim \text{Bin}(n, p)$, where n is the number of trials and p is the probability of success.

Exam hack

Always state the values of the **parameters** n and p in any binomial distribution question. This information may be written in the form $X \sim \text{Bin}(n, p)$.

CAS can calculate probabilities for a binomial distribution.

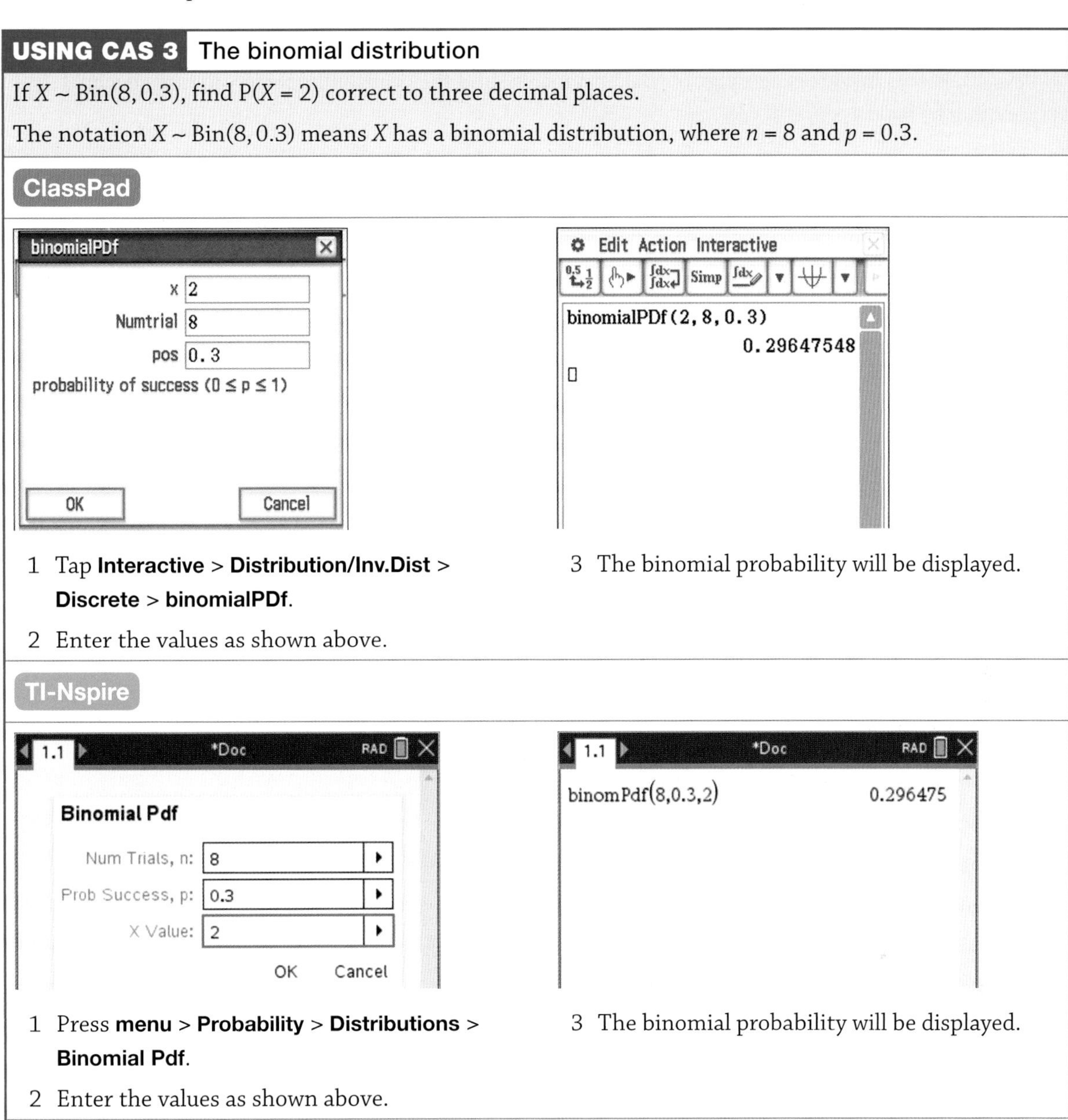

USING CAS 3 The binomial distribution

If $X \sim \text{Bin}(8, 0.3)$, find $P(X = 2)$ correct to three decimal places.

The notation $X \sim \text{Bin}(8, 0.3)$ means X has a binomial distribution, where $n = 8$ and $p = 0.3$.

ClassPad

1 Tap **Interactive** > **Distribution/Inv.Dist** > **Discrete** > **binomialPDf**.

2 Enter the values as shown above.

3 The binomial probability will be displayed.

TI-Nspire

1 Press **menu** > **Probability** > **Distributions** > **Binomial Pdf**.

2 Enter the values as shown above.

3 The binomial probability will be displayed.

$P(X = 2)$ is 0.296 to three decimal places.

Probabilities for a binomial distribution using CAS

Probabilities	ClassPad	TI-Nspire
$P(X = x)$ Probability of single outcome	binomialPDf	Binomial Pdf
$P(X \le x)$ $P(X \ge x)$ $P(x_1 \le X \le x_2)$ Probability of a range of outcomes	binomialCDf	Binomial Cdf

WORKED EXAMPLE 16 **Finding binomial probabilities**

In a football match, a full forward has a probability of 0.35 of kicking a goal from a free kick.

If he is awarded 8 free kicks, in range of goal, in a match, find the probability, correct to three decimal places, that he kicks

a 2 goals **b** at least 2 goals.

Steps	**Working**
a 1 The distribution is binomial because there are 2 outcomes on each trial: kick a goal or not kick a goal.	X = the number of goals scored from free kicks $X \sim \text{Bin}(8, 0.35)$

Exam hack

Before solving the binomial distribution problem, write the values of n and p.

Steps	**Working**
2 Write the probability using the formula: $\text{P}(X = x) = \binom{n}{x} p^x (1-p)^{n-x}$	$\text{P}(X = 2) = \binom{8}{2}(0.35)^2(0.65)^6$
3 Use CAS to calculate the answer. Use the binomial probability density function binomialPDf (ClassPad) or Binomial Pdf (TI-Nspire), as the probability is a single outcome.	$\text{P}(X = 2) = 0.259$
b 1 Write the required probability.	$\text{P}(X \geq 2)$
2 Use CAS to calculate the answer. Use the binomial cumulative distribution function binomialCDf (ClassPad) or Binomial Cdf (TI-Nspire) as the probability is a range of outcomes. The lower bound is 2 and the upper bound is 8. These bounds are inclusive.	

ClassPad

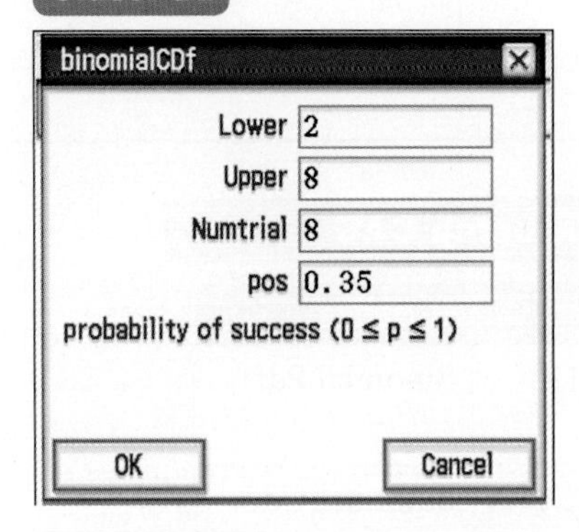

Tap **Interactive** > **Distribution/Inv.Dist** > **Discrete** > **binomialCDf** and enter the values shown.

TI-Nspire

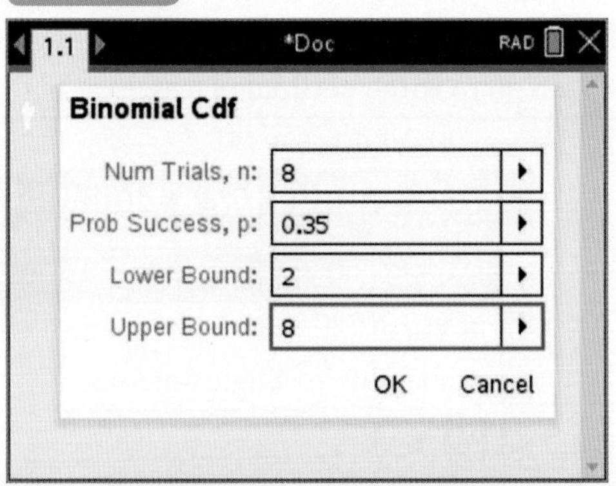

Press **menu** > **Probability** > **Distributions** > **Binomial Cdf** and enter the values shown.

Steps	**Working**
3 Write the probability correct to three decimal places.	$\text{P}(X \geq 2) = 0.831$

Technology-free binomial distribution problems

Binomial distribution probabilities can be calculated without CAS by using the binomial probability formula. This also requires a knowledge of **combinations,** which we learned in Year 11.

Combinations

$$\binom{n}{r} = \frac{n!}{(n-r)!r!}$$

To calculate, evaluate the fraction where the numerator is the product of r descending consecutive numbers starting with n and the denominator is $r!$

For example, $\binom{6}{3} = \dfrac{6!}{3! \times 3!} = \dfrac{6 \times 5 \times 4}{3 \times 2 \times 1}$.

WORKED EXAMPLE 17 Using the binomial distribution formula

A biased coin is tossed four times. The probability of a tail occurring on any toss is p.

a Find, in terms of p, the probability of obtaining

i four tails **ii** three tails.

b Find the value of p if the probability of obtaining four tails is equal to the probability of obtaining three tails.

Steps	Working
a i 1 The distribution is binomial because there are two possible outcomes on each trial: a head or a tail.	X represents the number of tails. $X \sim \text{Bin}(4, p)$
2 Calculate the probability of $X = 4$ using the formula: $\text{P}(X = x) = \binom{n}{x} p^x (1-p)^{n-x}$	$\text{P}(X = 4) = \binom{4}{4}(p)^4(1-p)^0$ $= p^4$
ii Calculate the probability of $X = 3$ using the formula: $\text{P}(X = x) = \binom{n}{x} p^x (1-p)^{n-x}$	$\text{P}(X = 3) = \binom{4}{3}(p)^3(1-p)^1$ $= 4p^3(1-p)$ $= 4p^3 - 4p^4$
b Equate the answers in part **a** and solve for p.	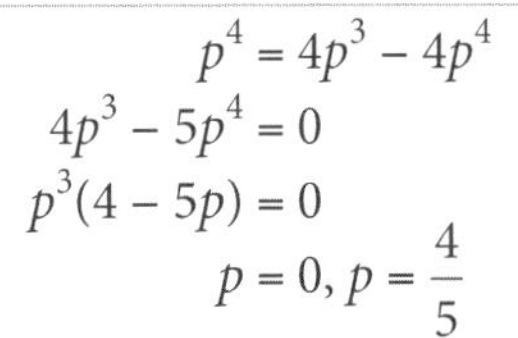 $p^4 = 4p^3 - 4p^4$ $4p^3 - 5p^4 = 0$ $p^3(4 - 5p) = 0$ $p = 0, p = \frac{4}{5}$ $p = \frac{4}{5}$ (p cannot be zero)

Exam hack

Practise writing binomial probabilities using the formula, as this is often necessary when the given probability is a variable.

The mean and variance of a binomial distribution

Worksheets
Binomial probability – Mean and standard deviation

Applications of binomial probability

Mean and variance of a binomial distribution

If X is a discrete random variable with a binomial distribution, then

$X \sim \text{Bin}(n, p)$

mean: $\mu = np$

variance: $\sigma^2 = np(1 - p)$

WORKED EXAMPLE 18 Finding the value of n and p for a binomial distribution

A binomial random variable has a mean of 12 and a variance of 9. Find the values of n and p.

Steps	Working
1 $\mu = np$ $\sigma^2 = np(1 - p)$	$np = 12$ [1] $np(1 - p) = 9$ [2]
2 Solve the equations by substitution to find the values of n and p.	Substitute [1] into [2]: $12(1 - p) = 9$ $1 - p = \frac{9}{12}$ $p = 1 - \frac{9}{12}$ $= \frac{1}{4}$ Substitute into [1]: $n \times \frac{1}{4} = 12$ $n = 48$ So $n = 48$, $p = \frac{1}{4}$.

Finding the number of trials, *n*

Use CAS to find the number of trials for a binomial distribution.

The problem is solved by creating a **discrete probability distribution** where the variable x represents the number of trials.

USING CAS 4 Finding the value of n for a binomial distribution

Amy plays basketball and practises shooting from the 3-point line. She finds during practice that her probability of making a 3-pointer is 0.2. Find the smallest number of shots Amy must take so that the probability she makes at least two 3-pointers is at least 0.72.

1 The problem is binomial because on each independent trial Amy will either hit or miss the 3-point shot. State the values of n and p and write the probability inequality.	Binomial, $n = x$, $p = 0.2$ X represents the number of 3-point shots scored. $P(X \geq 2) \geq 0.72$
2 The probability is a range of outcomes ($X \geq 2$), so use Binomial Cdf. $n = x$, $p = 0.2$, lower = 2, upper = x	$P(X \geq 2) = P(X = 2) + P(X = 3) + \ldots P(X = n)$ Using CAS notation $n = x$, $p = 0.2$, lower = 2 and upper = x. $\therefore$ Binomial Cdf $(x, 0.2, 2, x)$

ClassPad

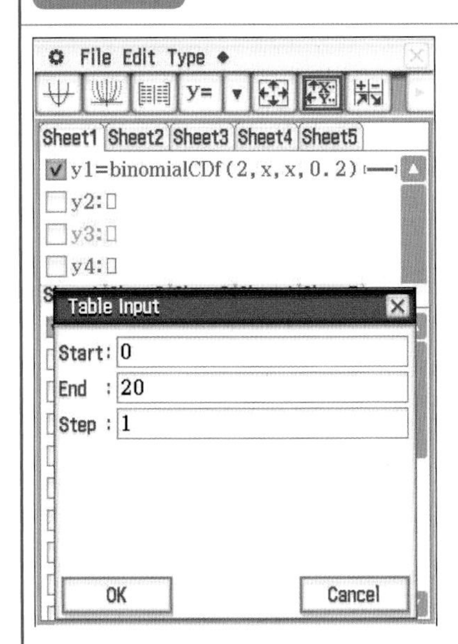

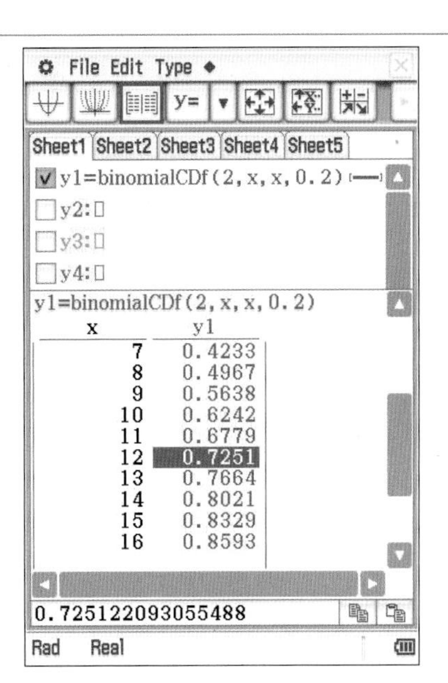

1 Tap **Menu** > **Graph&Table**.
2 Open the **Keyboard** > **down arrow** > **Catalog**.
3 Tap **B** and scroll down to select **binomialCDf**.
4 Enter the values **(2,x,x,0.2)** as shown above.
5 Tap **Table Input** to set a suitable range of values.
6 Tap **OK**.
7 Tap **Table** to view the values.
8 Scroll down the table to the first binomialCDf value greater than 0.72.

TI-Nspire

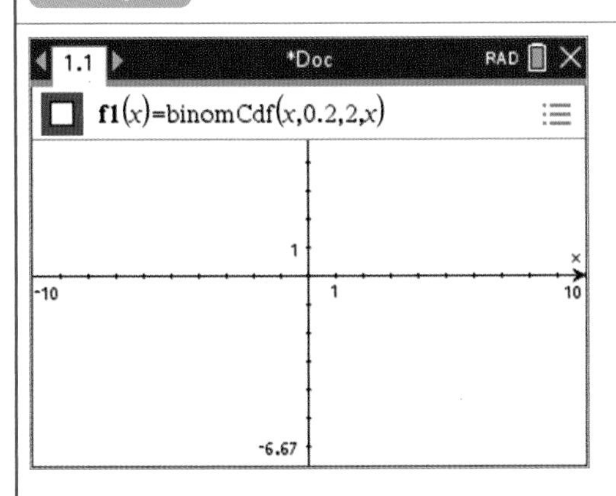

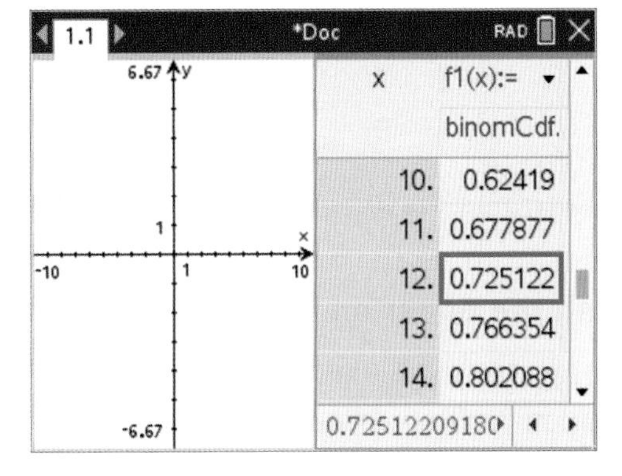

1 Add a **Graphs** page.
2 Press **catalogue** then **B** to jump to the functions starting with b.
3 Scroll down and select **binomCdf**.
4 Enter the values as shown above.
5 After pressing **enter**, no graph will appear.
6 Press **menu > Table > Split-screen Table**.
7 Scroll down the table on the right to the first binomCdf value greater than 0.72.

The number of attempts required is 12.

Video WACE question analysis: Discrete random variables

WACE QUESTION ANALYSIS

© SCSA MM2019 Q10 Calculator-assumed (7 marks)

A group of researchers conducted a study into the number of siblings of adult Australian citizens. They surveyed a total of 200 participants and recorded the number of siblings, X, of each participant.

A few days later the lead researcher discovered that the survey data had been misplaced. Fortunately, one of the research assistants had been doing some rough calculations on a whiteboard and the lead researcher was able to recover the following information about the probability distribution for X and the mean μ.

x	0	1	2	3
$P(X = x)$	0.2	a	b	0.1

$$\mu = 1.3$$

The letters a and b have been used to denote unknown probabilities.

a **i** Write **two** independent equations for a and b. (2 marks)

ii Hence solve for the unknown probabilities. (2 marks)

Later that day the research assistant found the complete probability distribution in their records and discovered that they had made an error in their original calculation of the mean. The correct probability distribution is given in the table below.

x	0	1	2	3
$P(X = x)$	0.2	0.3	0.4	0.1

b **i** Given that there were 200 participants in the study, complete the table below to show the number of participants N with 0, 1, 2 and 3 siblings. (1 mark)

x	0	1	2	3
$P(X = x)$	0.2	0.3	0.4	0.1
N	40			

ii Determine the correct mean and standard deviation of the number of siblings X. (2 marks)

Reading the question

- Highlight the definition of both random variables. The random variable is discrete if there are a countable number of options.
- Highlight the nature of the answer required in each part. This is particularly important in parts where more than one answer is required.
- Notice, in part **b ii**, that the mean and standard deviation are for the variable X not N.

Thinking about the question

- You will need to know the properties of a discrete probability distribution and how to calculate the mean.
- You will also need to be able to solve simultaneous linear equations.
- The question requires you to find the standard deviation from a discrete probability distribution table.
- Many probability questions can be answered using CAS; however, the number of marks in each part of the question will give an indication of the amount of working that must be shown. In part **b ii** there are only 2 marks allocated for two answers, which is an indication that CAS can be used.

Worked solution (✓ = 1 mark)

a **i** Probabilities add to 1.

$0.2 + a + b + 0.1 = 1$ ✓

$a + b = 0.7$ [1]

The mean is 1.3.

$0.2(0) + a(1) + b(2) + 0.1(3) = 1.3$ ✓

$a + 2b = 1$ [2]

ii From the first equation

$a = 0.7 - b$

Substituting into the second equation

$(0.7 - b) + 2b = 1$

$b = 0.3$

$a = 0.4$

$b = 0.3$ ✓

$a = 0.4$ ✓

b **i**

x	0	1	2	3
$P(X = x)$	0.2	0.3	0.4	0.1
N	40	**60**	**80**	**20**

determines all the correct N values ✓

ii μ **= 1.4** ✓

σ **= 0.9165** ✓

ClassPad

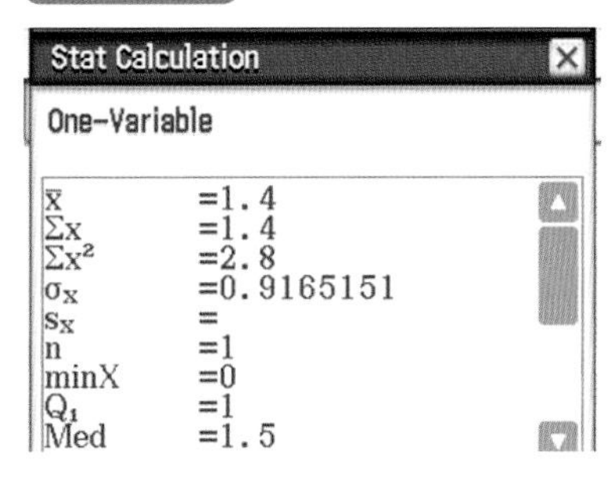

TI-Nspire

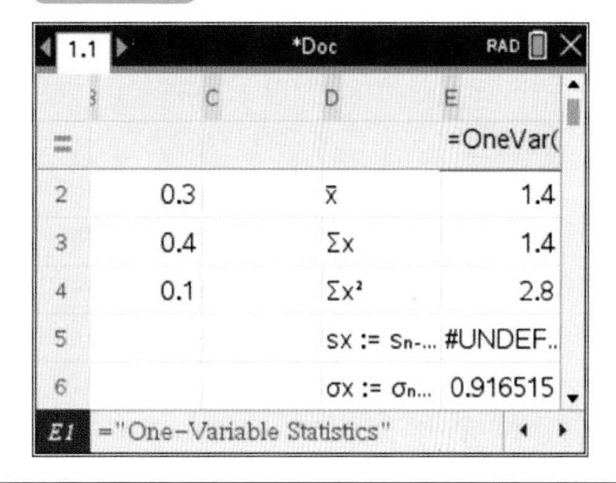

EXERCISE 5.4 The Bernoulli and binomial distributions

ANSWERS p. 398

Recap

1 The discrete random variable X has the following probability distribution.

X	0	1	2	3	4	5	6
$\mathrm{P}(X = x)$	0.05	0.13	0.27	0.1	0.25	0.14	0.06

The variance of X is

A 1.6091 **B** 2.5891 **C** 2.6125 **D** 3.03 **E** 11.77

2 The discrete random variable X has the following probability distribution, where $0 < p < \frac{1}{4}$.

X	0	1	2
$\mathrm{P}(X = x)$	$3p$	p	$1 - 4p$

The mean of X is

A 1 **B** $2 - 3p$ **C** $1 - 4p$ **D** $4 - 15p$ **E** $2 - 7p$

Mastery

3 WORKED EXAMPLE 15 A building has one alarm and the probability that the alarm fails overnight is 0.05. Let the random variable X denote the number of times the alarm fails overnight.

a State the distribution of X.

b Find the mean and variance of X.

4 Using CAS 3

a If $X \sim \mathrm{Bin}(15, 0.74)$, find $\mathrm{P}(X = 12)$ correct to three decimal places.

b If $X \sim \mathrm{Bin}(20, 0.625)$, find $\mathrm{P}(X < 11)$ correct to three decimal places.

c If $Y \sim \mathrm{Bin}(8, 0.4)$, find $\mathrm{P}(1 < Y \leq 6)$ correct to three decimal places.

5 WORKED EXAMPLE 16 A test has eight multiple-choice questions with five possible answers each. A student guesses the answer to each question.

a Find the probability, correct to three decimal places, that the student guesses half the questions correctly.

A cyclist has a probability of 0.05 of puncturing a tyre on a training ride. During the month of April, the cyclist completes one training ride each day and the probability of puncturing a tyre on one day is independent of a puncture on any other day.

b Find the probability, correct to four decimal places, that the rider has at most two punctures during April.

6 WORKED EXAMPLE 17 Imogen is a soccer player who practises her penalty kicks many times each day. Each time she takes a penalty kick her probability of scoring a goal is p, independent of any other penalty kick. In one game Imogen had 6 penalty kicks.

a Find, in terms of p, the probability of Imogen scoring

i 5 goals **ii** 6 goals.

b Find the value of p if the probability of scoring 5 goals is equal to the probability of scoring 6 goals.

7 WORKED EXAMPLE 18

a A binomial random variable, X, has an expected value of 90 and a variance of 36. Find the parameters n and p.

b A binomial random variable, X, has an expected value of 10 and a variance of 9. Find $P(X \geq 15)$ correct to two decimal places.

8 Using CAS 4

a The probability of a darts player hitting a bullseye is $\frac{1}{8}$. What is the smallest number of darts the player should throw so that the probability of hitting at least one bullseye is greater than 0.8?

b A transport company claims that there is a 0.75 probability that each delivery it makes will arrive on time or earlier. Assume that whether each delivery is on time or earlier is independent of other deliveries. If the company makes n deliveries in a day, find the minimum value of n such that there is at least a 0.95 probability that one or more deliveries will **not** arrive on time or earlier.

Calculator-free

9 © SCSA MM2018 Q1 (9 marks) A bag contains one red marble and four green marbles. A single marble is drawn from the bag. The random variable Y is defined as the number of green marbles drawn from the bag.

a Copy and complete the probability distribution for Y shown below. (2 marks)

y	0	1
$P(Y = y)$		

b State the distribution of Y. (1 mark)

c Determine the mean and standard deviation of the distribution. (2 marks)

The above process is repeated five times, with the marble being replaced every time. The random variable X is defined as the number of green marbles drawn from the bag in five attempts.

d State the distribution of X, including its parameters. (2 marks)

e Evaluate the probability of selecting exactly two green marbles. (2 marks)

10 (3 marks) A paddock contains 10 tagged sheep and 20 untagged sheep. Four times each day, one sheep is selected at random from the paddock, placed in an observation area and studied, and then returned to the paddock.

a What is the probability that the number of tagged sheep selected on a given day is zero? (1 mark)

b What is the probability that at least one tagged sheep is selected on a given day? (1 mark)

c What is the probability that no tagged sheep are selected on each of six consecutive days?

Express your answer in the form $\left(\frac{a}{b}\right)^c$, where a, b and c are positive integers. (1 mark)

11 (4 marks) A biased coin is tossed three times. The probability of a head from a toss of this coin is p.

a Find, in terms of p, the probability of obtaining

i three heads from the three tosses (1 mark)

ii two heads and a tail from the three tosses. (1 mark)

b If the probability of obtaining three heads equals the probability of obtaining two heads and a tail, find p. (2 marks)

12 (2 marks) It is known that 50% of the customers who enter a restaurant order a cup of coffee. If four customers enter the restaurant, what is the probability that more than two of these customers order coffee? (Assume that what any customer orders is independent of what any other customer orders.)

13 (5 marks) Records of the arrival times of trains at a busy station have been kept for a long period. The random variable X represents the number of minutes **after** the scheduled time that a train arrives at this station; that is, the lateness of the train. Assume that the lateness of one train arriving at this station is independent of the lateness of any other train. The distribution of X is given in the table below.

x	-1	0	1	2
$P(X = x)$	0.1	0.4	0.3	p

a Find the value of p. (1 mark)

b Find $E(X)$. (1 mark)

c Find $Var(X)$. (1 mark)

d A passenger catches a train at this station on five separate occasions. What is the probability that the train arrives **before** the scheduled time on exactly four of these occasions? (2 marks)

Calculator-assumed

14 © SCSA MM2019 Q18 (9 marks) A building has five alarms configured in such a way that the system functions if at least two of the alarms work. The probability that an alarm fails overnight is 0.05. Let the random variable X denote the number of alarms that fail overnight.

a State the distribution of X. (2 marks)

b What assumptions are required for the distribution in part **a** to be valid? (2 marks)

c What is the probability that the alarm system fails overnight? (2 marks)

One of the alarms is removed in the evening for maintenance and is not replaced.

d What is the probability that the alarm system still works in the morning? (3 marks)

15 (5 marks) FullyFit is an international company that owns and operates many fitness centres (gyms) in several countries. At every one of FullyFit's gyms, each member agrees to have his or her fitness assessed every month by undertaking a set of exercises called **S**. There is a five-minute time limit on any attempt to complete **S** and if someone completes **S** in less than three minutes, they are considered fit.

At FullyFit's Ascot gym, it has been found that the probability that any member will complete **S** in less than three minutes is $\frac{5}{8}$. This is independent of any other member.

In a particular week, 20 members of this gym attempt **S**.

a Find the probability, correct to four decimal places, that at least 10 of these 20 members will complete **S** in less than three minutes. (2 marks)

b Given that at least 10 of these 20 members complete **S** in less than three minutes, what is the probability, correct to three decimal places, that more than 15 of them complete **S** in less than three minutes? (3 marks)

16 (4 marks) A school has a class set of 22 new laptops kept in a recharging trolley. Provided each laptop is correctly plugged into the trolley after use, its battery recharges.

On a particular day, a class of 22 students uses the laptops. All laptop batteries are fully charged at the start of the lesson. Each student uses and returns exactly one laptop. The probability that a student does **not** correctly plug their laptop into the trolley at the end of the lesson is 10%. The correctness of any student's plugging-in is independent of any other student's correctness.

a Determine the probability that at least one of the laptops is **not** correctly plugged into the trolley at the end of the lesson. Give your answer correct to four decimal places. (2 marks)

b A teacher observes that at least one of the returned laptops is not correctly plugged into the trolley. Given this, find the probability that fewer than five laptops are **not** correctly plugged in. Give your answer correct to four decimal places. (2 marks)

17 (4 marks) Mani grows lemons, which are sold to a food factory. When a truckload of lemons arrives at the food factory, the manager randomly selects and weighs four lemons from the load. If one or more of these lemons is underweight, the load is rejected. Otherwise it is accepted.

It is known that 3% of Mani's lemons are underweight.

a Find the probability that a particular load of lemons will be rejected. Express the answer correct to four decimal places. (2 marks)

b Suppose that instead of selecting only four lemons, n lemons are selected at random from a particular load. Find the smallest integer value of n such that the probability of at least one lemon being underweight exceeds 0.5. (2 marks)

18 © SCSA MM2016 Q20ab (6 marks) A chocolate factory produces chocolates of which 80% are pink. Each box of chocolates contains exactly 30 pieces.

a Identify the probability distribution of X = the number of pink chocolates in a single box and also give the mean and standard deviation. (3 marks)

b Determine the probability, to three decimal places, that there are at least 27 pink chocolates in a randomly selected box. (3 marks)

19 (3 marks) Victoria Jones runs a small business making and selling statues.

The statues are made in a mould, then finished (smoothed and then hand-painted using a special gold paint) by Victoria herself. Victoria sends the statues **in order of completion** to an inspector, who classifies them as either 'Superior' or 'Regular', depending on the quality of their finish.

Victoria hears that another company, Shoddy Ltd, is producing similar statues (also classified as Superior or Regular), but its statues are entirely made by machines, on a construction line. The quality of any one of Shoddy's statues is independent of the quality of any of the others on its construction line. The probability that any one of Shoddy's statues is Regular is 0.8.

Shoddy Ltd wants to ensure that the probability that it produces at least two Superior statues in a day's production run is at least 0.9.

Calculate the minimum number of statues that Shoddy would need to produce in a day to achieve this aim.

5 Chapter summary

Probability of an event

$$P(A) = \frac{\text{number of ways } A \text{ can occur}}{\text{total number of possible outcomes}}$$

Tree diagrams

A tree diagram is a useful way of representing a series of simple events. Each simple event is represented by another stage of branches with the number of branches in each set equal to the number of possible outcomes for the event.

Selections without replacement

Selection without replacement occurs when an item is selected from a sample and is not replaced, then another item is selected. The events are dependent.

The addition rule for probability

$$P(A \cup B) = P(A) + P(B) - P(A \cap B)$$

Conditional probability formula

$$P(A|B) = \frac{P(A \cap B)}{P(B)}$$

If events A and B are independent then $P(A \mid B) = P(A)$.

The properties of a probability distribution

- All probabilities must be between 0 and 1 inclusive: $0 \leq p(x) \leq 1$.
- The sum of all probabilities must equal 1: $\Sigma p(x) = 1$.

The expected value (mean)

- The **expected value** of X is written in mathematical notation as $E(X)$ or μ.

  $E(X) = \Sigma x \cdot p(x)$

The variance and standard deviation

- The computational formula for the **variance** of a discrete probability distribution:

  $\sigma^2 = \text{Var}(X) = E(X^2) - \mu^2 = \Sigma(x - \mu)^2 p(x)$

  where $E(X^2) = \Sigma x^2 \cdot p(x)$ and $\mu = E(X)$.
- The **standard deviation** $\sigma = \text{SD}(X) = \sqrt{\text{Var}(X)}$.

The expected value and variance of $aX + b$

- $E(aX + b) = aE(X) + b$
- $\text{Var}(aX + b) = a^2 \text{Var}(X)$

Bernoulli distribution

The Bernoulli distribution is a discrete distribution having two possible outcomes, $x = 1$ and $x = 0$. The outcome $x = 1$, described as success, has probability p and the outcome $x = 0$, described as failure, has probability $1 - p$. The notation for a Bernoulli distribution is $X \sim \text{Bern}(p)$.

If $X \sim \text{Bern}(p)$ then $E(X) = p$ and $\text{Var}(X) = p(1 - p)$.

The binomial distribution

The **binomial distribution** is a special discrete probability distribution of the results of a series of Bernoulli trials.

- Every outcome has two possibilities, which are categorised as 'success' or 'failure'.
- There is a series of n independent trials.
- The probability of success, denoted as p, remains constant on each trial.
- Selections with replacement produce a binomial distribution.
- The probability of x successes is given by

$$P(X = x) = \binom{n}{x} p^x (1 - p)^{n-x}$$

- The notation for a discrete random variable X with a binomial distribution is $X \sim \text{Bin}(n, p)$, where n is the number of trials and p is the probability of success.

Probabilities	ClassPad	TI-Nspire
$P(X = x)$ Probability of single outcome	binomialPDf	Binomial Pdf
$P(X \le x)$ $P(X \ge x)$ $P(x_1 \le X \le x_2)$ Probability of a range of outcomes	binomialCDf	Binomial Cdf

The mean and variance of a binomial distribution

If X is a discrete random variable with a binomial distribution, then

mean: $\mu = np$

variance: $\sigma^2 = np(1 - p)$

Cumulative examination: Calculator-free

Total number of marks: 18 Reading time: 2 minutes Working time: 18 minutes

1 (2 marks) If $f(x) = (x^2 + 3)^2$, find $\dfrac{d^2y}{dx^2}$.

2 (3 marks) Let P be a point on the straight line $y = 2x - 4$ such that the length of OP, the line segment from the origin O to P, is a minimum. Find the coordinates of P.

3 © SCSA MM2020 Q1 (6 marks) Ashley and Xavier are playing a board game that requires them to use the spinner shown.

The player spins the arrowhead and the result is where the arrowhead is pointing when it stops moving. The diagram is showing a result of A.

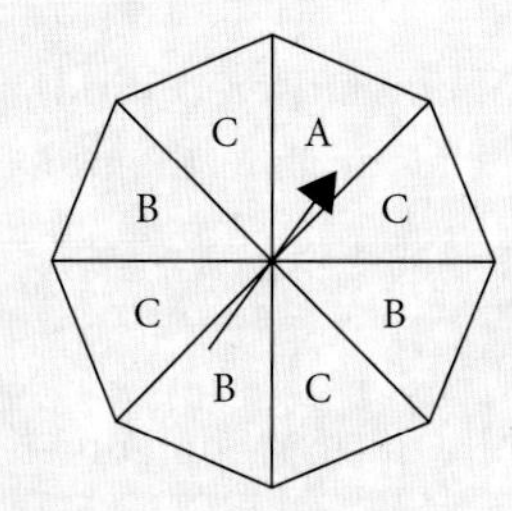

a If the spinner is spun three times, what is the probability that B is never a result? (1 mark)

Let the random variable X be defined as the number of times B is the result when the spinner is spun three times.

b Copy and complete the table below showing the probability distribution of X. (3 marks)

x	0	1	2	3
$P(X = x)$				

c Determine the mean and variance of the distribution. (2 marks)

4 (3 marks) A car manufacturer is reviewing the performance of its car model X. It is known that at any given six-month service, the probability of model X requiring an oil change is $\frac{17}{20}$, the probability of model X requiring an air filter change is $\frac{3}{20}$ and the probability of model X requiring both is $\frac{1}{20}$.

a State the probability that at any given six-month service model X will require an air filter change without an oil change. (1 mark)

b The car manufacturer is developing a new model, Y. The production goals are that the probability of model Y requiring an oil change at any given six-month service will be $\dfrac{m}{m+n}$, the probability of model Y requiring an air filter change will be $\dfrac{n}{m+n}$ and the probability of model Y requiring both will be $\dfrac{1}{m+n}$, where m and n are positive integers.

Determine m in terms of n if the probability of model Y requiring an air filter change without an oil change at any given six-month service is 0.05. (2 marks)

5 (4 marks) For a certain population the probability of a person being born with the specific gene SPGE1 is $\frac{3}{5}$. The probability of a person having this gene is independent of any other person in the population having this gene.

a In a randomly selected group of four people, what is the probability that three or more people have the SPGE1 gene? (2 marks)

b In a randomly selected group of four people, what is the probability that exactly two people have the SPGE1 gene, given that at least one of those people has the SPGE1 gene?

Express your answer in the form $\dfrac{a^3}{b^4 - c^4}$, where a, b and c are positive integers. (2 marks)

Cumulative examination: Calculator-assumed

Total number of marks: 30 Reading time: 3 minutes Working time: 30 minutes

1 © SCSA MM2019 Q7 (9 marks) A company's profit, in millions of dollars, over a five-year period can be modelled by the function:

$P(t) = 2t\sin(3t) \quad 0 \le t \le 5$ where t is measured in years.

The graph of $P(t)$ is shown below.

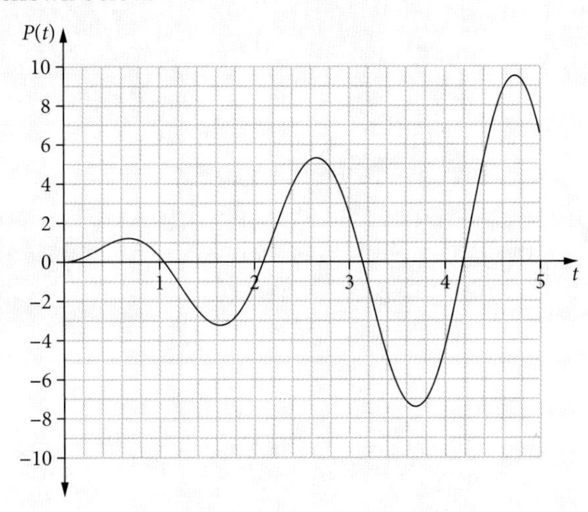

a Differentiate $P(t)$ to determine the marginal profit function, $P'(t)$. (2 marks)

b Calculate the rate of change of the marginal profit function when $t = \frac{\pi}{18}$ years. (4 marks)

c Use the increments formula at $t = \frac{7\pi}{6}$ to estimate the change in profit for a one month change in time. (3 marks)

2 © SCSA MM2017 Q13 (9 marks) Ravi runs a dice game in which a player throws two standard six-sided dice and the sum of the uppermost faces is calculated. If the sum is less than five, the player wins $20. If the sum is greater than eight, the player wins $10. Otherwise the player receives no money.

a Copy and complete the table below. (2 marks)

Amount won			
Probability			

b What is the expected amount of money won by a player each time they play? (2 marks)

c Liu Yang decides to play the game. If Ravi charges her $5 to roll two dice, who is likely to be better off in the long-term? Explain. (3 marks)

d If Ravi wants to make a long-term profit per game of 20% of what he charges, what should he charge a player to roll the two dice? (2 marks)

3 © SCSA MM2021 Q13abc (6 marks) A carnival game involves five buckets, each containing 5 blue balls and 15 red balls. A player blindly selects a ball from each bucket and wins the game if they select at least 4 blue balls. Let X denote the number of blue balls selected.

a State the distribution of X, including its parameters. (2 marks)

b What is the probability of a player winning the game on any given attempt? (2 marks)

c Players are charged \$2 for each attempt at the game and offered a \$150 prize if they win the game. By providing appropriate numerical justification, explain why this is not a good idea for the carnival organisers. (2 marks)

4 (6 marks) Doctors are studying the resting heart rate of adults in two neighbouring towns: Mathsland and Statsville. Resting heart rate is measured in beats per minute (bpm).

The doctors consider a person to have a slow heart rate if the person's resting heart rate is less than 60 bpm. The probability that a randomly chosen Mathsland adult has a slow heart rate is 0.1587. It is known that 29% of Mathsland adults play sport regularly. It is also known that 9% of Mathsland adults play sport regularly and have a slow heart rate. Let S be the event that a randomly selected Mathsland adult plays sport regularly and let H be the event that a randomly selected Mathsland adult has a slow heart rate.

a **i** Find $P(H|S)$, correct to three decimal places. (1 mark)

ii Are the events H and S independent? Justify your answer. (1 mark)

b Find the probability that a random sample of 16 Mathsland adults will contain exactly one person with a slow heart rate. Give your answer correct to three decimal places. (2 marks)

Every year at Mathsland Secondary College, students hike to the top of a hill that rises behind the school.

Students who take less than 15 minutes to get to the top of the hill are categorised as 'elite'.

The probability of a student at Mathland Secondary College being categorised as 'elite' is 0.0266.

The Year 12 students at Mathsland Secondary College make up $\frac{1}{7}$ of the total number of students at the school. Of the Year 12 students at Mathsland Secondary College, 5% are categorised as elite.

c Find the probability that a randomly selected non-Year 12 student at Mathsland Secondary College is categorised as elite. Give your answer correct to four decimal places. (2 marks)

LOGARITHMIC FUNCTIONS

CHAPTER

Syllabus coverage

TOPIC 4.1: THE LOGARITHMIC FUNCTION

Logarithmic functions

4.1.1 define logarithms as indices: $a^x = b$ is equivalent to $x = \log_a b$ i.e. $a^{\log_a b} = b$
4.1.2 establish and use the algebraic properties of logarithms
4.1.3 examine the inverse relationship between logarithms and exponentials: $y = a^x$ is equivalent to $x = \log_a y$
4.1.4 interpret and use logarithmic scales
4.1.5 solve equations involving indices using logarithms
4.1.6 identify the qualitative features of the graph of $y = \log_a x$ $(a > 1)$, including asymptotes, and of its translations $y = \log_a x + b$ and $y = \log_a (x - c)$
4.1.7 solve simple equations involving logarithmic functions algebraically and graphically
4.1.8 identify contexts suitable for modelling by logarithmic functions and use them to solve practical problems

Calculus of the natural logarithmic function

4.1.9 define the natural logarithm $\ln x = \log_e x$
4.1.10 examine and use the inverse relationship of the functions $y = e^x$ and $y = \ln x$

Mathematics Methods ATAR Course Year 12 syllabus pp. 12–13 © SCSA

Video playlists (5):

6.1 Logarithms
6.2 Exponential and logarithmic equations
6.3 The logarithmic function $y = \log_a (x)$
6.4 Applications of logarithmic functions
WACE question analysis Logarithmic functions

Worksheets (2):

6.1 Logarithm laws • Logarithms review

Puzzles (2):

6.2 Logarithms – Solving equations 1 • Logarithms – Solving equations 2

Nelson MindTap

To access resources above, visit **cengage.com.au/nelsonmindtap**

6.1 Logarithms

Logarithms have many uses in science. pH is logarithmic and is a measure of how acidic or basic a solution is. Other examples include the Richter scale for measuring earthquake strength and decibels (dB), which are used to measure sound intensity.

A **logarithm** (or log) is the power or **exponent** to which a base is raised to yield a certain number.

$3^4 = 81$

This means that when we multiply 3 by itself 4 times, we get 81.

Another way of writing this is by using a **logarithm**, which is abbreviated as **log**.

$\log_3(81) = 4$

This is read 'the logarithm of 81, to base 3, is 4'.

The logarithm of a number to base a is the **power** to which a must be raised to give that number.

For example: $\log_5(625)$ means the power to which 5 is raised to get 625.

The equation $5^4 = 625$, is equivalent to $\log_5(625) = 4$.

Video playlist
Logarithms

Worksheets
Logarithm laws

Logarithms review

Some special logarithms

Base 10: When the base is omitted from the logarithm, we assume it is a base 10 logarithm.
$\log(100)$ is interpreted as $\log_{10}(100)$ and has a value of 2.

Base e: When a logarithm is to base e, it is called a natural logarithm.
$\log_e(x) = \ln(x)$

Equations in logarithmic form and exponential form

The logarithmic equation $\log_a(b) = x$ can also be written in exponential form as $a^x = b$.

The relations $a^x = b$ and $\log_a(b) = x$ are equivalent as they show the same relationship between a, b and x, but they are written in different forms. The mathematical symbol for equivalence is $\Leftrightarrow$.

Logarithms		
Logarithmic form	$\Leftrightarrow$	Exponential form
$x = \log_a(b)$	$\Leftrightarrow$	$a^x = b$

WORKED EXAMPLE 1 Converting to logarithmic form

Write each statement in logarithmic form.

a $5^2 = 25$ **b** $2^{-3} = \frac{1}{8}$

Steps	Working
a Write as $\log_a(b) = x$, where a is the base and x is the power. base = 5, power = 2	$5^2 = 25$ $\log_5(25) = 2$
b base = 2, power = −3	$2^{-3} = \frac{1}{8}$ $\log_2\left(\frac{1}{8}\right) = -3$

WORKED EXAMPLE 2 Converting to exponential form

Write each statement in exponential form.

a $\log_2(64) = 6$ **b** $\log_7\left(\frac{1}{7}\right) = -1$

Steps	Working
a Write as $a^x = b$, where a is the base and x is the power. base = 2, power = 6	$\log_2(64) = 6$ $2^6 = 64$
b base = 7, power = −1	$\log_7\left(\frac{1}{7}\right) = -1$ $7^{-1} = \frac{1}{7}$

Exam hack

When you have to find the logarithm of a number to a particular base, you need to think 'What power of this will give the number?'

or

'How many times do I need to multiply the base to get the number?'

WORKED EXAMPLE 3 Evaluating logarithms

Evaluate each logarithm.

a $\log_4(64)$ **b** $\log_6\left(\frac{1}{216}\right)$ **c** $\log_9(1)$

Steps	Working
a **1** Think $4^x = 64$.	$4^3 = 64$
2 Evaluate the logarithm.	$\log_4(64) = 3$
b **1** Think $6^x = \frac{1}{216}$. The power must be negative.	$6^{-3} = \frac{1}{216}$
2 Evaluate the logarithm.	$\log_6\left(\frac{1}{216}\right) = -3$
c **1** Think $9^x = 1$. $\log_a(1) = 0$ always, because $a^0 = 1$.	$9^0 = 1$
2 Evaluate the logarithm.	$\log_9(1) = 0$

The algebraic properties of logarithms

The rules that apply to all logarithms can be established using the algebraic properties of exponentials (index laws).

6.1

Logarithm of a product

$$\log_a(mn) = \log_a(m) + \log_a(n)$$

This can be proved using the index law $a^x \times a^y = a^{x+y}$.

Let	$\log_a(m) = x$	and	$\log_a(n) = y$
	$\log_a(m) = x \Leftrightarrow a^x = m$	and	$\log_a(n) = y \Leftrightarrow a^y = n$
	$m \times n = a^x \times a^y$		
	$m \times n = a^{x+y}$	take $\log_a(\ )$ of both sides	
	$\log_a(m \times n) = \log_a(a^{x+y})$		
however,	$\log_a(a^b) = b$		
therefore,	$\log_a(m \times n) = x + y$		
	$\log_a(mn) = \log_a(m) + \log_a(n)$		

Logarithm of a quotient

$$\log_a\left(\frac{m}{n}\right) = \log_a(m) - \log_a(n)$$

This can be proved using the index law $a^x \div a^y = a^{x-y}$.

Let	$\log_a(m) = x$	and	$\log_a(n) = y$
	$\log_a(m) = x \Leftrightarrow a^x = m$	and	$\log_a(n) = y \Leftrightarrow a^y = n$
	$\frac{m}{n} = \frac{a^x}{a^y}$		
	$\frac{m}{n} = a^{x-y}$	take $\log_a(\ )$ of both sides	
	$\log_a\left(\frac{m}{n}\right) = \log_a(a^{x-y})$		
however,	$\log_a(a^b) = b$		
therefore,	$\log_a\left(\frac{m}{n}\right) = x - y$		
	$\log_a\left(\frac{m}{n}\right) = \log_a(m) - \log_a(n)$		

Logarithm of a power

$$\log_a(m^k) = k\log_a(m)$$

This can be proved using the index law $(a^x)^k = a^{kx}$.

Let	$\log_a(m) = x \Leftrightarrow a^x = m$
	$m^k = (a^x)^k$
	$m^k = a^{xk}$
	$\log_a(m^k) = \log_a(a^{kx})$
however,	$\log_a(a^b) = b$
	$\log_a(m^k) = kx$
therefore,	$\log_a(m^k) = k\log_a(m)$

These algebraic properties of logarithms are also called the **laws of logarithms**.

Laws of logarithms
$\log_a(mn) = \log_a(m) + \log_a(n)$
$\log_a\left(\frac{m}{n}\right) = \log_a(m) - \log_a(n)$
$\log_a(m^k) = k\log_a(m)$
$\log_a(1) = 0$
$\log_a(a^b) = b$ and $a^{\log_a(b)} = b$

WORKED EXAMPLE 4 Simplifying and evaluating logarithms using the laws of logarithms

Simplify each expression.

a $\log_4(32) - \log_4(2)$ **b** $\log_6(18) + \log_6(12)$ **c** $\log_2(144) - 2\log_2(3)$

Steps	Working
a 1 Use the logarithm law $\log_a\left(\frac{m}{n}\right) = \log_a(m) - \log_a(n)$.	$\log_4(32) - \log_4(2)$ $= \log_4(32 \div 2)$ $= \log_4(16)$
2 Evaluate the logarithm using $\log_a(a^b) = b$.	$= \log_4(4^2)$ $= 2$
b 1 Use $\log_a(mn) = \log_a(m) + \log_a(n)$.	$\log_6(18) + \log_6(12)$ $= \log_6(18 \times 12)$ $= \log_6(216)$
2 Write 216 as a power of 6 and simplify.	$= \log_6(6^3)$ $= 3$
c Use $\log_a(m^k) = k\log_a(m)$ then use the logarithm of a quotient law.	$\log_2(144) - 2\log_2(3)$ $= \log_2(144) - \log_2(3^2)$ $= \log_2(144 \div 9)$ $= \log_2(16)$ $= \log_2(2^4)$ $= 4$

WORKED EXAMPLE 5 Simplifying logarithms into a single expression

Simplify $3\log_2(x) + \log_2(y) - 4\log_2(x+3)$ to a single logarithm.

Steps	Working
1 Use $k\log_a(m) = \log_a(m^k)$.	$3\log_2(x) + \log_2(y) - 4\log_2(x+3)$ $= \log_2(x^3) + \log_2(y) - \log_2(x+3)^4$
2 Use $\log_a(mn) = \log_a(m) + \log_a(n)$.	$= \log_2(x^3y) - \log_2(x+3)^4$
3 Use $\log_a(m) - \log_a(n) = \log_a\left(\frac{m}{n}\right)$.	$= \log_2\left[\frac{x^3y}{(x+3)^4}\right]$

EXERCISE 6.1 Logarithms

ANSWERS p. 399

6.1

Mastery

1 WORKED EXAMPLE 1 Write the equivalent equation to each of the following, in logarithmic form.

a $7^2 = 49$
b $3^3 = 27$
c $2^4 = 16$
d $5^3 = 125$
e $11^0 = 1$
f $(2)^0 = 1$
g $5^{-2} = \frac{1}{25}$
h $4^{-2} = \frac{1}{16}$

2 WORKED EXAMPLE 2 Write the equivalent equation to each of the following, in exponential form.

a $\log_5(25) = 2$
b $\log_4(16) = 2$
c $\log_5(125) = 3$
d $\log_2(16) = 4$
e $\log_3(3) = 1$
f $\log_7(49) = 2$
g $\log_2(128) = 7$
h $\log_5(1) = 0$

3 WORKED EXAMPLE 3 Evaluate each logarithm.

a $\log_2(64)$
b $\log_9(81)$
c $\log_3(81)$
d $\log_7(343)$
e $\log_6(216)$
f $\log_5(1)$
g $\log_3(3)$
h $\log(100\,000)$
i $\log_3(243)$
j $\log_4(1024)$
k $\log_{\frac{1}{2}}\left(\frac{1}{16}\right)$
l $\log_5\left(\frac{1}{125}\right)$

4 WORKED EXAMPLE 4 Simplify and evaluate each expression.

a $\log_4(10) + \log_4(2) - \log_4(5)$
b $\log_5(25) + \log_5(125) - \log_5(625)$
c $\log_8\left(\frac{1}{8}\right) + \log_8(4)$
d $\log_2(16) + \log_2(4) + \log_2(8)$
e $\log(400) + \log(10) - \log(4)$
f $\log_5(8) - \log_5(4) - \log_5(2)$
g $\log_8(2) - \log_8\left(\frac{1}{4}\right)$
h $\log_4(256) - \log_4(32) + \log_4(2)$

5 WORKED EXAMPLE 5 Write each expression as a single logarithm.

a $5\log_4(x) + \log_4(x^2) - \log_4(x^3)$
b $3\log_7(x) - 5\log_7(x) + 4\log_7(x)$
c $4\log_6(x) - \log_6(x^2) - \log_6(x^3)$
d $\log_2(x+2) - \log_2(x+2)^2$
e $\log_4[(x-1)^3] - \log_4[(x-1)^2]$
f $\log_3(x-3) + \log_3(x+3) - \log_3(x^2-9)$

Calculator-free

6 (5 marks) Evaluate each logarithm.

a $\log_2(\sqrt{2})$ (1 mark)
b $\log_9(9\sqrt{9})$ (1 mark)
c $\log_4(\sqrt{64})$ (1 mark)
d $\log_7(\sqrt{343})$ (1 mark)
e $\log_6(\sqrt[3]{36})$ (1 mark)

Video playlist
Exponential and logarithmic equations

Puzzles
Logarithms – Solving equations 1

Logarithms – Solving equations 2

6.2 Exponential and logarithmic equations

Solving exponential equations

Every exponential equation has an equivalent **logarithmic equation**.

It is often necessary when solving an equation in exponential form to express it as an equation in logarithmic form. The two equations are equivalent ways of expressing the same relationship.

WORKED EXAMPLE 6 Solving exponential equations

Solve for x.

a $2^{x-1} = 3$ **b** $e^x = 7$

Steps	Working
a Express the exponential equation as a logarithmic equation and solve for x.	$2^{x-1} = 3$ $x - 1 = \log_2(3)$ $x = \log_2(3) + 1$
b Express the exponential equation as a logarithmic equation. Remember, $\log_e(x) = \ln(x)$.	$e^x = 7$ $x = \log_e(7)$ $x = \ln(7)$

WORKED EXAMPLE 7 Solving exponential equations using the null factor law

Solve $e^x(e^x - 6) = 0$ for x.

Steps	Working
1 Solve using the null factor law. There is only one solution as e^x is always positive.	$e^x - 6 = 0$ $e^x = 6$
2 Express the exponential equation as a logarithmic equation.	$x = \log_e(6)$ $x = \ln(6)$

Solving logarithmic equations

To solve logarithmic equations, we need to use the algebraic properties of logarithms to simplify expressions and be able to change expressions from logarithmic form ($\log_a(x) = b$) to exponential form ($a^b = x$). These laws of logarithms were covered in the previous section.

WORKED EXAMPLE 8 Solving logarithmic equations using the laws of logarithms

Solve $\log_2(x - 1) + 2\log_2(5) = 2$ for x.

Steps	Working
1 Use log laws to express the left-hand side of the equation as a single logarithm.	$\log_2(x-1) + 2\log_2(5) = 2$ $\log_2(x-1) + \log_2(5)^2 = 2$ $\log_2(x-1) + \log_2(25) = 2$ $\log_2(25(x-1)) = 2$
2 Change the equation from log form to exponential form.	$25(x-1) = 2^2$
3 Simplify and solve the equation.	$25x - 25 = 4$ $25x = 29$ $x = \dfrac{29}{25}$

WORKED EXAMPLE 9 Solving equations where every term is a logarithm

Solve $\log_7(x) = \log_7(3) + \log_7(6)$ for x.

Steps	Working
1 Use the laws of logarithms to express the right-hand side of the equation as a single logarithm.	$\log_7(x) = \log_7(3) + \log_7(6)$ $\log_7(x) = \log_7(3 \times 6)$ $\log_7(x) = \log_7(18)$
2 Equate the brackets.	$x = 18$

WORKED EXAMPLE 10 Solving equations using log form to exponential form transformations

a Given that $\log_5(x) = 2$ and $\log_2(y) = 6$, evaluate $2x + y$.

b Express y in terms of x given that $\log_3(x + y) - 1 = \log_3(x - y)$.

Steps	Working
a 1 Change each equation from logarithmic form to exponential form to solve for x and y.	$\log_5(x) = 2$ $x = 5^2$ $= 25$ $\log_2(y) = 6$ $y = 2^6$ $= 64$
2 Find the value of $2x + y$.	$2x + y = 2 \times 25 + 64$ $= 114$
b 1 Transpose the equation and simplify using log laws.	$\log_3(x + y) - 1 = \log_3(x - y)$ $\log_3(x + y) - \log_3(x - y) = 1$ $\log_3\left(\frac{x + y}{x - y}\right) = 1$
2 Change the equation into exponential form and make y the subject.	$\frac{x + y}{x - y} = 3^1$ $x + y = 3x - 3y$ $4y = 2x$ $y = \frac{x}{2}$

WORKED EXAMPLE 11 Solving simultaneous equations involving logarithms

The graph of $y = a\log_2(x - 4) + b$ passes through the points $(5, 8)$ and $(12, 17)$.

Find the values of a and b.

Steps	Working
1 Substitute the coordinate $(5, 8)$ into the equation and evaluate the logarithms to simplify. Remember, $\log_2(2^b) = b$.	$(5, 8)$ $8 = a\log_2(5 - 4) + b$ $8 = a\log_2(1) + b$ $8 = a \times 0 + b$ $b = 8$
2 Substitute the coordinate $(12, 17)$ into the equation and evaluate the logarithms to simplify.	$(12, 17)$ $y = a\log_2(x - 4) + 8$ $17 = a\log_2(12 - 4) + 8$ $17 = a\log_2(8) + 8$ $3a + 8 = 17$ $a = 3$

EXERCISE 6.2 Exponential and logarithmic equations

ANSWERS p. 399

Recap

1 Evaluate the following logarithms.

a $\log_2(32)$ **b** $\log_5(125)$ **c** $\log_3\left(\frac{1}{81}\right)$

2 Simplify each expression.

a $\log_2(96) - \log_2(3)$ **b** $\log_5(50) + \log_5(75) - \log_5(6)$

Mastery

3 WORKED EXAMPLE 6 Solve for x.

a $3^{x-5} = 7$ **b** $2^{x+3} - 5 = 7$ **c** $e^{3x} = 9$ **d** $e^{2x+3} = 2$

4 WORKED EXAMPLE 7 Solve for x.

a $e^x(e^x - 8) = 0$ **b** $5^x(5^x - 4) = 0$ **c** $7(2^{3x}) - 6 = 5(2^{3x})$ **d** $(3^x - 1)(3^{2x} - 2) = 0$

5 WORKED EXAMPLE 8 Solve each logarithmic equation for x.

a $\log_3(x+7) + \log_3(2) = 3$

b $2\log_2(3) + \log_2(x+1) = 4$

c $\log_2(3x-2) + 2\log_2(4) = 3$

d $\log_2(2x-4) + \log_2(5) = 1$

e $\ln(x-3) - \ln(4) = 0$

f $\log_2(x+2) - \log_2(3) = 3$

6 WORKED EXAMPLE 9 Solve each logarithmic equation for x.

a $\log_5(x) + \log_5(3) - \log_5(2) = \log_5(6)$

b $\log_2(x) + \log_2(6) = \log_2(3) + \log_2(x+7)$

c $\log_2(x) - 3\log_2(2) = \log_2(x+1) - 2\log_2(5)$

7 WORKED EXAMPLE 10

a Given that $\log_7(x) = 2$ and $\log_3(y) = 4$, evaluate $3y - 2x$.

b Express y in terms of x given that $\log_5(x-y) - 2 = \log_5(2y-x)$.

8 WORKED EXAMPLE 11

a The graph of $y = a\log_3(x+2) + b$ passes through the points $(-1, 10)$ and $(7, 14)$. Find the values of a and b.

b The graph of $y = a\log_2(x-7) + b$ passes through the points $(9, 26)$ and $(15, 36)$. Find the values of a and b.

Calculator-free

9 © SCSA MM2017 Q3 (4 marks) Solve $4e^{2x} = 81 - 5e^{2x}$ exactly for x.

10 © SCSA MM2016 Q1 (5 marks)

a Given that $\log_8(x) = 2$ and $\log_2(y) = 5$, evaluate $x - y$. (2 marks)

b Express y in terms of x given that $\log_2(x+y) + 2 = \log_2(x-2y)$. (3 marks)

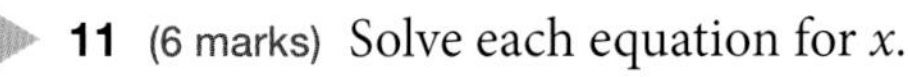

11 (6 marks) Solve each equation for x.

a $2\log_3(5) - \log_3(2) + \log_3(x) = 2$ (2 marks)

b $\log_e(3x + 5) + \log_e(2) = 2$ (2 marks)

c $\log_2(6 - x) - \log_2(4 - x) = 2$ (2 marks)

Calculator-assumed

12 (5 marks) The functions f and g are defined as $f(x) = \log_3(x + 1) - 3$ and $g(x) = \log_3(2)$. The graphs of the function intersect at the point (a, b).

a Show $\log_3\left(\frac{a+1}{2}\right) = 3$. (2 marks)

b Find the values of a and b. (3 marks)

6.3 The logarithmic function $y = \log_a(x)$

Video playlist The logarithmic function $y = \log_a(x)$

Graphing logarithmic functions

We can plot the **logarithmic function** for $y = \log_2(x)$ using a table of values. We substitute powers of 2 for x so that the logarithms are easier to calculate. This table of values and its graphical representation are shown below.

x	$\frac{1}{4}$	$\frac{1}{2}$	1	2	4
$y = \log_2(x)$	$\log_2\left(\frac{1}{4}\right) = -2$	$\log_2\left(\frac{1}{2}\right) = -1$	$\log_2(1) = 0$	$\log_2(2) = 1$	$\log_2(4) = 2$

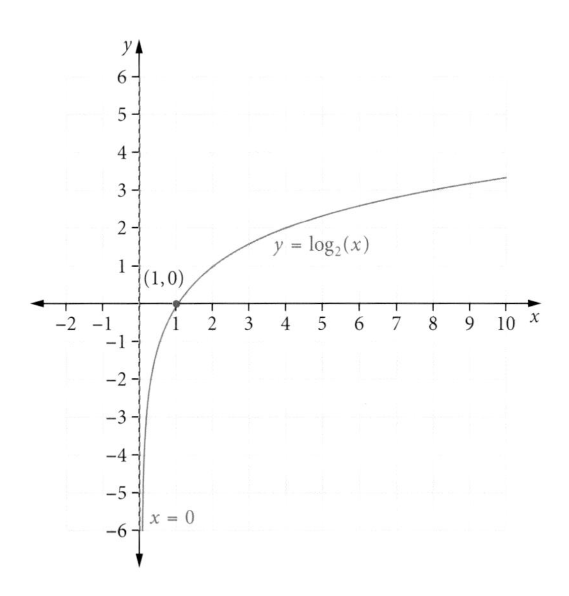

Properties of the logarithmic function $y = \log_a(x)$

- It is a strictly increasing function, increasing quickly at first, then more slowly.
- The gradient of the graph is always decreasing.
- x-intercept is at $(1, 0)$ as $\log_a(1) = 0$.
- The y-axis $(x = 0)$ is a vertical asymptote.

Changing the base of the logarithm does not alter the basic shape, x-intercept or the asymptote of the graph of the logarithmic function.

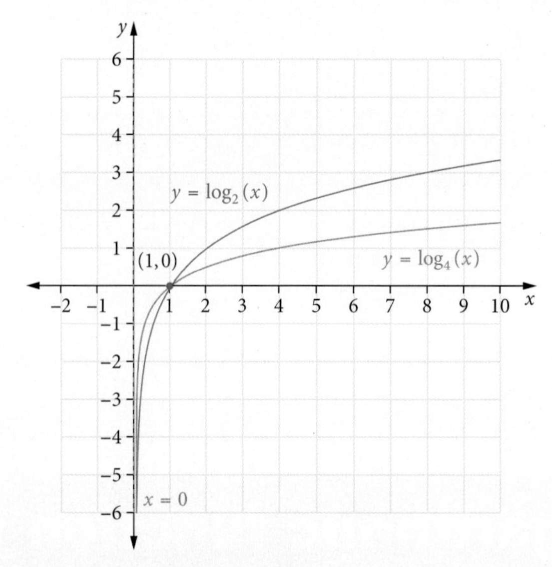

Translations of $y = \log_a(x)$

The graph of the function $y = \log_a(x - c) + b$ is a translation of $y = \log_a(x)$, c units right and b units up where b and c are positive real constants.

The horizontal translation of c units right produces a vertical asymptote at $x = c$.

The x-intercept $(1, 0)$ of $y = \log_a(x)$ translates to $(1 + c, b)$.

Properties of the logarithmic function $y = \log_a(x - c) + b$, where b and c are positive real constants

- It has the same shape as $y = \log_a(x)$.
- The horizontal translation is c units right and the vertical translation is b units up.
- $x = c$ is the vertical asymptote.
- Include the guiding point $(1 + c, b)$.

The graphs of $y = \log_2(x)$ and $y = \log_2(x - 1) + 2$ are shown below.

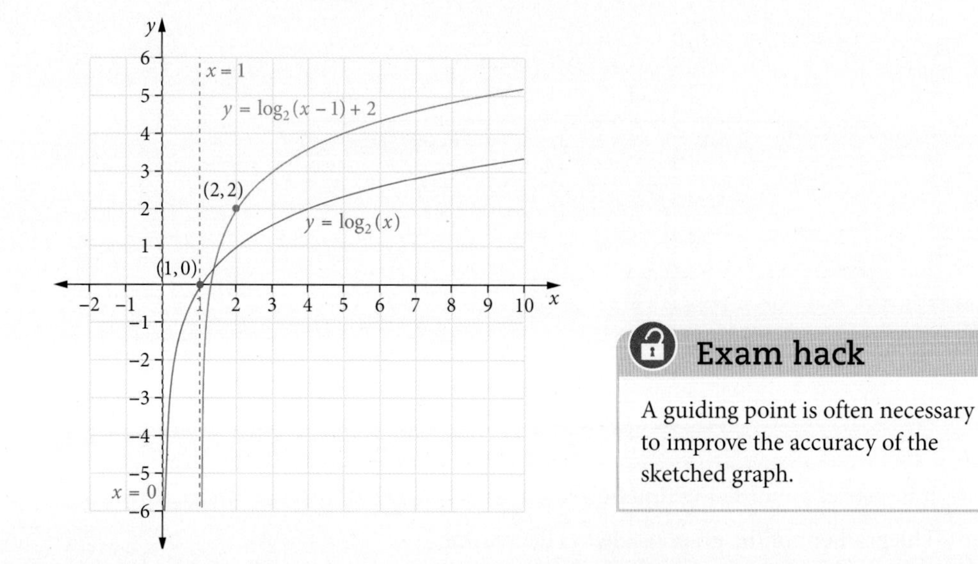

The graph of $y = \log_2(x)$ has been translated horizontally, 1 unit right, and vertically, 2 units up, to produce the graph of $y = \log_2(x - 1) + 2$. Note that the asymptote $x = 0$ and the x-intercept $(1, 0)$ on $y = \log_2(x)$ have translated to $x = 1$ and $(2, 2)$ respectively on $y = \log_2(x - 1) + 2$.

WORKED EXAMPLE 12 Using translations to sketch a logarithmic function

Sketch the graph of $f(x) = \log_2(x-3)$. Label the coordinates of the x-intercept and the asymptote with its equation.

Steps	Working
1 The graph of $y = \log_2(x)$ has a vertical asymptote at $x = 0$ and an x-intercept $(1, 0)$. Translate the function 3 units to the right.	The vertical asymptote is $x = 3$. x-intercept ($y = 0$): $\log_2(x-3) = 0$ $x - 3 = 2^0$ $x - 3 = 1$ $x = 4$ The x-intercept is $(4, 0)$.
2 Sketch the graph.	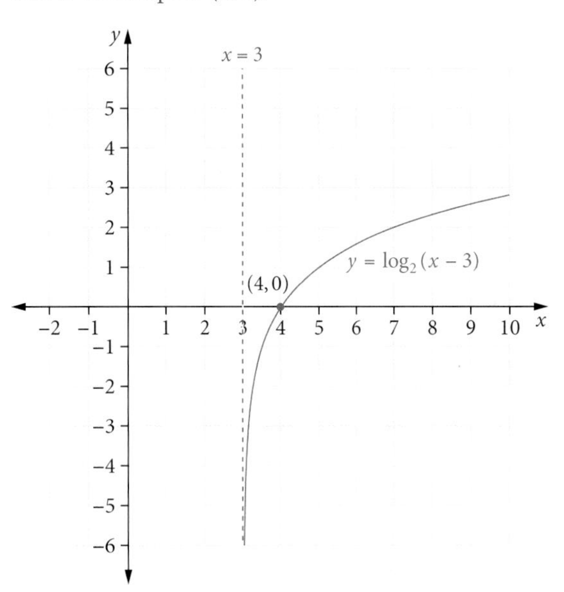

WORKED EXAMPLE 13 Using the intercept method to sketch a logarithmic function

Consider the function $f(x) = \log_3(x+1) - 2$. Describe the translations on $y = \log_3(x)$ and sketch $y = f(x)$. Label axes intercepts and the asymptote with its equation.

Steps	Working
1 State the translation.	The graph of $y = \log_3(x)$ is translated vertically 2 units down, and horizontally 1 unit left. The vertical asymptote is $x = -1$. The point $(1, 0)$ on $y = \log_3(x)$ is translated 1 unit left and 2 units down to $(0, -2)$.
2 Find the x-intercept and y-intercept.	x-intercept ($y = 0$): $\log_3(x+1) - 2 = 0$ $\log_3(x+1) = 2$ $x = 3^2 - 1 = 8$ The x-intercept is $(8, 0)$. y-intercept ($x = 0$): $y = \log_3(1) - 2$ $y = -2$ The y-intercept is $(0, -2)$.

3 Sketch the graph.

Exam hack

Include the asymptote equation $x = -1$, and the coordinates of any axes intercepts.

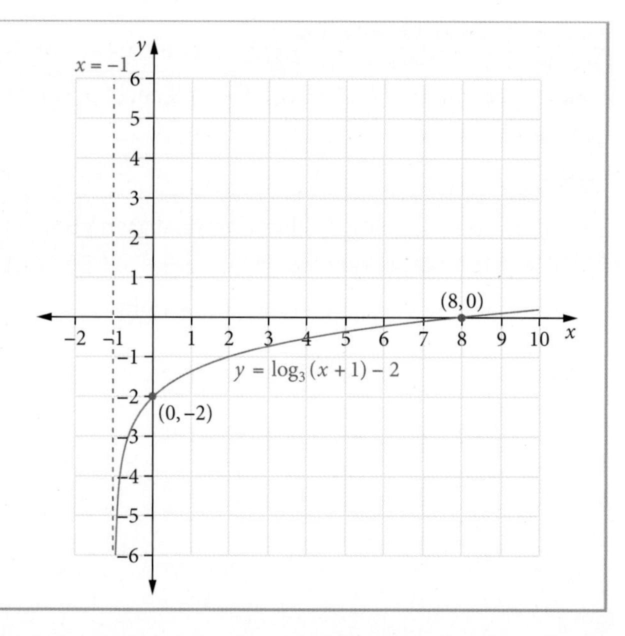

USING CAS 1 Graphing logarithmic functions

Graph $y = \ln(x - 2) + 3$.

ClassPad

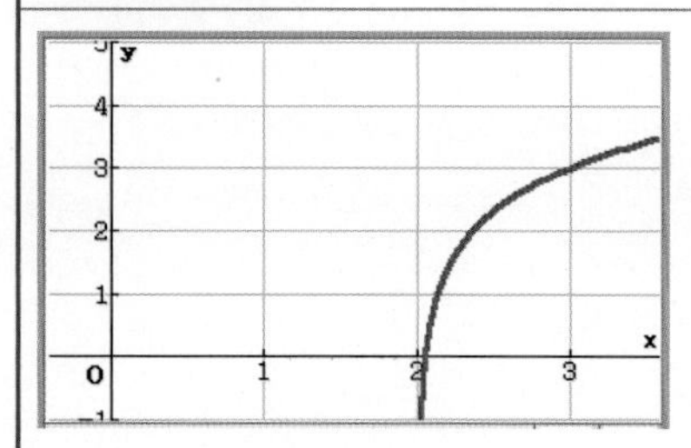

1 In **Main**, enter and highlight the equation using **ln** for $\log_e$ from **Math1**.

2 Tap **Graph** and drag the equation down into the Graph window.

3 Adjust the window settings to suit.

TI-Nspire

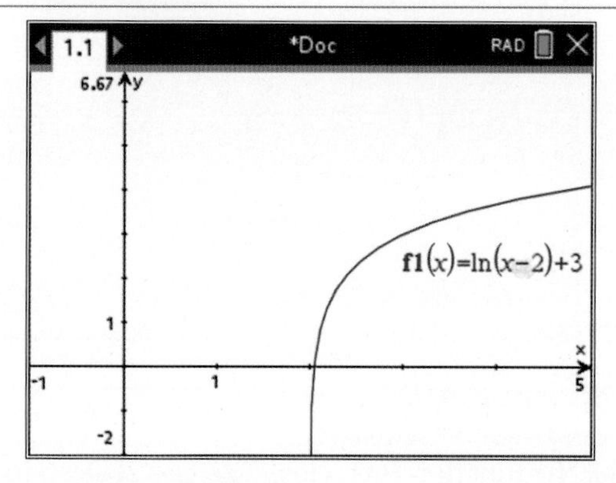

1 Add a **Graphs** page and enter the function as shown above. Press **ctrl** + $\mathbf{e}^{\mathbf{x}}$ for the **ln** function.

2 Adjust the window settings to suit.

Finding equations of logarithmic functions

Finding the equation of logarithmic functions is often a multi-step process, and every problem is different depending on the information and type of graph we are given.

For a logarithmic function $y = \log_a(x - c) + b$, the vertical asymptote is $x = c$.

WORKED EXAMPLE 14 Finding the rule of a logarithmic function

Find the rule for the logarithmic function $y = a\ln(x + b)$ if the vertical asymptote is $x = -3$ and the y-intercept is at the point $(0, 4\ln(9))$.

Steps	Working
1 The function $y = a\ln(x + b)$ has been translated b units to the left and will have an asymptote at $x = -b$.	Vertical asymptote is $x = -3$. $b = 3$ $y = a\ln(x + 3)$
2 Substitute $(0, 4\ln(9))$ into the equation.	$4\ln(9) = a\ln(0 + 3)$
3 Simplify $4\ln(9)$ using the log law $\ln(x^n) = n\ln(x)$.	$4\ln(3^2) = a\ln(3)$ $8\ln(3) = a\ln(3)$ $a = 8$
4 Write the function.	$y = 8\ln(x + 3)$

USING CAS 2 Finding rules for logarithmic functions

The graph of the logarithmic function $f(x) = \log_2(x - c) + b$ passes through the points $(3, 7)$ and $(11, 8)$. Find the values of b and c.

ClassPad

Define f(x)=log$_2$(x−c)+b
done
{f(3)=7, f(11)=8 | b, c
{b=4, c=−5}

1 In **Main**, enter and highlight the equation using **log** from **Math1**.
2 Tap **Interactive, Define**.
3 Enter the simultaneous equations as shown and solve for b and c.

TI-Nspire

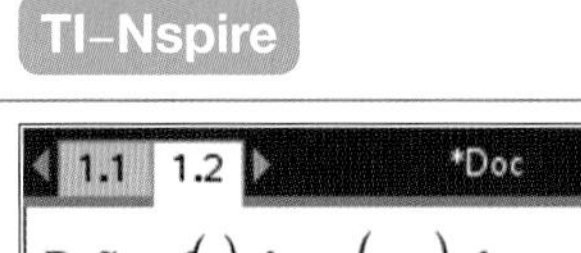
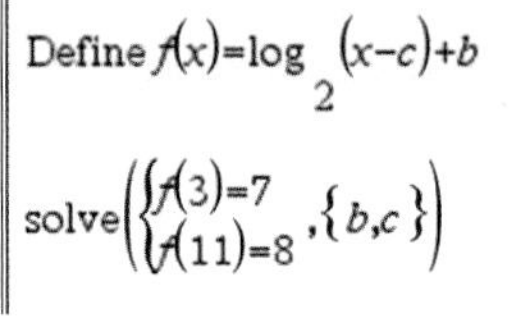
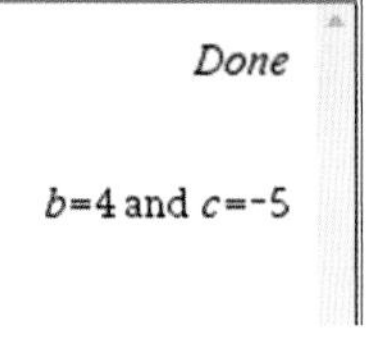

1 Add a **Calculator** page and define the function $f(x)$.
2 Press **menu** > **algebra** > **Solve system of equations**.
3 Enter the equations as shown.

$b = 4$, $c = -5$.

EXERCISE 6.3 The logarithmic function $y = \log_a(x)$

ANSWERS p. 399

Recap

1 Find the value of $x + y$ if $\log_2(x) = 3$ and $\log_3(y) = 2$.

2 Solve $e^{2x} + 16 = 2e^{2x}$ for x.

Mastery

3 WORKED EXAMPLE 12 Sketch the graph of $f(x) = \log_2(x + 4)$. Label the coordinates of the x- and y-intercepts and the asymptote with its equation.

4 Sketch the graph of $f(x) = \ln(x - 4)$. Label the coordinates of the x-intercept and the asymptote with its equation.

5 WORKED EXAMPLE 13 Consider the function $f(x) = \log_2(x + 2) - 1$. Describe the translations on $y = \log_2(x)$ and sketch $y = f(x)$, labelling axes intercepts and the asymptote.

6 Using CAS 1 Sketch the graph of $f(x) = \ln(x + 3) + 5$.

7 WORKED EXAMPLE 14 Find the rule for the logarithmic function $y = a\ln(x + b)$ if the vertical asymptote is $x = -4$ and the y-intercept is at the point $(0, 6\ln(2))$.

8 Using CAS 2 The graph of the logarithmic function $f(x) = \log_2(x - c) + b$ passes through the points $(3, 8)$ and $(33, 12)$ Find the values of b and c.

Calculator-free

9 © SCSA MM2019 Q4b (3 marks) Consider the graph of $y = \ln(x)$ shown below.

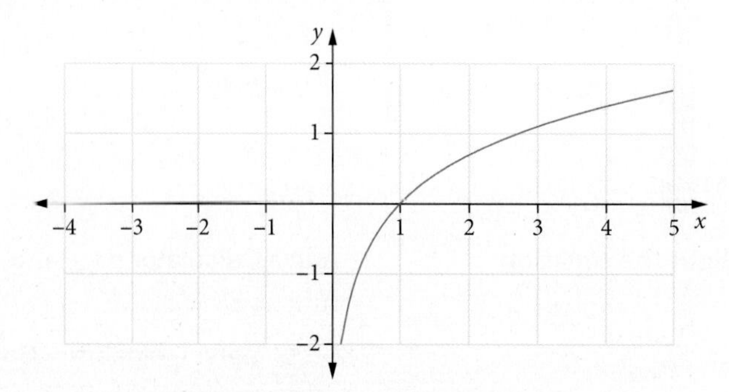

Copy the graph and on it sketch the graph of $y = \ln(x - 2) + 1$.

10 (6 marks) Determine the coordinates of the x- and y-intercepts and the equation of the asymptote for each of the logarithmic functions below.

a $f(x) = \log_3(x + 9) - 4$ (3 marks)

b $f(x) = \log_2(x + 8) - 3$ (3 marks)

11 (4 marks) Find the rule for the logarithmic function $y = \log_5(x - c) + b$, if the vertical asymptote is $x = 2$ and the graph passes through the point $(27, 10)$.

12 © SCSA MM2018 Q8 (8 marks) Consider the function $f(x) = \log_a(x - 1)$ where $a > 1$.

a Copy the axes below, and on it sketch the graph of $f(x)$, labelling important features. (3 marks)

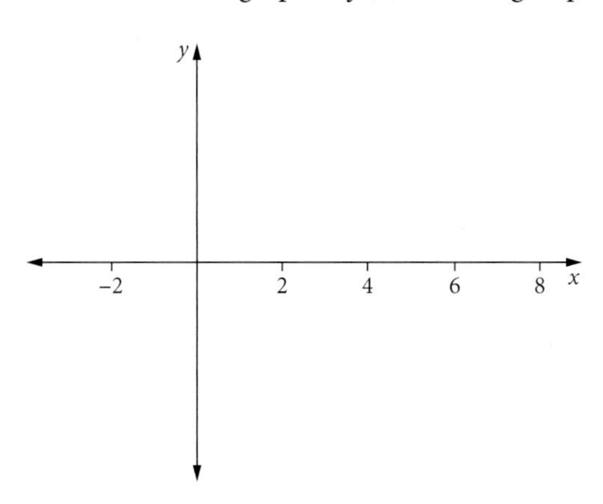

b Determine the value of m if $f(m) = 1$. (2 marks)

c Determine the coordinates of the x-intercept of $f(x + b) + c$, where b and c are positive real constants. (3 marks)

13 © SCSA MM2021 Q15b (2 marks) The graph of $y = m\log_3(x - p) + q$ has a vertical asymptote at $x = 5$. If this graph passes through the points $(6, 2)$ and $(14, -6)$, determine the values of m, p and q.

6.4 Applications of logarithmic functions

Video playlist
Applications of logarithmic functions

Modelling with logarithmic functions

Logarithms have many applications in the fields of finance and science. In this section we will look at some examples of those applications.

WORKED EXAMPLE 15 Using the graph of a logarithmic function to approximate logarithms and exponentials

Consider the graph of $y = \log_2(x)$ shown.

Use the graph to estimate the value of m in each of the following.

a $0.8 = \log_2(m)$

b $2^{m-2} - 5 = 0$

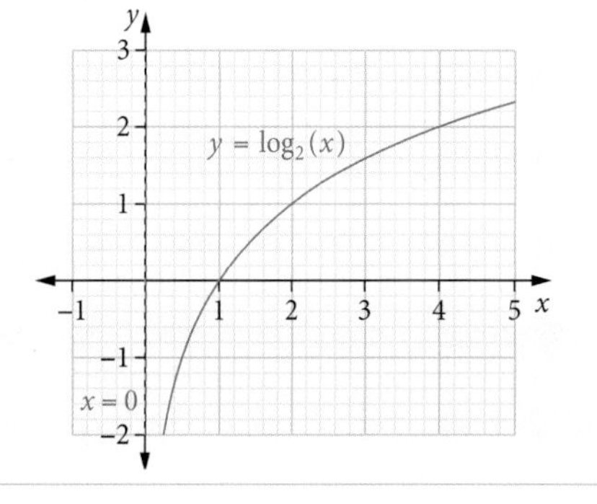

Steps	Working
a The values of $\log_2(x)$ are on the y-axis. Find the x value on the curve that corresponds to $y = 0.8$.	When $y = 0.8$, $x = 1.8$ $\log_2(1.8) = 0.8$ $m = 1.8$

b **1** Express the exponential equation as a logarithmic equation.

$$2^{m-2} - 5 = 0$$
$$2^{m-2} = 5$$
$$m - 2 = \log_2(5)$$

2 Use the graph to find y when $x = 5$.

Substitute into the equation and solve.

From the graph $\log_2(5) = 2.3$.

$$m - 2 = 2.3$$
$$m = 4.3$$

WORKED EXAMPLE 16 Applying logarithmic functions

The cost of manufacturing bicycle components depends on the number produced each day and this cost influences the profit. A company can manufacture a maximum of 80 components in one day. The daily profit from producing x components each day is given by the function $P(x) = (100 - x)\ln(3x + 1) - x$.

a Find the profit, to the nearest dollar, when 30 components are manufactured in a day.

b Sketch the graph of $P(x)$.

c Determine the number of components that result in the highest profit per day.

Steps	Working
a Define the $P(x)$ function on CAS and find $P(30)$.	$P(30) = 285.76$ \$286 profit is made when 30 components are manufactured.

ClassPad

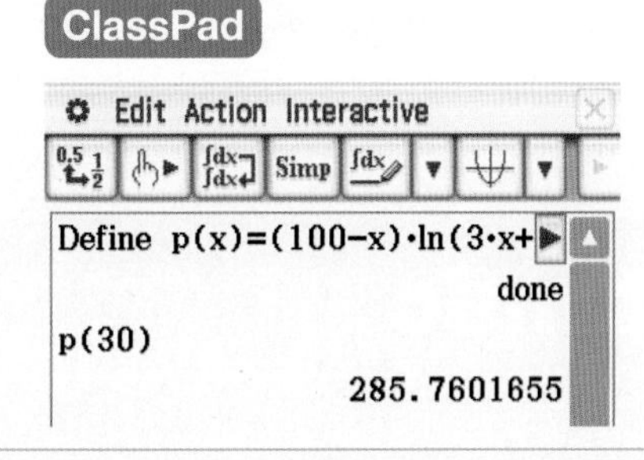

TI-Nspire

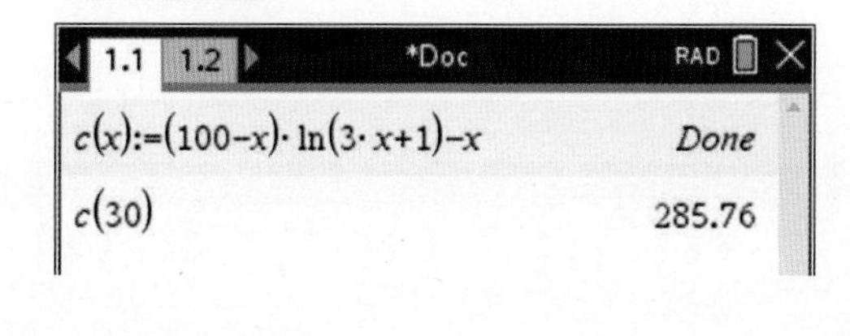

b Use CAS to graph the function over the domain $0 \le x \le 80$. Set the window to an appropriate scale and include the coordinates of the stationary point and the endpoints on the graph.

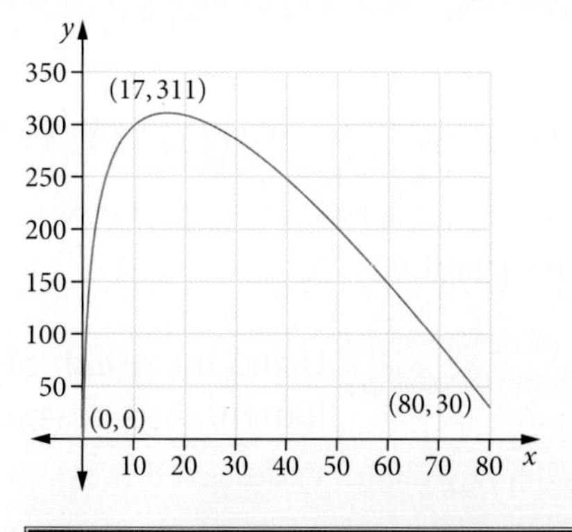

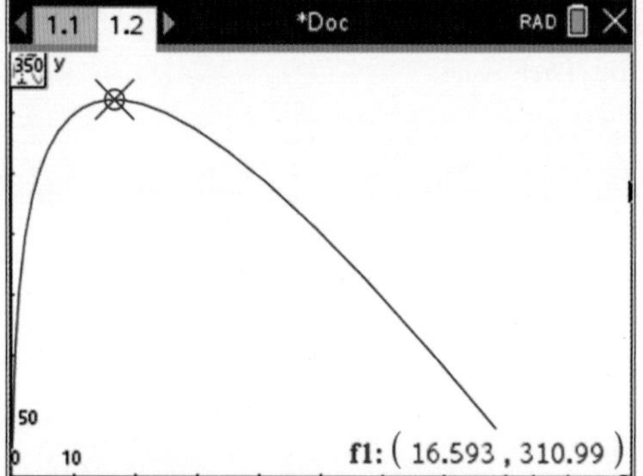

c Find the maximum function value using CAS and state the x-coordinate to the nearest integer.	$f(16) = 310.91$ and $f(17) = 310.95$ The maximum profit is made when 17 components are made each day.

6.4

Logarithmic scales

If a graph has a linear scale, we *add* the same number to move from one scale mark to the next. In the example on the right, the linear scale involves adding 10 each time.

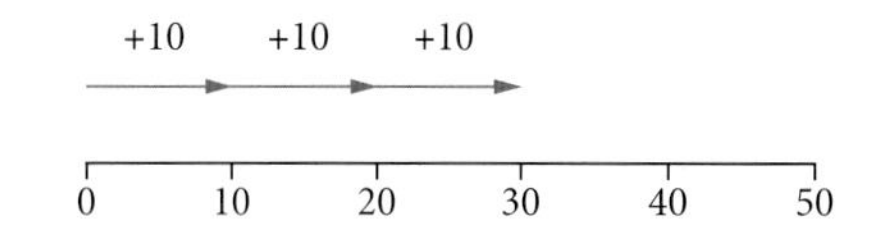

In situations where the values we need to plot cover a very large range, it is better to use a **logarithmic scale** (or **log scale**).

On a log scale, we *multiply* by the same number to move from one scale mark to the next. In the example on the right, the log scale involves multiplying by 10 each time.

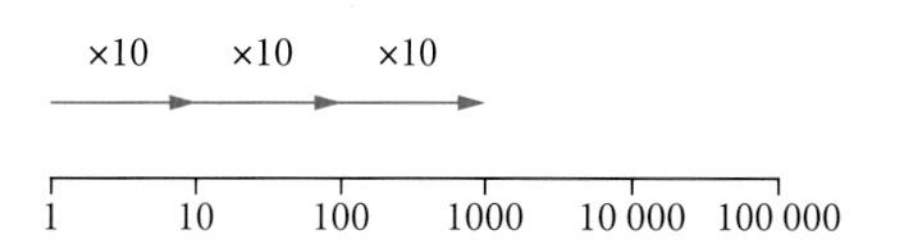

This is called a 'log base 10' or '$\log_{10}$' scale.

Measurements of acidity (pH), earthquake strength (Richter magnitude) and sound intensity (decibels – dB) are examples of logarithms.

WORKED EXAMPLE 17 Comparing values measured on a logarithm scale

© SCSA MM2016 Q12 MODIFIED

The Richter magnitude, M, of an earthquake is determined from the logarithm of the amplitude, A, of waves recorded by seismographs.

$M = \log_{10} \frac{A}{A_0}$, where A_0 is a reference value.

An earthquake in Double Spring Flat (NV, USA) in 1994 was estimated at 6.0 on the Richter scale, whereas the Java earthquake (Indonesia) in 2009 measured 7.0 on the same scale. How many times larger was the amplitude of the waves in Java compared to those of Double Spring Flat?

Steps	Working
1 Change the equation from logarithmic form to exponential form.	$M = \log_{10} \frac{A}{A_0}$ $\frac{A}{A_0} = 10^M$ $A = A_0 \times 10^M$
2 Substitute USA: $A = A_1, M_1 = 6$ and Indonesia: $A = A_2, M_2 = 7$ into: $A = A_0 \times 10^M$	USA: $A_1 = A_0 \times 10^6$ Indonesia: $A_2 = A_0 \times 10^7$
3 Calculate $\frac{A_2}{A_1}$.	$\frac{A_2}{A_1} = \frac{A_0 \times 10^7}{A_0 \times 10^6} = 10$ The amplitude of the waves in Java was 10 times larger than the waves in Double Spring Flat.

WORKED EXAMPLE 18 Problems involving log scales

© SCSA MM2019 Q17a–bi MODIFIED

A scientist is studying the growth of a certain type of bacteria under controlled laboratory conditions. A population of bacteria is incubated at a temperature of 30°C and the size of the population is measured at hourly intervals for five hours. The logarithm of the population size (N) appears to lie on a straight line when plotted against time (measured in hours) and the line of best fit, as shown on the axes below.

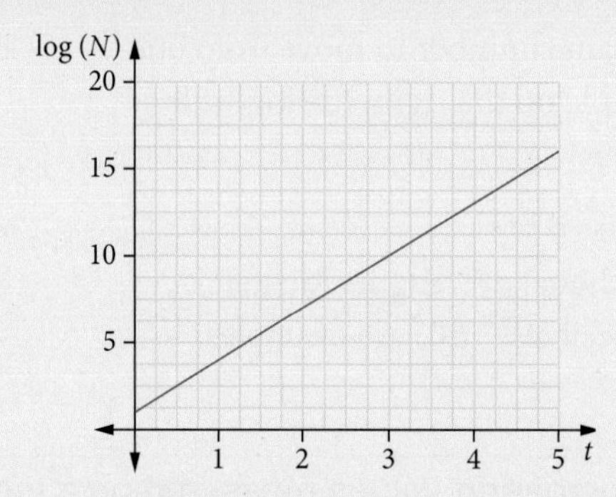

a On the basis of the graph above, what is the size of the bacteria population

i after one hour?

ii after two hours?

b The equation of the line can be written in the form $\log_{10}(N) = At + B$. Use the graph to determine the values of A and B.

c Use the equation to predict the number of bacteria after 20 minutes.

d Express the above equation in the form $N = a(10)^{bt}$.

Steps	Working
a **i** Find the value of $\log(N)$ when $t = 1$ and solve to find N.	$t = 1, \log_{10}(N) = 4$ Change to exponential form: $N = 10^4 = 10\,000$ bacteria
ii Repeat for $t = 2$. $\log(N) = \log_{10}(N)$	$t = 2, \log_{10}(N) = 7$ $N = 10^7 = 10\,000\,000$ bacteria
b Find the gradient and y-intercept from the graph. Use the linear form $y = mx + c$ to write the equation.	Use the points $(1, 4)$ and $(2, 7)$. $A = \frac{7-4}{2-1} = 3$ Vertical axis intercept, $B = 1$. $\log_{10}(N) = 3t + 1$
c Change 20 minutes into hours then substitute the value of t into $\log_{10}(N) = 3t + 1$.	$t = \frac{20}{60} = \frac{1}{3}$ $\log_{10}(N) = 3\left(\frac{1}{3}\right) + 1$ $\log_{10}(N) = 2$ $N = 10^2 = 100$
d Change the equation into exponential form and simplify using index laws.	$\log_{10}(N) = 3t + 1$ $N = 10^{3t+1}$ $N = 10^{3t} \times 10^1$ $N = 10(10)^{3t}$

9780170477536

WACE QUESTION ANALYSIS

 MM2019 Q17 Calculator-assumed (15 marks)

A microbiologist is studying the effect of temperature on the growth of a certain type of bacteria under controlled laboratory conditions. A population of bacteria is incubated at a temperature of 30°C and the size of the population measured at hourly intervals for six hours. The logarithm of the population size appears to lie on a straight line when plotted against time (measured in hours) and the line of best fit shown on the axes below.

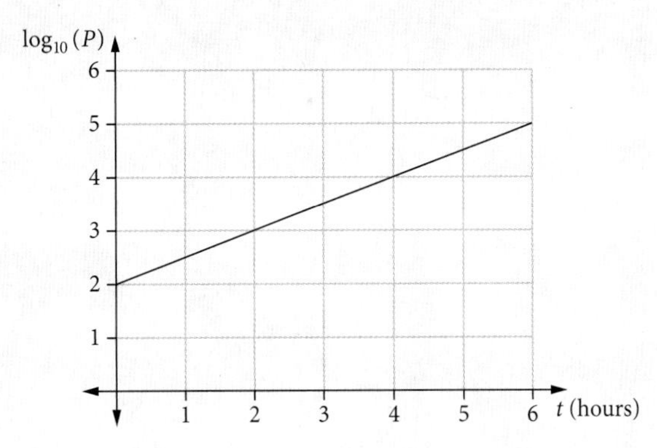

Video WACE question analysis: Logarithmic functions

a **i** On the basis of the graph above, what is the size of the bacteria population after two hours? (2 marks)

ii The equation of the line can be written in the form $\log_{10}(P) = At + B$. Use the graph to determine the values of A and B. (2 marks)

Another population of the same bacteria is cultured at 40°C. The size of the population, P, after t hours satisfies the equation

$$\log_{10}(P) = \frac{1}{3}t + 2.$$

b **i** Express the above equation in the form $P = A(10)^{Bt}$. (3 marks)

ii Determine the size of the population after exactly four hours to the nearest whole number. (1 mark)

iii Express the above equation in the form $t = C\log_{10}\left(\frac{P}{D}\right)$. (3 marks)

iv How many minutes does it take for the population to reach a size of 5000? Give your answer to the nearest minute. (2 marks)

c With reference to parts **a** and **b**, describe the effect of temperature on the population growth of this type of bacteria. (2 marks)

Reading the question

- The graph of the population is a log scale, where the log to base 10 of the population is on the vertical axis.
- Highlight the type, accuracy and units of the answer required in each part.
- Take note of the number of marks allocated to each part of the question. This will give an indication of the amount of working required.

Thinking about the question

- This question requires a knowledge of graphs using logarithmic scales.
- You will also need to be able to transform an equation from logarithmic form to exponential form.
- You will need to be able to find the equation of a straight line using a gradient and y-intercept. Remember, in this case, the subject of the linear equation is $\log_{10}(P)$.

Worked solution (✓ = 1 mark)

a i $\log_{10}(P) = 3$ ✓

$P = 10^3 = 1000$ ✓

ii gradient of $\frac{1}{2}$ and vertical axis intercept of 2

$A = \frac{1}{2}$ ✓ $B = 2$ ✓

$\log_{10}(P) = \frac{1}{2}t + 2$

b i $\log_{10}(P) = \frac{1}{3}t + 2$

$P = 10^{\frac{1}{3}t+2}$ ✓

$P = 10^2 \times 10^{\frac{1}{3}t}$ ✓

$P = 100(10)^{\frac{1}{3}t}$ ✓

ii $P = 100 \cdot 10^{\frac{4}{3}}$

$P = 2154$ ✓

iii

$$\begin{aligned}\log_{10}(P) &= \frac{1}{3}t + 2\\ \log_{10}(P) - 2 &= \frac{1}{3}t\\ \log_{10}(P) - \log_{10}(100) &= \frac{1}{3}t\\ \log_{10}\left(\frac{P}{100}\right) &= \frac{1}{3}t\\ t &= 3\log_{10}\left(\frac{P}{100}\right)\end{aligned}$$

expresses 2 in terms of a log of base 10 ✓

applies appropriate log law to arrive at single log expression (second last line) ✓

determines correct expression ✓

iv $t = 3\log_{10}\left(\frac{5000}{100}\right)$

$t = 5.0969$ **hours** ✓

$t = 306$ **minutes** ✓

c The equation at 30°C has a greater slope than that of the 40°C equation, which indicates a greater growth rate. Parts **a** and **b** would seem to indicate that the lower temperature incubation results in a higher growth rate.

identifies features of the equations in parts a and b that relate to growth ✓

states lower temperature has higher growth ✓

EXERCISE 6.4 Applications of logarithmic functions

ANSWERS p. 400

Recap

1 The graph of the logarithmic function $f(x) = \log_2(x - c) + b$ has a vertical asymptote at $x = 10$ and passes through the point $(12, 21)$. Find the values of b and c.

2 The graph of the logarithmic function $f(x) = \log_3(x - c) + b$ passes through the points $(17, 4)$ and $(11, 3)$. Find the values of b and c.

Mastery

3 WORKED EXAMPLE 15 Consider the graph of $y = \log_2(x)$ shown below.

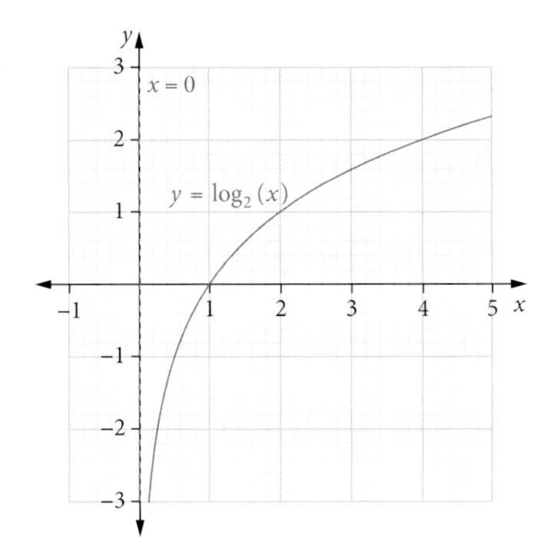

Use the graph to estimate the value of n in each of the following.

a $\log_2(n) = 1.9$

b $2^{n-3} - 0.7 = 0$

4 WORKED EXAMPLE 16 A farmer finds that the cost of raising sheep for wool, meat and breeding stock varies depending on the number of sheep. The farm can support a maximum of 400 sheep. The monthly cost $C(x)$ of raising x sheep is given by the function

$$C(x) = (x - 200)\ln(0.5x + 1) - x + 1000.$$

a Find the monthly cost, to the nearest dollar, when the farmer raises 100 sheep.

b Determine the number of sheep the farmer must raise to produce the least cost.

c How many sheep should the farmer raise to keep costs below \$600 per month?

5 A small colony of black peppered moths live on a small isolated island. In summer, the population begins to increase. If t is the number of days after 12 midnight on 1 January, the equation that best models the number of moths in the colony at any given time is

$$N(t) = 500\ln(21t + 3).$$

a What is the population of the species on 1 January?

b What is the population of moths after 30 days?

c On which day is the population first greater than 2000?

6 WORKED EXAMPLE 17 The Richter magnitude, M, of an earthquake is determined from the logarithm of the amplitude, A, of waves recorded by seismographs.

$$M = \log_{10} \frac{A}{A_0}, \text{ where } A_0 \text{ is a reference value.}$$

An earthquake in México City (Mexico) in 1985 was estimated at 8.0 on the Richter scale, while the San Francisco Bay Area earthquake (CA, USA) in 1989 measured 6.9 on the same scale. How many times larger was the amplitude of the waves in México City compared to the waves in San Francisco Bay Area?

7 WORKED EXAMPLE 18 The number of cases (N) of a virus is recorded daily. The logarithm of the number of cases appears to lie on a straight line when plotted against time (measured in days) and the line of best fit, as shown on the axes below.

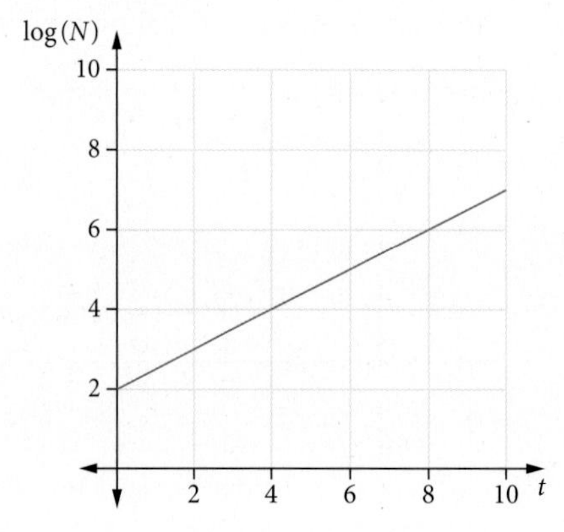

a On the basis of the graph above, how many virus cases are recorded

  i at the start of the outbreak

  ii after two days.

b The equation of the line can be written in the form $\log_{10}(N) = At + B$. Use the graph to determine the values of A and B.

c Use the equation to predict the number of cases after six days.

d Express the above equation in the form $N = a(10)^{bt}$.

Calculator-free

8 © SCSA MM2019 Q4a (3 marks) Consider the graph of $y = \ln(x)$ shown below.

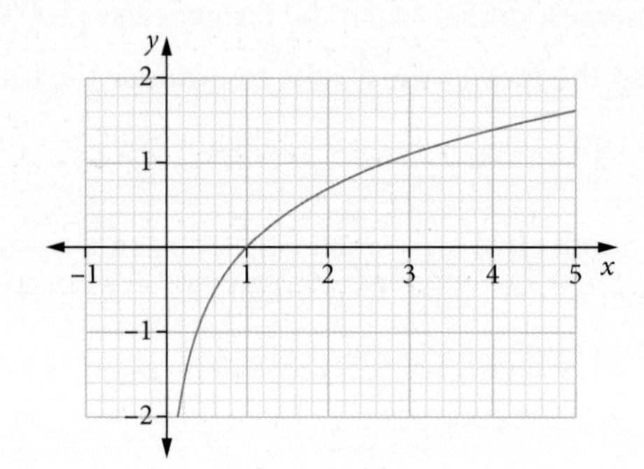

Use the graph to estimate the value of p in each of the following.

a $1.4 = \ln(p)$ (1 mark)

b $e^{p+1} - 3 = 0$ (2 marks)

9 © SCSA MM2019 Q12 (6 marks) Part of Josie's workout at her gym involves a 10 minute run on a treadmill. The treadmill's program makes her run at a constant 12.3 km/h for the first 2 minutes and then her speed, $s(t)$, is determined by the equation below, where t is the time in minutes after she began running.

$$s(t) = 10 - \frac{\ln(t - 1.99)}{t} \text{ km/h}$$

a Sketch the graph of her speed during this run versus time. (3 marks)

b At what time(s) is Josie's speed 10 km/h? (1 mark)

c At what time(s) during her run is Josie's acceleration zero? (2 marks)

10 © SCSA MM2016 Q12 (3 marks) The Richter magnitude, M, of an earthquake is determined from the logarithm of the amplitude, A, of waves recorded by seismographs.

$M = \log_{10} \frac{A}{A_0}$, where A_0 is a reference value.

An earthquake in a town in New Zealand in November 2015 was estimated at 5.5 on the Richter scale, while the earthquake just north of Hayman Island measured 3.4 on the same scale. How many times larger was the amplitude of the waves in New Zealand compared to those at Hayman Island?

11 © SCSA MM2018 Q18ab (4 marks) The ear has the remarkable ability to handle an enormous range of sound levels. In order to express levels of sound meaningfully in numbers that are more manageable, a logarithmic scale is used, rather than a linear scale. This scale is the decibel (dB) scale.

The sound intensity level, L, is given by the formula below:

$$L = 10\log\left(\frac{I}{I_0}\right) \text{ dB}$$

where I is the sound intensity and I_0 is the reference sound intensity.

I and I_0 are measured in watt/m^2.

a Listening to a sound intensity of 5 billion times that of the reference intensity ($I = 5 \times 10^9 I_0$) for more than 30 minutes is considered unsafe. To what sound intensity level does this correspond? (2 marks)

b The reference sound intensity, I_0, has a sound intensity level of 0 dB. If a household vacuum cleaner has a sound intensity $I = 1 \times 10^{-5}$ watt/m^2 and this corresponds to a sound intensity level $L = 70$ dB, determine I_0. (2 marks)

6 Chapter summary

Logarithms

A **logarithm** (or log) is the power or **exponent** to which a base is raised to yield a certain number.

When a logarithm is to base e it is called a natural logarithm.

$\log_e(x) = \ln(x)$

Solving logarithmic and exponential equations

It is often necessary when solving an equation in exponential form to express it as an equation in logarithmic form.

Logarithmic form	$\Leftrightarrow$	Exponential form
$x = \log_a(b)$	$\Leftrightarrow$	$a^x = b$

Laws of Logarithms

$\log_a(mn) = \log_a(m) + \log_a(n)$

$\log_a\left(\frac{m}{n}\right) = \log_a(m) - \log_a(n)$

$\log_a(m^k) = k\log_a(m)$

$\log_a 1 = 0$

$\log_a(a^b) = b$ and $a^{\log_a(b)} = b$

Properties of the logarithmic function $y = \log_a(x)$

- It is a strictly increasing function, increasing quickly at first, then more slowly.
- The gradient of the graph is always decreasing.
- The x-intercept is at $(1, 0)$ as $\log_a(1) = 0$.
- The y-axis ($x = 0$) is a vertical asymptote.

Changing the base of the logarithm does not alter the basic shape, x-intercept or the asymptote of the graph of the logarithmic function.

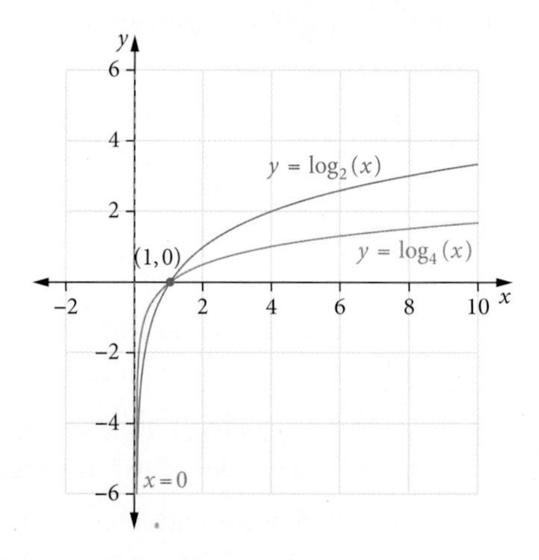

Properties of the logarithmic function $y = \log_a(x - c) + b$, where b and c are positive real constants

- It has the same shape as $y = \log_a(x)$.
- The horizontal translation is c units right and the vertical translation is b units up.
- $x = c$ is the vertical asymptote.
- Include the guiding point $(1 + c, b)$.

$f(x) = \ln(x)$ is called the **natural logarithmic function**, and is also written as $f(x) = \log_e(x)$.

Logarithmic scales

On a log scale, we *multiply* by the same number to move from one scale mark to the next. In the example on the right, the log scale involves multiplying by 10 each time.

This is called a 'log base 10' or '$\log_{10}$' scale.

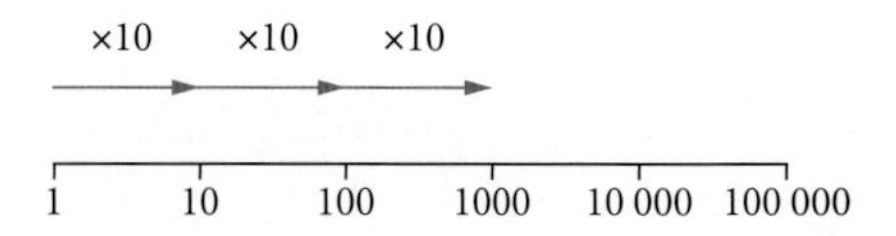

Cumulative examination: Calculator-free

Total number of marks: 22 Reading time: 3 minutes Working time: 22 minutes

1 (5 marks) Given $y = x + \sqrt{x^2 - 4}$, show that $(x^2 - 4)\dfrac{d^2y}{dx^2} + x\dfrac{dy}{dx} - y = 0$.

2 © SCSA MM2016 Q5 (6 marks) Consider the graph of $y = f(x)$ which is drawn below.

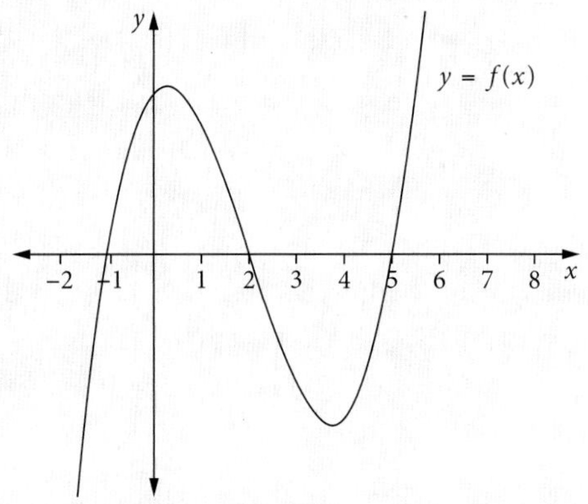

Let $A(x)$ be defined by the integral $A(x) = \int_{-1}^{x} f(t)\,dt$ for $-1 \le x \le 6$.

It is known that $A(2) = 15$, $A(5) = 0$ and $A(6) = 8$.

Copy the axes below and on them sketch the function of $A(x)$ for $-1 \le x \le 6$, labelling clearly key features such as x-intercepts, turning points and inflection points if any.

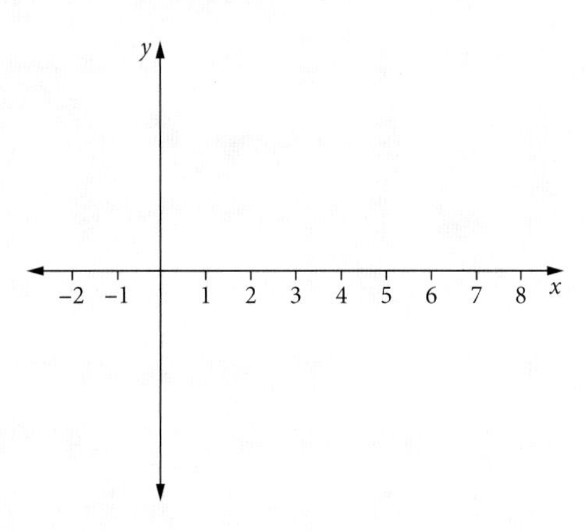

3 (8 marks) Solve each equation for x.

a $e^{2x+3} = 11$ (2 marks)

b $5e^{2x} = 27 + 2e^{2x}$ (2 marks)

c $\log_2(3x - 2) = 4$ (2 marks)

d $\ln(8x + 4) - \ln(2) = 3$ (2 marks)

4 (3 marks) Find the values of a and b if the logarithmic function $f(x) = \ln(x - a) + b$ has a vertical asymptote with equation $x = -6$ and passes through the point $(-5, 4)$.

Cumulative examination: Calculator-assumed

Total number of marks: 30 Reading time: 3 minutes Working time: 30 minutes

1 © SCSA MM2020 Q11 (9 marks) The line $y = x + c$ is tangent to the graph of $f(x) = e^x$.

a Obtain the coordinates of the point of intersection of the tangent with the graph of $f(x)$. (2 marks)

b What is the value of c? (1 mark)

c Sketch the graph of $f(x)$ and the tangent on the same axes. (1 mark)

d Evaluate the exact area between the graph of $f(x)$, the tangent line, and the line $x = \ln 2$. (3 marks)

e Given that $g(x)$ is the inverse function of $f(x)$, write a definite integral that could be used to determine the area between the graph of $g(x)$, the x-axis, and the line $x = \ln 2$. (2 marks)

2 (5 marks) A roulette wheel has thirty-seven slots, numbered 0 to 36. Slot 0 is green. Eighteen of the remaining slots are black, and the other eighteen are red. In a single game, a person spins the wheel and at the same time rolls a ball around the wheel in the opposite direction. As the wheel slows, the ball falls into one of the slots. The wheel is carefully balanced so that the ball is equally likely to fall into any of the slots.

a In a single game, what is the probability, correct to three decimal places, that the ball falls into a black slot? (1 mark)

Several games are played one after the other. Assume that the result of each game is independent of the result of any other game.

b What is the probability, correct to three decimal places, that the first time that the ball falls into a black slot is in the sixth game? (2 marks)

c What is the probability, correct to three decimal places, that the ball falls into a black slot three times in the first six games? (2 marks)

3 © SCSA MM2020 Q13 (7 marks) A company manufactures small machine components. They can manufacture up to 200 of a particular component in one day. The total cost, in hundreds of dollars, incurred in manufacturing the components is given by

$$C(x) = \frac{x \ln(2x + 1)}{3} - 2x + 120$$

where x is the number of components that will be produced on that day.

a Determine the total cost of manufacturing 20 components in one day. (1 mark)

b Sketch the graph of $C(x)$. (3 marks)

c With reference to your graph in part **b**, explain how many components the company should manufacture per day if the total cost is to be as low as possible. (3 marks)

4 © SCSA MM2021 Q16 MODIFIED (9 marks) An analyst was hired by a large company at the beginning of 2021 to develop a model to predict profit. At that time, the company's profit was \$4 million. The model developed by the analyst was:

$$P(x) = \frac{20 \ln (x + a)}{x + 5}$$

where $P(x)$ is the profit in millions of dollars after x weeks and a is a constant.

a Show that $a = e$. (2 marks)

b What does the model predict the profit will be after five weeks? (1 mark)

c What is this maximum profit and during which week will it occur? (2 marks)

d According to the model, during which week will the company's profit fall below its value at the beginning of 2021? (1 mark)

The model proved accurate and after 10 weeks the company implemented some changes. From this time the analyst used a new model to predict the profit:

$$N(y) = 2e^{b(10+y)}$$

where $N(y)$ is the profit in millions of dollars y weeks from this point in time and b is a constant.

e The company is projecting its profit to exceed \$5 million. During which week does the new model suggest this will happen? (3 marks)

CUMULATIVE EXAMINATION: CALCULATOR-ASSUMED

CHAPTER

CALCULUS OF THE NATURAL LOGARITHMIC FUNCTION

Syllabus coverage

TOPIC 3.1 FURTHER DIFFERENTIATION AND APPLICATIONS

The second derivative and applications of differentiation

3.1.10 use the increments formula: $\delta y = \frac{dy}{dx} \times \delta x$ to estimate the change in the dependent variable y resulting from changes in the independent variable x

3.1.11 apply the concept of the second derivative as the rate of change of the first derivative function

3.1.12 identify acceleration as the second derivative of position with respect to time

3.1.13 examine the concepts of concavity and points of inflection and their relationship with the second derivative

3.1.14 apply the second derivative test for determining local maxima and minima

3.1.15 sketch the graph of a function using first and second derivatives to locate stationary points and points of inflection

3.1.16 solve optimisation problems from a wide variety of fields using first and second derivatives

TOPIC 3.2 INTEGRALS

Applications of integration

3.2.19 calculate the area under a curve

3.2.20 calculate the area between curves determined by functions of the form $y = f(x)$

3.2.21 determine displacement given velocity in linear motion problems

3.2.22 determine positions given linear acceleration and initial values of position and velocity

TOPIC 4.1 THE LOGARITHMIC FUNCTION

Calculus of the natural logarithmic function

4.1.9 define the natural logarithm $\ln x = \log_e x$

4.1.10 examine and use the inverse relationship of the functions $y = e^x$ and $y = \ln x$

4.1.11 establish and use the formula $\frac{d}{dx}\ln x = \frac{1}{x}$

4.1.12 establish and use the formula $\int \frac{1}{x}\, dx = \ln x + c$, for $x > 0$

4.1.13 determine derivatives of the form $\frac{d}{dx}(\ln f(x))$ and integrals of the form $\int \frac{f'(x)}{f(x)}\, dx$, for $f(x) > 0$

4.1.14 use logarithmic functions and their derivatives to solve practical problems

Video playlist Differentiating natural logarithmic functions

Worksheets Derivatives of logarithmic functions

Exponential and logarithmic functions

7.1 Differentiating natural logarithmic functions

The first derivative of ln (*x*)

The first derivative of ln (x) can be found using the algebraic property $e^{\ln x} = x$.

Differentiate both sides. $$\frac{d}{dx}\left(e^{\ln x}\right) = \frac{d}{dx}(x)$$

Using the chain rule: $$\frac{d}{dx}(\ln x)e^{\ln x} = 1$$

$$\frac{d}{dx}(\ln x) \times x = 1$$

Therefore, $$\frac{d}{dx}(\ln x) = \frac{1}{x}$$

The first derivative of ln (*x*)

$$\frac{d}{dx}(\ln(x)) = \frac{1}{x}$$

The first derivative of ln (*ax*)

Using the logarithm law: $$\ln(ax) = \ln(a) + \ln(x)$$

$$\frac{d}{dx}(\ln(ax)) = \frac{d}{dx}(\ln(a)) + \frac{d}{dx}(\ln(x))$$

and as ln (a) is a constant, $$\frac{d}{dx}(\ln(a)) = 0$$

$$\frac{d}{dx}(\ln(ax)) = \frac{1}{x}$$

The first derivative of ln (*ax*)

$$\frac{d}{dx}(\ln(ax)) = \frac{1}{x}$$

The first derivative of ln (*f* (*x*))

We can use the chain rule to create a formula for the derivative of the natural logarithm of a function $f(x)$.

If $y = \ln(f(x))$,

then $\frac{dy}{dx} = \frac{1}{f(x)} \times f'(x) = \frac{f'(x)}{f(x)}$.

The chain rule for natural logarithmic functions

If $y = \ln(f(x))$, then $\frac{d}{dx}(\ln(f(x)) = \frac{f'(x)}{f(x)}$.

WORKED EXAMPLE 1	Finding the derivative of $y = \ln(f(x))$
Find the first derivative of each logarithmic function. **a** $y = \ln(3x - 7)$ **b** $y = \ln(9x^2 - x)$	
Steps	**Working**
a 1 Use the rule $\frac{d}{dx}(\ln(f(x))) = \frac{f'(x)}{f(x)}$.	$f(x) = 3x - 7$ $f'(x) = 3$
2 Let $f(x) = \ln(3x - 7)$ and differentiate.	$\frac{d}{dx}(\ln(3x - 7)) = \frac{3}{3x - 7}$
b Let $f(x) = 9x^2 - x$ and differentiate.	$f(x) = 9x^2 - x$ $f'(x) = 18x - 1$ $\frac{d}{dx}(\ln(9x^2 - x)) = \frac{18x - 1}{9x^2 - x}$

Some derivatives of natural logarithmic functions are much easier if the logarithm is first simplified using the laws of logarithms.

The laws of logarithms for natural logarithms

$\ln(xy) = \ln(x) + \ln(y)$

$\ln\left(\frac{x}{y}\right) = \ln(x) - \ln(y)$

$\ln(x^n) = n\ln(x)$

Also remember,

$\ln(1) = 0$

$\ln(e^n) = n$

WORKED EXAMPLE 2	Using the laws of logarithms to find the first derivative of natural logarithmic functions
Find the first derivative of each logarithmic function. **a** $y = \ln(5x)$ **c** $y = \ln(\sqrt{x})$	 **b** $y = \ln((2x - 5))^2$ **d** $y = \ln((x + 2)(x + 5))$
Steps	**Working**
a Use the rule $\frac{d}{dx}(\ln(ax)) = \frac{1}{x}$.	$\frac{d}{dx}(\ln(5x)) = \frac{1}{x}$
b 1 Simplify using the logarithm law $\ln(x^n) = n\ln(x)$.	$y = \ln((2x - 5)^2)$ $y = 2\ln(2x - 5)$
2 Use the chain rule $\frac{dy}{dx} = \frac{dy}{du} \times \frac{du}{dx}$.	$u = 2x - 5$ $y = 2\ln(u)$ $\frac{du}{dx} = 2$ $\frac{dy}{du} = \frac{2}{u} = \frac{2}{2x - 5}$ $\frac{dy}{dx} = 2 \times \frac{2}{2x - 5} = \frac{4}{2x - 5}$

c Simplify using the logarithm laws and differentiate.	$y = \ln(\sqrt{x}) = \ln\left(x^{\frac{1}{2}}\right)$ $y = \frac{1}{2}\ln(x)$ $\frac{dy}{dx} = \frac{1}{2} \times \frac{1}{x}$ $\frac{dy}{dx} = \frac{1}{2x}$
d Simplify using logarithm laws and differentiate.	$y = \ln((x+2)(x+5))$ $y = \ln(x+2) + \ln(x+5)$ $\frac{dy}{dx} = \frac{1}{x+2} + \frac{1}{x+5}$

WORKED EXAMPLE 3 **Finding the first derivative using the product rule**

Find the first derivative of $y = x^2 \ln(x)$.

Steps	**Working**
1 Identify u and v.	$u = x^2$ $\quad v = \ln(x)$
2 Differentiate to obtain $\frac{du}{dx}$ and $\frac{dv}{dx}$.	$\frac{du}{dx} = 2x$ $\quad \frac{dv}{dx} = \frac{1}{x}$
3 Use the product rule $\frac{dy}{dx} = u\frac{dv}{dx} + v\frac{du}{dx}$ and simplify.	$\frac{dy}{dx} = u\frac{dv}{dx} + v\frac{du}{dx}$ $= x^2 \times \frac{1}{x} + \ln(x) \times 2x$ $= x + 2x\ln(x)$

WORKED EXAMPLE 4 **Finding the first derivative using the quotient rule**

Find the first derivative of $f(x) = \frac{\ln(x)}{x^2}$.

Steps	**Working**
1 Let $\frac{u}{v} = \frac{\ln(x)}{x^2}$.	$u = \ln(x)$ $\quad v = x^2$ $\frac{du}{dx} = \frac{1}{x}$ $\quad \frac{dv}{dx} = 2x$
2 Differentiate using the quotient rule $\frac{d}{dx}\left(\frac{u}{v}\right) = \frac{v\frac{du}{dx} - u\frac{dv}{dx}}{v^2}$.	$f'(x) = \frac{x^2 \times \frac{1}{x} - 2x\ln(x)}{(x^2)^2}$ $= \frac{x - 2x\ln(x)}{x^4}$ $= \frac{1 - 2\ln(x)}{x^3}$

Finding the second derivative of a natural logarithmic function

The second derivative of a function is the rate of change of the first derivative. This can be used to find the rate at which the gradient of a function is changing. It is also used to find points of inflection and to determine the nature of a stationary point.

WORKED EXAMPLE 5 Finding the second derivative of a natural logarithmic function

Find the second derivative of $y = \ln(2x - 3)$.

Steps	Working
1 Find the first derivative.	Let $f(x) = 2x - 3$. $f'(x) = 2$ $\frac{d}{dx}(\ln(2x-3)) = \frac{2}{2x-3}$
2 Write $\frac{2}{2x-3}$ as $2(2x-3)^{-1}$ and differentiate to find the second derivative.	$\frac{2}{2x-3} = 2(2x-3)^{-1}$ $\frac{d^2y}{dx^2} = -2(2x-3)^{-2} \times 2$ $= -4(2x-3)^{-2}$ $= \frac{-4}{(2x-3)^2}$

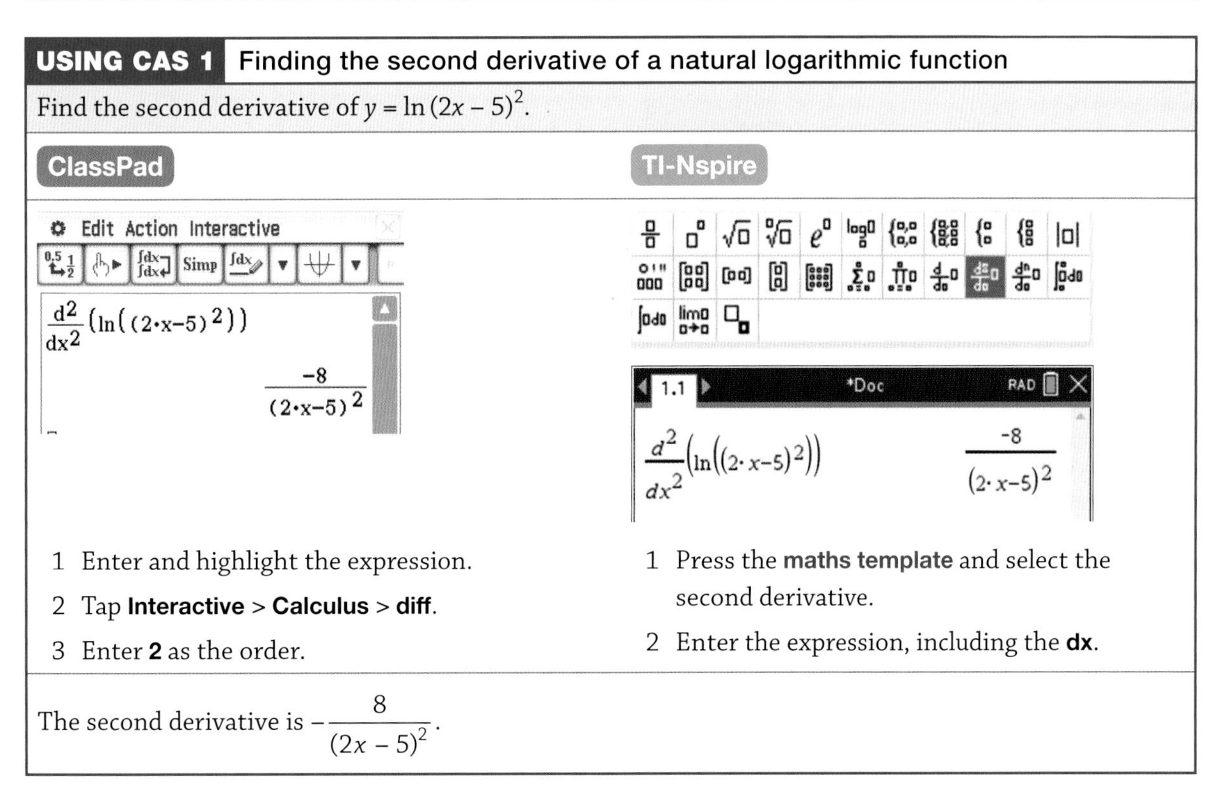

USING CAS 1 Finding the second derivative of a natural logarithmic function

Find the second derivative of $y = \ln(2x - 5)^2$.

ClassPad

1 Enter and highlight the expression.
2 Tap **Interactive** > **Calculus** > **diff**.
3 Enter **2** as the order.

TI-Nspire

1 Press the **maths template** and select the second derivative.
2 Enter the expression, including the **dx**.

The second derivative is $-\frac{8}{(2x-5)^2}$.

EXERCISE 7.1 Differentiating natural logarithmic functions

ANSWERS p. 401

Mastery

1 WORKED EXAMPLE 1 Find $\frac{dy}{dx}$ for each of the natural logarithm functions below.

a $y = \ln(8x - 5)$ **b** $y = \ln(3x^2 + 6x)$ **c** $y = 3\ln(x^4 + 8x)$

2 WORKED EXAMPLE 2 Find the derivative of each function using the laws of logarithms to simplify where necessary.

a $y = \ln(2x)$ **b** $y = 3\ln(5x)$ **c** $y = 2\ln(4x - 3)$

d $y = \ln(\sqrt[4]{x - 4})$ **e** $y = \ln((2x + 1)^3)$

3 WORKED EXAMPLE 3 Find $f'(x)$ for each function.

a $f(x) = (x^2 - 2x)\ln(x)$ **b** $f(x) = x^3\ln(x^3)$ **c** $f(x) = \frac{1}{x}\ln(x)$

4 WORKED EXAMPLE 4 Find $f'(x)$ if $f(x) = \frac{\ln(2x)}{x^3}$.

5 WORKED EXAMPLE 5 Find the second derivatives of the following natural logarithmic functions.

a $y = \ln(5x + 4)$ **b** $y = 2\ln((4x + 1)^2)$

6 Using CAS 1 Given $f(x) = \ln(4x - 3)$, find

a $f'(x)$ **b** $f''(x)$.

Calculator-free

7 (2 marks) Find the first derivative of $f(x) = \sin(\ln(x^2))$ at $x = e$.

8 (4 marks)

a Show that $\ln\sqrt{\frac{3x + 3}{3x - 2}} = \frac{1}{2}\ln(3x + 3) - \frac{1}{2}\ln(3x - 2)$. (3 marks)

b Hence find the first derivative of $f(x) = \ln\left(\sqrt{\frac{3x + 3}{3x - 2}}\right)$ at $x = 2$. (1 mark)

9 (2 marks) If $y = x^2\ln(x)$, find $\frac{dy}{dx}$.

10 (2 marks) Differentiate $x\ln(x)$ with respect to x.

11 (3 marks) Let $f(x) = \frac{\ln(x)}{x^2}$.

a Find $f'(x)$. (2 marks)

b Evaluate $f'(1)$. (1 mark)

12 (2 marks) For $f(x) = \log_e(x^2 + 1)$, find $f'(2)$.

Calculator-assumed

13 © SCSA MM2016 Q13a MODIFIED (2 marks) Determine $\frac{d}{dx}(x^3\ln(2x))$.

Applications of derivatives of the natural logarithmic function

Stationary points and their nature

Local maxima occur when $\frac{dy}{dx} = 0$ and $\frac{d^2y}{dx^2} < 0$.

Local minima occur when $\frac{dy}{dx} = 0$ and $\frac{d^2y}{dx^2} > 0$.

Stationary points of inflection occur when $\frac{dy}{dx} = 0$, $\frac{d^2y}{dx^2} = 0$ and the curve either changes from concave up to concave down or concave down to concave up. A point on a curve where the concavity changes is called a point of inflection and satisfies the same conditions as a stationary point of inflection; however, it is not a stationary point so $\frac{dy}{dx}$ is NOT equal to zero.

Video playlist Applications of derivatives of the natural logarithmic function

WORKED EXAMPLE 6 Finding the coordinates and nature of a local maximum

The function $f(x) = \ln(10x - x^2)$ has a stationary point in the interval $0 < x < 10$.

a Find the coordinates of the stationary point.

b Use the second derivative to determine the nature of the stationary point.

Steps	Working
a 1 Find $f'(x)$. Use the rule $\frac{d}{dx}(\ln(f(x))) = \frac{f'(x)}{f(x)}$.	$f'(x) = \frac{10 - 2x}{10x - x^2}$
2 Solve $f'(x) = 0$.	Stationary when $f'(x) = 0$. $\frac{10 - 2x}{10x - x^2} = 0$ $10 - 2x = 0$ $x = 5$
3 Find $f(5)$ and state the coordinates of the stationary point.	$f(5) = \ln(10 \times 5 - 5^2) = \ln(25)$ Stationary point is $(5, \ln(25))$.
b 1 Find the second derivative by differentiating $f'(x)$ using the quotient rule.	$u = 10 - 2x$ $v = 10x - x^2$ $\frac{du}{dx} = -2$ $\frac{dv}{dx} = 10 - 2x$ $f''(x) = \frac{-2(10x - x^2) - (10 - 2x)(10 - 2x)}{(10x - x^2)^2}$ $= \frac{-20x + 2x^2 - 100 + 40x - 4x^2}{(10x - x^2)^2}$ $= \frac{-2x^2 + 20x - 100}{(10x - x^2)^2}$

2 Find the nature of the stationary point by finding $f''(5)$.	$f''(5) = \dfrac{-2(5)^2 + 20(5) - 100}{(10(5) - (5)^2)^2}$ $= \dfrac{-50}{625} = -\dfrac{2}{25}$ The stationary point $(5, \ln 25)$ is a local maximum as $f''(5) < 0$.

The first derivative can also be used to find an optimum solution for a function which may be the minimum production cost or maximum population number.

WORKED EXAMPLE 7 Finding the optimum solution for a natural logarithmic function

The population of tadpoles in a dam is recorded each week for eight weeks. The number of tadpoles N, after t weeks, is modelled by the function $N(t) = 100\ln(-t^2 + 8t + 9)$.

Find

a $N'(t)$

b the number of weeks when the population of tadpoles is a maximum

c the maximum population of tadpoles.

Steps	Working
a Find $N'(t)$. Use the rule $\dfrac{d}{dx}(\ln(f(x)) = \dfrac{f'(x)}{f(x)}$.	$N'(t) = \dfrac{100(8 - 2t)}{-t^2 + 8t + 9}$
b Solve $N'(t) = 0$.	Stationary when $N'(t) = 0$. $100(8 - 2t) = 0$ $t = 4$ The population of tadpoles is a maximum at 4 weeks.
c Find $N(4)$ and round to the nearest integer.	$N(4) = 100\ln(-(4)^2 + 8(4) + 9)$ $N(4) = 100\ln(25) \approx 322$ tadpoles

WORKED EXAMPLE 8 Finding the equation of the tangent

Find the equation of the tangent to the curve $f(x) = \ln(2x + e)$ at $x = 0$.

Steps	Working
1 Differentiate $f(x)$.	$f'(x) = \dfrac{2}{2x + e}$
2 Find $f'(0)$ and $f(0)$.	$f'(0) = \dfrac{2}{2(0) + e} = \dfrac{2}{e}$ $f(0) = \log_e(2(0) + e) = \log_e(e) = 1$
3 Use the formula $y - y_1 = m(x - x_1)$ to find the equation of the tangent.	$m = \dfrac{2}{e}$ The point $(0, 1)$ is on the curve $f(x)$. $y - 1 = \left(\dfrac{2}{e}\right)(x - 0)$ $y = \left(\dfrac{2}{e}\right)x + 1$

The increments formula

The increments formula can be used to approximate the increase in the y value for a corresponding small increase in the x value.

$$\delta y \approx \frac{dy}{dx} \times \delta x$$

For a given function $y = f(x)$, we can use δy to find an approximation for $f(x + \delta x)$.

$$f(x + \delta x) \approx f(x) + \delta y$$

WORKED EXAMPLE 9 **Applying the increments formula**

Given that $\ln(5) \approx 1.609$, use the increments formula to determine an approximation for $\ln(5.01)$.

Steps	Working
1 Find the first derivative of $\ln(x)$.	$y = \ln(x)$ $\frac{dy}{dx} = \frac{1}{x}$
2 Find the values of x and δx.	x increases from 5 to 5.01, therefore, $x = 5$ and $\delta x = 0.01$.
Find the value of $\frac{dy}{dx}$ at the given x value.	When $x = 5$ $\frac{dy}{dx} = \frac{1}{5}$.
3 Substitute into $\delta y \approx \frac{dy}{dx} \times \delta x$.	$\delta y \approx \frac{1}{5} \times 0.01 = 0.002$
4 Substitute into $f(x + \delta x) \approx f(x) + \delta y$.	$f(x) = \ln(x)$ $f(x + \delta x) \approx f(x) + \delta y$ $f(5.01) \approx f(5) + 0.002$ $\approx 1.609 + 0.002$ Therefore, $\ln(5.01) \approx 1.611$.

Straight line motion and the natural logarithmic function

Displacement: $x(t)$

Velocity: $v(t) = \frac{dx}{dt}$

Acceleration: $a(t) = \frac{dv}{dt} = \frac{d^2x}{dt^2}$

WORKED EXAMPLE 10 Straight line motion

The distance covered by Ali on her morning run is given by the function $x(t) = 5\ln(4t + 1)$, where x is the number of kilometres travelled in t hours. Find

a Ali's running speed at time t hours

b the number of hours Ali has been running when her speed is 10 km/h

c Ali's acceleration after she has been running for 2 hours.

Steps	Working
a Find the first derivative. $v(t) = x'(t)$	$x'(t) = \dfrac{5 \times 4}{4t + 1}$ $v(t) = \dfrac{20}{4t + 1}$ km/h
b Solve $x'(t) = 10$.	$\dfrac{20}{4t + 1} = 10$ $20 = 10(4t + 1)$ $4t + 1 = 2$ $t = \dfrac{1}{4}$ h
c **1** Find the second derivative of $x(t)$. $a(t) = v'(t) = x''(t)$	$v(t) = 20(4t + 1)^{-1}$ $v'(t) = -20(4t + 1)^{-2} \times 4$ $a(t) = \dfrac{-80}{(4t + 1)^2}$
2 Substitute $t = 2$ into $a(t)$.	$a(2) = \dfrac{-80}{(4(2) + 1)^2}$ $a(2) = \dfrac{-80}{81}$ km/h^2

EXERCISE 7.2 Applications of derivatives of the natural logarithmic function

ANSWERS p. 401

Recap

1 Find $\dfrac{dy}{dx}$ for each of the natural logarithmic functions below.

a $y = \ln(5 - 2x)$

b $y = \ln(x^3 + x^2)$

2 Find the second derivative of the function $y = \ln(x + 6)$.

Mastery

3 WORKED EXAMPLE 6 The function $f(x) = \ln(8x - x^2)$ has a stationary point in the interval $0 < x < 8$.

a Find the coordinates of the stationary point.

b Use the second derivative to determine the nature of the stationary point.

4 WORKED EXAMPLE 7 The population of frogs in a wetland is recorded each week for ten weeks. The number of frogs N, after t weeks, is modelled by the function

$$N(t) = 200\ln(-t^2 + 12t + 13).$$

Find

a $N'(t)$

b the number of weeks when the population of frogs is a maximum

c the maximum population of frogs.

5 WORKED EXAMPLE 8 Find the equation of the tangent to the curve $f(x) = \ln(x + e^2)$ at $x = 0$.

6 Find the equation of the tangent to the graph of $y = \ln(x)$ at the point $(3, \ln(3))$.

7 Find the equation of the tangent to the graph of $y = 3\ln(x - 2)$ at the point where the curve crosses the x-axis.

8 WORKED EXAMPLE 9 Given that $\ln(3) \approx 1.0986$, use the increments formula to determine an approximation for $\ln(3.003)$.

9 WORKED EXAMPLE 10 Simon rows a straight stretch of river, for three hours each evening. The distance covered by Simon is given by the function $x(t) = 8\ln(2t + 1)$, where x is the number of kilometres travelled in t hours. Find

a Simon's rowing speed at time t hours

b the number of hours Simon has been rowing when his speed is 4 km/h

c Simon's acceleration when $t = 1$ hour.

Calculator-free

10 © SCSA MM2018 Q6 (8 marks) A company manufactures and sells an item for $\$x$. The profit, $\$P$, made by the company per item sold is dependent on the selling price and can be modelled by the function

$$P(x) = \frac{50\ln\left(\frac{x}{2}\right)}{x^2} \text{ where } 1.5 \le x \le 10$$

The graph of $P(x)$ is shown below:

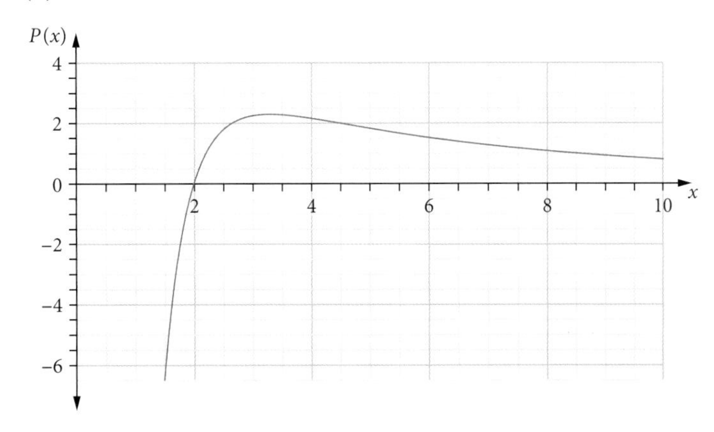

a Describe how the profit per item sold varies as the selling price changes. (3 marks)

b Determine the exact price that should be charged for the item if the company wishes to maximise the profit per item sold. (5 marks)

11 © SCSA MM2021 Q1b (3 marks) Let $f'(x) = x\ln(2x)$. Determine a simplified expression for the rate of change of $f'(x)$.

12 © SCSA MM2021 Q3 (3 marks) Given that $\ln(2) \approx 0.693$, use the increments formula to determine an approximation for $\ln(2.02)$.

Calculator-assumed

13 © SCSA MM2016 Q13ab (5 marks)

a Determine $\frac{d}{dx}(x^2 \ln x)$. (2 marks)

b Using your answer from part **a**, show that the graph of $y = x^2 \ln x$ has only one stationary point. (3 marks)

14 (9 marks) The distance covered by a marathon runner in a training run is given by the function $x(t) = \frac{18\ln(2t+1)}{5}$, where x is the number of kilometres travelled in t hours.

Find

a the speed in terms of t (2 marks)

b the acceleration in terms of t (2 marks)

c the runner's acceleration after 2 hours (2 marks)

d after how many hours will the runner be slowing down at a rate of 1 km/h. Give your answer in hours and minutes, to the nearest minute. (3 marks)

Video playlist Integrals producing natural logarithmic functions

Worksheet Integration of $y = \frac{1}{x}$

7.3 Integrals producing natural logarithmic functions

Integration of reciprocal functions

A **reciprocal function** is a fraction where the variable x, appears only in the denominator.

$$\frac{d}{dx}\ln(x) = \frac{1}{x}$$

So, if we integrate both sides of the equation

$$\int \frac{d}{dx}\ln(x)\,dx = \int \frac{1}{x}\,dx$$

$$\int \frac{1}{x}\,dx = \ln(x) + c$$

Note that this integral is only defined for $x > 0$ because this is the domain of $\ln(x)$.

For the case where $x < 0$, $-x > 0$ so $\ln(-x)$ is defined.

$$\frac{d}{dx}(\ln(-x)) = \frac{1}{-x} \times -1 \quad \text{by chain rule}$$

$$\frac{d}{dx}(\ln(-x)) = \frac{1}{x}$$

$$\int \frac{1}{x}\,dx = \ln(-x) + c, \text{ where } x < 0.$$

We can summarise this as $\int \frac{1}{x}\,dx = \begin{cases} \ln(x) + c, & x > 0 \\ \ln(-x) + c, & x < 0 \end{cases}$

In the Methods course we only consider the case where the denominator is positive.

The integral of $\frac{1}{x}$

$$\int \frac{1}{x}\,dx = \ln(x) + c, \text{ where } x > 0$$

WORKED EXAMPLE 11 Integrating a simple reciprocal function

Find $\int \frac{4}{7x}\,dx$, $x > 0$.

Steps	Working
1 Factorise by taking out the constant $\frac{4}{7}$.	$\int \frac{4}{7x}\,dx = \frac{4}{7}\int \frac{1}{x}\,dx$
2 Use $\int \frac{1}{x}\,dx = \ln(x) + c$.	$= \frac{4}{7}\ln(x) + c$

Integrating $y = \frac{f'(x)}{f(x)}$

In section 7.1, we used the chain rule to find the derivative below.

$$\frac{d}{dx}\ln(f(x)) = \frac{f'(x)}{f(x)}$$

If we integrate both sides

$$\int \frac{d}{dx}\ln(f(x))\,dx = \int \frac{f'(x)}{f(x)}\,dx$$

$$\int \frac{f'(x)}{f(x)}\,dx = \ln(f(x)) + c$$

Integral of $\frac{f'(x)}{f(x)}$

$$\int \frac{f'(x)}{f(x)}\,dx = \ln(f(x)) + c \text{ for } f(x) > 0$$

When integrating a function of the form $\frac{g(x)}{f(x)}$, test whether the numerator, $g(x)$, is a multiple of the derivative of the denominator, $f'(x)$. If this is the case, the integral will be a natural logarithmic function.

WORKED EXAMPLE 12 Integrals of the form $\int \frac{f'(x)}{f(x)}\,dx$ where $f(x) > 0$

Find each integral.

a $\int \frac{2x-3}{x^2-3x+5}\,dx$ where $x^2 - 3x + 5 > 0$

b $\int \frac{12x}{3x^2-7}\,dx$ where $3x^2 - 7 > 0$

Steps	Working
a 1 Find the derivative of the denominator, $f(x)$.	$f(x) = x^2 - 3x + 5$ $f'(x) = 2x - 3$
2 As this derivative is equal to the numerator, write the integral in the form $\int \frac{f'(x)}{f(x)}\,dx = \ln(f(x)) + c$.	$\int \frac{2x-3}{x^2-3x+5}\,dx = \ln(x^2 - 3x + 5) + c$ The restriction $x^2 - 3x + 5 > 0$ ensures the natural logarithmic function is defined.
b 1 Find the derivative of the denominator, $f(x)$.	$f(x) = 3x^2 - 7$ $f'(x) = 6x$
2 As this derivative is equal to a multiple of the numerator, write the integral in the form $\int \frac{f'(x)}{f(x)}\,dx = \ln(f(x)) + c$.	$\int \frac{6x}{3x^2-7}\,dx = \ln(3x^2 - 7)$
3 Multiply both sides by 2 and include the constant in your answer.	$\int \frac{12x}{3x^2-7}\,dx = 2\ln(3x^2 - 7) + c$

USING CAS 2 Finding integrals that produce a natural logarithmic function

Find $\int \frac{12x}{3x^2-7}\,dx$.

ClassPad

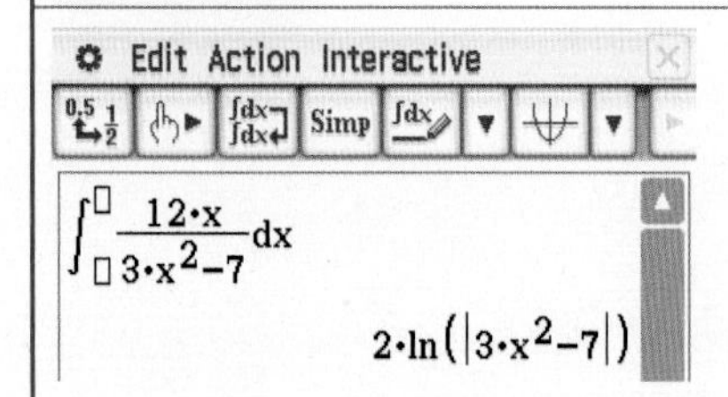

1 Enter and highlight the expression.

2 Tap **Interactive** > **Calculus** > ∫.

3 Tap **OK**.

TI-Nspire

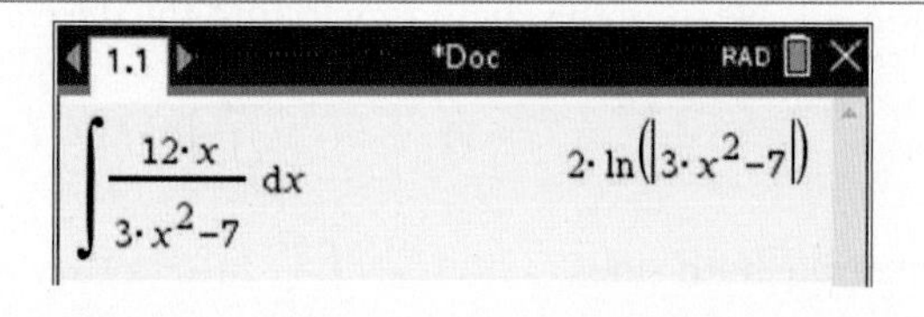

1 Press **menu > calculus > integral**.

2 Enter the expression, including the **dx**.

$$\int \frac{12x}{3x^2-7}\,dx = 2\ln(3x^2 - 7) + c$$

Exam hack

$|3x^2 - 7|$ means the absolute value or modulus of $3x^2 - 7$.

The modulus of a value is its magnitude and can never be negative. This ensures that the natural logarithmic function is always defined, as $\ln(x)$ is not defined when x is negative. It is not necessary to include this modulus sign in your exam answers as this is beyond the scope of the course.

Integration by recognition

Integration by recognition uses the derivative of a function to find the anti-derivative.

WORKED EXAMPLE 13 Integration by recognition

Find the first derivative of $2x\ln(2x)$ and hence find $\int \ln(2x)\,dx$ where $x > 0$.

Steps	Working
1 Find the first derivative of $2x\ln(2x)$ using the product rule.	$u = 2x$ $v = \ln(2x)$ $\frac{du}{dx} = 2$ $\frac{dv}{dx} = \frac{1}{x}$ $\frac{dy}{dx} = 2x \times \frac{1}{x} + 2\ln(2x)$ $\frac{dy}{dx} = 2 + 2\ln(2x)$
2 Write as a derivative equation and integrate both sides.	$\frac{d}{dx}(2x\ln(2x)) = 2 + 2\ln(2x)$ $\int \frac{d}{dx}(2x\ln(2x))\,dx = \int 2 + 2\ln(2x)\,dx$
3 Simplify the equation.	$2x\ln(2x) = \int 2\,dx + 2\int \ln(2x)\,dx$ $2x\ln(2x) = 2x + 2\int \ln(2x)\,dx$
4 Transpose so that $\int \ln(2x)\,dx$ is the subject of the equation.	$2\int \ln(2x)\,dx = 2x\ln(2x) - 2x$ $\int \ln(2x)\,dx = x\ln(2x) - x + c$

WORKED EXAMPLE 14 Integrating $\frac{1}{ax+b}$ where $x > -\frac{b}{a}$

Find $\int \frac{4}{12x+5}\,dx$ where $x > -\frac{5}{12}$.

Steps	Working
1 Find the derivative of the denominator, $f(x)$.	$f(x) = 12x + 5$ $f'(x) = 12$
2 As this derivative is equal to a multiple of the numerator, write the integral in the form $\int \frac{f'(x)}{f(x)}\,dx = \ln(f(x)) + c$.	$\int \frac{12}{12x+5}\,dx = \ln(12x+5)$
3 Multiply both sides by $\frac{1}{3}$.	$\frac{1}{3}\int \frac{12}{12x+5}\,dx = \frac{1}{3}\ln(12x+5) + c$ $\int \frac{4}{12x+5}\,dx = \frac{1}{3}\ln(12x+5) + c$

WORKED EXAMPLE 15	Evaluating definite integrals
Evaluate $\int_3^5 \frac{1}{x-2}\,dx$.	
Steps	**Working**
1 Find the integral.	$\int_3^5 \frac{1}{x-2}\,dx = [\ln(x-2)]_3^5$
2 Evaluate the integral. Remember, $\ln(1) = 0$.	$= \ln(5-2) - \ln(3-2)$ $= \ln(3) - \ln(1)$ $= \ln(3)$

WORKED EXAMPLE 16	Finding $f(x)$ given $f'(x)$ and a point
Find the equation of the curve $f(x)$ given that $f'(x) = \frac{2}{2x+7}$ where $x > -\frac{7}{2}$ and the curve passes through $(1, 0)$.	
Steps	**Working**
1 Integrate $f'(x)$ to find $f(x)$.	$f'(x) = \frac{2}{2x+7}$ $\int f'(x)\,dx = \int \frac{2}{2x+7}\,dx$ $f(x) = \int \frac{2}{2x+7}\,dx$
2 Use the formula $\int \frac{f'(x)}{f(x)}\,dx = \ln(f(x)) + c$.	$f(x) = \ln(2x+7) + c$
3 Substitute the coordinates $(1, 0)$ to find the constant, c. Write the function $f(x)$.	$0 = \ln(2 \times 1 + 7) + c$ $c = -\ln(9)$ $f(x) = \ln(2x+7) - \ln(9)$ $f(x) = \ln\left(\frac{2x+7}{9}\right),\ x > -\frac{7}{2}$

WORKED EXAMPLE 17	Finding an unknown pronumeral
Given that $\int_2^k \left(\frac{2}{2x+5}\right) dx = 7$, find the value of k.	
Steps	**Working**
1 Find the integral.	$\int_2^k \left(\frac{2}{2x+5}\right) dx = [\log_e(2x+5)]_2^k$ $= \log_e(2k+5) - \log_e(2(2)+5)$ $= \log_e(2k+5) - \log_e(9)$ $= \log_e\left(\frac{2k-5}{9}\right)$
2 Make this equal to 7 and solve.	$\log_e\left(\frac{2k+5}{9}\right) = 7$ $\frac{2k+5}{9} = e^7$ $2k+5 = 9e^7$ $2k = 9e^7 - 5$ $k = \frac{9e^7 - 5}{2}$

EXERCISE 7.3 Integrals producing natural logarithmic functions

ANSWERS p. 402

7.3

Recap

1 Find $f'(x)$ given $f(x) = e^x \ln(x)$.

2 For $f(x) = \log_e(x^3 + 1)$, find $f'(2)$.

Mastery

3 WORKED EXAMPLE 11 Find each integral for $x > 0$.

a $\int \frac{2}{x} dx$ **b** $\int \frac{6}{5x} dx$ **c** $\int \frac{1}{3x} dx$

4 WORKED EXAMPLE 12 Find each integral.

a $\int \frac{2x + 11}{x^2 + 11x - 15} dx$ for $x^2 + 11x - 15 > 0$ **b** $\int \frac{15x^2}{x^3 - 13} dx$ for $x^2 - 13 > 0$

c $\int \frac{18x^2 + 16x}{3x^3 + 4x^2 + 1} dx$ for $3x^3 + 4x^2 + 1 > 0$

5 Using CAS 2 Find each integral.

a $\int \frac{1}{x^2 - 11x + 30} dx$ for $x^2 - 11x + 30 > 0$ **b** $\int \frac{3}{4x^2 - 25} dx$ for $4x^2 - 25 > 0$

6 WORKED EXAMPLE 13

a Find the first derivative of $f(x) = x^2 \log_e(2x)$ where $x > 0$.

b Hence, find $\int x \log_e(2x)\, dx$.

7 a Find the first derivative of $f(x) = x \log_e(x^3)$ where $x > 0$.

b Hence, find $\int \log_e(x^3)\, dx$.

8 Find each integral.

a $\int \frac{1}{5x + 3} dx$ for $x > -\frac{3}{5}$ **b** $\int \frac{3}{2x - 5} dx$ for $x > \frac{5}{2}$

9 WORKED EXAMPLE 15 Evaluate each definite integral.

a $\int_1^5 \frac{1}{x} dx$ **b** $\int_2^9 \frac{1}{x - 1} dx$ **c** $\int_6^7 \frac{1}{3x - 2} dx$ **d** $\int_2^4 \frac{1}{20 - 3x} dx$ **e** $\int_e^{4e} \frac{1}{x} dx$

10 WORKED EXAMPLE 16 Find the equation of the curve $f(x)$ given that $f'(x) = \frac{7}{3x - 5}$ where $x > \frac{5}{3}$ and $f(2) = 7$.

11 Find the equation of the curve $f(x)$ given that $f'(x) = \frac{9}{x - 3} + 4$ where $x > 3$ and $f(4) = 5$.

12 WORKED EXAMPLE 17 Given that $\int_2^m \frac{3}{3x - 1} dx = 7$, find the value of m.

13 Given that $\int_k^4 \frac{-1}{5 - x} dx = \ln(2)$, find the value of k.

Calculator-free

14 (4 marks) Let $y = x\log_e(3x)$ where $x > 0$.

a Find $\frac{dy}{dx}$. (2 marks)

b Hence, calculate $\int_1^2 (\log_e(3x) + 1)\,dx$. Express your answer in the form $\log_e(a)$, where a is a positive integer. (2 marks)

15 (5 marks)

a Let $\int_4^5 \frac{2}{2x - 1}\,dx = \log_e(b)$. Find the value of b. (2 marks)

b Find p given that $\int_2^3 \frac{1}{1 - x}\,dx = \log_e(p)$. (3 marks)

16 © SCSA MM2018 Q7 (6 marks)

a Determine a simplified expression for $\frac{d}{dx}(x\ln(x))$. (2 marks)

b Use your answer from part **a** to show that $\int \ln(x)\,dx = x\ln(x) - x + c$, where c is a constant. (4 marks)

Calculator-assumed

17 (7 marks) The function $f(x)$ has the first derivative $f'(x) = \frac{x + 5}{x - 1}$, where $x > 1$ and $f(2) = 1$.

a If $f'(x) = a + \frac{b}{x - 1}$, show that $a = 1$ and $b = 6$. (1 mark)

b Find $f(x)$. (3 marks)

c Find the gradient of $f(x)$ at $x = 2$. (1 mark)

d Find the equation of the tangent to $f(x)$ at $x = 2$. (2 marks)

Video playlist
Applications of anti-differentiation involving natural logarithms

7.4 Applications of anti-differentiation involving natural logarithms

The area between a curve and the x-axis

The area bounded by the curve $y = \frac{1}{x}$, the x-axis and the lines $x = a$ and $x = b$ is given by the integral equation:

$$\text{area} = \int_a^b \frac{1}{x}\,dx$$

$$\text{area} = [\ln(x)]_a^b = \ln(b) - \ln(a) \text{ units}^2$$

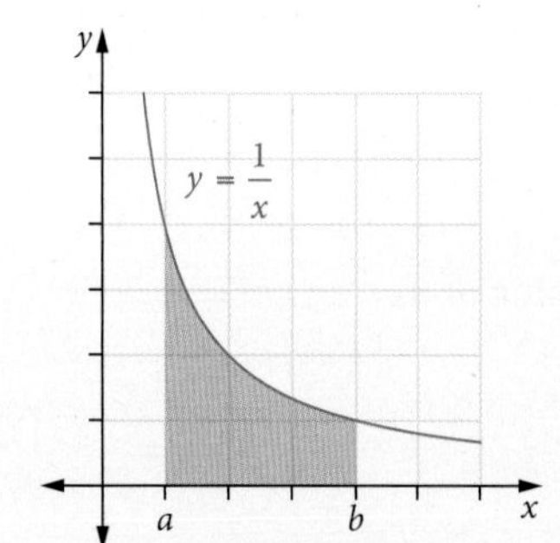

Exam hack

Always sketch the graph of the function when calculating the area and write square units or units^2 after evaluating the integral.

WORKED EXAMPLE 17 Calculating the area between a reciprocal function and the x-axis

Find the area bounded by the curve $f(x) = \dfrac{1}{2x - 4}$, the x-axis and the lines $x = 3$ and $x = 6$.

Steps	Working
1 Sketch the graph of the function and shade the area described. The vertical asymptote occurs where: $2x - 4 = 0$, $x = 2$	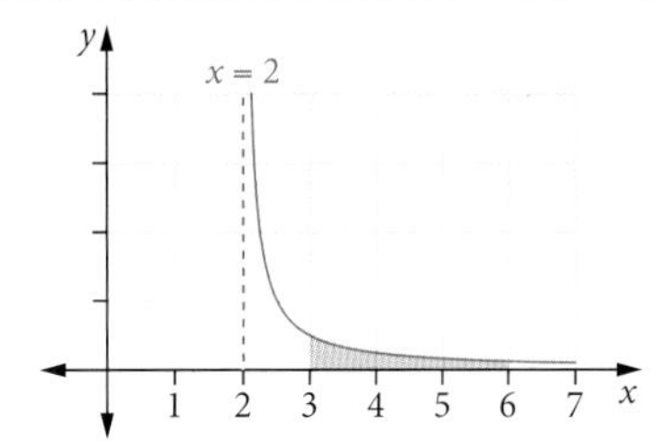
2 Write an integral equation for the area and evaluate. Use the laws of logarithms to simplify the answer.	$\text{area} = \int_3^6 \frac{1}{2x-4}\,dx$ $= \frac{1}{2}\left[\ln(2x-4)\right]_3^6$ $= \frac{1}{2}(\ln(2 \times 6 - 4) - \ln(2 \times 3 - 4))$ $= \frac{1}{2}(\ln(8) - \ln(2))$ $= \frac{1}{2}\ln(4) = \ln(2)\text{ units}^2$

The area bounded by two curves

In the formula given below, $f(x)$ is the upper function and $g(x)$ the lower function. The area is bounded by the functions between the intersection points $x = a$ and $x = b$.

Areas between curves

If $f(x) > g(x)$ for $a < x < b$, then the upper function is $f(x)$ and the lower function is $g(x)$.

$$\text{bounded area} = \int_a^b (\text{upper} - \text{lower})\,dx = \int_a^b [f(x) - g(x)]\,dx$$

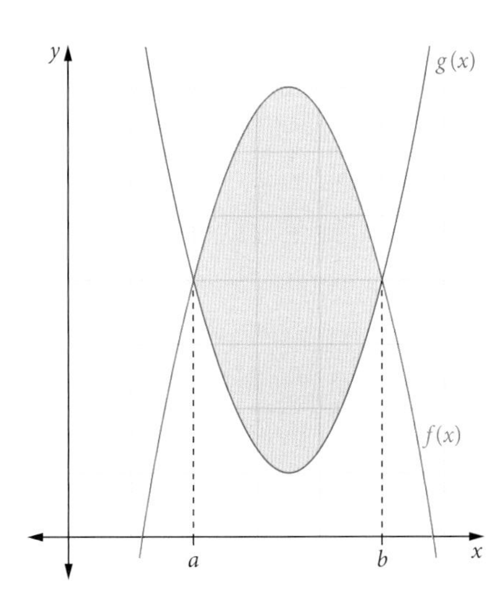

WORKED EXAMPLE 18 Calculating the area bounded by two functions

The functions $f(x) = e^x$ and $g(x) = 3$ intersect at the point $(b, 3)$.

Find

a the exact value of b

b the area bounded by the $f(x)$, $g(x)$ and the y-axis.

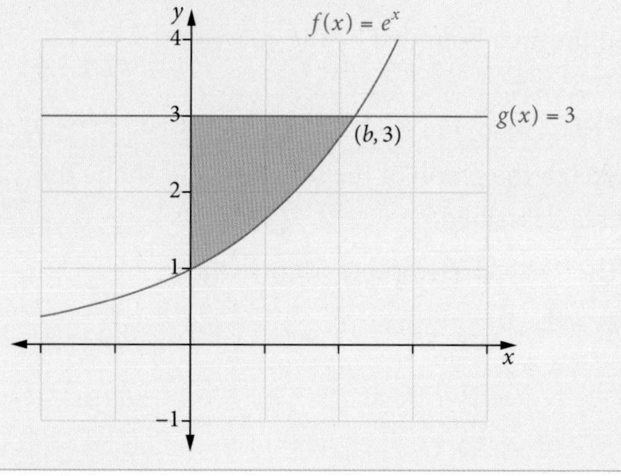

Steps	Working
a Solve $f(x) = g(x)$. Change the equation into exponential form.	$e^x = 3$ $x = \ln(3)$ $b = \ln(3)$
b 1 Write an integral equation for the area and evaluate.	area $= \int_0^{\ln 3} (3 - e^x)\,dx$ $= \left[3x - e^x\right]_0^{\ln 3}$
2 Use the laws of logarithms to simplify the answer.	$= 3\ln(3) - e^{\ln 3} - (0 - e^0)$
3 Use the logarithm law $a^{\log_a b} = b$ to simplify $e^{\ln 3}$.	$= 3\ln(3) - 3 + 1$ $= 3\ln(3) - 2$ units^2

WACE QUESTION ANALYSIS

Video WACE question analysis: Calculus of the natural logarithmic function

© SCSA MM2020 Q7 Calculator-free (13 marks)

Consider the function $f(x) = e^{2x} - 4e^x$.

a Determine the coordinates of the x-intercept(s) of f. You may wish to consider the factorised version of f: $f(x) = e^x(e^x - 4)$. (3 marks)

b Show that there is only one turning point on the graph of f, which is located at $(\ln(2), -4)$. (3 marks)

c Determine the coordinates of the point(s) of inflection of f. (3 marks)

d Copy the axes on the right and on them sketch the function f, labelling clearly all intercepts, the turning point and point(s) of inflection. Some approximate values of the natural logarithmic function provided in the table below may be helpful.

x	1	2	3	4
$\ln(x)$	0	0.7	1.1	1.4

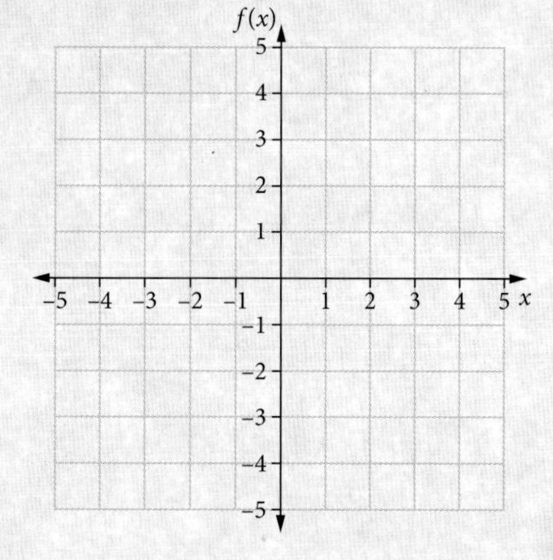

(4 marks)

Reading the question

- Highlight the type of answer required in each part. Where the coordinates are required, you need to find both x and y.
- 'Show' in part **b** indicates that you need to have all working shown. Three marks are allocated, so your working must have at least three parts.
- Highlight the information you should label on your sketched graph. The table of values will assist in getting a better shape.

Thinking about the question

- This function is exponential so solutions to this equation will be natural logarithms.
- You will need to be able to find a first and second derivative.
- You will also need to use the first derivative to find stationary points and the second derivative to find the point(s) of inflection.
- Make sure all the required coordinates are labelled on the graph. You will need to use the table of values to approximate some of your coordinates.

Worked solution (✓ = 1 mark)

a $f(x) = e^x(e^x - 4)$

$\mathbf{e^x(e^x - 4) = 0}$ ✓

$e^x - 4 = 0$

$e^x = 4$

$\mathbf{x = \ln(4)}$ ✓

x-intercept is $(\ln(4), 0)$. ✓

b $f'(x) = 2e^{2x} - 4e^x$

Solve $f'(x) = 0$.

$0 = 2e^{2x} - 4e^x$

$= 2e^x(e^x - 2)$

$e^x = 2$

$x = \ln(2)$

Substitute $x = \ln(2)$ into $f(x)$:

$f(\ln(2)) = e^{2\ln(2)} - 4e^{\ln(2)}$

$= e^{\ln(4)} - 4e^{\ln(2)}$

$= 4 - 8$

$= -4$

Turning point is at $(\ln(2), -4)$.

differentiates $f(x)$ correctly and equates to 0 ✓

shows the steps required to solve for x ✓

demonstrates the use of log laws to determine the y-coordinate ✓

c $\mathbf{f''(x) = 4e^{2x} - 4e^x}$ ✓

$f''(x) = 4e^x(e^x - 1)$

Point of inflection occurs when $f''(x) = 0$.

$e^x = 1$

$\mathbf{x = \ln(1) = 0}$ ✓

$f(0) = e^0 - 4e^0 = -3$

Point of inflection is at $(0, -3)$. ✓

d

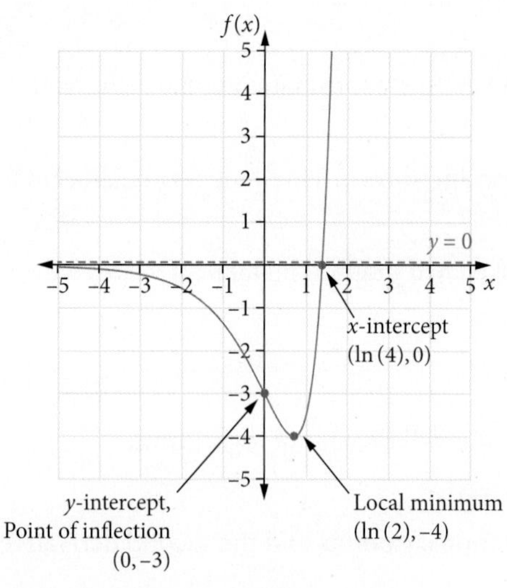

intercepts correct and labelled ✓

turning point and inflection point correct and labelled ✓

concavity correct ✓

limiting behaviour correct ✓

EXERCISE 7.4 Applications involving natural logarithms

ANSWERS p. 402

Recap

1 Given that $\int_0^m \frac{4}{4x+1}\,dx = \ln(13)$ find the value of m.

2 Find the equation of the curve $f(x)$ given that $f'(x) = \frac{1}{x+3}$, $x > -3$ and $f(-2) = 12$.

Mastery

3 WORKED EXAMPLE 17 Find the area bounded by the curve $f(x) = \frac{1}{3x-9}$, the x-axis and the lines $x = 4$ and $x = 5$.

4 Find the area bounded by the curve $f(x) = \frac{4}{x}$, the x-axis and the lines $x = 1$ and $x = e^3$.

5 WORKED EXAMPLE 18 The functions $f(x) = e^x$ and $g(x) = 7$ intersect at the point $(b, 7)$.

Find

a the exact value of b

b the area bounded by the $f(x)$, $g(x)$ and the y-axis.

6 The functions $f(x) = e^x$ and $g(x) = e^2$ intersect at the point (a, e^2).

Find

a the value of a

b the area bounded by $f(x)$, the x-axis, the y-axis and the line $x = a$.

Calculator-free

7 (4 marks) Part of the graph of $f: f(x) = x\log_e(x)$ is shown.

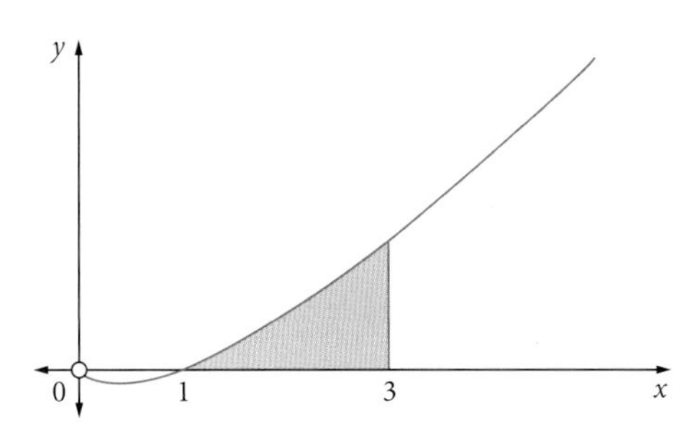

a Find the derivative of $x^2\log_e(x)$. (1 mark)

b Use your answer to part **a** to find the area of the shaded region in the form $a\log_e(b) + c$, where a, b and c are non-zero real constants. (3 marks)

Exam hack

You will need to use integration by recognition to find the required integral.

8 © SCSA MM2017 Q5 (8 marks)

a Consider the shaded area shown between the graph of $y = e^x$, the y axis and the line $y = 2$.

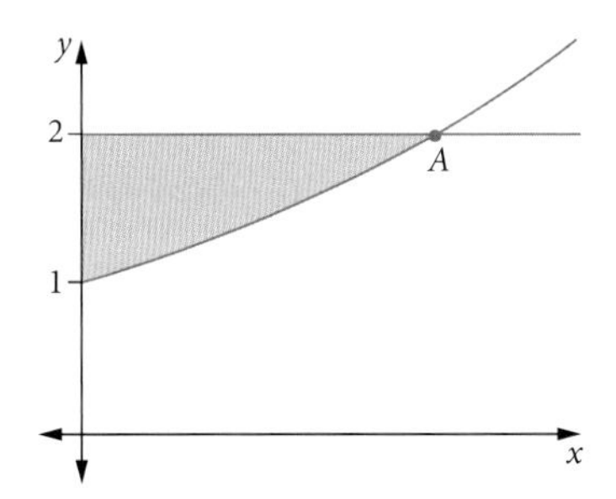

i Determine the coordinates of the point A. (1 mark)

ii Hence or otherwise determine the area between the graph of $y = e^x$, the y axis and the line $y = 2$. (3 marks)

b If the area between the graph of $y = e^x$, the y axis, the x axis and the line $x = k$, where $k \geq 0$, is to be equal to 2 square units, determine the exact value of k. (4 marks)

9 © SCSA MM2019 Q5 (8 marks)

a Determine the area bound by the graph of $f(x) = e^x$ and the x-axis between $x = 0$ and $x = \ln 2$. (3 marks)

b Hence, determine the area bound by the graph of $f(x) = e^x$, the line $y = 2$ and the y-axis. (2 marks)

c Determine the area bound by the graph of $f(x) = e^x$, the line $y = a$ and the y-axis, where a is a positive constant. (3 marks)

Calculator-assumed

10 (4 marks) A small colony of black peppered moths live on a small isolated island. In summer, the population begins to increase. If t is the number of days after 12 midnight on 1 January, the equation that best models the number of moths in the colony at any given time is

$N = 500\ln(21t + 3)$.

a What is the population of black peppered moths on 1 January? (1 mark)

b What is the population of moths after 30 days? (1 mark)

c On which day is the population first greater than 2000? (2 marks)

11 (8 marks) Harrison is training for a 100 m swimming race and wants to get under 50 seconds. At the start, his best time was 1 minute. After 12 days of intensive training, his time has reduced to 55 seconds. Harrison's swim times, T minutes after t days, are modelled using the function $T = 60 - a\ln(t + 1)$.

a Find the value of a, correct to three decimal places. (2 marks)

b How many days will it take him to get under 50 seconds? (2 marks)

c At what rate (in seconds per day, correct to three decimal places) is Harrison's time decreasing at this point? (2 marks)

d How long would it take him to be an Olympic champion contender (under 46 s), assuming his body could stand the training regime? (2 marks)

12 (8 marks) David can currently make about 5 skateboards in a day. He starts to improve his productivity and after two weeks has increased his productivity to 7 skateboards per day. David's daily productivity is modelled using the function $N = k + a\ln(t + 1)$, where t is the number of weeks after starting.

a Find the value of a, correct to three decimal places. (2 marks)

b How long will it take him to get his productivity up to 10 skateboards per day? (2 marks)

c What will be his rate of productivity increase (in skateboards/day) after four weeks? (2 marks)

d What will be his rate of productivity increase after ten weeks? (2 marks)

13 © SCSA MM2016 Q13cd (5 marks)

a Sketch the graph of $y = x^2 \ln x$, showing all features. (3 marks)

b Calculate the area bounded by the graph of $y = x^2 \ln x$, the x axis, $x = 1$ and $x = e$. (2 marks)

14 (12 marks) The diagram shows part of the graph of the function $f(x) = \frac{7}{x}$.

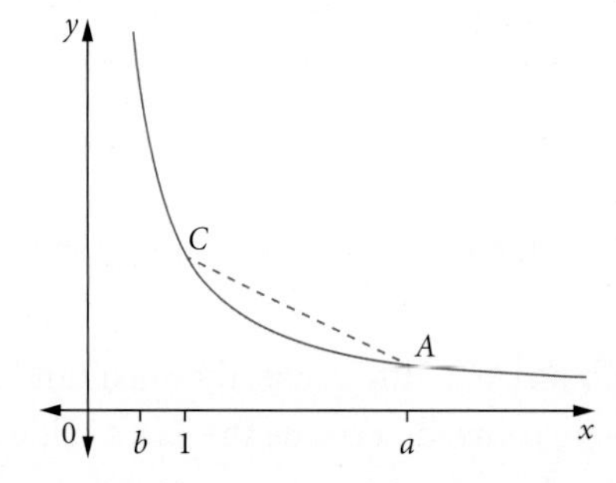

The line segment CA is drawn from the point $C(1, f(1))$ to the point $A(a, f(a))$, where $a > 1$.

a **i** Calculate the gradient of CA in terms of a. (1 mark)

ii At what value of x between 1 and a does the tangent to the graph of f have the same gradient as CA? (2 marks)

b **i** Calculate $\int_1^e f(x)\,dx$. (1 mark)

ii Let b be a positive real number less than one. Find the exact value of b such that $\int_b^1 f(x)\,dx$ is equal to 7. (2 marks)

c **i** Express the area of the region bounded by the line segment CA, the x-axis, the line $x = 1$ and the line $x = a$ in terms of a. (2 marks)

ii For what exact value of a does this area equal 7? (1 mark)

iii Using the value for a determined in **c ii**, explain in words, without evaluating the integral, why $\int_1^a f(x)\,dx < 7$. Use this result to explain why $a < e$. (1 mark)

d Find the exact values of m and n such that $\int_1^{mn} f(x)\,dx = 3$ and $\int_1^{\frac{m}{n}} f(x)\,dx = 2$. (2 marks)

7 Chapter summary

The first derivative of ln (*x*)

$$\frac{d}{dx}(\ln(x)) = \frac{1}{x}$$

$$\frac{d}{dx}(\ln(ax)) = \frac{1}{x}$$

$$\frac{d}{dx}(\ln(f(x)) = \frac{f'(x)}{f(x)}$$

The laws of logarithms for natural logarithms

$\ln(xy) = \ln(x) + \ln(y)$

$\ln\left(\frac{x}{y}\right) = \ln(x) - \ln(y)$

$\ln(x^n) = n\ln(x)$

Also remember,

$\ln(1) = 0$

$\ln(e^n) = n$

Stationary points and their nature

- Local maxima occur when $\frac{dy}{dx} = 0$ and $\frac{d^2y}{dx^2} < 0$.
- Local minima occur when $\frac{dy}{dx} = 0$ and $\frac{d^2y}{dx^2} > 0$.
- Stationary points of inflection occur when $\frac{dy}{dx} = 0$ and $\frac{d^2y}{dx^2} = 0$ and the concavity of the curve changes from concave up to concave down or from concave down to concave up..

The increments formula

- The increments formula can be used to approximate the increase in the y value for a corresponding small increase in the x value.

  $$\delta y \approx \frac{dy}{dx} \times \delta x$$

  For a given function $y = f(x)$, we can use δy to find an approximation for $f(x + \delta x)$.

  $f(x + \delta x) \approx f(x) + \delta y$

Integration of reciprocal functions

- $\int \frac{1}{x}\,dx = \ln(x) + c$ for $x > 0$
- $\int \frac{f'(x)}{f(x)}\,dx = \ln(f(x)) + c$ for $f(x) > 0$

Areas between curves

- If $f(x) > g(x)$ for $a < x < b$, then the upper function is $f(x)$ and the lower function is $g(x)$.

  bounded area $= \int_a^b (\text{upper} - \text{lower})\,dx = \int_a^b [f(x) - g(x)]\,dx$

Cumulative examination: Calculator-free

Total number of marks: 36 Reading time: 4 minutes Working time: 36 minutes

1 (4 marks) The diameter d, in centimetres, of a species of gum tree after t years is given by the rule $d(t) = d_0 e^{mt}$. The diameter is 2 cm when the tree is planted, and 10 cm after 2 years.

a Write two equations that can be used to find the constants d_0 and m. (2 marks)

b Calculate the exact values of the constants d_0 and m. (2 marks)

2 (7 marks) A coin is biased so that the probability of tossing a head is p and the probability of tossing a tail is $\frac{2}{3}$. The coin is tossed three times. The discrete random variable X represents the number of tails that occur.

a Find the value of p. (1 mark)

b List the probability distribution of the discrete random variable X. (3 marks)

c Find $P(X \geq 1)$. (1 mark)

d Find $P(X = 2 \mid X \geq 1)$. (2 marks)

3 (3 marks) Find the coordinates of the x-intercepts of $f(x) = 2e^{2x} - 7e^x + 6$, if the factors of $2e^{2x} - 7e^x + 6$ are $(2e^x - 3)(e^x - 2)$.

4 © SCSA MM2018 Q3cii (3 marks) Evaluate $\int_0^1 \frac{3x + 1}{3x^2 + 2x + 1}\, dx$.

5 © SCSA MM2021 Q7 (9 marks)

a Consider the function, $f(x) = \frac{1}{x}$ graphed twice below.

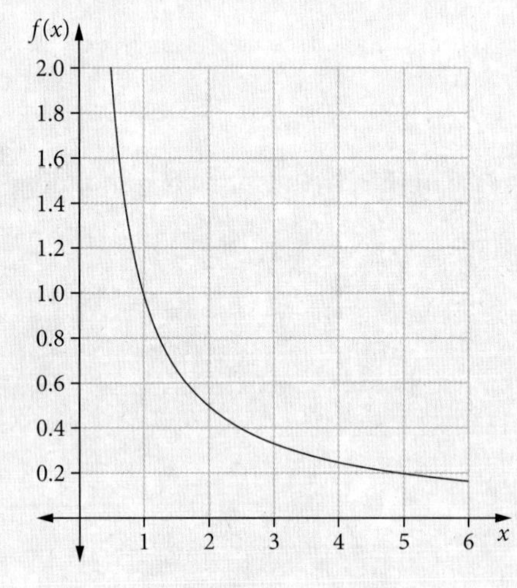

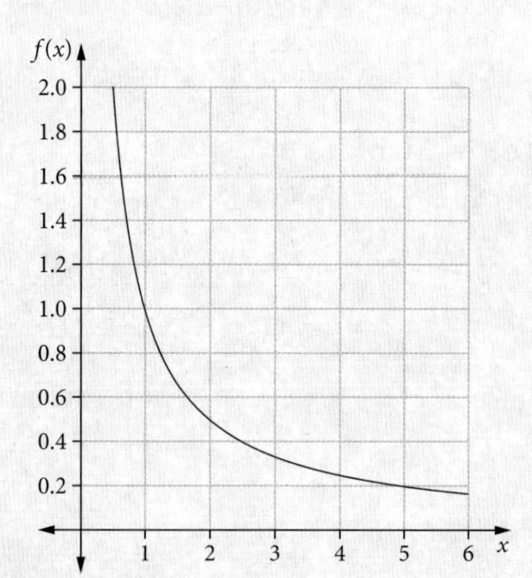

i Copy the graphs and on them shade **two** different regions (one on each graph) each with area exactly ln (2). (2 marks)

ii Given that $\int_a^b \frac{1}{x}\, dx = \ln(3)$, what is the relationship between a and b? (2 marks)

b Another graph of $f(x) = \frac{1}{x}$ is shown below.

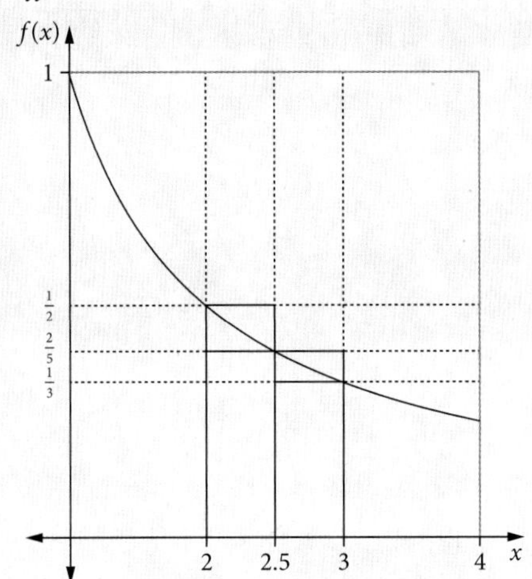

i By considering the areas of the rectangles shown, demonstrate and explain why $\frac{11}{30} < \int_2^3 \frac{1}{x}\,dx < \frac{9}{20}$. (3 marks)

ii Hence show that $\frac{11}{30} < \ln(1.5) < \frac{9}{20}$. (2 marks)

6 (3 marks) The derivative with respect to x of the function $f(x)$ has the rule $f'(x) = \frac{1}{2} - \frac{1}{2x-2}$. Given that $f(2) = 0$, find $f(x)$ in terms of x.

7 (7 marks) Let $f(x) = x^2e^{kx}$, where k is a positive real constant.

a Show that $f'(x) = xe^{kx}(kx + 2)$. (1 mark)

b Find the value of k for which the graphs of $y = f(x)$ and $y = f'(x)$ have exactly one point of intersection. (2 marks)

Let $g(x) = -\frac{2xe^{kx}}{k}$. The diagram below shows sections of the graphs of f and g for $x \geq 0$.

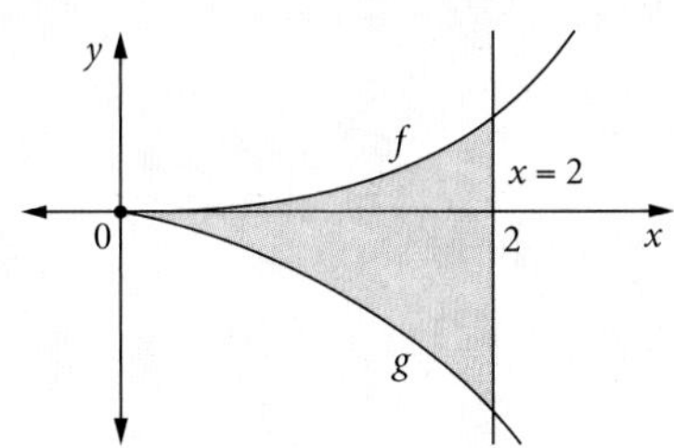

Let A be the area of the region bounded by the curves $y = f(x)$, $y = g(x)$ and the line $x = 2$.

c Write down a definite integral that gives the value of A. (1 mark)

d Using your result from part **a**, or otherwise, find the value of k such that $A = \frac{16}{k}$. (3 marks)

Cumulative examination: Calculator-assumed

Total number of marks: 29 Reading time: 3 minutes Working time: 29 minutes

1 (10 marks)

Consider the function $f(x) = \frac{1}{27}(ax - 1)^3(b - 3x) + 1$, where a and b are real constants.

a Write down, in terms of a and b, the possible values of x for which $(x, f(x))$ is a stationary point of $f(x)$. (3 marks)

b For what value of a does $f(x)$ have no stationary points? (1 mark)

c Find a in terms of b given that $f(x)$ has one stationary point. (2 marks)

d What is the maximum number of stationary points that $f(x)$ can have? (1 mark)

e Assume that there is a stationary point at $(1, 1)$ and another stationary point (p, p) where $p \neq 1$. Find the value of p. (3 marks)

2 (1 mark) If $\int_1^{12} g(x)\,dx = 5$ and $\int_{12}^{5} g(x)\,dx = -6$, then determine the value of $\int_1^{5} g(x)\,dx$.

3 (9 marks) Consider the function $f(x) = x^4 \ln(4x)$.

a Use the product rule to find $f'(x)$. (2 marks)

b Hence find $\int x^3 \ln(4x)\,dx$. (3 marks)

c Use the result of part **b** to find $\int_{0.25}^{1} x^3 \ln(4x)\,dx$. (2 marks)

An object moves in a straight line with a velocity given by the $v(t) = t^3 \ln(4t)$ m/s.

d Find the distance travelled by the object between $t = 0.25$ s and $t = 1$ s. Give your answer to the nearest centimetre. (2 marks)

4 © SCSA MM2020 Q11 (9 marks)

The line $y = x + c$ is tangent to the graph of $f(x) = e^x$.

a Obtain the coordinates of the point of intersection of the tangent with the graph of $f(x)$. (2 marks)

b What is the value of c? (1 mark)

c Sketch the graph of $f(x)$ and the tangent on the same axes. (1 mark)

d Evaluate the exact area between the graph of $f(x)$, the tangent line, and the line $x = \ln 2$. (3 marks)

e Given that $g(x)$ is the inverse function of $f(x)$, write a definite integral that could be used to determine the area between the graph of $g(x)$, the x-axis, and the line $x = \ln 2$. (2 marks)

9780170477536

CONTINUOUS RANDOM VARIABLES AND THE NORMAL DISTRIBUTION

Syllabus coverage

TOPIC 4.2: CONTINUOUS RANDOM VARIABLES AND THE NORMAL DISTRIBUTION

General continuous random variables

4.2.1 use relative frequencies and histograms obtained from data to estimate probabilities associated with a continuous random variable

4.2.2 examine the concepts of a probability density function, cumulative distribution function, and probabilities associated with a continuous random variable given by integrals; examine simple types of continuous random variables and use them in appropriate contexts

4.2.3 identify the expected value, variance and standard deviation of a continuous random variable and evaluate them using technology

4.2.4 examine the effects of linear changes of scale and origin on the mean and the standard deviation

Normal distributions

4.2.5 identify contexts, such as naturally occurring variation, that are suitable for modelling by normal random variables

4.2.6 identify features of the graph of the probability density function of the normal distribution with mean μ and standard deviation σ and the use of the standard normal distribution

4.2.7 calculate probabilities and quantiles associated with a given normal distribution using technology, and use these to solve practical problems

Mathematics Methods ATAR Course Year 12 syllabus p. 13 © SCSA

Video playlists (5):

8.1 General continuous random variables
8.2 Measures of centre and spread
8.3 Uniform and triangular distribution
8.4 The normal distribution
WACE question analysis Continuous random variables and the normal distribution

Worksheets (10):

8.1 Probability density functions
8.4 The normal curve • The standard normal curve • Areas under the normal curve • *z*-scores • The standard normal curve • The normal distribution • Applying the normal distribution • Normal distribution – Worded problems 1 • Normal distribution – Worded problems 2

To access resources above, visit **cengage.com.au/nelsonmindtap**

8.1 General continuous random variables

Recall that in Chapter 5, we defined a random variable, X, as a set of numerical quantities with elements defined as the outcomes of a random chance experiment. However, there are two types of random variables: discrete and continuous. Chapter 5 dealt with discrete random variables, whose possible values are any countable set of numbers, such as the number of pets in a household. In this chapter, we will explore continuous random variables.

A continuous random variable can take the value of any real number over a given continuous interval. Examples of continuous random variables include the height of students in a class, the minimum daily temperatures or the mass of pet dogs.

Relative frequencies, estimates and histograms

Video playlist
General continuous random variables

Worksheet
Probability density functions

As with discrete random variables, the idea of a continuous random variable is not new to you. You may remember constructing **frequency histograms** to display the frequencies of the outcomes of a continuous numerical data set in which the variable being measured is on the horizontal axis and the frequency is on the vertical axis. For example, suppose a local vet recorded the mass of the next 100 pet dogs that visited the clinic and displayed the results in a histogram, such as the one below.

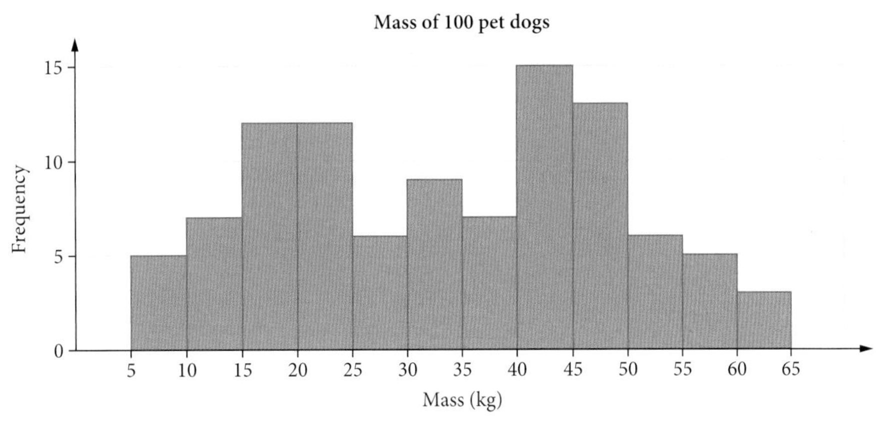

In this example, the continuous random variable is the mass of a pet dog visiting the clinic, recorded in kilograms. Let this variable be M. From this experimental data, we can then calculate a **relative frequency** of an outcome, which is its frequency as a proportion of the total number of observations. This relative frequency is also called an **experimental probability**. Given that the probability of an outcome is coming from the collection of data, it cannot be considered a theoretical probability as we would expect there to be differences in the data collected with every set of experiments conducted. As a result, based on this single set of data, we can use the relative frequencies of each outcome as estimates of the theoretical probabilities and then carry out calculations involving the same chance experiment. For example, we could estimate that 15 out of every 100 pet dogs that visit this particular vet clinic have a mass between 40 and 45 kilograms. This can be written using the following probability notation $P(40 \le M < 45) = \frac{15}{100} = 0.15$ and all relative frequencies could be displayed in a **relative frequency histogram**.

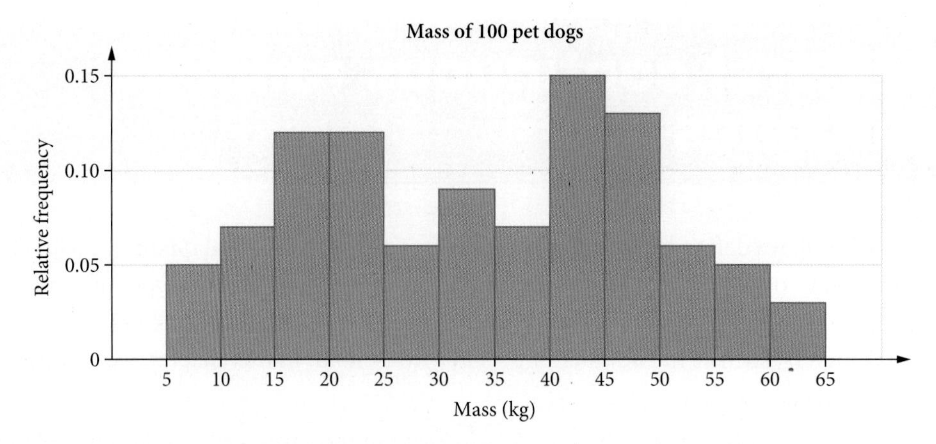

Now without the specific dataset, we do not know whether this interval with a relative frequency of $\frac{15}{100}$ is actually $40 \le M < 45$, $40 \le M \le 45$, $40 < M \le 45$, or even $40 < M < 45$. This brings up an important question for continuous random variables:

What is the probability that a pet dog has a mass of EXACTLY 40 kg?

Or similarly,

What is the probability that a pet dog has a mass of EXACTLY 45 kg?

Because continuous random variables can take on any real numerical value, there is an underlying assumption that the probability that the mass of a dog is an exact discrete value such as 45 kg is negligible; that is, $P(M = 45) = 0$ because the degree of accuracy of mass can never be measured exactly. It could be 45.0001 kg or 44.999 987 4 kg.

Probability at a point for a continuous random variable

For a continuous random variable, X, $P(X = k) = 0$, where k is a discrete outcome.

WORKED EXAMPLE 1 **Probabilities from a histogram**

The histogram below shows the results of a survey of 80 workers who were asked how long it took them to arrive at work that morning.

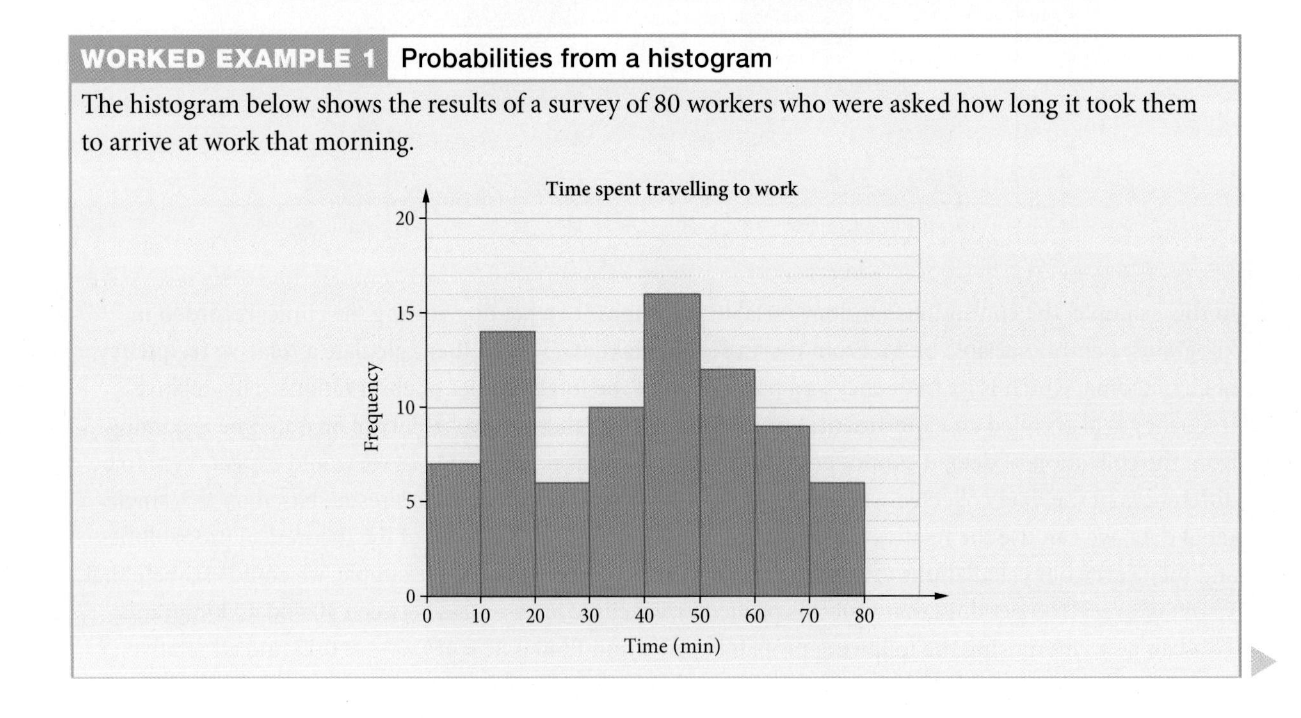

Let T be the continuous random variable representing the time taken to arrive at work, measured in minutes.

a Complete the frequency table below.

Time	Frequency
$0 \le t < 10$	7
$10 \le t < 20$	14
$20 \le t < 30$	6
$30 \le t < 40$	
$40 \le t < 50$	16
$50 \le t < 60$	12
$60 \le t < 70$	9
$70 \le t < 80$	
Total	80

b Use the frequencies from this data set to estimate the following probabilities:

i $P(10 \le T < 20)$ **ii** $P(T \ge 10 \mid T \le 20)$ **iii** $P(T < 50)$

iv $P(T \ge 30)$ **v** $P(25 \le T \le 45)$

Steps	Working
a 1 Read the frequencies from the histogram to complete the table.	$n(30 \le T < 40) = 10$ $n(70 \le T < 80) = 6$
2 Check that the frequencies total 80.	$7 + 14 + 6 + 10 + 16 + 12 + 9 + 6 = 80$
b i Express the frequency for $10 \le t < 20$ as a relative frequency and use it as an estimate of the probability.	$P(10 \le T < 20) = \frac{14}{80}$
ii Use the conditional probability formula $P(A \mid B) = \frac{P(A \cap B)}{P(B)}$ and recognise that $P(T \le 20) = P(T < 20)$ for the continuous random variable.	$P(T \ge 10 \mid T \le 20) = \frac{P(10 \le T < 20)}{P(T < 20)}$ $= \frac{14}{21}$
iii Add up all frequencies for $t < 50$ and express the **cumulative relative frequency** as an estimate of the probability.	$P(T < 50) = \frac{7 + 14 + 6 + 10 + 16}{80}$ $= \frac{53}{80}$
iv Use the complement rule to $P(T \ge 30) = 1 - P(T < 30)$ to obtain an estimate of the probability.	$P(T \ge 30) = 1 - \frac{27}{80}$ $= \frac{53}{80}$

Exam hack

Unless you are specifically asked to simplify a fraction in the exam, there's no need to.

v 1 Assume a uniform distribution of the times within each interval to estimate the proportions from $20 \le T < 30$ and $40 \le T < 50$.

$$P(25 \le T < 30) \approx \frac{5}{10} \times \frac{6}{80}$$
$$\approx \frac{3}{80}$$
$$P(40 \le T < 45) \approx \frac{5}{10} \times \frac{16}{80}$$
$$\approx \frac{8}{80}$$

2 Express the cumulative relative frequency as an estimate of the probability.

$$P(25 \le T \le 45) \approx \frac{3 + 10 + 8}{80}$$
$$\approx \frac{21}{80}$$

Probability density functions and integrals

Suppose the vet clinic collected the mass of pet dogs over an entire year. With significantly more data, the width of the columns would become more precise and there would be significantly more columns such that the tops of the columns would start to form a smooth curve representing the range of relative frequencies over the entire interval.

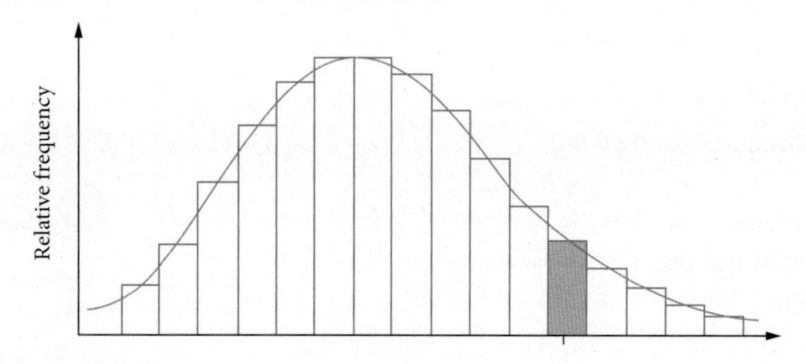

This smooth curve that is formed, which can also sometimes be defined by a rule $f(x)$, is called the **probability density function (pdf)** of a continuous random variable, X. Given that the area under the curve represents the columns of a relative frequency histogram with infinitely small width, we can say that the probability of an interval $a \le X \le b$ is given by the definite integral of the probability density function $f(x)$.

$$P(a \le X \le b) = \int_a^b f(x)\,dx$$

Based on the earlier result that $P(X = k) = 0$, then

$$P(a \le X \le b) = P(a < X < b) = P(a < X \le b) = P(a \le X < b) = \int_a^b f(x)\,dx.$$

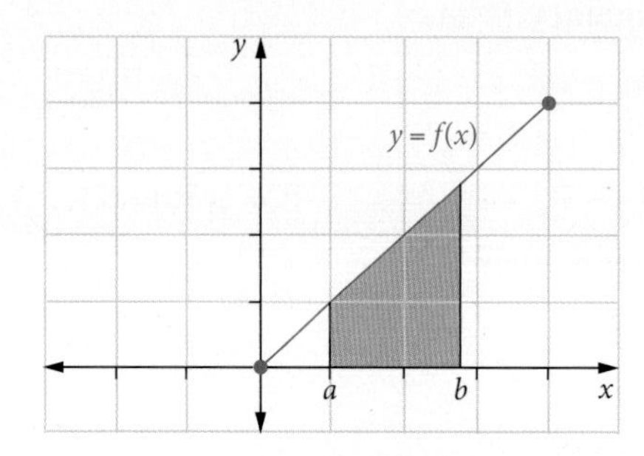

Much like the two conditions of the probability distribution of a discrete random variable, there are two required properties of a probability density function of a continuous random variable for it to be considered valid:

1. $f(x) \ge 0$ for all values of x.
2. The total area under the curve from the lowest value of x to the highest value of x is 1. This is often represented as the definite integral, $\int_{-\infty}^{\infty} f(x)\,dx = 1$, where $\pm\infty$ represent the lower and upper bounds of x.

WORKED EXAMPLE 2 Finding an unknown in $f(x)$ for a valid continuous random variable

The probability density function for a continuous random variable X is given by

$$f(x) = \begin{cases} kx & 1 \le x \le 7 \\ 0 & \text{otherwise} \end{cases}$$

Find the value of k.

Steps	Working
1 Establish a definite integral representing the total area under the curve.	$\int_1^7 kx\,dx = 1$
2 Evaluate the definite integral and solve for k.	$\left[\frac{kx^2}{2}\right]_1^7 = 1$ $\frac{49k}{2} - \frac{k}{2} = 1$ $24k = 1$ $k = \frac{1}{24}$

In some simple cases, like the one above, an integral calculation may not be required, as the area under the curve could be calculated using simpler area formulas, such as:

area of a rectangle = lw

area of a triangle = $\frac{1}{2}bh$

area of a trapezium = $\frac{1}{2}(a+b)h$

area of a sector = $\frac{\theta}{360°}\pi r^2$

WORKED EXAMPLE 3 Probabilities from the graph of a probability density function

The graph of the probability density function $f(x)$ for a continuous random variable X is shown below.

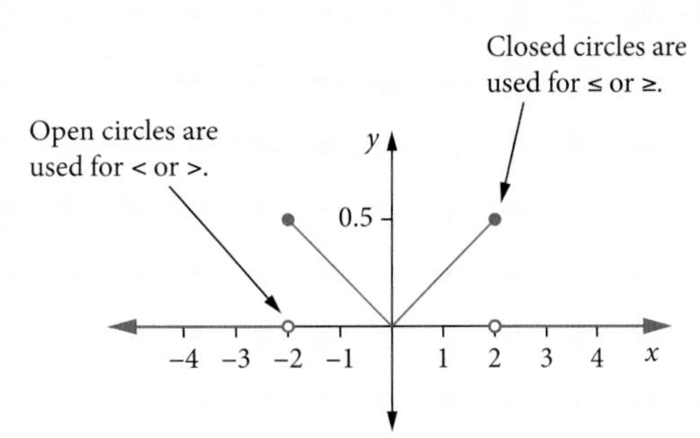

Determine the following probabilities:

a $P(X = 1)$ **b** $P(0 \le X \le 1)$ **c** $P(-1 < X < 1)$ **d** $P(X > -1 \mid X < 1)$

Steps	Working
a Recognise that $P(X = k) = 0$.	$P(X = 1) = 0$
b Use the area of a triangle formula to calculate the probability, interpreting the height of the triangle using the gradient of the line.	$P(0 \le X \le 1) = \frac{1}{2}(1)\left(\frac{1}{4}\right)$ $= \frac{1}{8}$

c **1** Use the symmetry of the diagram to recognise that $P(0 \le X \le 1) = P(-1 \le X \le 0)$.	$P(-1 < X < 1) = 2\left(\frac{1}{8}\right)$
2 Recognise that $P(X = k) = 0$ and so $P(-1 \le X \le 1) = P(-1 < X < 1)$.	$= \frac{1}{4}$
d **1** Use the conditional probability formula $P(A \mid B) = \frac{P(A \cap B)}{P(B)}$.	$P(X > -1 \mid X < 1)$ $= \frac{P(-1 < X < 1)}{P(X < 1)}$
2 Recognise that $P(X \le 1) = P(X < 0) + P(0 \le X < 1)$.	$= \frac{\frac{1}{4}}{\frac{1}{2} + \frac{1}{8}}$ $= \frac{\left(\frac{2}{8}\right)}{\left(\frac{5}{8}\right)} = \frac{2}{5}$

Exam hack

In cases where the graph of the probability density function is not given to you, if you recognise the shape of the graph from the rule, always draw a quick sketch of it to help represent the bounds of the definite integral.

WORKED EXAMPLE 4 Probabilities from the rule of a probability density function

The probability density function for a continuous random variable X is given by

$$f(x) = \begin{cases} \frac{3}{32}x(4 - x) & 0 \le x \le 4 \\ 0 & \text{otherwise} \end{cases}$$

Determine the following probabilities:

a $P(X < 4)$ **b** $P(0 \le X \le 2)$ **c** $P(X \ge 3)$ **d** $P(X > 0 \mid X < 2)$

Steps	**Working**
a Recognise that all values of x are less than 4 and that $P(X = 4) = 0$.	$P(X < 4) = P(0 \le X \le 4) = 1$
b **1** Establish a definite integral for the area under the curve.	$\int_0^2 \frac{3}{32}x(4 - x)\,dx$ $= \frac{3}{32}\int_0^2 4x - x^2 dx$
2 Evaluate the definite integral. **Exam hack** If you can recognise critical features of a probability density function such as this function having roots at $x = 0$ and $x = 4$, with a line of symmetry at $x = 2$, then pay attention to the number of marks. For 2 marks or less, it could be inferred that $P(0 \le X \le 2) = \frac{1}{2}$.	$= \frac{3}{32}\left[2x^2 - \frac{x^3}{3}\right]_0^2$ $= \frac{3}{32}\left(8 - \frac{8}{3}\right)$ $= \frac{3}{32}\left(\frac{16}{3}\right)$ $= \frac{1}{2}$

c Recognise the probability statement has an upper bound at 4 and establish the definite integral.

$$P(3 \le X < 4) = \int_3^4 \frac{3}{32}x(4-x)\,dx$$

$$= \frac{3}{32}\left[2x^2 - \frac{x^3}{3}\right]_3^4$$

$$= \frac{3}{32}\left(32 - \frac{64}{3} - 18 + 9\right)$$

$$= 1 - \frac{27}{32} = \frac{5}{32}$$

d 1 Use the conditional probability formula $P(A \mid B) = \dfrac{P(A \cap B)}{P(B)}$.

$$P(X > 0 \mid X < 2) = \frac{P(0 < X < 2)}{P(X < 2)}$$

2 Recognise that $P(0 < X < 2) = P(0 \le X < 2)$ (from part **b**) and $P(X < 2) = P(X \le 2)$.

$$= \frac{\left(\frac{1}{2}\right)}{\left(\frac{1}{2}\right)} = 1$$

Exam hack

Questions such as this could be worth 1 mark. If a conditional probability question is only allocated 1 mark, there will usually be something to notice about the intervals given. For example, in this case it is certain that $x > 0$ if $x < 2$ as the function starts from $x = 0$.

USING CAS 1 Finding an unknown in the domain of $f(x)$ for a valid continuous random variable

The probability density function for a continuous random variable X is given by

$$f(x) = \begin{cases} \dfrac{3x(10-x)}{500} & \text{if } 0 \le x \le a \\ 0 & \text{otherwise} \end{cases}$$

Find the value of a.

ClassPad

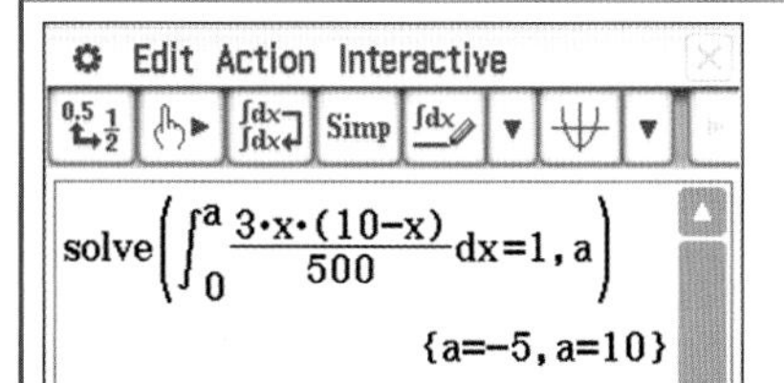

1 Enter the definite integral with lower and upper limits of **0** and **a** and set equal to **1**.

2 Highlight the integral equation and solve for a, as shown above.

3 Select the solution that suits the domain, i.e. in this case greater than 0.

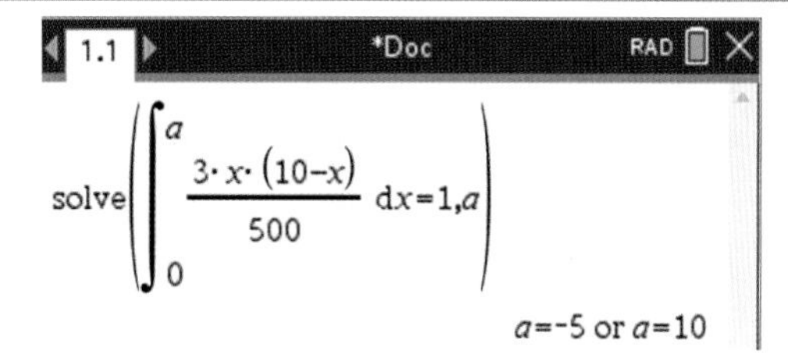

1 Set the definite integral equal to 1 and solve for **a**, as shown above.

2 Select the positive solution.

$a \ne -5$ as this is outside the domain.

$a = 10$

Exam hack

In the Calculator-assumed section, if you can see that a probability density function is being used multiple times throughout a question, it may be beneficial for time efficiency to use CAS to define the probability density function. It is especially helpful when the probability calculation goes across two separate functions in the piece-wise definition.

USING CAS 2 Defining probability density functions

Consider the following probability density function $f(x)$ for the continuous random variable X.

$$f(x) = \begin{cases} \dfrac{x}{4} & 0 \le x < 2 \\ 1 - \dfrac{x}{4} & 2 \le x \le 4 \\ 0 & \text{otherwise} \end{cases}$$

Calculate the following probabilities:

a $P(X < 3)$ **b** $P(X \ge 1 \mid X < 2)$ **c** $P(X > 3.5 \mid X > 1.5)$

ClassPad **TI-Nspire**

a

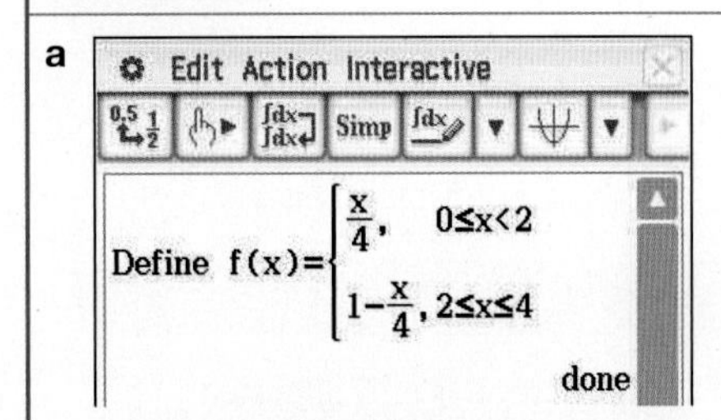

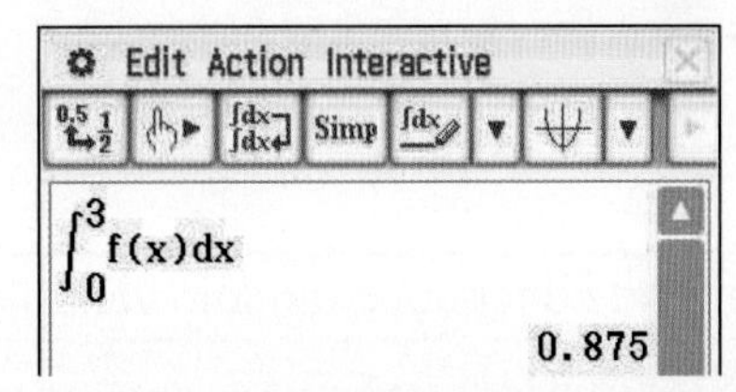

1 Using the **Math3** keyboard, **Define f(x)** piece-wise-defined function template.

2 Find the definite integral of **f(x)** from 0 to 3.

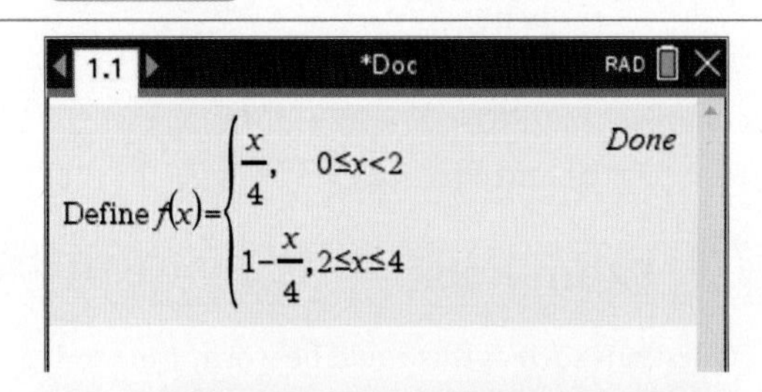

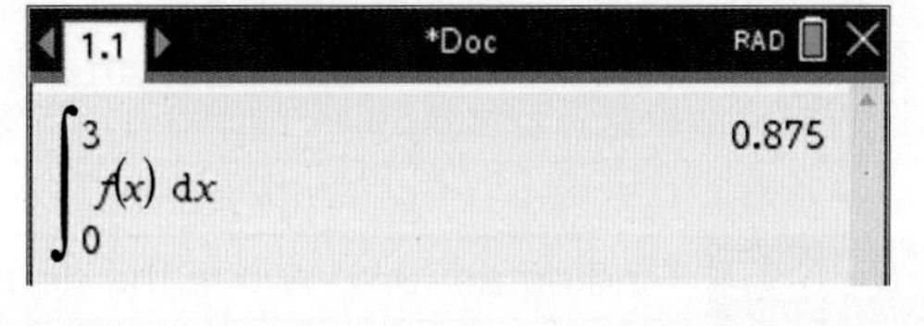

1 Define the piece-wise function **f(x)**.

2 Find the definite integral of **f(x)** from 0 to 3.

b

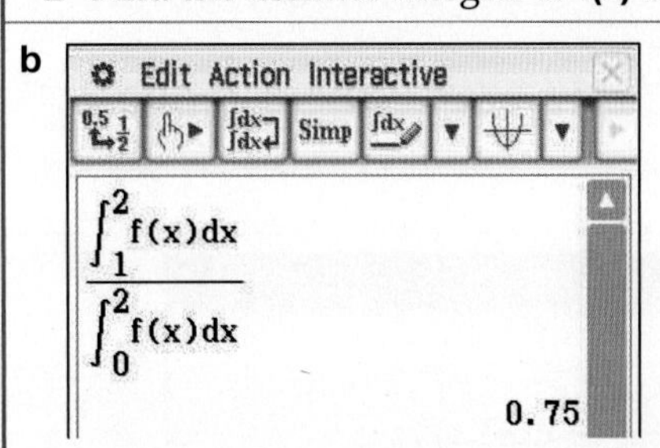

Use the conditional probability formula to find the definite integral of **f(x)** from 1 to 2, divided by the definite integral of **f(x)** from 0 to 2.

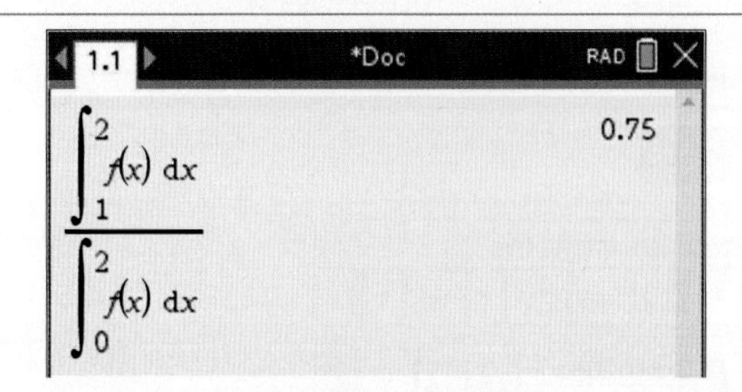

Use the conditional probability formula to find the definite integral of **f(x)** from 1 to 2, divided by the definite integral of **f(x)** from 0 to 2.

c

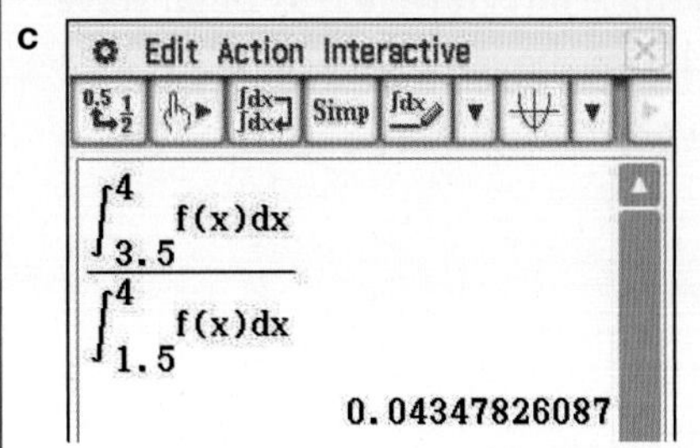

Use the conditional probability formula* to find the definite integral of **f(x)** from 3.5 to 4, divided by the definite integral of **f(x)** from 1.5 to 4.

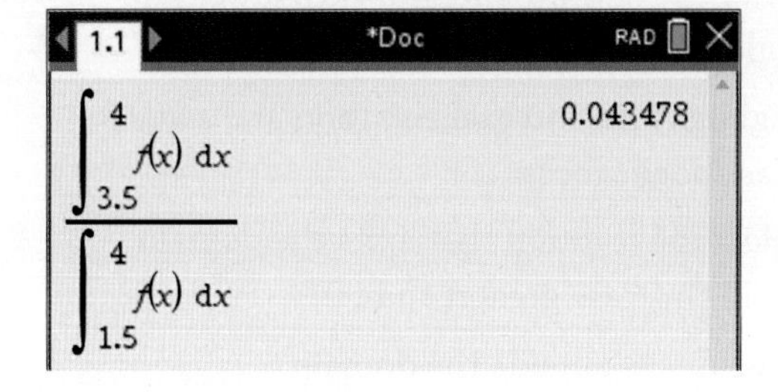

Use the conditional probability formula* to find the definite integral of **f(x)** from 3.5 to 4, divided by the definite integral of **f(x)** from 1.5 to 4.

a 0.875 **b** 0.75 **c** 0.0435

*Note: In this case $P(A \cap B) = P(A)$ because the interval $X > 3.5$ (i.e. set A) is completely contained within the interval $X > 1.5$ (i.e. set B).

Cumulative probability distributions

In many of the calculations so far, we have had to use the idea of a cumulative probability; that is, the addition of probabilities for a continuous random variable X from consecutive intervals up to a certain value of x, $P(X \le x)$. We have also seen that $P(X > x)$ can also be expressed as a cumulative probability using the complement rule:

$$P(X > x) = 1 - P(X \le x)$$

Now suppose we wanted to expressed the cumulative probabilities generally, either as a table of values, graphically or as a function.

Recall the histogram and corresponding frequency table that showed the results of a survey of 80 workers who were asked how long it took them to arrive at work that morning, where T was the continuous random variable representing the time taken to arrive at work, measured in minutes.

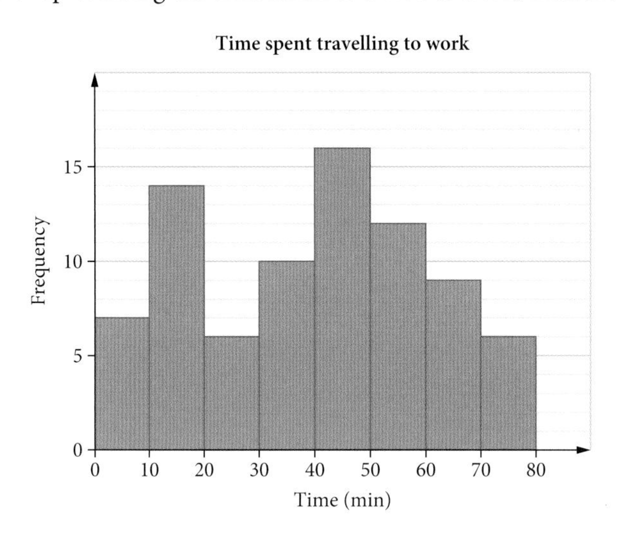

Time	Frequency
$0 \le t < 10$	7
$10 \le t < 20$	14
$20 \le t < 30$	6
$30 \le t < 40$	10
$40 \le t < 50$	16
$50 \le t < 60$	12
$60 \le t < 70$	9
$70 \le t < 80$	6
Total	80

Instead of displaying this data in a frequency table, we can display it in a cumulative frequency table, which then makes calculations of the form $P(T \ge t)$ easier.

Time	Cumulative frequency	Cumulative relative frequency
$t < 10$	7	$\frac{7}{80}$
$t < 20$	21	$\frac{21}{80}$
$t < 30$	27	$\frac{27}{80}$
$t < 40$	37	$\frac{37}{80}$
$t < 50$	53	$\frac{53}{80}$
$t < 60$	65	$\frac{65}{80}$
$t < 70$	74	$\frac{74}{80}$
$t < 80$	80	$\frac{80}{80}$
Total	80	1

Note that the final row of the cumulative frequency column should give the same value as the total, and the final row of the cumulative relative frequency column should give 1.

Suppose now that this cumulative data was used to create a smooth curve through the tops of the cumulative frequency histogram columns. We would obtain a function such as the one below, with the horizontal axis representing T and the vertical axis representing the cumulative relative frequencies whereby $P(T < 80) = 1$. This is called the **cumulative probability distribution** of T.

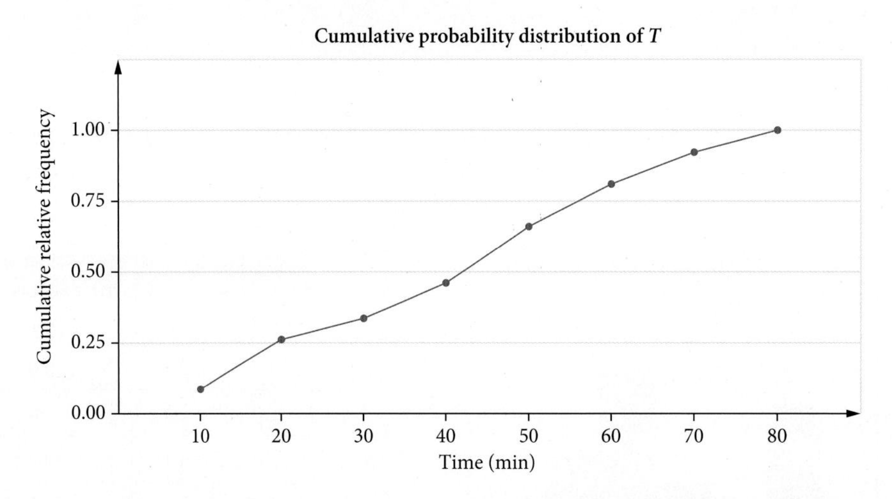

In cases where a rule for the function of the cumulative probability distribution can be obtained, then it is called the **cumulative distribution function (cdf)** and typically denoted using $F(x)$.

$$P(X \le x) = F(x)$$

For a continuous random variable X with a probability density function f defined over $a \le x \le b$, then the cumulative distribution function F is summing all areas under the curve of f from $x = a$ to some value of x. Let f be defined in terms of the dummy variable t, and so using the fundamental theorem of calculus, we can say that $F(x) = \int_a^x f(t)\,dt$ and so $\frac{d}{dx}(F(x)) = \frac{d}{dx}\left(\int_a^x f(t)\,dt\right) = f(x)$.

Cumulative distribution function

The probability density function $f(x)$ of a continuous random variable X is the derivative of the cumulative distribution function $F(x)$ and, hence, the cumulative distribution function $F(x)$ is the integral of $f(x)$ such that

$$F(x) = \begin{cases} 0 & x < a \\ \int_a^x f(t)\,dt & a \le x \le b \\ 1 & x > b \end{cases}$$

WORKED EXAMPLE 5 Finding a cumulative distribution function given a probability density function

The probability density function for a continuous random variable X is given by

$$f(x) = \begin{cases} \frac{x}{48} & 2 \le x \le 10 \\ 0 & \text{otherwise} \end{cases}$$

a Determine the cumulative distribution function of X.

b Hence, calculate $P(X \le 5)$.

Steps	Working
a **1** Replace the x in $f(x)$ with a dummy variable t and establish the definite integral $F(x) = \int_a^x f(t)\,dt$.	$f(t) = \frac{t}{48}$
2 Evaluate the definite integral in terms of x and establish the piece-wise definition of $F(x)$.	$F(x) = \int_2^x \frac{t}{48}\,dt$ $= \left[\frac{t^2}{96}\right]_2^x$ $= \frac{x^2 - 4}{96}$ $F(x) = \begin{cases} 0 & x < 2 \\ \frac{x^2-4}{96} & 2 \le x \le 10 \\ 1 & x > 10 \end{cases}$
b Evaluate $F(5)$.	$F(5) = \frac{5^2 - 4}{96} = \frac{21}{96}$

Exam hack

When a probability density function has more than two components in its piece-wise definition, write out the addition of the definite integrals that would give you a total area under the curve of 1.

WORKED EXAMPLE 6 Finding a probability density function given a cumulative distribution function

The cumulative distribution function for a continuous random variable X is given by

$$F(x) = \begin{cases} 0 & x < 0 \\ \frac{x^2}{8} & 0 \le x < 2 \\ x - \frac{x^2}{8} - 1 & 2 \le x \le 4 \\ 1 & x > 4 \end{cases}$$

a Find $P(X > 3)$.

b Determine the probability density function of X, $f(x)$.

c Sketch the graph of $f(x)$.

d Hence, show how to obtain the rule $F(x) = x - \frac{x^2}{8} - 1$ for $2 \le x \le 4$ from $f(x)$.

Steps	Working
a **1** Express $P(X > 3)$ in terms of a cumulative probability.	$P(X > 3) = 1 - P(X \le 3)$
2 Use $F(x)$ to evaluate the probability.	$P(X > 3) = 1 - F(3)$ $= 1 - \left(3 - \frac{9}{8} - 1\right)$ $= 1 - \frac{7}{8}$ $= \frac{1}{8}$

b 1 Obtain $f(x)$ by differentiating each component of $F(x)$.	For $0 \le x < 2$, $f(x) = \frac{d}{dx}\left(\frac{x^2}{8}\right) = \frac{x}{4}$ For $2 \le x \le 4$, $f(x) = \frac{d}{dx}\left(x - \frac{x^2}{8} - 1\right) = 1 - \frac{x}{4}$
2 Express $f(x)$ in the piece-wise definition.	$f(x) = \begin{cases} \frac{x}{4} & 0 \le x < 2 \\ 1 - \frac{x}{4} & 2 \le x \le 4 \\ 0 & \text{otherwise} \end{cases}$
c 1 Find the end points of each component of the piece-wise function.	$f(0) = 0$ $f(2) = \frac{1}{2}$ $f(4) = 0$
2 Sketch the graph of $y = f(x)$ ensuring all critical features are shown.	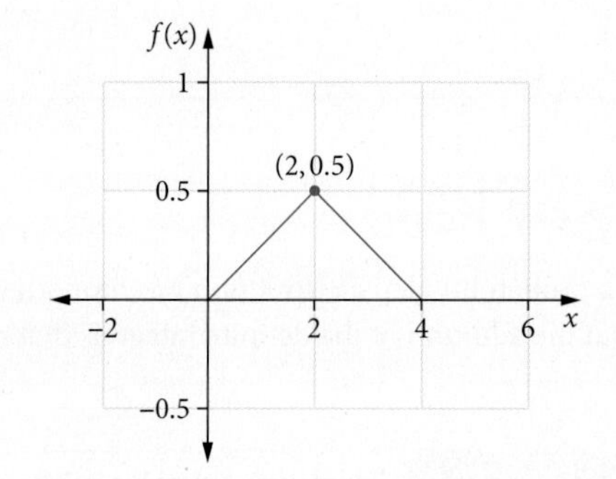
d 1 Establish an expression using definite integrals for the area under the curve of $f(x)$ from 0 to any value of x between $2 \le x \le 4$.	For $2 \le x \le 4$, $F(x) = \int_0^2 \frac{x}{4}\,dx + \int_2^x 1 - \frac{t}{4}\,dt$ $= \frac{1}{2}(2)\left(\frac{1}{2}\right) + \left[t - \frac{t^2}{8}\right]_2^x$
2 Use geometry, where efficient, to assist with any calculations.	$= \frac{1}{2} + \left(x - \frac{x^2}{8} - 2 + \frac{1}{2}\right)$ $= x - \frac{x^2}{8} - 1$

EXERCISE 8.1 General continuous random variables

ANSWERS p. 403

Mastery

1 WORKED EXAMPLE 1 The histogram below shows the amount of time, in minutes, that Jared spends completing homework each day, over the period of 28 school days. He does some homework on every school day.

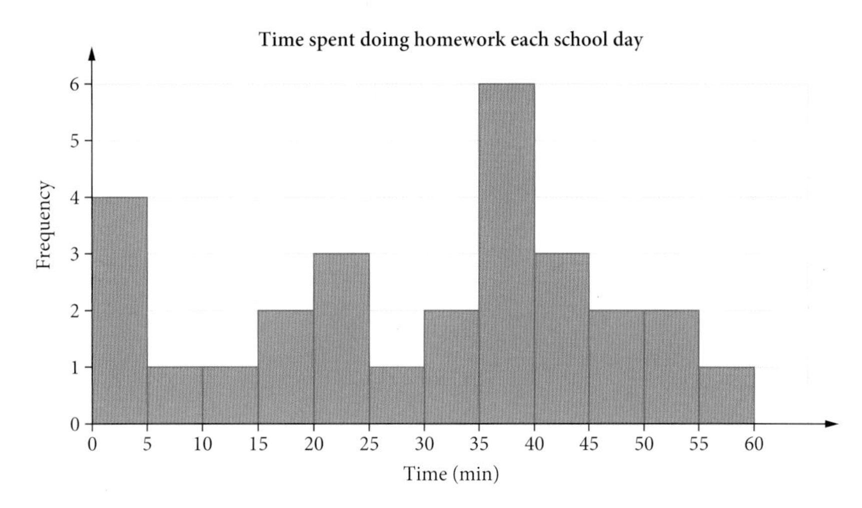

Let T be the continuous random variable representing the time spent completing homework on a school day, measured in minutes.

a Copy and complete the frequency table below.

Time	Frequency
$0 \le t < 5$	4
$5 \le t < 10$	1
$10 \le t < 15$	1
$15 \le t < 20$	2
$20 \le t < 25$	
$25 \le t < 30$	1
$30 \le t < 35$	2
$35 \le t < 40$	
$40 \le t < 45$	3
$45 \le t < 50$	2
$50 \le t < 55$	2
$55 \le t < 60$	1
Total	28

b Use the frequencies from this data set to estimate the following probabilities:

i $P(15 \le T \le 20)$ **ii** $P(T \ge 15 \mid T \le 20)$ **iii** $P(T < 40)$

iv $P(T \ge 20)$ **v** $P(42 \le T \le 48)$

2 WORKED EXAMPLE 2 The probability density function for a continuous random variable X is given by

$$f(x) = \begin{cases} 5kx - kx^2 & 0 \le x \le 5 \\ 0 & \text{otherwise} \end{cases}$$

Find the value of k.

3 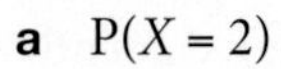 WORKED EXAMPLE 3 The graph of the probability density function $f(x)$ for a continuous random variable X is shown.

Determine the following probabilities:

a $P(X = 2)$

b $P(0 \le X \le 2)$

c $P(1 < X < 2)$

d $P(X > 1 \mid X < 2)$

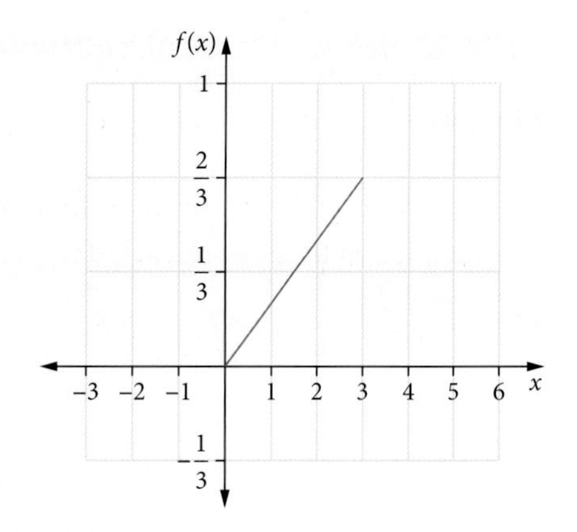

4 WORKED EXAMPLE 4

a The probability density function for a continuous random variable X is given by

$$f(x) = \begin{cases} 4x^3 & 0 \le x \le 1 \\ 0 & \text{otherwise} \end{cases}$$

Determine $P\left(X > \frac{1}{2}\right)$.

b The probability density function for a continuous random variable Y is given by

$$f(y) = \begin{cases} 0.01e^{-0.01y} & y \ge 0 \\ 0 & \text{otherwise} \end{cases}$$

Determine $P(50 \le Y \le 80)$, leaving your answer exact.

c The probability density function for a continuous random variable Z is given by

$$f(z) = \begin{cases} \pi \sin(2\pi z) & 0 \le z \le \frac{1}{2} \\ 0 & \text{otherwise} \end{cases}$$

Determine $P\left(Z > \frac{1}{4} \mid Z < \frac{1}{3}\right)$, leaving your answer exact.

5 Using CAS 1 The probability density function for a continuous random variable X is given by

$$f(x) = \begin{cases} \frac{4x^3}{625} & 0 \le x \le a \\ 0 & \text{otherwise} \end{cases}$$

Find the value of a.

6 Using CAS 2 Consider the following probability density function $f(x)$ for the continuous random variable X.

$$f(x) = \begin{cases} \frac{5-x}{25} & 0 \le x < 5 \\ \frac{x-5}{25} & 5 \le x \le 10 \\ 0 & \text{otherwise} \end{cases}$$

Calculate the following probabilities:

a $P(X < 8)$

b $P(X \ge 2 \mid X < 5)$

c $P(X > 8 \mid X > 5)$

7 WORKED EXAMPLE 5 The probability density function for a continuous random variable X is given by

$$f(x) = \begin{cases} \dfrac{2x}{9} & 0 \le x \le 3 \\ 0 & \text{otherwise} \end{cases}$$

a Determine the cumulative distribution function of X.

b Hence, calculate $P\left(X \le \dfrac{5}{2}\right)$.

8 WORKED EXAMPLE 6 The cumulative distribution function for a continuous random variable X is given by

$$F(x) = \begin{cases} 0 & x < 0 \\ \dfrac{x}{4} - \dfrac{x^2}{32} & 0 \le x < 4 \\ \dfrac{x^2}{32} - \dfrac{x}{4} + 1 & 4 \le x \le 8 \\ 1 & x > 8 \end{cases}$$

a Find $P(X \ge 6)$.

b Determine the probability density function of X, $f(x)$.

c Sketch the graph of $f(x)$.

d Hence, show how to obtain the rule $F(x) = \dfrac{x^2}{32} - \dfrac{x}{4} + 1$ for $4 \le x \le 8$ from $f(x)$.

Calculator-free

9 © SCSA MM2017 Q1 (5 marks) Anastasia is a university student. She records the time it takes for her to get from home to her campus each day. The histogram of relative frequencies below shows the journey times she recorded.

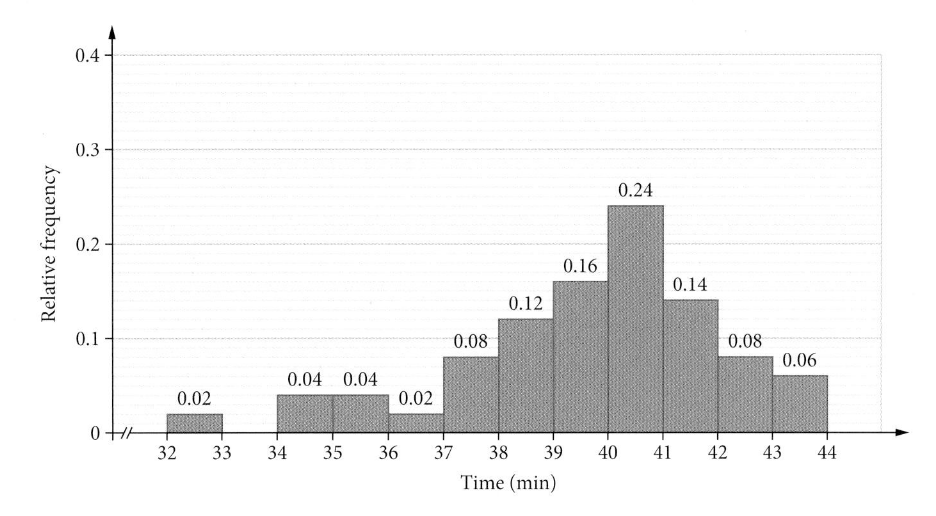

Use the above data to estimate the probability of her next journey from home to her university campus

a taking her less than 36 minutes (1 mark)

b taking at least 35 minutes but no more than 39 minutes. (2 marks)

On three consecutive days, Anastasia needs to be on campus no later than 10 am.

c If she leaves her home at 9:22 am each day, use the above data to estimate the probability that she makes it on or before time on all three days. (2 marks)

10 © SCSA MM2020 Q4 (9 marks) The heights reached by a species of small plant at maturity are measured by a team of biologists. The results are shown in the histogram of relative frequencies below.

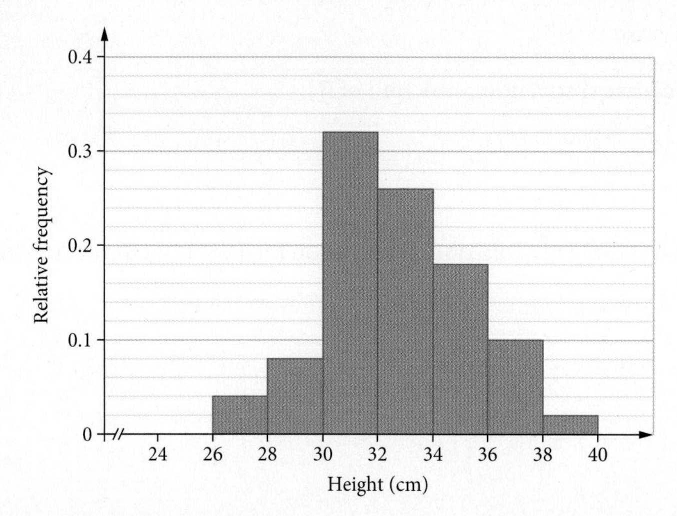

a Determine the probability that a mature plant of this species reaches no higher than 30 cm. (1 mark)

b If a mature plant reaches a height of at least 32 cm, what is the probability that its height reaches above 38 cm? (2 marks)

Another team of biologists is studying the mature heights of a species of hedge. The height, h metres, has a probability density function, $d(h)$, as given below.

$$d(h) = \begin{cases} \dfrac{h-1}{5} & 1 \leq h \leq 2 \\ kh^2 & 2 < h \leq 4 \\ 0 & \text{otherwise} \end{cases}$$

c What percentage of hedges from this study reaches a mature height less than 2 m? (3 marks)

d Determine the value of k. (3 marks)

11 (7 marks) The probability density function $f(x)$ of a continuous random variable X is given by

$$f(x) = \begin{cases} \dfrac{x+1}{k} & 0 \leq x \leq 4 \\ 0 & \text{otherwise} \end{cases}$$

a Show that $k = 12$. (2 marks)

b Find the value b of such that $P(X \leq b) = \dfrac{5}{8}$. (3 marks)

c Hence, determine $P(X > 2 \mid X < 3)$. (2 marks)

12 (4 marks) A continuous random variable X has a probability density function given by

$$f(x) = \begin{cases} \dfrac{a}{x^2} & x \geq a \\ 0 & \text{otherwise} \end{cases}$$

a Use the expression $\lim_{x \to \infty}\left(\int_a^k \frac{a}{x^2}\,dx\right)$ to justify why $f(x)$ is a valid probability density function. (2 marks)

b Determine $P(X > 2a)$. (2 marks)

Calculator-assumed

13 © SCSA MM2017 Q11a (2 marks) A pizza shop estimates that the time X hours to deliver a pizza from when it is ordered is a continuous random variable with probability density function given by

$$f(x) = \begin{cases} \frac{4}{3} - \frac{2}{3}x & 0 < x < 1 \\ 0 & \text{otherwise} \end{cases}$$

Determine the probability of a pizza being delivered within half an hour of being ordered.

14 © SCSA MM2018 Q10ab (5 marks) The following function is a probability density function on the given interval:

$$f(x) = \begin{cases} ax^2(x-2) & \text{for } 0 \le x \le 2 \\ 0 & \text{otherwise} \end{cases}$$

a Find the value of a. (3 marks)

b Find the probability that $x \ge 1.2$. (2 marks)

15 (7 marks) Each night Kim goes to the gym or the pool. When Kim goes to the gym, the time, T hours, that she spends working out is a continuous random variable with probability density function given by

$$f(t) = \begin{cases} 4t^3 - 24t^2 + 44t - 24 & \text{for } 1 \le t \le 2 \\ 0 & \text{otherwise} \end{cases}$$

a Sketch the graph of $y = f(t)$. Label any stationary points with their coordinates, correct to two decimal places. (3 marks)

b What is the probability, correct to three decimal places, that she spends less than 75 minutes working out when she goes to the gym? (2 marks)

c Kim calls her longest workout sessions 'super sessions'. If Kim spends more than k minutes in the gym she will have done a 'super session' and the probability of this occurring is 0.1. Find the value of k, to the nearest minute. (2 marks)

16 (4 marks) Sharelle is the goal shooter for her netball team. The time in hours that Sharelle spends training each day is a continuous random variable with probability density function given by

$$f(x) = \begin{cases} \frac{1}{64}(6-x)(x-2)(x+2) & \text{for } 2 \le x \le 6 \\ 0 & \text{otherwise} \end{cases}$$

a Sketch the probability density function, and label the local maximum with its coordinates, correct to two decimal places. (2 marks)

b What is the probability, correct to four decimal places, that Sharelle spends less than 3 hours training on a particular day? (2 marks)

17 (6 marks)

a The continuous random variable, X, has a probability density function given by

$$f(x) = \begin{cases} \frac{1}{4}\cos\left(\frac{x}{2}\right) & 3\pi \leq x \leq 5\pi \\ 0 & \text{otherwise} \end{cases}$$

Determine the value of a to two decimal places such that $P(X < a) = \frac{\sqrt{3}+2}{4}$. (3 marks)

b A probability density function $f(t)$ for a continuous random variable T is given by

$$f(t) = \begin{cases} \cos(t) + 1 & k \leq t \leq (k+1) \\ 0 & \text{otherwise} \end{cases}$$

where $0 < k < 2$. Show that the exact value of k is $\frac{\pi - 1}{2}$. (3 marks)

18 (7 marks) In a chocolate factory, the time, Y seconds, taken to produce a chocolate has the following probability density function.

$$f(y) = \begin{cases} 0 & y < 0 \\ \frac{y}{16} & 0 \leq y \leq 4 \\ 0.25e^{-0.5(y-4)} & y > 4 \end{cases}$$

a Explain, with the appropriate working, why $f(y)$ is a valid probability density function for a continuous random variable Y. (4 marks)

b Find, correct to four decimal places, $P(3 \leq Y \leq 5)$. (3 marks)

8.2 Measures of centre and spread

Video playlist
Measures of centre and spread

Expected value (mean) as a measure of centre

Recall from Chapter 5 that the expected value of a random variable X is the summation of all outcomes x multiplied by their corresponding probabilities $p(x)$. In discrete cases, we used the summation notation below as the values of $p(x)$ were defined for discrete values of x.

$$E(X) = \sum_{i=1}^{n} x_i p(x_i)$$

However, now that X is continuous, $p(x)$ is also a continuous curve and so the summation takes the form of a definite integral.

The expected value (mean) of a continuous random variable X

Let a continuous random variable X have a probability density function $f(x)$ defined over the interval $a \leq x \leq b$. Then the expected value of X is given by

$$E(X) = \mu = \int_a^b x f(x)\,dx$$

In some cases, the integral is written as

$$E(X) = \mu = \int_{-\infty}^{\infty} x f(x)\,dx$$

where $-\infty$ represents the lowest possible value of x and ∞ represents the highest possible value of x.

WORKED EXAMPLE 7 Finding the expected value of a continuous random variable with one component

The probability density function for a continuous random variable X is given by

$$f(x) = \begin{cases} 4x^3 & \text{if } 0 \le x \le 1 \\ 0 & \text{otherwise} \end{cases}$$

Find the expected value of X.

Steps	Working
1 Write the integral for the formula $E(X) = \int_{-\infty}^{\infty} x f(x)\,dx$ and simplify.	$E(x) = \int_0^1 x \times 4x^3\,dx$ $= \int_0^1 4x^4\,dx$
2 Evaluate the integral.	$E(X) = \left[\frac{4x^5}{5}\right]_0^1$ $= \frac{4(1)^5}{5} - 0$ $= \frac{4}{5}$

Exam hack

When a probability density function $f(x)$ has a piece-wise definition with two or more components, you will be required to write two or more separate integrals to calculate the value of $E(X)$.

WORKED EXAMPLE 8 Finding the expected value of a continuous random variable with two components

The probability density function for a continuous random variable X is given by

$$f(x) = \begin{cases} \frac{x}{4} & 0 \le x < 2 \\ 1 - \frac{x}{4} & 2 \le x \le 4 \\ 0 & \text{otherwise} \end{cases}$$

Find $E(X)$.

Steps	Working
1 Write the two integrals for the formula $E(X) = \int_{-\infty}^{\infty} x f(x)\,dx$ and simplify.	$E(X) = \int_0^2 x \times \frac{x}{4}\,dx + \int_2^4 x \times \left(1 - \frac{x}{4}\right)dx$ $= \int_0^2 \frac{x^2}{4}\,dx + \int_2^4 \left(x - \frac{x^2}{4}\right)dx$
2 Evaluate the integrals. The graph of the pdf shows by symmetry that the mean is 2. 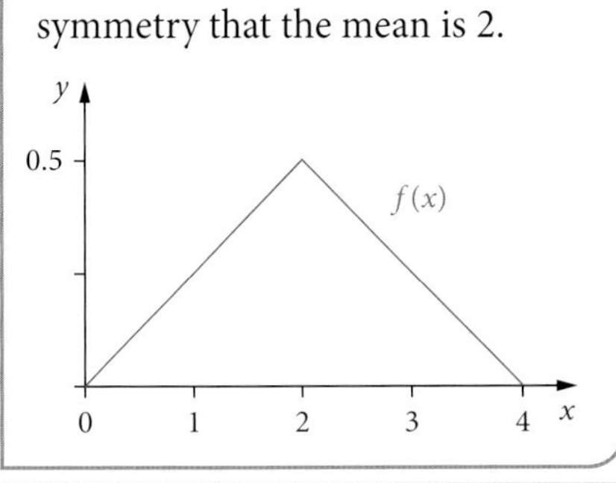	$E(X) = \left[\frac{x^3}{12}\right]_0^2 + \left[\frac{x^2}{2} - \frac{x^3}{12}\right]_2^4$ $= \left[\frac{2^3}{12} - 0\right] + \left[\left(\frac{4^2}{2} - \frac{4^3}{12}\right) - \left(\frac{2^2}{2} - \frac{2^3}{12}\right)\right]$ $= \frac{8}{12} + \left[\left(8 - \frac{64}{12}\right) - \left(2 - \frac{8}{12}\right)\right]$ $= 2$

In some cases, $xf(x)$ will produce a function that we do not have the skills and techniques required to integrate by hand. This is where CAS can be useful.

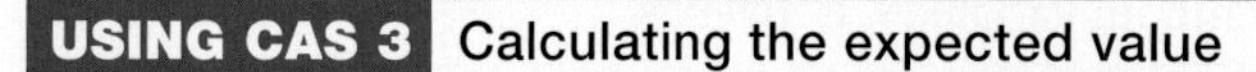

A continuous random variable X is defined by the following probability density function.

$$f(x) = \begin{cases} k\sin(2\pi x) & 0 \le x \le \dfrac{1}{2} \\ 0 & \text{otherwise} \end{cases}$$

a Show that the value of k is π.

b Determine $\text{E}(X)$.

ClassPad

a

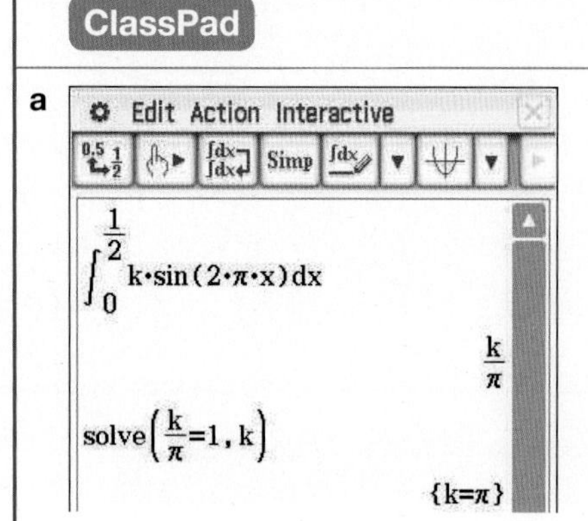

1 Establish the definite integral giving a total area under the curve.

2 Solve the result equal to 1.

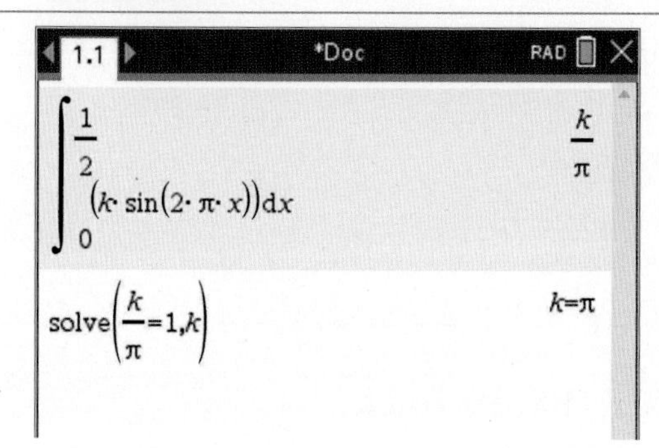

1 Establish the definite integral giving a total area under the curve.

2 Solve the result equal to 1.

b

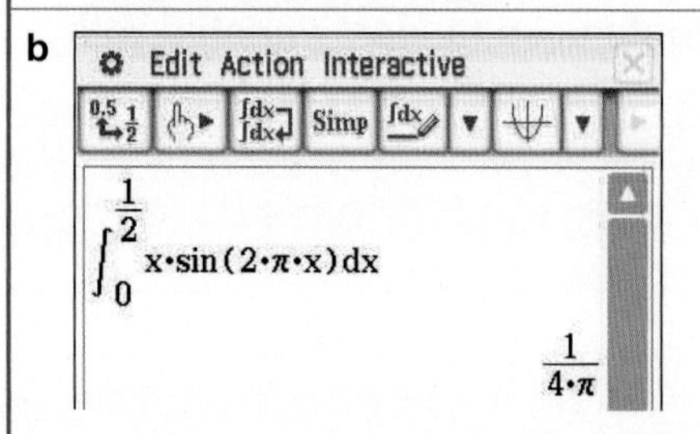

Establish and evaluate the definite integral for the expected value using $\text{E}(X) = \int_{-\infty}^{\infty} xf(x)\,dx$.

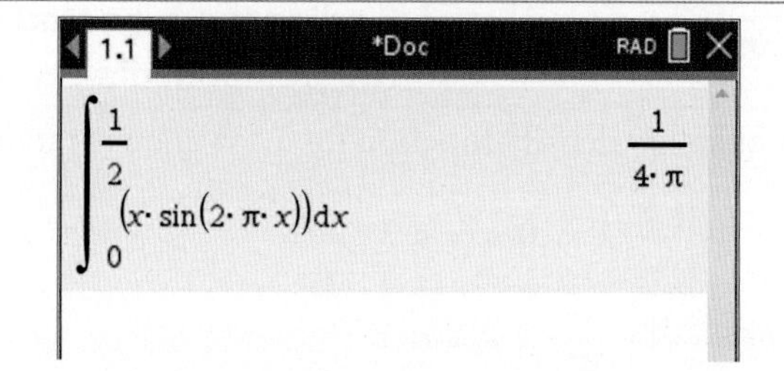

Establish and evaluate the definite integral for the expected value using $\text{E}(X) = \int_{-\infty}^{\infty} xf(x)\,dx$.

This gives the expected value of $\dfrac{1}{4\pi}$.

The median and other percentiles

In addition to the expected value (or mean) being used to measure the centre of a continuous random variable, we can also encounter problems that involve calculating the **median** of a continuous random variable. Recall that the **median** is the middle score of a random variable, when the outcomes are in ascending or descending order.

The median of a continuous random variable X

For a continuous random variable X, let $x = m$ be the median. Then

$$\text{P}(X < m) = \text{P}(X \le m) = \text{P}(X > m) = \text{P}(X \ge m) = 0.5$$

The value of m can be solved for using an appropriate definite integral of the probability density function $f(x)$ defined over the interval $a \le x \le b$.

$$\int_a^m f(x)\,dx = \int_m^b f(x)\,dx = 0.5$$

WORKED EXAMPLE 9 **Finding the median of a continuous random variable**

The probability density function for a continuous random variable X is given by

$$f(x) = \begin{cases} \dfrac{x}{32} & 0 \le x \le 8 \\ 0 & \text{otherwise} \end{cases}$$

Find the median, m, of X.

Steps	Working
1 Sketch the probability density function.	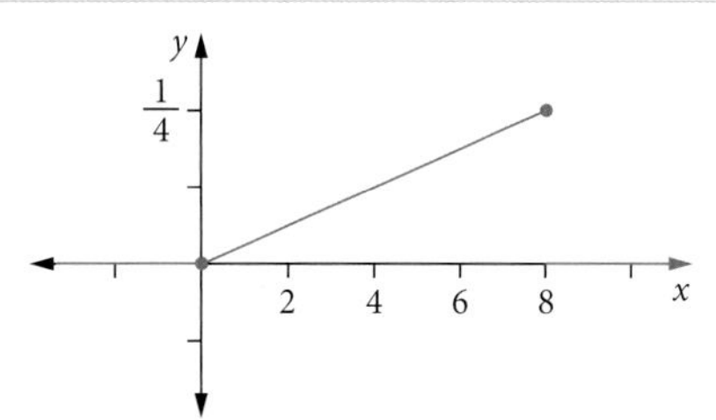
2 Establish and solve the integral equation of the 50th percentile.	$\int_0^m \frac{x}{32}\,dx = 0.50$ $\left[\frac{x^2}{64}\right]_0^m = \frac{1}{2}$ $\frac{m^2}{64} = \frac{1}{2}$ $m^2 = 32$ $m = \pm\sqrt{32} = \pm 4\sqrt{2}$ $\therefore m = 4\sqrt{2}$ (as $m > 0$)

The median is a specific example of a **percentile** of a continuous random variable X, which is a value of x for which a certain percentage of scores, p%, fall below that value. That is, the median is the 50th percentile. Other common percentiles include

- the 25th percentile or lower quartile
- the 75th percentile or upper quartile.

A percentile of a continuous random variable X

For a continuous random variable X with a probability density function $f(x)$, the value of the pth percentile ($x = k$) is determined by the integral equation

$$P(X \le k) = \frac{p}{100} \Rightarrow \int_{-\infty}^k f(x)\,dx = \frac{p}{100}.$$

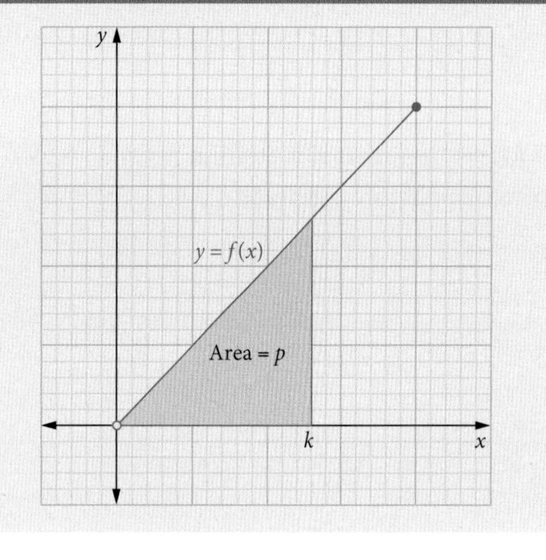

WORKED EXAMPLE 10 **Finding a percentile of a continuous random variable**

The probability density function for a continuous random variable X is given by

$$f(x) = \begin{cases} \dfrac{x}{32} & 0 \leq x \leq 8 \\ 0 & \text{otherwise} \end{cases}$$

Find the value of k such that $P(X < k) = 0.75$.

Steps	Working
1 Sketch the probability density function.	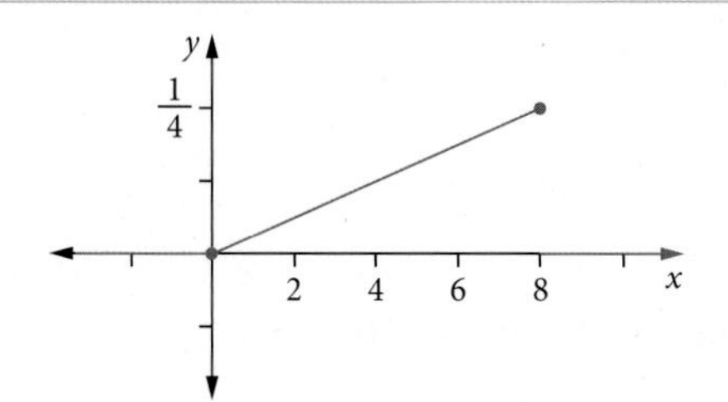
2 Establish and solve the integral equation of the 75th percentile.	$\int_0^k \frac{x}{32}\,dx = 0.75$ $\left[\frac{x^2}{64}\right]_0^k = \frac{3}{4}$ $\frac{k^2}{64} - 0 = \frac{3}{4}$ $k^2 = 48$ $k = \pm\sqrt{48}$ $= \pm 4\sqrt{3}$ $\therefore\ k = 4\sqrt{3} \quad (\text{as } k > 0)$

Variance and standard deviation as measures of spread

Similarly, recall from Chapter 5 that the variance and standard deviation were useful measures of spread of a random variable X about its mean, whereby

- the variance is the weighted average of the squared deviations from the mean
- the standard deviation is the square root of the variance.

Once again, in discrete cases we used the summation notation below as the values of $p(x)$ were defined for discrete values of x.

$$\text{Var}(x) = \sum_{i=1}^{n}(x_i - \mu)^2 p(x_i)$$

However, now that X is continuous, $p(x)$ is also a continuous curve and so the summation again takes the form of a definite integral.

8.2

The variance and standard deviation of a continuous random variable X

Let a continuous random variable X have a probability density function $f(x)$ defined over the interval $a \le x \le b$. Then the variance of X is given by

$$\text{Var}(X) = \sigma^2 = \int_a^b (x - \mu)^2 f(x)\,dx.$$

In some cases, the integral is written as

$$\text{Var}(X) = \sigma^2 = \int_{-\infty}^{\infty} (x - \mu)^2 f(x)\,dx$$

where $-\infty$ represents the lowest possible value of x and ∞ represents the highest possible value of x.

Remember, the more efficient computational formula for variance can also be used:

$$\text{Var}(X) = \text{E}(X^2) - \text{E}(X)^2 = \int_a^b x^2 f(x)\,dx - \mu^2$$

The standard deviation is then $\text{SD}(X) = \sqrt{\text{Var}(X)}$.

WORKED EXAMPLE 11 Finding the variance and standard deviation of a continuous random variable

The probability density function for a continuous random variable X is given by

$$f(x) = \begin{cases} 4x^3 & \text{if } 0 \le x \le 1 \\ 0 & \text{otherwise} \end{cases}$$

Find the variance and standard deviation of X.

Steps	Working
1 Write the formula for the variance and calculate $\text{E}(X)$ and $\text{E}(X^2)$ using $\text{E}(X^2) = \int_{-\infty}^{\infty} x^2 f(x)\,dx$. Note: μ or $\text{E}(X)$ was calculated in Worked example 7.	$\text{Var}(X) = \text{E}(X^2) - \mu^2$ $\text{E}(X) = \int_0^1 x \times 4x^3\,dx = \frac{4}{5}$ $\text{E}(X^2) = \int_0^1 x^2 \times 4x^3\,dx$ $= \int_0^1 4x^5\,dx$ $\text{E}(X^2) = \left[\frac{4x^6}{6}\right]_0^1$ $= \frac{2(1^6)}{3} - 0$ $= \frac{2}{3}$
2 Substitute into the variance formula.	$\text{Var}(X) = \text{E}(X^2) - \mu^2$ $= \frac{2}{3} - \left(\frac{4}{5}\right)^2$ $= \frac{2}{3} - \frac{16}{25}$ $= \frac{2}{75}$
3 Use the fact $\text{SD}(X) = \sqrt{\text{Var}(X)}$.	$\text{SD}(X) = \sqrt{\frac{2}{75}} = \frac{\sqrt{6}}{15}$

Linear changes of scale and origin

Recall the ideas of a linear change of scale and origin from Chapter 5, such that

- a change of scale of a random variable X multiplies all values of x by a scalar multiple a, i.e. $X \rightarrow aX$
- a change of origin of a random variable X adds a value b (or subtracts b from) all values of x, i.e. $X \rightarrow X \pm b$.

The effects of these changes of scale and origin have the same effect on the mean, variance and standard deviation of a continuous random variable as they did to a discrete random variable due to the properties of definite integration.

Expected value, variance and standard deviation of $Y = aX + b$

For a linear transformation $Y = aX + b$ of a continuous random variable X:

$$E(aX + b) = a\,E(X) + b$$

$$\text{Var}(aX + b) = a^2\,\text{Var}(X)$$

$$\text{SD}(aX + b) = |a|\,\text{SD}(X)$$

WORKED EXAMPLE 12 **Finding the expected value, variance and standard deviation of $aX + b$**

A continuous random variable X has a mean of 4 and variance of 16. A continuous random variable Y is defined such that $Y = -\frac{1}{2}X + 5$. Determine

a $E(Y)$

b $\text{Var}(Y)$

c $\text{SD}(Y)$.

Steps	Working
a Apply the formula $E(aX + b) = aE(X) + b$.	$E(Y) = E\left(-\frac{1}{2}X + 5\right)$ $= -\frac{1}{2}(4) + 5$ $= 3$
b Apply the formula $\text{Var}(aX + b) = a^2\,\text{Var}(X)$.	$\text{Var}(Y) = \text{Var}\left(-\frac{1}{2}X + 5\right)$ $= \frac{1}{4}(16)$ $= 4$
c Apply the formula $\text{SD}(aX + b) = |a|\,\text{SD}(X)$.	$\text{SD}(Y) = \frac{1}{2}(4)$ $= 2$ or $\sqrt{\text{Var}(Y)} = \sqrt{4}$ $= 2$

EXERCISE 8.2 Measures of centre and spread

ANSWERS p. 404

Recap

1 The probability density function for the continuous random variable X is given by

$$f(x) = \begin{cases} 3x^2 & 0 \le x \le 1 \\ 0 & \text{otherwise} \end{cases}$$

The value of $P\left(X > \frac{1}{2}\right)$ is

A $\frac{1}{27}$ **B** $\frac{1}{8}$ **C** $\frac{2}{3}$ **D** $\frac{7}{8}$ **E** $\frac{26}{27}$

2 If X is a continuous random variable such that $P(X > 5) = a$ and $P(X > 8) = b$ then $P(X < 5 \mid X < 8)$ is

A $\frac{a}{b}$ **B** $\frac{a-b}{1-b}$ **C** $\frac{1-b}{1-a}$ **D** $\frac{ab}{1-b}$ **E** $\frac{a-1}{b-1}$

Mastery

3 WORKED EXAMPLE 7 The probability density function for a continuous random variable X is given by

$$f(x) = \begin{cases} \frac{1}{288}(12x - x^2) & 0 \le x \le 12 \\ 0 & \text{otherwise} \end{cases}$$

Find the expected value of X.

4 WORKED EXAMPLE 8 A continuous random variable X has a probability density function given by

$$f(x) = \begin{cases} \frac{3x}{8} & 0 \le x \le 2 \\ 3 - \frac{9}{8}x & 2 < x \le \frac{8}{3} \\ 0 & \text{otherwise} \end{cases}$$

Find $E(X)$.

5 Using CAS 3 A continuous random variable X is defined by the following probability density function.

$$f(x) = \begin{cases} k\sin(\pi x) & 0 \le x \le 1 \\ 0 & \text{otherwise} \end{cases}$$

a Show that the value of k is $\frac{\pi}{2}$.

b Determine $E(X)$.

6 WORKED EXAMPLE 9 The probability density function for a continuous random variable X is given by

$$f(x) = \begin{cases} \frac{x}{12} & 1 \le x \le 5 \\ 0 & \text{otherwise} \end{cases}$$

Find the median, m, of X.

7 WORKED EXAMPLE 10 The probability density function for a continuous random variable X is given by

$$f(x) = \begin{cases} \frac{x}{8} & 0 \le x \le 4 \\ 0 & \text{otherwise} \end{cases}$$

Find the value of k such that $P(X < k) = 0.3$.

8 WORKED EXAMPLE 11 A continuous random variable X has a probability density function given by

$$f(x) = \begin{cases} 2x & 0 \le x \le 1 \\ 0 & \text{otherwise} \end{cases}$$

Find the variance and standard deviation of X.

9 WORKED EXAMPLE 12 A continuous random variable X has a mean of 3 and variance of 9. A continuous random variable Y is defined such that $Y = \frac{2}{3}X - 1$. Determine

a $E(Y)$ **b** $\text{Var}(Y)$ **c** $SD(Y)$.

Calculator-free

10 (6 marks) The continuous random variable X, with probability density function $p(x)$ defined over $a \le x \le b$, has mean 2 and variance 5.

a Determine the value of $\int_a^b x^2 p(x)\,dx$. (2 marks)

b Determine $E(3X + 2)$. (2 marks)

c Determine $\text{Var}\left(-\frac{1}{5}X - 1\right)$. (2 marks)

11 (4 marks) A continuous random variable X has the probability density function f given by

$$f(x) = \begin{cases} \frac{k}{x^2} & 1 \le x \le 2 \\ 0 & \text{otherwise} \end{cases}$$

a Show that $k = 2$. (2 marks)

b Determine $E(X)$. (2 marks)

12 (3 marks) A continuous random variable, X, has a probability density function given by

$$f(x) = \begin{cases} \frac{1}{5}e^{-\frac{x}{5}} & x \ge 0 \\ 0 & x < 0 \end{cases}$$

The median of X is m such that $P(X < m) = 0.5$. Determine the exact value of m.

13 (7 marks) The probability density function for a continuous random variable X is given by

$$f(x) = \begin{cases} ax(5 - x) & 0 \le x \le 5 \\ 0 & \text{otherwise} \end{cases}$$

where a is a positive constant.

a Show that $a = \frac{6}{125}$. (3 marks)

b Explain why $E(X) = \frac{5}{2}$. (1 marks)

c Show that $SD(X) = \frac{\sqrt{5}}{2}$. (3 marks)

Calculator-assumed

14 © SCSA MM2018 Q10c (2 marks) The following function is a probability density function on the given interval

$$f(x) = \begin{cases} ax^2(x-2) & \text{for } 0 \le x \le 2 \\ 0 & \text{otherwise} \end{cases}$$

Find the median of the distribution.

15 © SCSA MM2017 Q11bc (7 marks) A pizza shop estimates that the time X hours to deliver a pizza from when it is ordered is a continuous random variable with probability density function given by

$$f(x) = \begin{cases} \frac{4}{3} - \frac{2}{3}x & 0 < x < 1 \\ 0 & \text{otherwise} \end{cases}$$

a Calculate the mean delivery time to the nearest minute. (3 marks)

b Calculate the standard deviation of the delivery time to the nearest minute. (4 marks)

16 (10 marks) A continuous random variable X has a probability density function given by

$$f(x) = \begin{cases} \frac{1}{8}e^{-\frac{x}{8}} & x \ge 0 \\ 0 & x < 0 \end{cases}$$

a Show that $f(x)$ is a valid probability density function for a continuous random variable. (3 marks)

b Determine

i $E(X)$ (2 marks)

ii $Var(X)$. (3 marks)

c Find the value of k, correct to three decimal places, for which $P(X < k) = 0.25$. (2 marks)

17 (10 marks) For the continuous random variable X with probability density function

$$f(x) = \begin{cases} \log_e(x) & 1 \le x \le e \\ 0 & \text{otherwise} \end{cases}$$

determine the following correct to four decimal places, where appropriate.

a $P(X > 1 \mid X < 2)$ (2 marks)

b k such that $P(X < k) = 0.8$ (2 marks)

c $E(X)$ (2 marks)

d $Var(X)$ (2 marks)

e $SD(Y)$ where $Y = -2X$ (2 marks)

18 (5 marks) FullyFit is an international company that owns and operates many fitness centres (gyms) in several countries. At every one of FullyFit's gyms, each member agrees to have his or her fitness assessed every month by undertaking a set of exercises called S. There is a five-minute time limit on any attempt to complete S and if someone completes S in less than three minutes, they are considered fit.

When FullyFit surveyed all its gyms throughout the world, it was found that the time taken by members to complete S is a continuous random variable X, with a probability density function g, as defined below.

$$g(x) = \begin{cases} \dfrac{(x-3)^3 + 64}{256} & 1 \le x \le 3 \\ \dfrac{x+29}{128} & 3 < x \le 5 \\ 0 & \text{otherwise} \end{cases}$$

a Find $E(X)$, correct to four decimal places. (2 marks)

b In a random sample of 200 FullyFit members, how many members would be expected to take more than four minutes to complete S? Give your answer to the nearest integer. (3 marks)

19 (8 marks) Rebecca's Robotics manufactures three types of components for robots: sensors, motors and controllers.

The weight, w, in grams, of controllers is modelled by the following probability density function.

$$C(w) = \begin{cases} \dfrac{3}{640000}(330 - w)^2(w - 290) & 290 \le w \le 330 \\ 0 & \text{otherwise} \end{cases}$$

a Determine the mean weight, in grams, of the controllers. (2 marks)

b Determine the probability that a randomly selected controller weighs less than the mean weight of the controllers. (2 marks)

c Determine the standard deviation of the weight, in grams, of the controllers. (2 marks)

d Determine the probability that a randomly selected controller weighs more than one standard deviation greater than the mean weight of the controllers. (2 marks)

8.3 Uniform and triangular distributions

Uniform continuous random variables

In Chapter 5 we examined simple examples of discrete random variables that had the same probability for each outcome $x \in X$. Suppose that is now the case for a continuous random variable X defined over the interval $a \leq x \leq b$ such that each value of x has an equally likely chance of occurring. This random variable can be described as a **uniform continuous random variable**.

Continuous uniform distribution

For a continuous random variable X, if X is uniformly distributed over the interval $a \leq x \leq b$, then the probability density function of X is defined as

$$f(x) = \begin{cases} \dfrac{1}{b-a} & a \leq x \leq b \\ 0 & \text{otherwise} \end{cases}$$

A uniform continuous random variable can be denoted using $X \sim U[a, b]$.

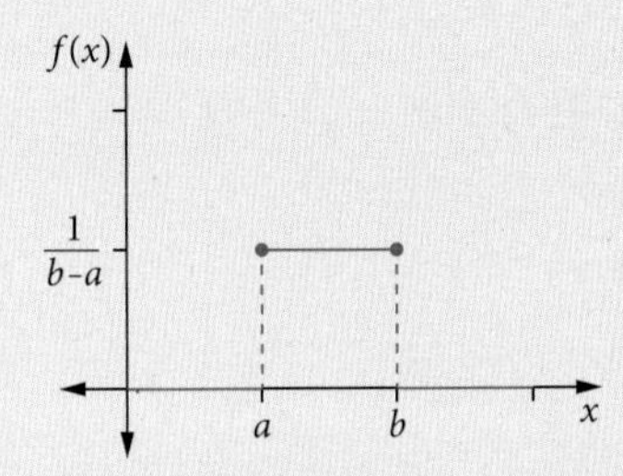

The probability density function creates a rectangular-shaped graph.

Video playlist
Uniform continuous random variables

Exam hack

In most cases, probabilities from a continuous uniform distribution can be found simply using the area of rectangles rather than definite integration! Regardless, always be sure to show the appropriate working, whether it is the area of the rectangles used or the definite integrals used.

WORKED EXAMPLE 13 Calculating probabilities from a continuous uniform distribution

A uniformly distributed continuous random variable X has the probability density function

$$f(x) = \begin{cases} \dfrac{1}{8} & 12 \leq x \leq 20 \\ 0 & \text{otherwise} \end{cases}$$

Determine

a $P(X = 13)$

b $P(X > 15)$

c $P(X \leq 18 \mid X > 15)$

Steps	Working
a Recognise that $P(X = k) = 0$ for a continuous random variable.	$P(X = 13) = 0$
b **1** Draw a sketch of $f(x)$ and shade the appropriate region under the curve.	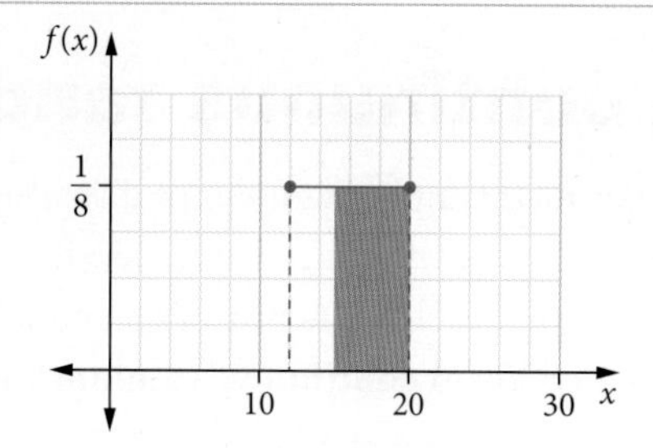
2 Calculate the area of the rectangle.	$P(X > 15) = P(15 < X \le 20)$ $= 5\left(\frac{1}{8}\right)$ $= \frac{5}{8}$
c **1** Establish the conditional probability formula $P(A\|B) = \frac{P(A \cap B)}{P(B)}$.	$(X \le 18 \| X > 15) = \frac{P(15 < X \le 18)}{P(X > 15)}$
2 Represent the probability diagrammatically.	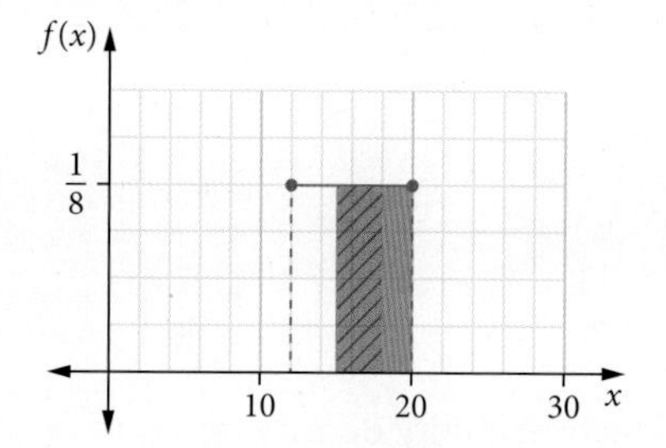
3 Calculate the areas of the rectangles and, hence, the probability.	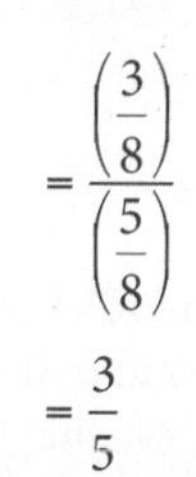 $= \frac{\left(\frac{3}{8}\right)}{\left(\frac{5}{8}\right)}$ $= \frac{3}{5}$

The cumulative distribution function of a uniform continuous random variable can also be obtained using integration, such that for $a \le x \le b$,

$$F(x) = \int_a^x \frac{1}{b-a}\,dt$$

$$= \left[\frac{t}{b-a}\right]_a^x$$

$$= \frac{x}{b-a} - \frac{a}{b-a}$$

$$= \frac{x-a}{b-a}$$

Cumulative distribution function of a continuous uniform distribution

For a uniformly distributed continuous random variable $X \sim U[a, b]$, the cumulative distribution function of X is defined as

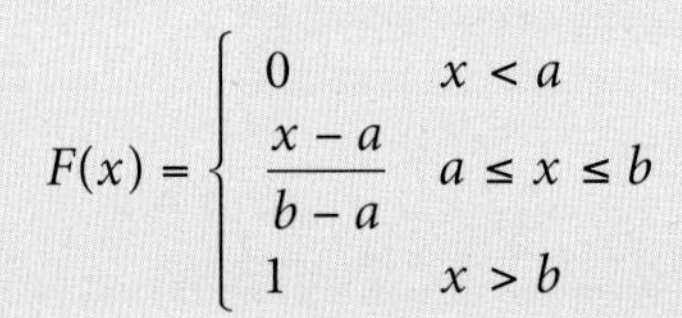

$$F(x) = \begin{cases} 0 & x < a \\ \dfrac{x-a}{b-a} & a \le x \le b \\ 1 & x > b \end{cases}$$

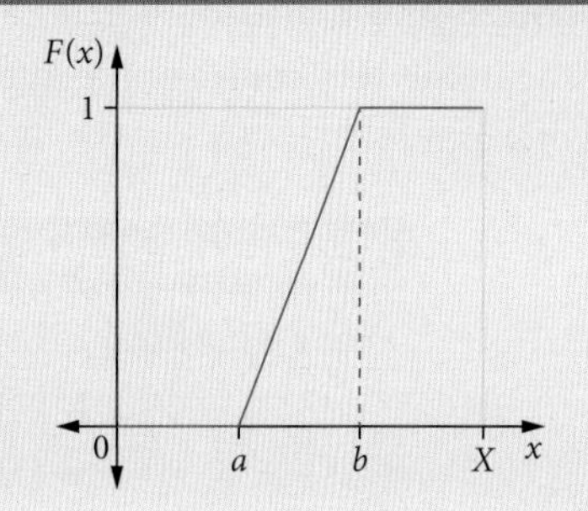

WORKED EXAMPLE 14	**Determining the probability density function from a cumulative distribution function**

A uniformly distributed continuous random variable Y has a cumulative distribution function given by

$$F(y) = \begin{cases} 0 & y < a \\ \dfrac{y-a}{10} & a \le y \le 15 \\ 1 & y > 15 \end{cases}$$

a Determine the value of a.

b State the probability density function of Y.

c Hence, or otherwise, calculate $P(Y < 12 \mid Y < 14)$.

Steps	**Working**
a Recognise that $b - a = 10$ and solve for a.	$15 - a = 10$ $a = 5$
b Establish the piece-wise-defined function for the uniform continuous random variable either using $f(x) = \dfrac{1}{b-a}$ or $f(x) = \dfrac{d}{dx}(F(x))$.	$f(y) = \begin{cases} \dfrac{1}{10} & 5 \le y \le 15 \\ 0 & \text{otherwise} \end{cases}$
c 1 Recognise that in $P(A \mid B)$, set A is contained within set B and so $P(A \mid B) = \dfrac{P(A)}{P(B)}$.	$P(Y < 12 \mid Y < 14) = \dfrac{P(Y < 12)}{P(Y < 14)}$
2 Use the cumulative distribution function to evaluate $P(Y < 12)$ and $P(Y < 14)$, or consider the area of rectangles.	$= \dfrac{\left(\dfrac{12-5}{10}\right)}{\left(\dfrac{14-5}{10}\right)}$ $= \dfrac{7}{9}$

When considering the expected value of a random variable with a symmetrical distribution such as the continuous uniform distribution, note that the symmetry of the distribution means that the expected value should lie in the very centre of the interval of x values and, hence, can be calculated using the average of the values of a and b.

Expected value of a uniform continuous random variable

For a uniformly distributed continuous random variable, $X \sim U[a, b]$,

$$\text{E}(X) = \frac{a+b}{2}$$

This can also be proven using integration, i.e. $\text{E}(X) = \int_a^b x\left(\frac{1}{b-a}\right)dx$.

Note that due to the symmetry of the distribution, the median of a uniform continuous random variable is the same as the mean.

We can then use integration and the computational formula for variance, $\text{Var}(X) = \text{E}(X^2) - \text{E}(X)^2$ to derive a formula for the variance and standard deviation of a uniformly distributed continuous random variable.

$$\begin{aligned}
\text{E}(X^2) &= \int_a^b x^2\left(\frac{1}{b-a}\right)dx \\
&= \frac{1}{b-a}\int_a^b x^2\,dx \\
&= \frac{1}{b-a}\left[\frac{x^3}{3}\right]_a^b \\
&= \frac{1}{b-a} \times \frac{b^3-a^3}{3}
\end{aligned}$$

Using the factorisation $b^3 - a^3 = (b-a)(b^2 + ab + a^2)$,

$$\text{E}(X^2) = \frac{b^2 + ab + a^2}{3}$$

$$\begin{aligned}
\text{Var}(X) &= \frac{b^2 + ab + a^2}{3} - \left(\frac{a+b}{2}\right)^2 \\
&= \frac{b^2 + ab + a^2}{3} - \frac{a^2 + 2ab + b^2}{4} \\
&= \frac{4b^2 + 4ab + 4a^2 - 3a^2 - 6ab - 3b^2}{12} \\
&= \frac{b^2 - 2ab + a^2}{12}
\end{aligned}$$

Using the factorisation $b^2 - 2ab + a^2 = (b-a)^2$,

$$\text{Var}(X) = \frac{(b-a)^2}{12}$$

Variance and standard deviation of a uniform continuous random variable

For a uniformly distributed continuous random variable, $X \sim U[a, b]$,

$$\text{Var}(X) = \frac{(b-a)^2}{12}$$

and

$$\text{SD}(X) = \frac{b-a}{\sqrt{12}}.$$

Exam hack

The formulas for the expected value, variance and standard deviation of uniform continuous random variable are not on the formula sheet. Write them on your notes, but be sure to pay attention to the number of marks in a question to calculate mean, variance or standard deviation (more than 2 marks per statistic) or any specific instructions to *use integration*. In these cases, the formulas may not be helpful!

WORKED EXAMPLE 15 Finding the expected value, variance and standard deviation of a uniform continuous random variable

A uniformly distributed continuous random variable X is defined such that $X \sim U[-10, 10]$.

Find the

a probability density function of X

b expected value of X

c variance of X

d standard deviation of X.

Steps	Working
a **1** Identify the values of a and b.	$a = -10, b = 10$
2 Use the piece-wise definition of $f(x)$. $f(x) = \begin{cases} \frac{1}{b-a} & a \le x \le b \\ 0 & \text{otherwise} \end{cases}$	$f(x) = \begin{cases} \frac{1}{20} & -10 \le x \le 10 \\ 0 & \text{otherwise} \end{cases}$
b Use the formula $\text{E}(X) = \frac{a+b}{2}$.	$\text{E}(X) = \frac{-10+10}{2}$ $= 0$
c Use the formula $\text{Var}(X) = \frac{(b-a)^2}{12}$.	$\text{Var}(X) = \frac{20^2}{12}$ $= \frac{400}{12}$ $= \frac{100}{3}$
d Use the formula $\text{SD}(X) = \sqrt{\text{Var}(X)}$.	$\text{SD}(X) = \sqrt{\frac{100}{3}}$ $= \frac{10\sqrt{3}}{3}$

Triangular continuous random variables

In some previous examples, such as Worked example 8, we were exposed to a particular type of probability density function for a special continuous random variable. You may recall seeing a probability density function with a graph in the shape of a triangle. Continuous random variables with triangular-shaped graphs are called **triangular continuous random variables.**

Triangular distributions can be both symmetrical and asymmetrical. Suppose a continuous random variable X is triangularly distributed with a probability density function $f(x)$ defined over the interval $a \leq x \leq b$. Let the 'peak' of the triangle occur at $x = c$. This peak can be considered the **mode** of the triangular distribution as it is the value of x with the highest value of $f(x)$.

If symmetrical, then the mode occurs halfway between a and b and, hence, is the mean and median of the distribution, i.e. $c = \text{E}(X) = \dfrac{a+b}{2}$.

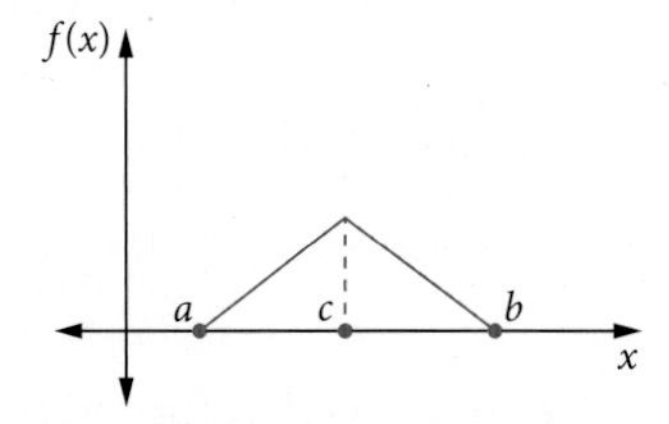

If asymmetrical, then the mean will **not** be the same as the mode.

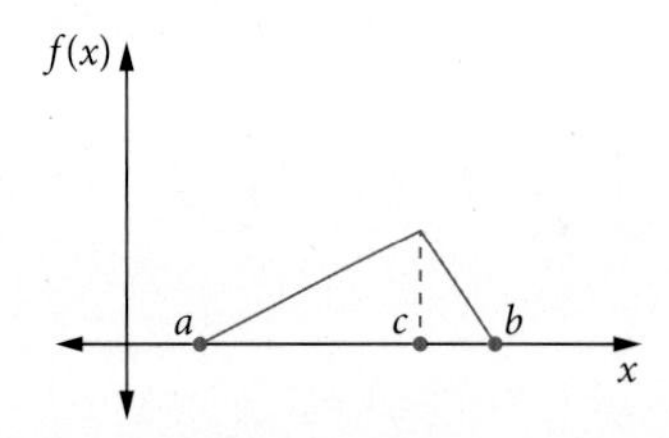

Maximum value of $f(x)$ in a triangular distribution

Given that the area of the triangle with base $(b - a)$ and a perpendicular height $h = f(c)$ needs to equal 1, then

$$\frac{1}{2}(b-a)f(c) = 1$$

$$f(c) = \frac{2}{b-a}$$

For a triangular continuous random variable X, then the maximum value of $f(x)$ at the mode $x = c$ is

$$f(c) = \frac{2}{b-a}$$

Although the probability density function, expected value and variance of a triangular distribution can be generalised regardless of the symmetry, it is not an explicit part of this course. As a result, it is often more useful to consider the function as a piece-wise-defined function with linear components, as we saw in previous examples, and use the appropriate geometric and integration techniques to solve any related problems.

8.3

WORKED EXAMPLE 16 Using a symmetrical triangular distribution

Consider the following probability density function for the continuous random variable X.

$$f(x) = \begin{cases} x & 0 \le x \le 1 \\ 2 - x & 1 < x \le 2 \\ 0 & \text{otherwise} \end{cases}$$

a Sketch the graph of $y = f(x)$ labelling the coordinates of all critical features. Hence, describe the shape of the distribution.

b Determine

i $P(X < 1.5)$ **ii** $P(X < 2 \mid X > 1.5)$ **iii** $E(X)$ **iv** $Var(X)$.

Steps	**Working**
a 1 Sketch each linear component over the given domains. **2** Label all axes intercepts and the 'peak'.	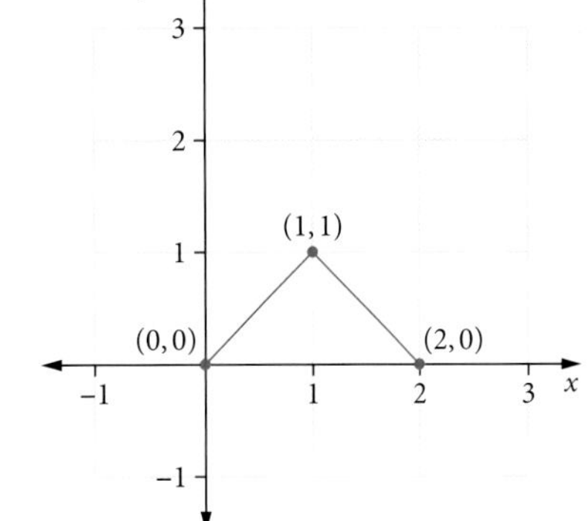
3 Describe the shape of the distribution, including the parameters.	It is a symmetrical triangular distribution over the interval $0 \le x \le 2$.
b i 1 Identify the region on the graph.	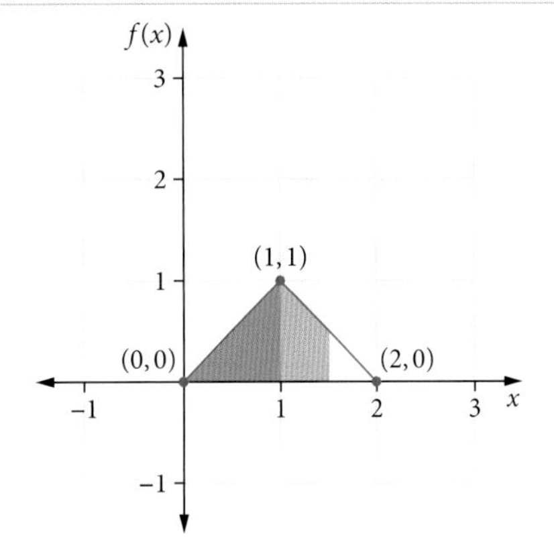
2 Use a geometric approach (i.e. triangle and trapezium) to calculate the probability as the area under the curve.	$P(X < 1.5) = \frac{1}{2} + \frac{1}{2}\left(1 + \frac{1}{2}\right)\left(\frac{1}{2}\right)$ $= \frac{1}{2} + \frac{1}{4}\left(\frac{3}{2}\right)$ $= \frac{7}{8}$

ii **1** Use the conditional probability formula $P(A \mid B) = \frac{P(A \cap B)}{P(B)}$.

$$P(X < 2 \mid X > 1.5) = \frac{P(1.5 < X < 2)}{P(X > 1.5)}$$

2 Use the complement to find $P(X > 1.5)$ given part **a**.

$$= \frac{\frac{1}{2}\left(\frac{1}{2}\right)\left(\frac{1}{2}\right)}{1 - \frac{7}{8}}$$

$$= 1$$

Exam hack

Be sure to look out for 'trick' questions like this one. Of course it is certain that $x < 2$ if $x > 1.5$, as the upper bound is 2.

iii Use the symmetry to identify the mean.

$$E(X) = 1$$

iv **1** Find $E(X^2)$ using integration.

$$E(X^2) = \int_0^1 x^2(x)\,dx + \int_1^2 x^2(2 - x)\,dx$$

$$= \int_0^1 x^3\,dx + \int_1^2 2x^2 - x^3 dx$$

2 Simplify the result for ease of computation.

$$= \left[\frac{x^4}{4}\right]_0^1 + \left[\frac{2x^3}{3} - \frac{x^4}{4}\right]_1^2$$

$$= \frac{1}{4} + \left(\frac{16}{3} - \frac{16}{4} - \frac{2}{3} + \frac{1}{4}\right)$$

$$= -\frac{14}{4} + \frac{14}{3}$$

$$= \frac{-42 + 56}{12}$$

$$= \frac{14}{12}$$

$$= \frac{7}{6}$$

3 Use $\text{Var}(X) = E(X^2) - E(X)^2$.

$$\text{Var}(X) = \frac{7}{6} - 1^2$$

$$= \frac{1}{6}$$

WORKED EXAMPLE 17 Using an asymmetrical triangular distribution

Consider the following probability density function for the continuous random variable X.

$$f(x) = \begin{cases} \frac{3x}{8} & 0 \le x \le 2 \\ 3 - \frac{9}{8}x & 2 < x \le \frac{8}{3} \\ 0 & \text{otherwise} \end{cases}$$

a Show that it is a valid probability density function for X.

b Determine the 30th percentile.

c Show the use of integration to determine $E(X)$.

Steps	Working
a 1 Identify the two conditions of a valid pdf. • $f(x) \geq 0$ for all x in $a \leq x \leq b$ • $\int_a^b f(x)\,dx = 1$	The end points of the line segment with equation $\frac{3x}{8}$ are $(0,0)$ and $\left(2,\frac{3}{4}\right)$. The end points of the line segment with equation $3-\frac{9}{8}x$ are $\left(2,\frac{3}{4}\right)$ and $\left(\frac{8}{3},0\right)$.
2 Show the first condition graphically or numerically.	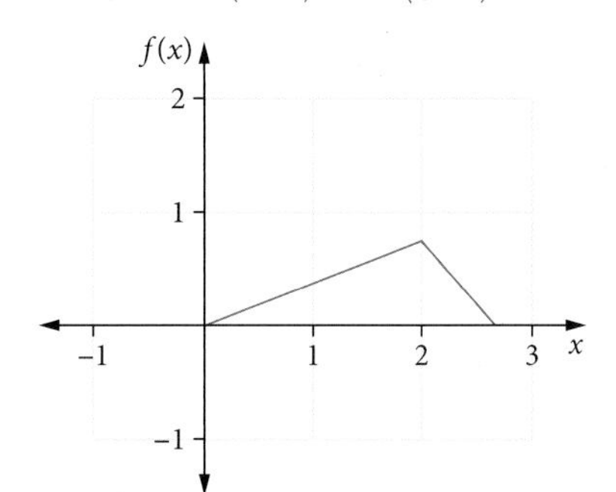
3 Show the second condition geometrically using the area of triangles or algebraically using integration.	And so $f(x) \geq 0$ for all values of x in the domain $0 \leq x \leq \frac{8}{3}$. $\int_0^{\frac{8}{3}} f(x)\,dx = \frac{1}{2}(2)\left(\frac{3}{4}\right)+\frac{1}{2}\left(\frac{2}{3}\right)\left(\frac{3}{4}\right)$ $= \frac{3}{4}+\frac{1}{4}$ $= 1$ Therefore, $f(x)$ is a valid pdf.
b 1 Recognise that $P(X < 2) = 0.75$ and so the 30th percentile lies in the first triangular region. Assign a variable to the value of the percentile.	Let the 30th percentile be $x = k$.
2 Establish the appropriate integral (or area formula of a triangle) to calculate the 30th percentile.	$\int_0^k \frac{3x}{8}\,dx = \frac{3}{10}$ $\left[\frac{3x^2}{16}\right]_0^k = \frac{3}{10}$ $\frac{3k^2}{16} = \frac{3}{10}$ Alternatively, $A = \frac{1}{2}bh$ $\frac{3}{10} = \frac{1}{2}k\left(\frac{3k}{8}\right)$ $\frac{3k^2}{16} = \frac{3}{10}$ $k^2 = \frac{8}{5}$ $k = \pm\frac{2\sqrt{10}}{5}$ $k = \frac{2\sqrt{10}}{5}$ (as $k > 0$)

c **1** Establish the sum of two definite integrals for E(X).	$\mathrm{E}(X) = \int_0^2 x\left(\frac{3x}{8}\right)dx + \int_2^{\frac{8}{3}} x\left(3 - \frac{9x}{8}\right)dx$ $= \int_0^2 \frac{3x^2}{8}dx + \int_2^{\frac{8}{3}} 3x - \frac{9}{8}x^2 dx$
2 Evaluate the definite integrals.	$= \left[\frac{x^3}{8}\right]_0^2 + \left[\frac{3x^2}{2} - \frac{9}{24}x^3\right]_2^{\frac{8}{3}}$ $= 1 + \left(\frac{3}{2}\left(\frac{64}{9}\right) - \frac{9}{24}\left(\frac{512}{27}\right) - 6 + 3\right)$ $= 1 + \frac{32}{3} - \frac{64}{9} - 3$ $= \frac{14}{9}$

Problems involving the uniform, triangular and binomial distributions

In some practical problems where continuous random variables are modelled by uniform or triangular distributions, we may be asked to solve a binomial problem for which the probability of success is to be determined from the uniform or triangular distribution. The features of a binomially distributed random variable were explored in Chapter 5.

Features of a binomial distribution

For $X \sim \mathrm{Bin}(n, p)$, then

$$\mathrm{P}(X = x) = \binom{n}{x} p^x (1 - p)^{n-x}$$

$$\mathrm{E}(X) = np$$

$$\mathrm{Var}(X) = np(1 - p)$$

$$\mathrm{SD}(X) = \sqrt{np(1 - p)}$$

In these problem types, be sure to define a new random variable when you notice the context change to a binomial situation.

WORKED EXAMPLE 18 **Modelling using the uniform and binomial distributions**

The local supermarket packages potatoes in bags with a labelled weight of 4 kg. The weight of the bags of potatoes is found to be uniformly distributed, with weights ranging from 3980 g to 4040 g. Let the weight of a bag of potatoes be represented by the continuous random variable W.

a Determine the mean and standard deviation of W.

b Determine the probability that a randomly selected bag of potatoes weighs less than the labelled weight of 4 kg.

Suppose 50 different bags of potatoes are to be sampled.

c How many of the 50 bags are expected to weigh less than 4 kg? Answer to the nearest whole bag.

d Determine the probability, correct to four decimal places, that at least 20 of the 50 bags will have a weight less than 4 kg.

Steps	Working
a Use the formulas $\text{E}(X) = \dfrac{a+b}{2}$ and $\text{SD}(X) = \dfrac{b-a}{\sqrt{12}}$.	$\text{E}(W) = \dfrac{3980 + 4040}{2}$ $= \dfrac{8020}{2}$ $= 4010\,\text{g}$ $\text{SD}(W) = \dfrac{4040 - 3980}{\sqrt{12}} = \dfrac{60}{\sqrt{12}}$ $= 10\sqrt{3}$ or $17.32\,\text{g}$
b 1 Determine the value of $f(x)$ using $\dfrac{1}{b-a}$.	$f(w) = \dfrac{1}{4040 - 3980} = \dfrac{1}{60}$
2 Consider the area of the rectangle as the probability.	$\text{P}(W < 4000) = 20\left(\dfrac{1}{60}\right) = \dfrac{1}{3}$
c 1 Define a new binomial random variable.	Let X be the number of bags out of 50 with a weight less than 4kg. Then $X \sim \text{Bin}\left(50, \dfrac{1}{3}\right)$.
2 Determine $\text{E}(X) = np$.	$\text{E}(X) = 50\left(\dfrac{1}{3}\right) = \dfrac{50}{3} = 16\dfrac{2}{3}$
3 Interpret your answer as a whole number of bags.	≈ 17 bags
d 1 Write the appropriate probability statement.	$\text{P}(X \geq 20) = \text{P}(20 \leq X \leq 50)$
2 Use CAS to evaluate the probability using **binomialCDf** for ClassPad and **binom Cdf** for TI-Nspire.	

ClassPad

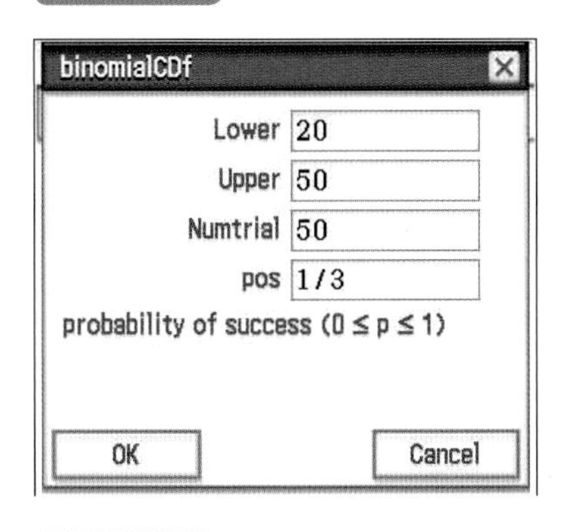

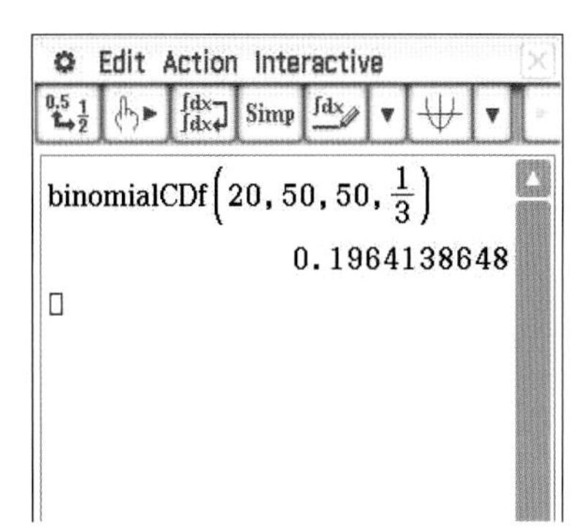

TI-Nspire

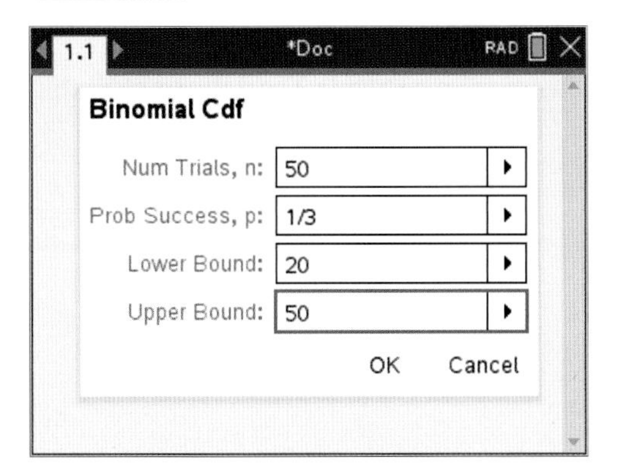

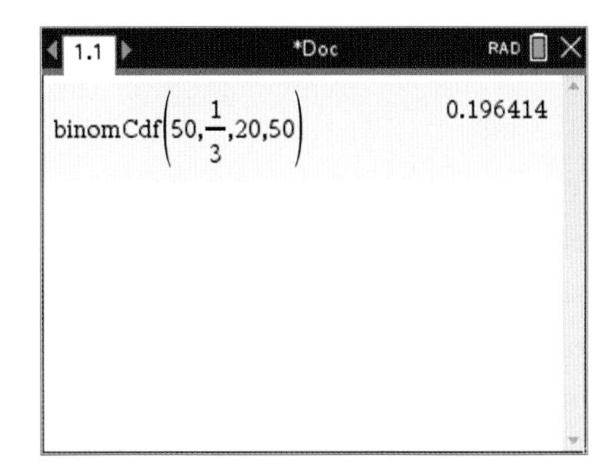

3 Round to four decimal places.	$\text{P}(20 \leq X \leq 50) = 0.1964$

EXERCISE 8.3 Uniform and triangular distributions

ANSWERS p. 405

Recap

1 A continuous random variable has a probability density function given by

$$f(t) = \begin{cases} 0.05e^{-0.05t} & \text{for } t > 0 \\ 0 & \text{for } t \leq 0 \end{cases}$$

The mean value of T is

A 0.05 **B** 2 **C** 5 **D** 10 **E** 20

2 A continuous random variable X has a probability density function given by

$$f(x) = \begin{cases} \frac{\pi}{2}\sin(\pi x) & 0 \leq x \leq 1 \\ 0 & \text{otherwise} \end{cases}$$

The value of the 50th percentile is

A $\frac{1}{2}$ **B** 1 **C** 1.5 **D** $\frac{\pi}{2}$ **E** π

Mastery

3 WORKED EXAMPLE 13 A uniformly distributed continuous random variable X has the probability density function

$$f(x) = \begin{cases} \frac{1}{6} & 9 \leq x \leq 15 \\ 0 & \text{otherwise} \end{cases}$$

Determine

a $\text{P}(X = 10)$ **b** $\text{P}(X > 10)$ **c** $\text{P}(X \leq 12 \mid X > 10)$.

4 WORKED EXAMPLE 14 A uniformly distributed continuous random variable T has a cumulative distribution function given by

$$F(t) = \begin{cases} 0 & t < 7 \\ \frac{t-17}{13} & 7 \leq t \leq b \\ 1 & t > b \end{cases}$$

a Determine the value of b.

b State the probability density function of T.

c Hence, or otherwise, calculate $\text{P}(T > 15 \mid T > 10)$.

5 WORKED EXAMPLE 15 A uniformly distributed continuous random variable X is defined such that $X \sim U[-2, 48]$.

Find the

a probability density function of X

b expected value of X

c variance of X

d standard deviation of X.

8.3

6 WORKED EXAMPLE 16 Consider the following probability density function for the continuous random variable X.

$$f(x) = \begin{cases} \frac{1}{4}x - \frac{1}{2} & 2 \leq x \leq 4 \\ \frac{3}{2} - \frac{1}{4}x & 4 < x \leq 6 \\ 0 & \text{otherwise} \end{cases}$$

a Sketch the graph of $y = f(x)$ labelling the coordinates of all critical features. Hence, describe the shape of the distribution.

b Determine

i $P(X < 3)$ **ii** $P(X < 3 \mid X < 4)$ **iii** $E(X)$ **iv** $Var(X)$.

You may use your calculator for any calculations.

7 WORKED EXAMPLE 17 Consider the following probability density function for the continuous random variable X.

$$f(x) = \begin{cases} 2x & 0 \leq x < \frac{1}{2} \\ -\frac{2}{3}x + \frac{4}{3} & \frac{1}{2} \leq x \leq 2 \\ 0 & \text{otherwise} \end{cases}$$

a Show that it is a valid probability density function for X.

b Determine the 70th percentile.

c Show the use of integration to determine $E(X)$.

8 WORKED EXAMPLE 18 The lengths of plastic pipes that are cut by a particular machine are a uniformly distributed random variable, L, with lengths ranging from 245 mm to 251 mm. The machine is calibrated to cut pipes at a length of 250 mm.

a Determine the mean and standard deviation of L.

b Determine the probability that a randomly selected pipe has a length greater than the calibrated length of 250 mm.

Suppose 60 different cut pipes are to be sampled.

c How many of the 60 pipes are expected to have a length greater than 250 mm?

d Determine the probability, correct to four decimal places, that at least 10 of the 60 pipes will have a length of more than 250 mm.

Calculator-free

9 © SCSA MM2019 Q3 MODIFIED (8 marks) Waiting times, T hours, for patients at a hospital emergency department can be up to four hours. The associated probability density function is shown below.

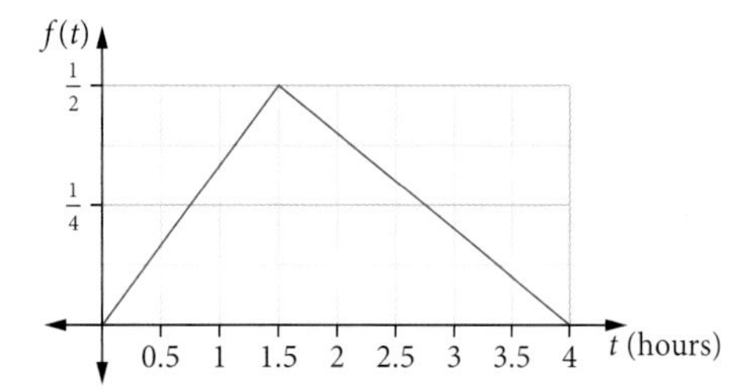

a Determine the equation of the probability density function, $f(t)$. (3 marks)

b What is the probability a patient will wait less than one hour? (2 marks)

c What is the probability a patient will wait between one hour and three hours? (3 marks)

10 © SCSA MM2019 Q6 MODIFIED (11 marks) The error X in digitising a communication signal has a distribution with probability density function given by

$$f(x) = \begin{cases} 1 & -0.5 < x < 0.5 \\ 0 & \text{otherwise} \end{cases}$$

a Sketch the graph of $f(x)$ and, hence, describe the distribution. (3 marks)

b What is the probability that the error is at least 0.35? (1 mark)

c If the error is negative, what is the probability that it is less than −0.35? (2 marks)

d An engineer is more interested in the square of the error. What is the probability that the square of the error is less than 0.09? (2 marks)

e Calculate the variance of the error. (3 marks)

Calculator-assumed

11 (10 marks) The time Jennifer spends on her homework each day varies, but it is known that she does some homework every day. The continuous random variable T, which models the time, t, in minutes, that Jennifer spends each day on her homework, has a probability density function f, where

$$f(t) = \begin{cases} \frac{1}{625}(t - 20) & 20 \le t < 45 \\ \frac{1}{625}(70 - t) & 45 \le t \le 70 \\ 0 & \text{otherwise} \end{cases}$$

a Sketch the graph of $y = f(t)$, labelling the coordinates of all critical features. (3 marks)

b Hence, describe the shape of the distribution of T. (1 mark)

c Find

i $P(25 \le T \le 55)$ (2 marks)

ii $P(T \le 25 \mid T \le 55)$ (2 marks)

iii k such that $P(T \ge k) = 0.7$, correct to four decimal places. (2 marks)

12 (8 marks) In the MaxFun amusement park there is a small train called ChooChooCharlie which does a circuit of the park. The continuous random variable T, the time in minutes for a circuit to be completed, has a probability density function f with rule

$$f(t) = \begin{cases} \frac{1}{100}(t - 10) & \text{if } 10 \le t < 20 \\ \frac{1}{100}(30 - t) & \text{if } 20 \le t \le 30 \\ 0 & \text{otherwise} \end{cases}$$

a Sketch the graph of $y = f(t)$, labelling the coordinates of all critical features. (3 marks)

b Hence, describe the shape of the distribution of T. (1 mark)

c Find the exact probability that the time taken by ChooChooCharlie to complete a full circuit is

i less than 25 minutes (2 marks)

ii less than 15 minutes, given that it is less than 25 minutes. (2 marks)

13 © SCSA MM2016 Q16 (10 marks) An automated milk bottling machine fills bottles uniformly to between 247 mL and 255 mL. The label on the bottle states that it holds 250 mL.

a Determine the probability that a bottle selected randomly from the conveyor belt of this machine contains less than the labelled amount. (3 marks)

b Calculate the mean and standard deviation of the amount of milk in the bottles. (4 marks)

A worker selects bottles from the conveyor belt, one at a time.

c Determine the probability that in a selection of 15 bottles, five bottles containing less than the labelled amount have been selected. (3 marks)

14 (9 marks) Black Mountain coffee is sold in packets labelled 250 grams. Let the weight of a packet of Black Mountain coffee be represented by the continuous random variable W. The packing process produces packets whose weights form a uniform distribution over the interval $245 \leq w \leq 253$ grams.

a Determine the mean and standard deviation of W. (2 marks)

b Determine the probability that a randomly selected packet of Black Mountain coffee weighs more than the labelled weight of 250 g. (2 marks)

Suppose 100 different packets of Black Mountain coffee are to be sampled.

c How many of the 100 packets are expected to weigh more than 250 g? Answer to the nearest whole packet. (3 marks)

d Determine the probability, correct to four decimal places, that at least a quarter of the 100 packets weigh more than 250 g. (2 marks)

15 (8 marks) The distribution of a continuous random variable, X, is defined by the probability density function $p(x)$ over $-a \leq x \leq b$ where a and b are positive real constants. The graph of the function $y = p(x)$ is shown below.

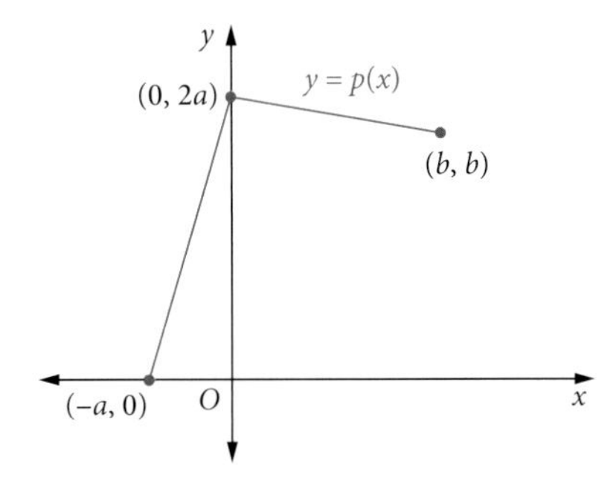

It is known that the average value of y (i.e. the average *height* of the function) over the interval $-a \leq x \leq b$ is $\frac{3}{4}$.

a Write two equations in terms of a and b using the probability density function of X. (4 marks)

b Determine the exact values of a and b. (2 marks)

c Hence, determine $P(X > 0)$. (2 marks)

Video playlist
The normal distribution

Worksheet
The normal curve

8.4 The normal distribution

Contexts suitable for the normal distribution

The **normal distribution** is one of the most frequently used probability distributions when modelling continuous random variables in both the natural and physical worlds, largely due to its critical features. A **normally distributed random variable** has

- a symmetrical distribution about its mean value, μ, meaning the mean, median and mode of a normal random variable are all equal
- a bell-shaped distribution, whereby the oblique points of inflection of the curve occur at values that are one standard deviation, σ, on either side of the mean.

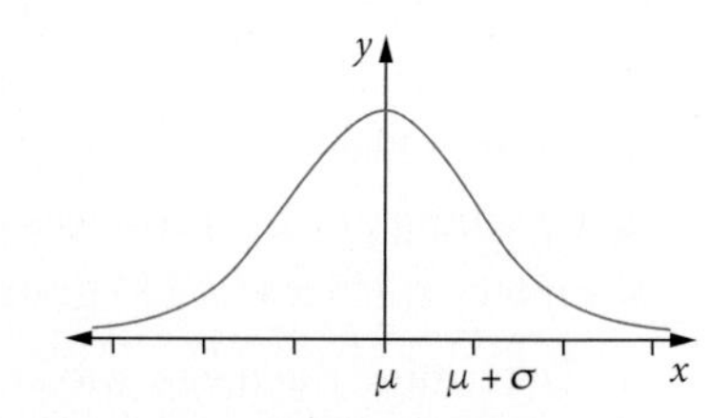

Probability density function of a normally distributed random variable

For a normal random variable, X, with a mean μ and standard deviation σ, the probability density function is given by the equation

$$f(x) = \frac{1}{\sigma\sqrt{2\pi}}e^{-\frac{1}{2}\left(\frac{x-\mu}{\sigma}\right)^2}$$

The function $f(x)$ satisfies the two conditions of a continuous random variable:

- $f(x) > 0$ for all $-\infty < x < \infty$
- $\int_{-\infty}^{\infty} f(x)\,dx = 1$.

A normal random variable can be denoted by $X \sim N(\mu, \sigma^2)$, where the mean and variance are the parameters of the distribution.

The probability density function of a normal random variable can be used to show another critical feature of the normal distribution. This feature is known as the **68–95–99.7% rule**. These proportions indicate the probability that a value of x lies within one, two and three standard deviations from the mean, respectively.

The 68–95–99.7% rule

For $X \sim N(\mu, \sigma^2)$, then

- $P(\mu - \sigma \le X \le \mu + \sigma) \approx 0.68$, i.e.

  $$\int_{\mu-\sigma}^{\mu+\sigma} f(x)\,dx \approx 0.68$$

- $P(\mu - 2\sigma \le X \le \mu + 2\sigma) \approx 0.95$, i.e.

  $$\int_{\mu-2\sigma}^{\mu+2\sigma} f(x)\,dx \approx 0.95$$

- $P(\mu - 3\sigma \le X \le \mu + 3\sigma) \approx 0.997$, i.e.

  $$\int_{\mu-3\sigma}^{\mu+3\sigma} f(x)\,dx \approx 0.997$$

This rule is useful to carry out approximate calculations without a calculator.

Exam hack

Sketch a bell-curve with a maximum of three standard deviations labelled either side of the mean to assist with probability calculations.

WORKED EXAMPLE 19 Using the 68–95–99.7% rule

A continuous random variable X is normally distributed with a mean of 120 and a standard deviation of 20. Use the 68–95–99.7% rule to approximate the following probabilities.

a $P(100 \leq X \leq 140)$

b $P(X > 140)$

c $P(X \leq 80)$

Steps	**Working**
a **1** 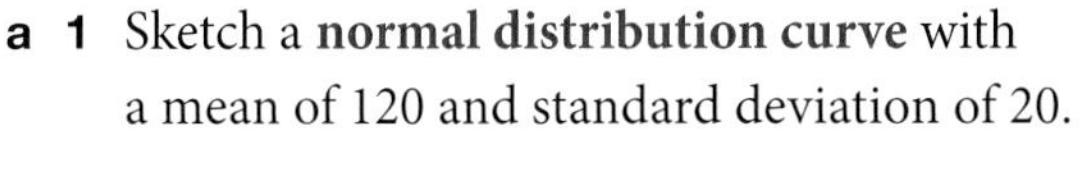 Sketch a **normal distribution curve** with a mean of 120 and standard deviation of 20.	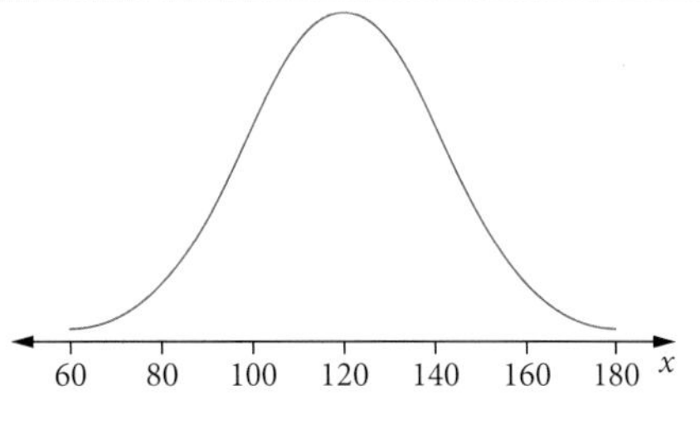
2 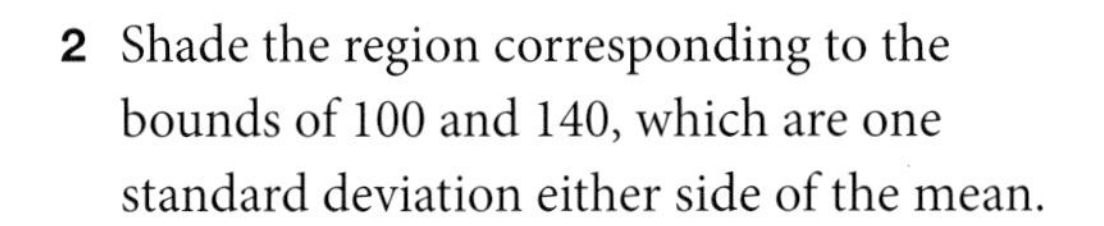 Shade the region corresponding to the bounds of 100 and 140, which are one standard deviation either side of the mean.	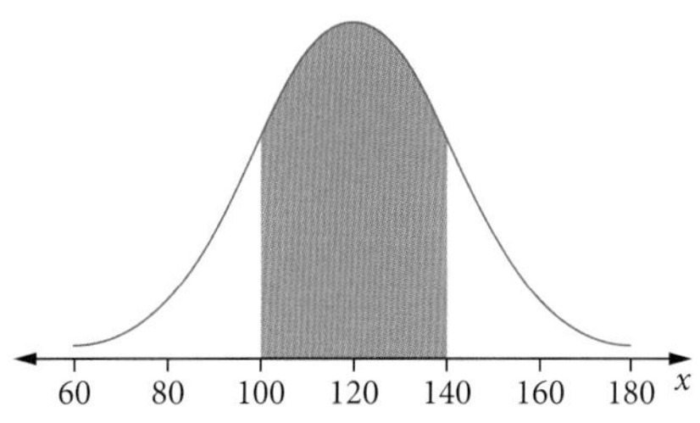
3 Approximate the probability using the 68–95–99.7% rule.	$P(100 \leq X \leq 140) \approx 0.68$
b **1** Shade the region corresponding to the lower bound of 140, which is one standard deviation above the mean.	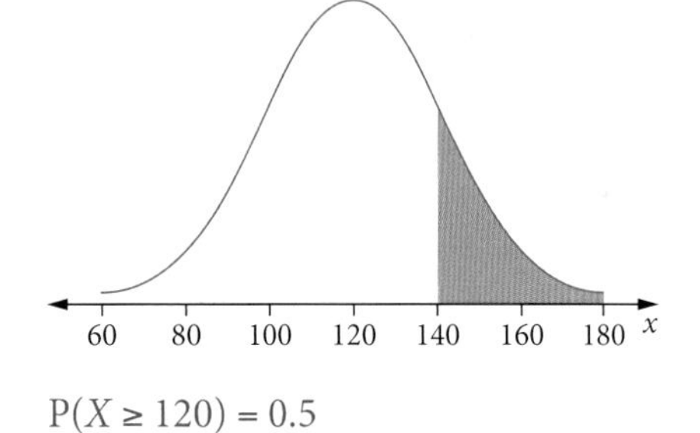
2 Approximate the probability using the 68–95–99.7% rule.	$P(X \geq 120) = 0.5$ $P(120 \leq X \leq 140) \approx \frac{0.68}{2} \approx 0.34$ $P(X > 140) \approx 0.5 - 0.34 \approx 0.16$
c **1** Shade the region corresponding to the upper bound of 80, which is two standard deviations below the mean.	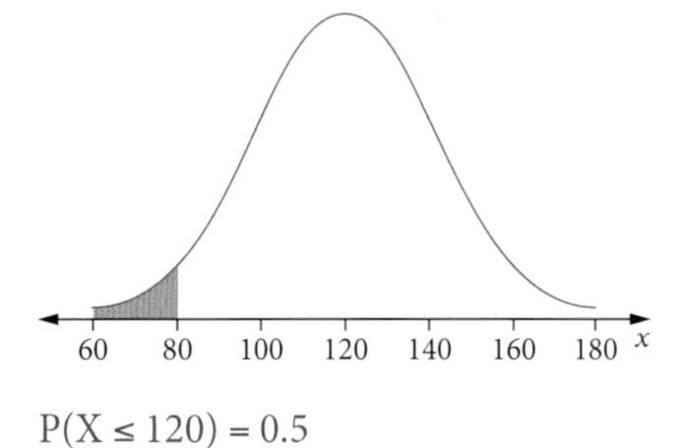
2 Approximate the probability using the 68–95–99.7% rule.	$P(X \leq 120) = 0.5$ $P(80 \leq X \leq 120) \approx \frac{0.95}{2} \approx 0.475$ $P(X < 80) \approx 0.5 - 0.475 \approx 0.025$

Together, the properties of

1 being a continuous variable, with

2 a symmetrical shape about the mean, such that

3 the proportions of scores either side of the mean follow the 68–95–99.7% rule

form the three conditions that can be used to determine whether a practical context can be suitably modelled by a normal distribution.

WORKED EXAMPLE 20 Justifying contexts suitable for the normal distribution

For each of the following situations, give one reason why it would **not** be appropriate to model the distribution of the continuous random variable with a normal distribution.

a The time taken (in minutes) for 100 students to complete an online road safety quiz has a mean of 25 minutes and standard deviation of 8 minutes. Seventy-five students completed the quiz within 17 to 33 minutes.

b A gardener has a nursery containing 2000 basil plants with a mean height of 14 cm and standard deviation of 4 cm. Approximately 1500 of the basil plants have a height, H, smaller than 14 cm.

c Results of a survey show that the number of electronic devices, D, owned by Australian teenagers is symmetrically distributed, with a mean number of devices of 2.

Steps	Working
Use the three properties of 1 continuous random variable 2 symmetrical shape 3 68–95–99.7% rule to find a reason why the context cannot be suitably modelled by a normal distribution.	**a** 1 Is time a continuous random variable? Yes 2 Is the distribution symmetrical? Insufficient information 3 Does the distribution satisfy the 68–95–99.7% rule? No $\mu - \sigma = 17,\ \mu + \sigma = 33$ $P(17 \le T \le 33) \approx 0.75 \ne 0.68$ Therefore, not appropriate as too many times are within one standard deviation from the mean.
	b 1 Is height a continuous random variable? Yes 2 Is the distribution symmetrical? No $\mu = 14$ $P(H < 14) = \frac{1500}{2000} = 0.75 \ne 0.5$ Therefore, not appropriate as the distribution is not symmetrical, i.e. it is skewed to the right (positively skewed), as there are more plants with a height less than the mean.
	c 1 Is number of devices a continuous random variable? No Therefore, not appropriate as the random variable is discrete.

The standard normal distribution

Often a useful way to think about a normally distributed random variable, X, is to consider how many standard deviations a particular score x lies above or below the mean of the distribution $\mathrm{E}(X) = \mu$. When normal distributions are thought about in this way, it does not actually matter what the values of the mean μ and standard deviation σ are, as we can set the mean as a default value of 0 and the standard deviation as 1 to obtain what is known as the **standard normal distribution**.

The standard normal distribution and z-scores

Let a normal random variable have a mean of 0 and a standard deviation of 1. This normal random variable is the **standard normal random variable** denoted by the distribution $Z \sim N(0, 1)$ and is used to describe the number of standard deviations a score is from the mean. This value is called a ***z*-score**.

The probability density function for a standard normal random variable is given by

$$f(z) = \frac{1}{\sqrt{2\pi}} e^{-\frac{1}{2}z^2}$$

Worksheets
The standard normal curve

Areas under the normal curve

z-scores

From the 68–95–99.7% rule, we can then deduce that for $Z \sim N(0, 1)$:

- $\mathrm{P}(-1 \le Z \le 1) \approx 0.68$. That is, $\int_{-1}^{1} f(z)\,dz \approx 0.68$.
- $\mathrm{P}(-2 \le Z \le 2) \approx 0.95$. That is, $\int_{-2}^{2} f(z)\,dz \approx 0.95$.
- $\mathrm{P}(-3 \le Z \le 3) \approx 0.997$. That is, $\int_{-3}^{3} f(z)\,dz \approx 0.997$.

The symmetrical properties of the standard normal distribution

Let $\mathrm{P}(Z \le c) = A$, then $\mathrm{P}(Z > c) = 1 - A$, as the total area under the curve is 1.

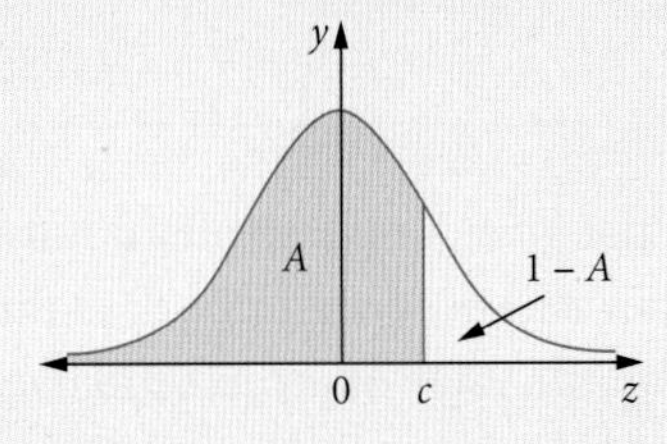

By symmetry, $\mathrm{P}(Z \ge -c) = A$ and so, $\mathrm{P}(Z < -c) = 1 - A$.

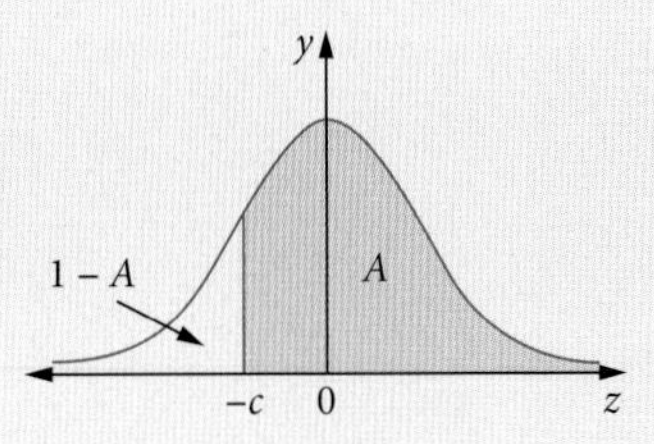

WORKED EXAMPLE 21 Solving problems involving $Z \sim N(0, 1)$

Let the continuous random variable Z be the standard normal random variable.

a Determine $P(-3 < Z < 2)$.

b If $P(Z \leq 1.5) = 0.933$ to three decimal places, determine $P(-1.5 \leq Z \leq 1.5)$ to three decimal places.

Steps	Working
a 1 Sketch the **standard normal distribution curve** and shade the region under the curve for $-3 < z < 2$.	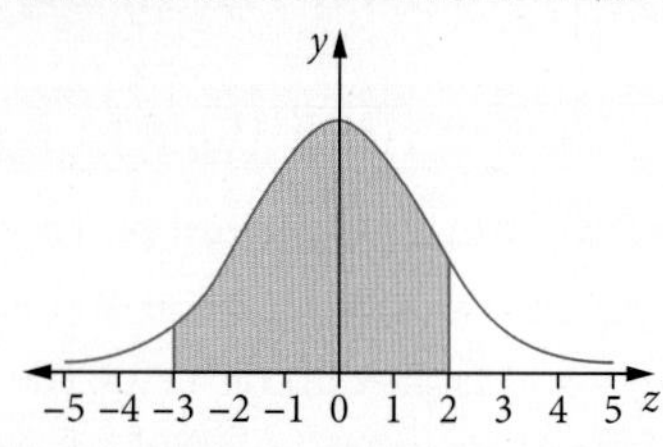
2 Approximate the probability using the 68–95–99.7% rule.	$P(-3 < Z < 3) \approx 0.997$ $P(-3 < Z < 0) \approx \frac{0.997}{2} \approx 0.4985$ $P(-2 < Z < 2) \approx 0.95$ $P(0 < Z < 2) \approx \frac{0.95}{2} \approx 0.475$ $P(-3 < Z < 2) \approx 0.4985 + 0.475 \approx 0.9735$
b 1 Draw normal distribution curves that illustrate $P(Z \leq 1.5) = 0.933$ and $P(-1.5 \leq Z \leq 1.5)$.	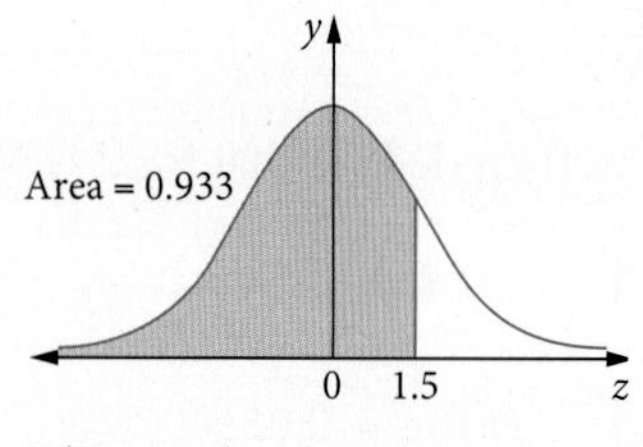 $P(Z \leq 1.5) = 0.933$ 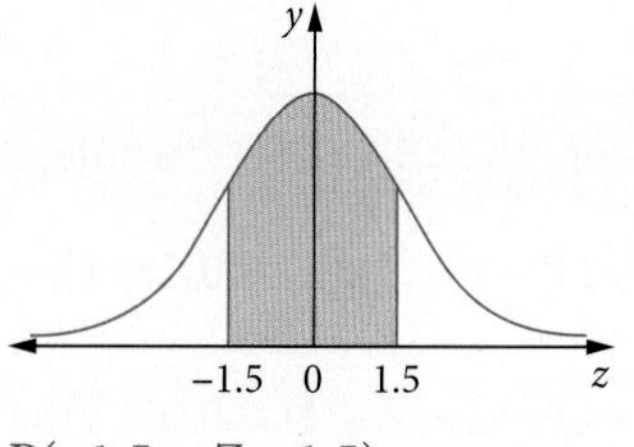 $P(-1.5 \leq Z \leq 1.5)$
2 To find $P(-1.5 \leq Z \leq 1.5)$ (the region shaded under the second curve), first find $P(Z \geq 1.5)$ (the region unshaded under the first curve).	$P(Z \geq 1.5) = 1 - 0.933$ $= 0.067$
3 By symmetry, subtract double the value found above from 1.	$P(-1.5 \leq Z \leq 1.5) = 1 - 2 \times 0.067$ $= 0.866$

Whenever a normal random variable is not given in the standard normal form, that is, it is given as $X \sim N(\mu, \sigma^2)$ where $\mu \neq 0$ and $\sigma \neq 1$, it can always be scaled back to the standard normal distribution. That is, the value of μ can be made to be 0 and the value of σ can be made to be 1 such that all values of x are scaled back to the corresponding z-score (which is called its **standard score**). This process is called the **standardisation** of a normal random variable.

To do so, let's use the example $X \sim N(120, 20^2)$.

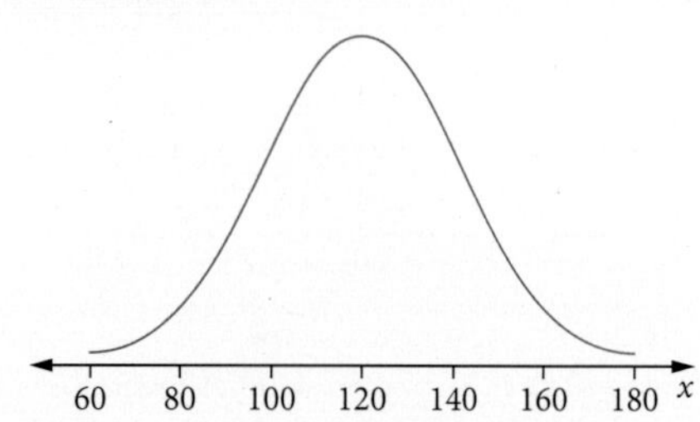

Imagine starting with the value of x that is the mean, $x = 120$. To make the mean 0, we would need to subtract the mean $\mu = 120$ from x, and to be consistent to the shape of the distribution, we would need to do it to every other value of x. That is, we apply a linear change of origin $X \to X - 120$.

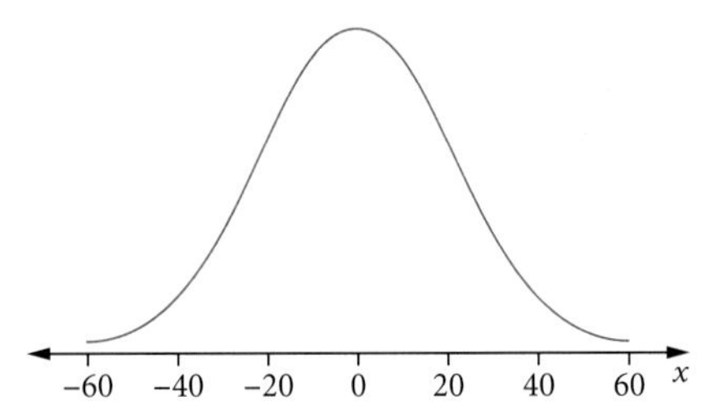

Then we need the distance between the scores to reduce by a scale factor of the standard deviation $\sigma = 20$. That is, we need to apply a linear change of scale $X - 120 \to \frac{1}{20}(X - 120)$.

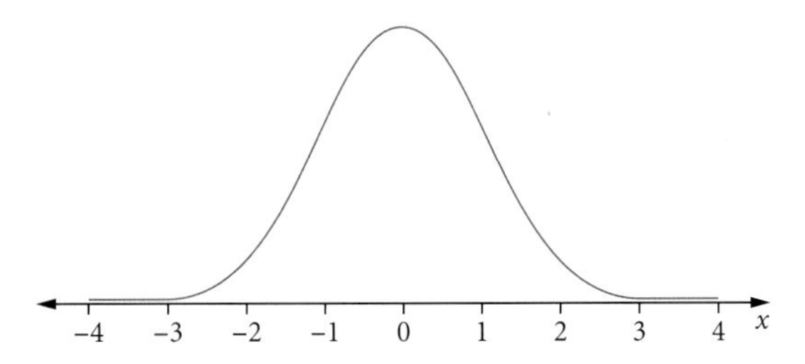

So, the standard normal random variable in this case would be defined as $Z = \frac{X - 120}{20}$.

Standardising a normal random variable

For a normal random variable $X \sim N(\mu, \sigma^2)$, the standard normal distribution $Z \sim N(0, 1)$ can be obtained using the linear transformation $Z = \frac{X - \mu}{\sigma}$.

To standardise a single value of x, that is, to obtain its z-score (standard score), use the formula

$$z = \frac{x - \mu}{\sigma}$$

where z is the standard score

x is the value of the score from the distribution $X \sim N(\mu, \sigma^2)$

μ is the mean of X

σ is the standard deviation of X.

It can then be shown that

$$\begin{aligned} \mathrm{E}(Z) &= \mathrm{E}\left(\frac{1}{\sigma}X - \frac{\mu}{\sigma}\right) \\ &= \frac{1}{\sigma}\mathrm{E}(X) - \frac{\mu}{\sigma} \\ &= \frac{\mu}{\sigma} - \frac{\mu}{\sigma} \\ &= 0 \end{aligned}$$

$$\begin{aligned} \mathrm{Var}(Z) &= \mathrm{Var}\left(\frac{1}{\sigma}X - \frac{\mu}{\sigma}\right) \\ &= \frac{1}{\sigma^2}\mathrm{Var}(X) \\ &= \frac{1}{\sigma^2}(\sigma^2) \\ &= 1 \end{aligned}$$

When standardising a normal random variable, it is important to note that the corresponding areas under the curve **do not** change.

WORKED EXAMPLE 22 Standardising a normal random variable

A continuous random variable X is normally distributed with a mean of 80 and a standard deviation of 15. Let $Z \sim N(0, 1)$.

a Determine the standard score of the following values of x.

i $x = 65$

ii $x = 117.5$

iii $x = 77$

b If $P(Z \leq 1) = 0.841$, find $P(X < 65)$.

Steps	Working
a Use the z-score formula $z = \dfrac{x - \mu}{\sigma}$ for each of the values of x.	**i** $z = \dfrac{65 - 80}{15} = -\dfrac{15}{15} = -1$ **ii** $z = \dfrac{117.5 - 80}{15} = \dfrac{37.5}{15} = 2.5$ **iii** $z = \dfrac{77 - 80}{15} = \dfrac{-3}{15} = -0.2$
b 1 Draw a normal distribution curve that illustrates $P(Z \leq 1) = 0.841$.	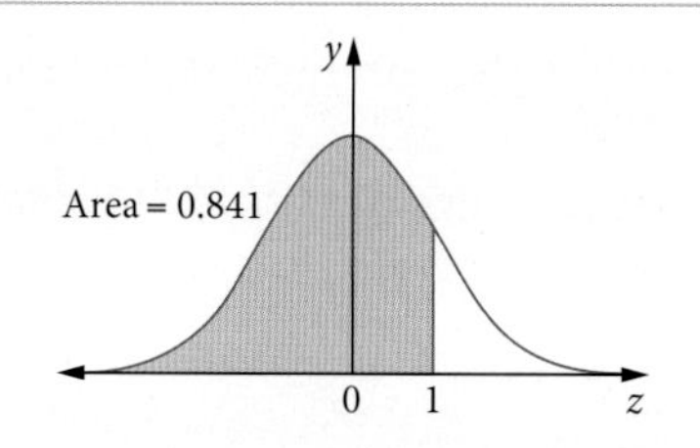 $P(Z \leq 1) = 0.841$
2 Use the standard score for $x = 65$ to draw a corresponding diagram of $P(X < 65)$ in terms of Z.	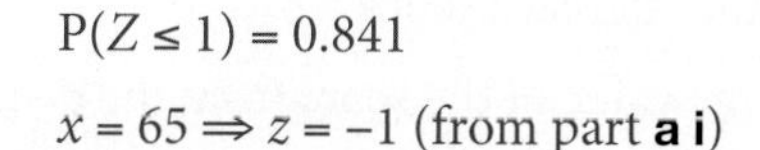 $x = 65 \Rightarrow z = -1$ (from part **a i**) 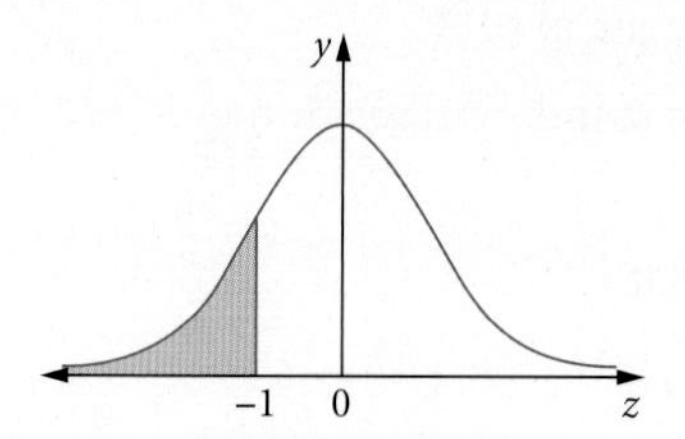
3 Calculate the probability using symmetry.	$P(Z > -1) = P(Z \leq 1) = 0.841$ $P(X < 65) = P(Z < -1) = 1 - P(Z > -1) = 1 - 0.841 = 0.159$

Calculating probabilities and quantiles with the normal distribution

Although the 68–95–99.7% rule and z-scores are useful in helping us to calculate probabilities that are a 'nice' number of standard deviations from the mean, we will typically need the assistance of CAS to compute any other probabilities.

The ClassPad and the TI-Nspire use the **normCDf** and **Normal Cdf** respectively function to carry out probabilities of the form $P(a \le X \le b)$. When dealing with problems such as $P(X \le b)$ or $P(X \ge a)$, you will need to think about these in terms of $\pm\infty$ for your calculator input. That is:

- $P(X \le b) = P(-\infty < X \le b)$
- $P(X \ge a) = P(a \le X < \infty)$.

Worksheets
The standard normal curve

The normal distribution

USING CAS 4 Probabilities for a normally distributed random variable

A continuous random variable X is normally distributed with mean 20 and standard deviation 5. Find the following probabilities, correct to four decimal places.

a $P(X \le 28)$

b $P(X > 19)$

c $P(15 < X < 22)$

ClassPad

1 Tap **Interactive** > **Distribution/Inv. Dist** > **Continuous** > **normCDf**.

2 In the dialogue box, enter the corresponding lower, upper, σ and μ values as shown.

3 Tap **OK** and the probability will be displayed.

(Note: always use $-\infty$ or ∞ from the **Math2 keyboard** for the 'ends' of the normal distribution.)

a

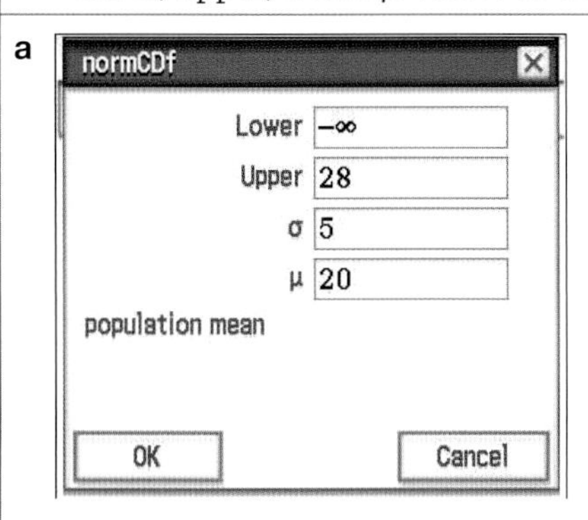

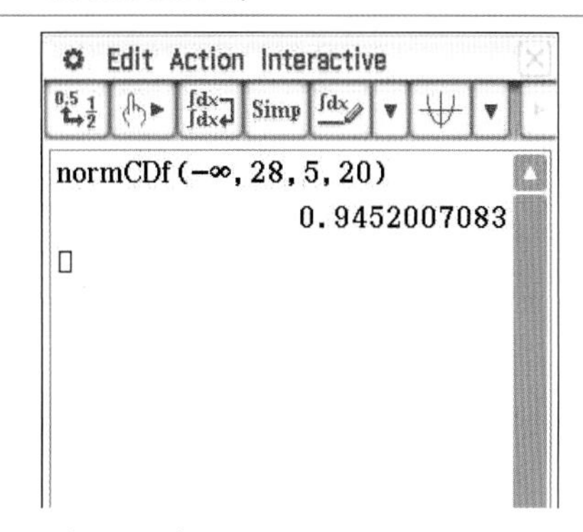

$P(X \le 28) = 0.9452$

b

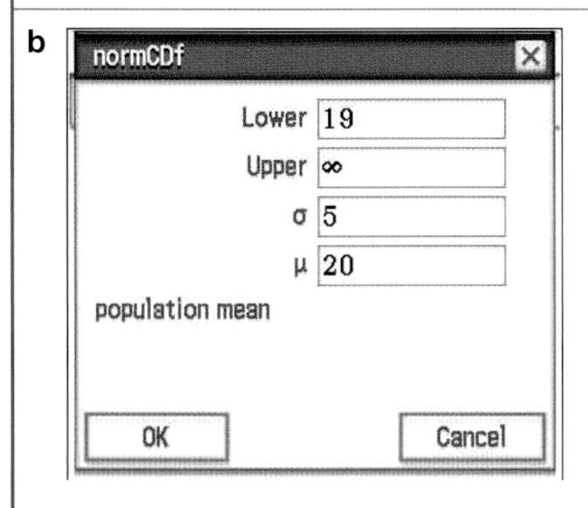

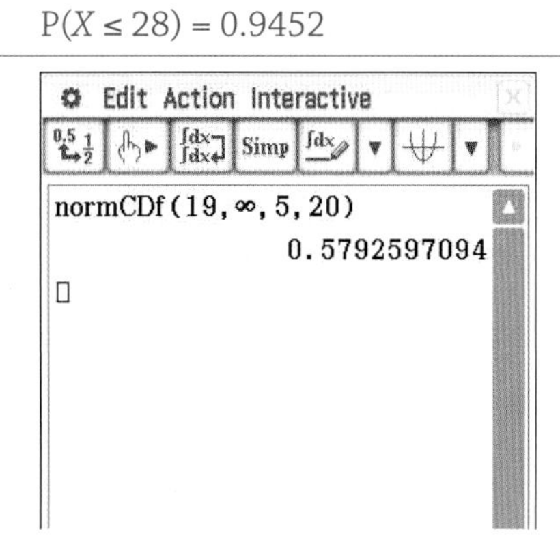

$P(X > 19) = 0.5793$

c

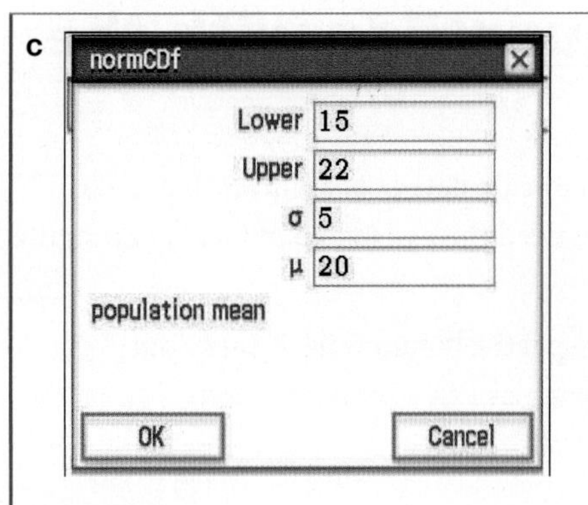

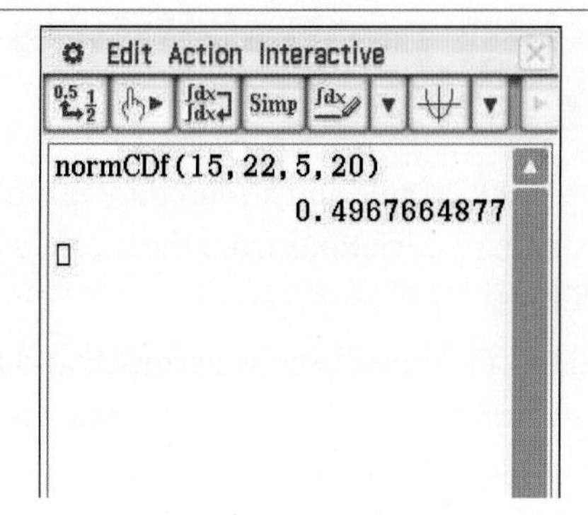

$P(15 < X < 22) = 0.4968$

TI-Nspire

1 Press **menu** > **Probability** > **Distributions** > **Normal Cdf**.

2 In the dialogue box, enter the corresponding lower bound, upper bound, μ and σ values as shown.

3 The probability will be displayed.

a

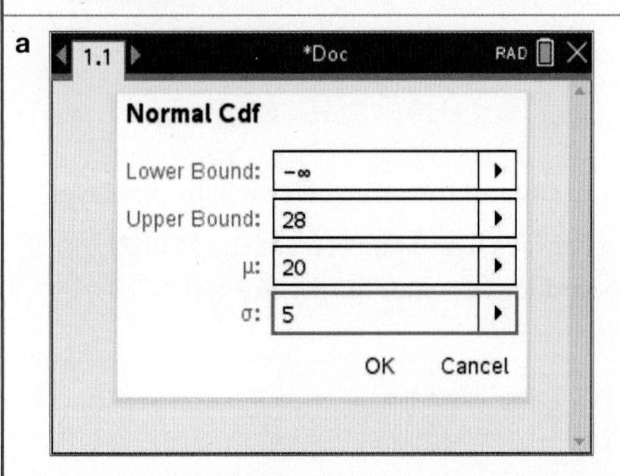

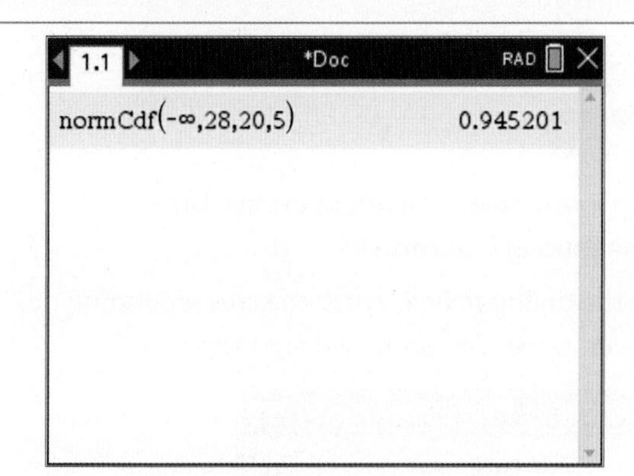

$P(X \leq 28) = 0.9452$

b

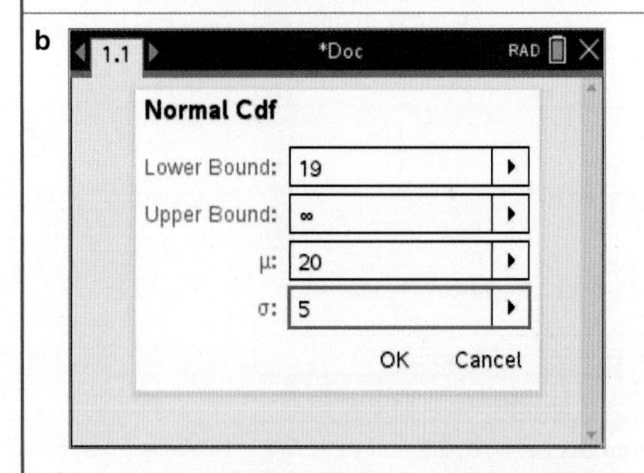

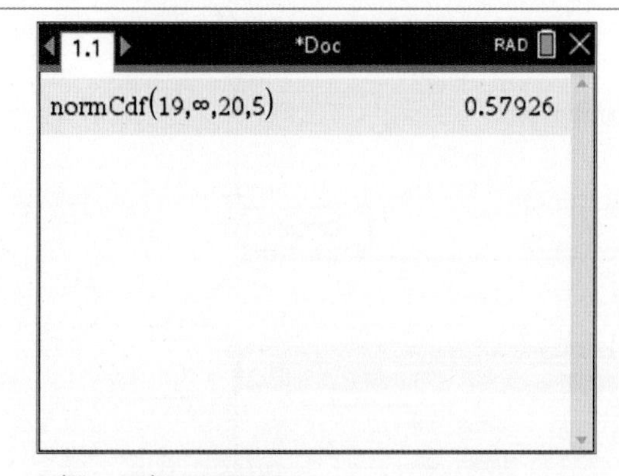

$P(X > 19) = 0.5793$

c

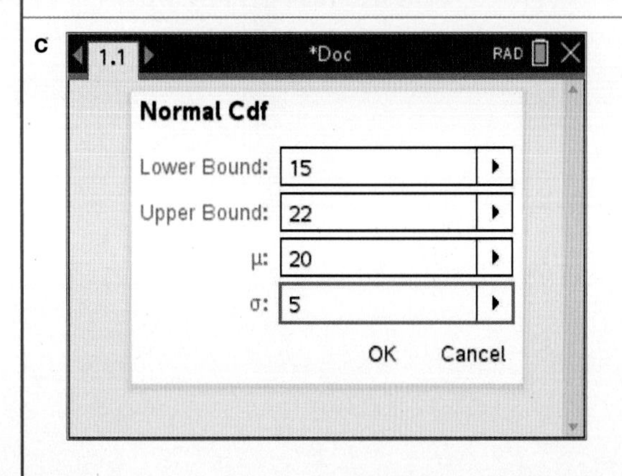

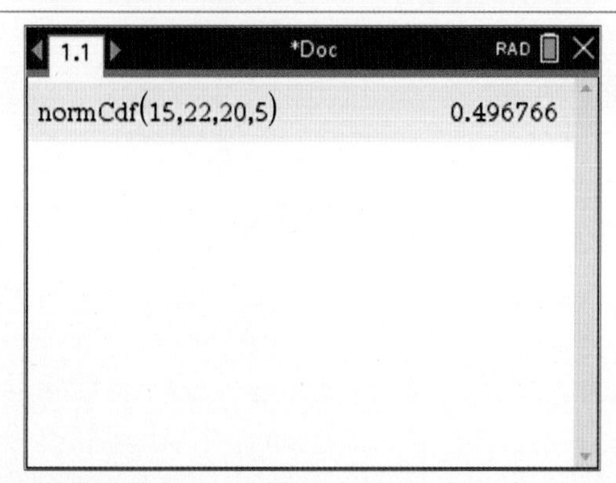

$P(15 < X < 22) = 0.4968$

As with general continuous random variables, we can solve problems of the form $P(X \le k) = p$ with normal random variables. Previously we saw the language of the percentile to describe such a problem. For example, the 75th percentile is the value of k such that $P(X \le k) = 0.75$. This value of k can also be referred to as the 0.75 **quantile**.

A quantile is simply the decimal form of a percentile, such that the p quantile is the score below which that $100p\%$ of the variable lies, where $0 < p < 1$.

Once again, we can use CAS and the **inverse normal distribution** function to solve these problems, but the ClassPad and TI-Nspire use slightly different conventions.

USING CAS 5 Finding quantiles using the inverse normal distribution

A continuous random variable X is normally distributed with a mean of 50 and a standard deviation of 10. Find the value of c, correct to two decimal places, if

a $P(X \le c) = 0.72$ **b** $P(X \ge c) = 0.8$.

Steps	Working
a 1 Draw the normal distribution curve, label c on the x-axis, and shade the given area. $P(X \le c) = 0.72$ indicates that 72% of the values are less than c. **2** Use CAS to solve for the 0.72 quantile.	$\mu = 50,\ \sigma = 10$ 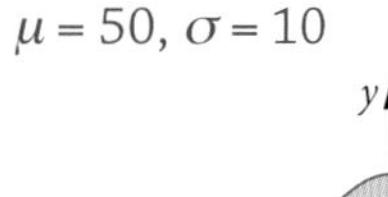 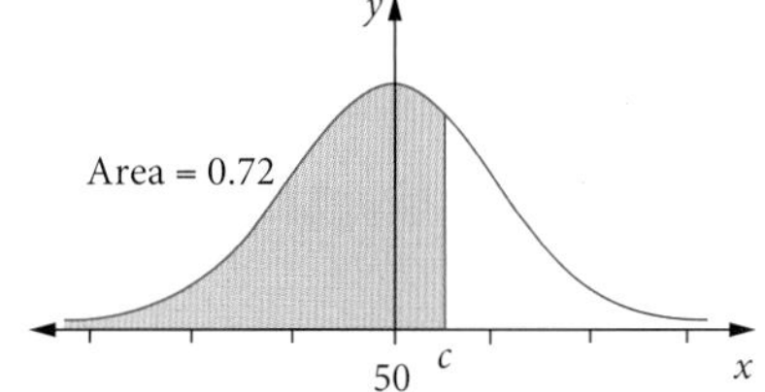
b 1 Draw the normal distribution curve, label c on the x-axis, and shade the given area. $P(X \ge c) = 0.8$ indicates that 80% of the values are greater than c. **2** Use CAS to solve for the 0.20 quantile.	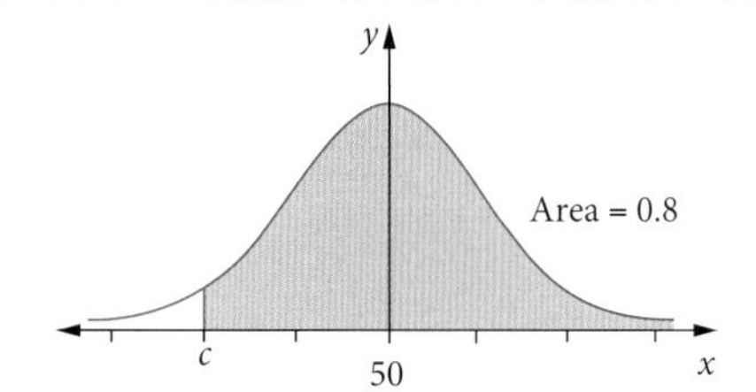

ClassPad

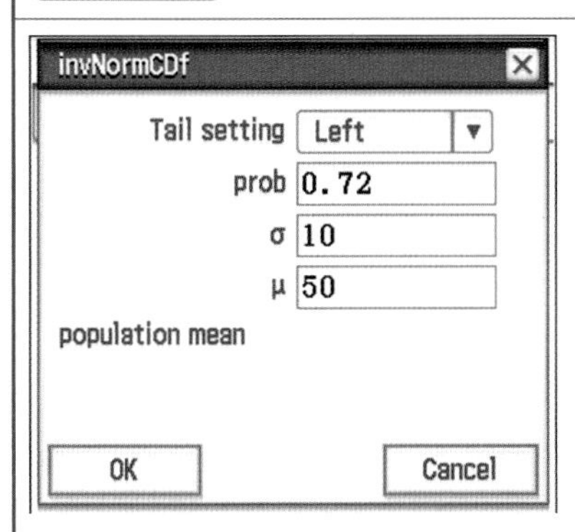

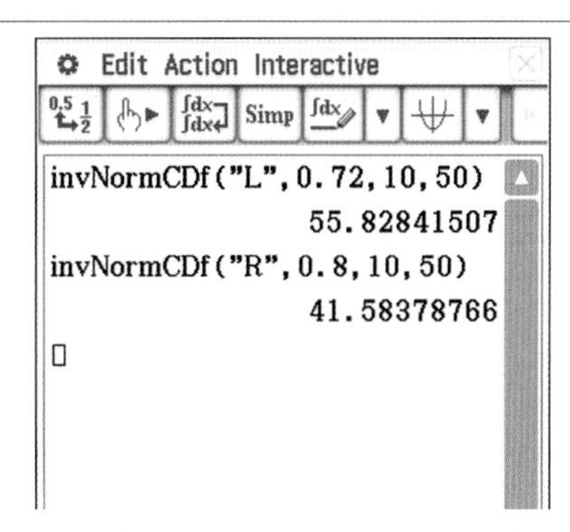

1 Tap **Interactive** > **Distribution/Inv.Dist** > **Inverse** > **invNormCDf**.
2 In the dialogue box, use the default **Tail setting** as **Left** for < or ≤.
3 Enter the values as shown above.
4 The answer for part **a** will be displayed.
5 Repeat for part **b** but change the **Tail** setting to **Right**.

a $c \approx 55.83$

b $c \approx 41.58$

TI-Nspire

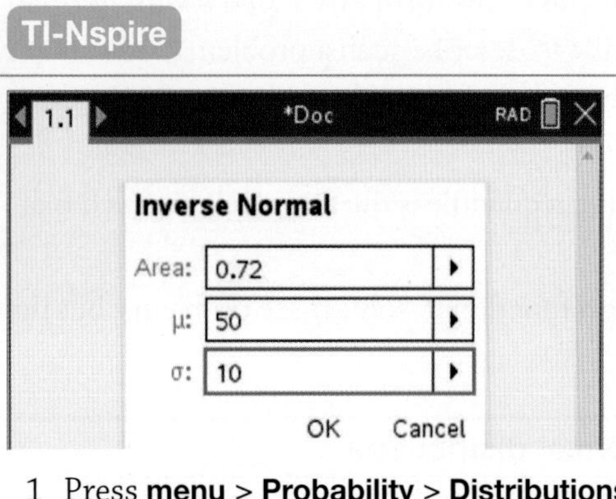

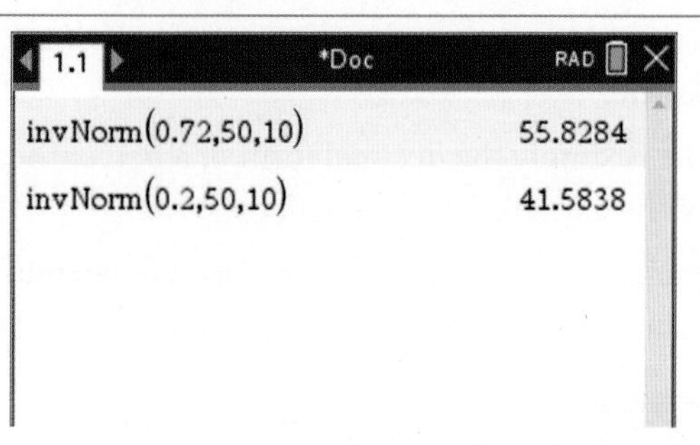

1. Press **menu** > **Probability** > **Distributions** > **Inverse Normal**.
2. In the dialogue box, enter the values as shown above.
3. The answer for part **a** will be displayed.
4. Repeat for part **b** using **area = 1 − 0.8 = 0.2**.

a $c \approx 55.83$

b $c \approx 41.58$

Worksheets
Applying the normal distribution
Normal distribution – Worded problems 1
Normal distribution – Worded problems 2

Problems involving the normal and binomial distributions

In some problems, we may not have all the information about the parameters of a normal random variable; that is, either μ or σ^2 or both may be unknown. In those cases, we must have sufficient information about probabilities or corresponding z-scores in order to determine the unknown parameters.

Typically for these problems, we should look to connect the values of μ and σ with x and its corresponding z-score using the formula $z = \dfrac{x - \mu}{\sigma}$.

WORKED EXAMPLE 23 Finding a parameter of a normal distribution

A continuous random variable X has the distribution $X \sim N(42, \sigma^2)$. If $P(X \geq 36) = 0.85$, find the standard deviation of X, correct to two decimal places.

Steps	Working
1 Draw the normal distribution curve for X and the corresponding normal distribution curve for Z.	$\mu = 42$ 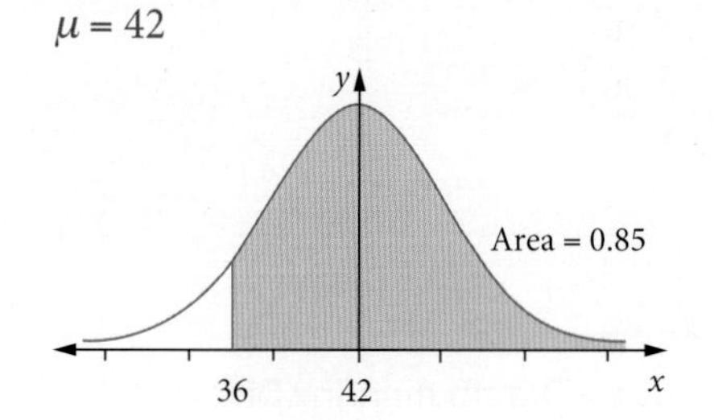 $P(X \geq 36) = 0.85$ 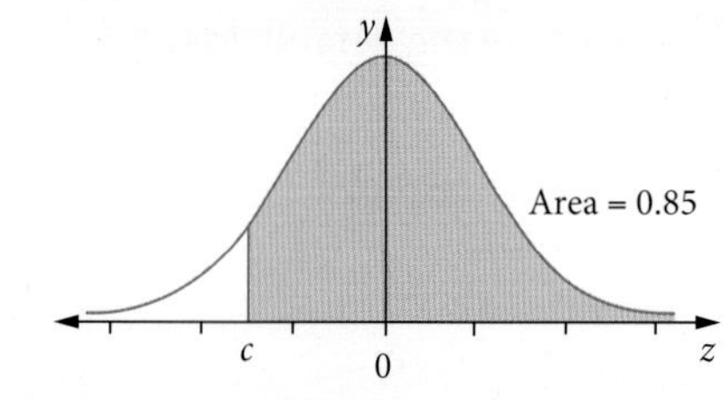 $P(Z \geq c) = 0.85$ $P(Z < c) = 1 - 0.85 = 0.15$
2 Use the CAS inverse normal distribution to find c (see Step 3 on the following page).	$c = -1.036$

3 Substitute into $z = \dfrac{x - \mu}{\sigma}$ and solve for σ algebraically or using CAS.	$\mu = 42, x = 36, z = -1.036$ $z = \dfrac{x - \mu}{\sigma}$ $-1.036 = \dfrac{36 - 42}{\sigma}$ $-1.036\sigma = -6$ $\sigma = 5.79$

ClassPad

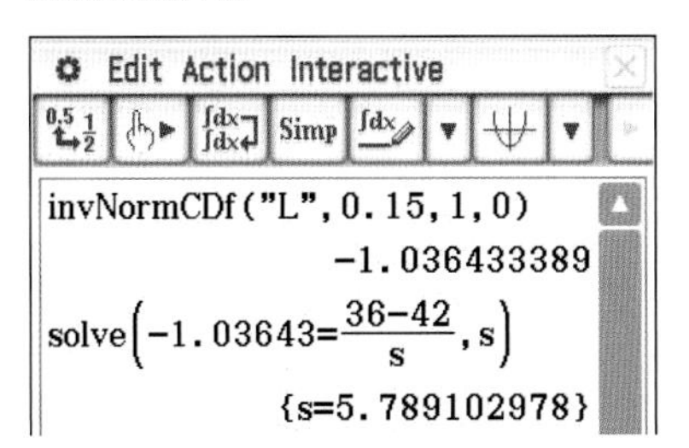

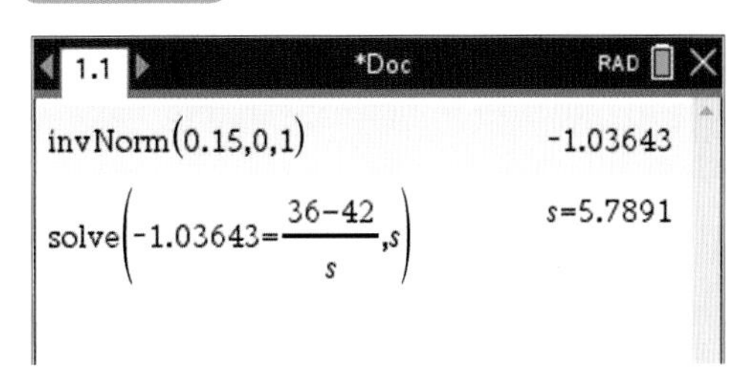

Be prepared to solve problems involving normal random variables in contextual situations, incorporating all the knowledge and skills you have learnt in this exercise. Like with the uniform and triangular distributions, be prepared to revisit the use of the binomial distribution in situations whereby you have n trials and a probability of success p being obtained from a normal distribution.

WORKED EXAMPLE 24 Applying the normal distribution in context

Year 10 students complete a fitness endurance task. The times taken to complete the task are normally distributed with a mean of 15 minutes and a standard deviation of 2 minutes.

a Find, correct to four decimal places, the proportion of students who complete the task in less than 14 minutes.

b Students who complete the task in a time between 13.5 minutes and 17 minutes are classified as having average fitness levels. Find, correct to four decimal places, the probability of a student selected at random being classified as having average fitness.

c Find, correct to four decimal places, the probability of a student with average fitness completing the task in less than 14 minutes.

Suppose a sample of Year 10 students was to be taken, and 30 of these students classified as having average fitness.

d Determine the probability, correct to four decimal places, that exactly 3 of the 30 students of average fitness completed the task in less than 14 minutes.

Steps	**Working**
a 1 Define an appropriate random variable for the context.	Let X represent the time taken to complete the fitness task in minutes, such that $X \sim N(15, 4)$.
2 Draw the normal distribution curve and show the three standard deviations above and below the mean on the horizontal axis scale.	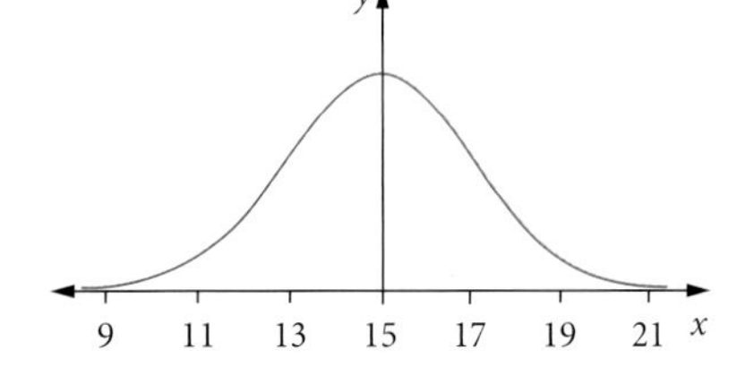
3 Calculate $P(X < 14)$ using CAS.	$P(X < 14) = 0.3085$

ClassPad

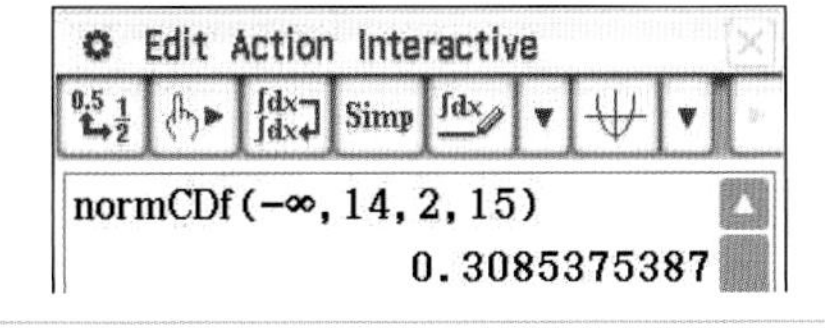

TI-Nspire

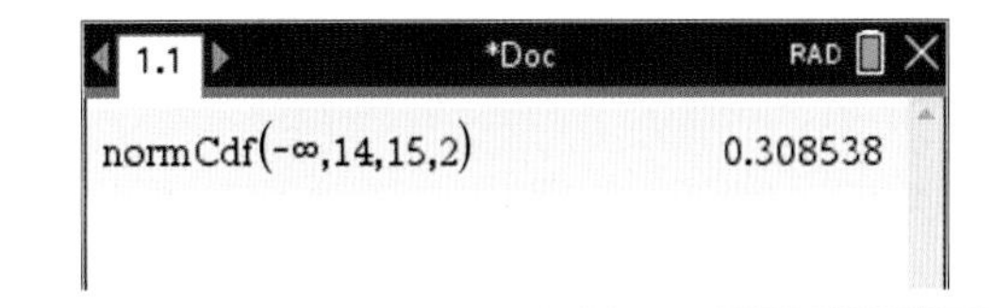

b Calculate $P(13.5 < X < 17)$ using CAS.

$P(13.5 < X < 17) = 0.6147$

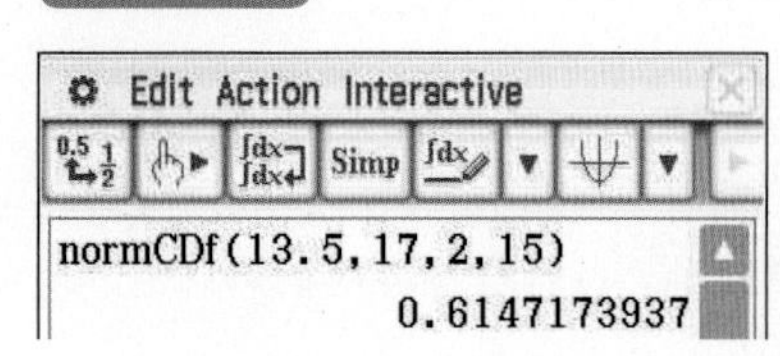

TI-Nspire

normCdf(13.5,17,15,2) 0.614717

c 1 Recognise the conditional language 'student with average fitness completing the task in less than 14 minutes' as 'a student completes a task in under 14 minutes given they have average fitness'.

2 Write and use the conditional probability formula $P(A \mid B) = \dfrac{P(A \cap B)}{P(B)}$.

$P(X < 14 \mid 13.5 < X < 17)$

$= \dfrac{P(13.5 < X < 14)}{P(13.5 < X < 17)}$

3 Calculate using CAS.

$= \dfrac{P(13.5 < X < 14)}{P(13.5 < X < 17)}$

$= \dfrac{0.081\,9102\ldots}{0.614\,7174\ldots}$

$= 0.1332$

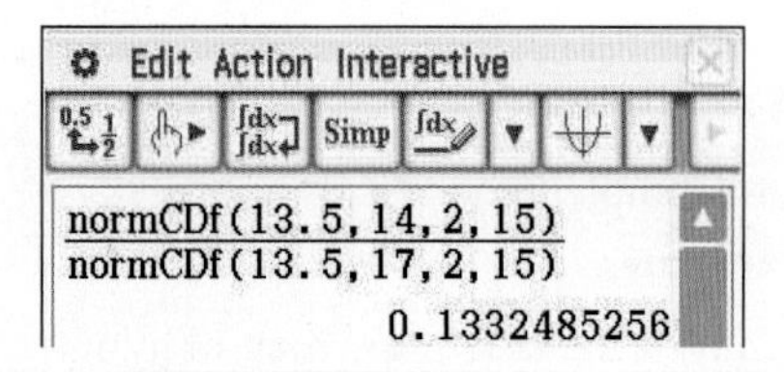

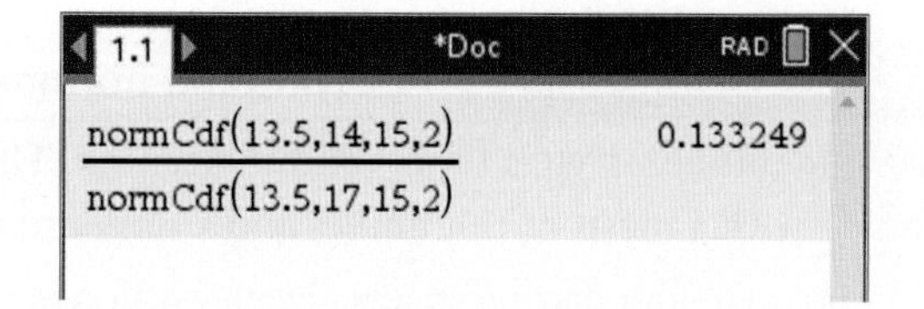

d 1 Define a new binomial random variable and its distribution.

Let Y be the number of students out of 30 who completed the task in less than 14 minutes, given that they classified as having average fitness.

$Y \sim \text{Bin}(30, 0.1332)$

2 Establish the **binomialPDf** (ClassPad) or **Binomial Pdf** (TI-Nspire) calculation and use CAS to solve.

$P(Y = 3) = \binom{30}{3}(0.1332)^3(1 - 0.1332)^{27}$

$= 0.2021$

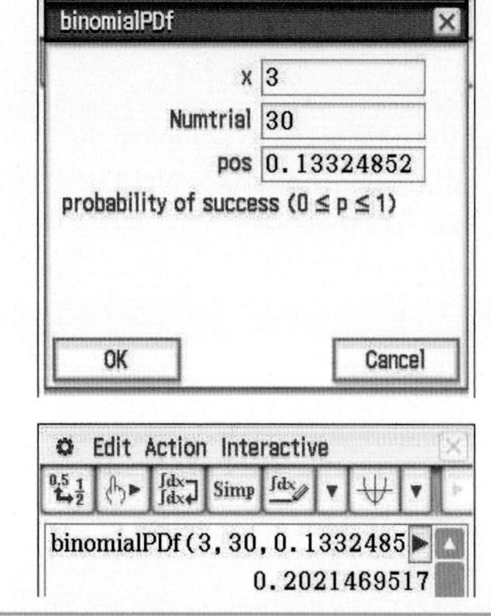

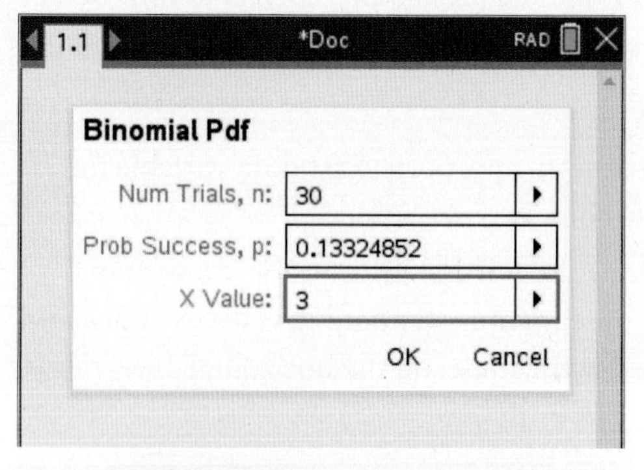

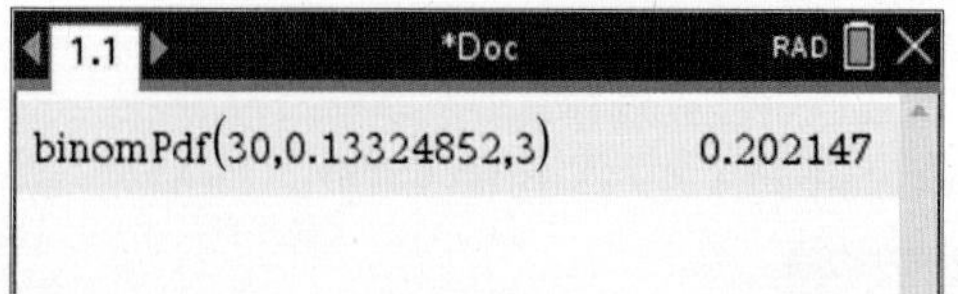

Exam hack

In questions involving probabilities that carry through multiple parts, be sure to use the full values from CAS.

WACE QUESTION ANALYSIS

© SCSA MM2020 Q16 Calculator-assumed (7 marks)

A large refrigerator in a scientific laboratory is always required to maintain a temperature between 0°C and 1°C to preserve the integrity of biological samples stored inside. A scientist working in the laboratory suspects that the refrigerator is not maintaining the required temperature and decides to record the temperature every hour for seven days. Based on these measurements, the scientist concludes that the temperature, T, in the refrigerator is normally distributed with a mean of 0.8°C and a standard deviation of 0.4°C.

a Temperature in degrees Fahrenheit, T_f, is given by $T_f = \frac{9}{5}T + 32$. Determine the mean and standard deviation of the refrigerator temperature in degrees Fahrenheit. (2 marks)

b Determine the probability that the refrigerator temperature is above 1°C. Give your answer rounded to four decimal places. (1 mark)

Video
WACE question analysis: Continuous random variables and the normal distribution

The histogram of data gathered by the scientist is shown below. N denotes the number of observations in each temperature interval.

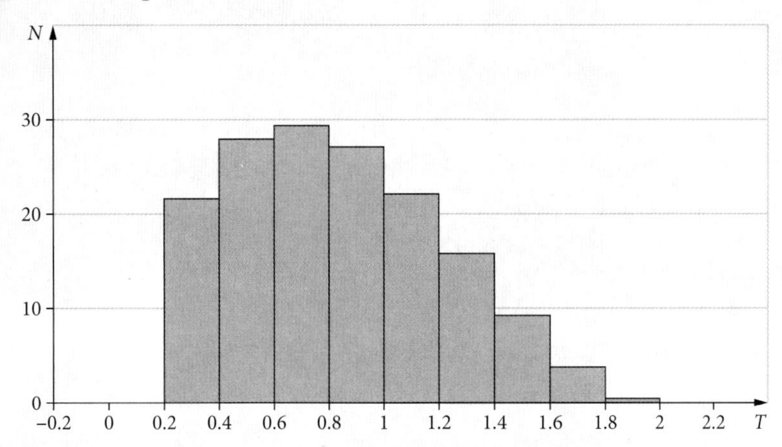

c Do you agree that the normal distribution was an appropriate model to use? Provide a reason to justify your response. (2 marks)

An alternative probability density function proposed to model the refrigerator temperature, in degrees Celsius, is given by:

$$p(t) = \frac{3}{4}t^3 - 3t^2 + 3t, \; 0 \le t \le 2$$

d Determine the probability that the refrigerator temperature is above 1°C using the new model. (2 marks)

Reading the question

- Identify and highlight the type of distribution and its corresponding parameters.
- Highlight any rounding instructions for probability questions.
- Highlight any key command words, e.g. *justify*.

Thinking about the question

- Consider the properties of change of scale and origin for $E(aX + b)$ and $SD(aX + b)$.
- Make a mental list of the conditions that make a normal distribution a suitable model for a continuous random variable.

Worked solution (✓ = 1 mark)

a The mean of T_f is

$$\mu_{T_f} = \frac{9}{5}\mu_T + 32$$
$$= \frac{9}{5}\left(\frac{4}{5}\right) + 32$$
$$= \frac{836}{25} = 33\frac{11}{25} = 33.44$$

The standard deviation of T_f is

$$\sigma_{T_f} = \frac{9}{5}\sigma_T$$
$$= \frac{9}{5}\left(\frac{2}{5}\right)$$
$$= \frac{18}{25} = 0.72$$

correctly determines mean using a change of scale and origin ✓

correctly determines standard deviation using a change of scale ✓

b $T \sim N(0.8, 0.4^2)$

$P(T > 1) = 0.3085$

determines the correct probability using CAS ✓

c No. The distribution appears to be skewed to the right (non-symmetric).

states that it is not an appropriate model ✓

justifies conclusion based on the lack of symmetry in the histogram ✓

d
$$P(T \geq 1) = \int_1^2 p(t)\,dt$$
$$= \int_1^2 \left(\frac{3}{4}t^3 - 3t^2 + 3t\right)dt$$
$$= \left[\frac{3}{16}t^4 - t^3 + \frac{3}{2}t^2\right]_1^2$$
$$= 1 - \frac{11}{16}$$
$$= \frac{5}{16} \quad \{0.3125\}$$

Exam hack

Be sure to pay attention to rounding instructions within questions, as they form part of the marking behaviours! In all answers requiring a written response, be sure to communicate clearly with an appropriate use of mathematical terminology; for example, correct use of the terms skewness and/or symmetry.

establishes the correct integral to determine the probability ✓

evaluates the integral to obtain the correct probability ✓

EXERCISE 8.4 The normal distribution

ANSWERS p. 407

Recap

Questions 1 and 2 relate to the context below.

The lifetime T, in hours, of a particular type of light globe can be modelled by a symmetrical triangular distribution over the interval $50 \le t \le 150$.

1 The mean lifetime of the light globes in hours is

A 1 **B** 100 **C** 500 **D** 800 **E** 1000

2 The maximum value of the probability density function $f(t)$ is

A 0.01 **B** 0.02 **C** 0.2 **D** 1 **E** 2

Mastery

3 WORKED EXAMPLE 19 A continuous random variable X is normally distributed with a mean of 62 and a standard deviation of 8. Use the 68–95–99.7% rule to approximate the following probabilities.

a $P(46 \le X \le 78)$ **b** $P(X > 70)$ **c** $P(X \le 38)$

4 WORKED EXAMPLE 20 For each of the following situations, give one reason why it would **not** be appropriate to model the distribution of the continuous random variable with a normal distribution.

a The mass, M, of a collection of different vehicles has a mean of 1500 kg and a standard deviation of 700 kg.

b The number of followers, F, that students in a class have on their Snaptagram accounts has a mean of 7 and a standard deviation of 1.9.

5 WORKED EXAMPLE 21 Let the continuous random variable Z be the standard normal random variable.

a Determine $P(-1 < Z < 2)$.

b If $P(Z \le a) = 0.82$ and $P(Z \le b) = 0.18$ to two decimal places, determine $P(b \le Z \le a)$.

6 WORKED EXAMPLE 22 A continuous random variable X is normally distributed with a mean of 35 and a standard deviation of 7. Let $Z \sim N(0, 1)$.

a Determine the standard score of the following values of x.

i $x = 14$ **ii** $x = 49$ **iii** $x = 59.5$

b If $P(Z \ge -2) = 0.975$, determine $P(X \ge 49)$.

7 Using CAS 4 A continuous random variable X is normally distributed with mean 150 and standard deviation 15. Find the following probabilities, correct to four decimal places.

a $P(X > 188)$ **b** $P(X \le 140)$ **c** $P(132 < X < 159)$

8 Using CAS 5 A continuous random variable X is normally distributed with a mean of 2400 and a standard deviation of 400. Find the value of c, correct to two decimal places, if

a $P(X \le c) = 0.28$ **b** $P(X \ge c) = 0.65$.

9 WORKED EXAMPLE 23 A continuous random variable X has the distribution $X \sim N(\mu, 50^2)$. If $P(X \leq 170) = 0.2743$, find the mean of X correct to the nearest integer.

10 WORKED EXAMPLE 24 The Clucky Hen Egg Farm produces eggs whose weights are normally distributed with a mean of 78 g and a standard deviation of 6 g.

a Find the probability, correct to four decimal places, that a randomly selected egg weighs more than 69 g.

b Eggs that have a weight larger than 80 g are considered 'Jumbo' eggs. Find, correct to four decimal places, the probability of an egg selected at random being classified as a 'Jumbo' egg.

c Find, correct to four decimal places, the probability of an egg weighing more than 69 g being classified as 'Jumbo'.

Suppose a sample of eggs was taken, with 80 eggs found to weigh more than 69 g.

d Determine the probability, correct to four decimal places, that exactly half of these 80 eggs will be classified as 'Jumbo'.

Calculator-free

11 (3 marks) Let X be a normally distributed random variable with mean 5 and variance 9.

a State $P(X > 5)$. (1 mark)

Let Z be the standard normal random variable.

b Find the value of b such that $P(X > 7) = P(Z < b)$. (2 marks)

12 (5 marks) Let the random variable X be normally distributed with mean 2.5 and standard deviation 0.3. Let Z be defined as $Z \sim N(0, 1)$.

a Find b such that $P(X > 3.1) = P(Z < b)$. (2 marks)

b Using the fact that $P(Z < -1) = 0.16$ correct to two decimal places, find $P(X < 2.8 \mid X > 2.5)$, rounding your answer correct to two decimal places. (3 marks)

13 (6 marks) Let X be a normally distributed random variable with a mean of 72 and a standard deviation of 8. Let Z be the standard normal random variable. Use the result that $P(Z < 1) = 0.84$ correct to two decimal places, to find

a the probability that X is greater than 80 (2 marks)

b the probability that X is between 64 and 72 (2 marks)

c the probability that X is less than 64, given that it is less than 72. (2 marks)

14 (2 marks) The random variable X is normally distributed with mean 100 and standard deviation 4. If $P(X < 106) = q$, express $P(94 < X < 100)$ in terms of q.

15 © SCSA MM2018 Q2 MODIFIED (6 marks) The heights of a large group of women are normally distributed with a mean $\mu = 163$ cm and standard deviation $\sigma = 7$ cm.

a A statistician says that almost all of the women have heights in the range 142 cm to 184 cm. Comment on the validity of her statement. Justify your answer. (2 marks)

b Approximately what percentage of women in the group has a height greater than 170 cm? (2 marks)

c Approximately 2.5% of the women are shorter than what height? (2 marks)

16 © SCSA MM2021 Q6 (7 marks)

a The graphs of three normal distributions are displayed below. The distributions have been labelled A, B and C.

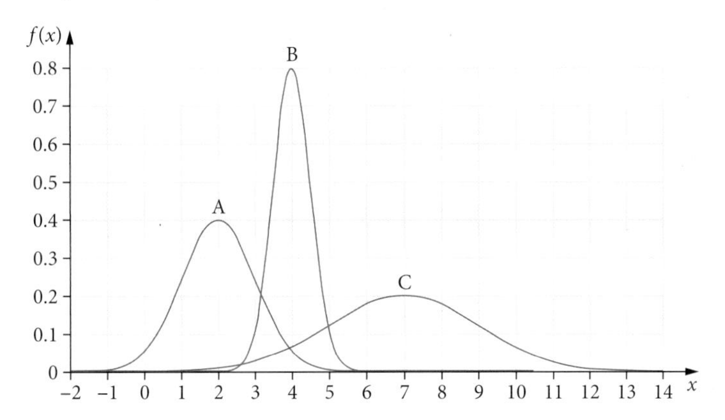

i What is the mean of distribution A? (1 mark)

ii Which of the distributions has the largest standard deviation? Justify your answer. (1 mark)

b A random variable X is normally distributed. The distribution of X is graphed below.

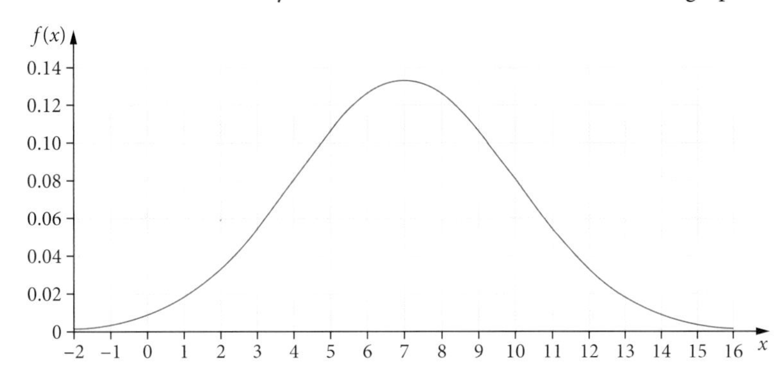

i Copy the graph and on it shade the region with area corresponding to $P(6 \leq X \leq 9)$. (1 mark)

ii Is $P(6 \leq X \leq 9) \geq 0.5$? Justify your answer. (2 marks)

c A random variable Y has probability $P(Y \geq 2) > P(Y > 2)$. Explain whether it is possible for the distribution of Y to be normal or binomial. (2 marks)

Calculator-assumed

17 (5 marks) The weights of packets of lollies are normally distributed with a mean of 200 g. It is known that 97% of these packets of lollies have a weight of more than 190 g.

a Determine the standard deviation of the distribution, correct to one decimal place. (3 marks)

b Hence, determine the probability that a randomly selected packet of lollies will have a weight between 195 g and 205 g. (2 marks)

18 © SCSA MM2016 Q18 (6 marks) The waiting times at a Perth Airport departure lounge have been found to be normally distributed. It is observed that passengers wait for less than 55 minutes, 5% of the time, while there is a 13% chance that the waiting times will be greater than 100 minutes.

a Determine the mean and standard deviation for the waiting times at Perth Airport departure lounge. (5 marks)

b Determine the probability that the waiting time will be between 75 and 90 minutes. (1 mark)

19 © SCSA MM2020 Q8 MODIFIED (7 marks) The weight, X, of chicken eggs from a farm is normally distributed with mean 60 g and standard deviation 5 g. Eggs with a weight of more than 67 g are classed as 'large'.

a What proportion of eggs from the farm are 'large'? (2 marks)

b What proportion of 'large' eggs are less than 75 g in weight? (3 marks)

c The heaviest 0.05% of eggs fetch a higher price. What is the minimum weight of these eggs? (2 marks)

20 © SCSA MM2019 Q11 (8 marks) A pizza company runs a marketing campaign based on the delivery times of its pizzas. The company claims that it will deliver a pizza in a radius of 5 km within 30 minutes of ordering or it is free. The manager estimates that the actual time, T, from order to delivery is normally distributed with mean 25 minutes and standard deviation 2 minutes.

a What is the probability that a pizza is delivered free? (1 mark)

b On a busy Saturday evening, a total of 50 pizzas are ordered. What is the probability that more than three are delivered free? (2 marks)

The company wants to reduce the proportion of pizzas that are delivered free to 0.1%.

c The manager suggests this can be achieved by increasing the advertised delivery time. What should the advertised delivery time be? (2 marks)

After some additional training the company was able to maintain the advertised delivery time as 30 minutes but reduce the proportion of pizzas delivered free to 0.1%.

d Assuming that the original mean of 25 minutes is maintained, what is the new standard deviation of delivery times? (3 marks)

21 © SCSA MM2021 Q8 (9 marks) The weights W (in grams) of carrots sold at a supermarket have been found to be normally distributed with a mean of 142.8 g and a standard deviation of 30.6 g.

a Determine the percentage of carrots sold at the supermarket that weigh more than 155 g. (2 marks)

Carrots sold at the supermarket are classified by weight, as shown in the table below.

Classification	Small	Medium	Large	Extra large
Weight W (grams)	$W \leq 110$	$110 < W \leq 155$	$155 < W \leq 210$	$W > 210$
$P(W)$		0.5131	0.3310	

b Copy and complete the table above, providing the missing probabilities. (2 marks)

c Of the carrots being sold at the supermarket that are **not** of medium weight, what proportion is small? (2 marks)

The supermarket sells bags of mixed-weight carrots, with 12 randomly-selected carrots placed in each bag.

d If a customer purchases a bag of mixed-weight carrots, determine the probability that there will be at most two small carrots in the bag. (3 marks)

22 (12 marks) A transport company has detailed records of all its deliveries. The number of minutes a delivery is made before or after its scheduled delivery time can be modelled as a normally distributed random variable, T, with a mean of zero and a standard deviation of four minutes. A graph of the probability distribution of T is shown below.

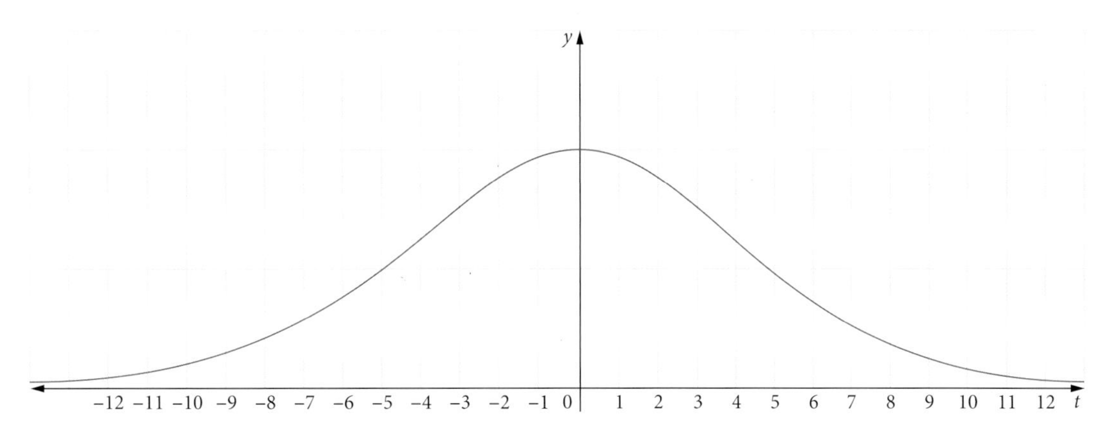

a If $P(T \le a) = 0.6$, find a to the nearest minute. (1 mark)

b Find the probability, correct to four decimal places, of a delivery being no later than three minutes after its scheduled delivery time, given that it arrives after its scheduled delivery time. (2 marks)

Using the model described, the transport company can make 46.48% of its deliveries over the interval $-3 \le t \le 2$.

c With an improvement to the delivery model, 46.48% of the transport company's deliveries can be made over the interval $-4.5 \le t \le 0.5$ such that the mean of T is k, while the standard deviation stays as four minutes. Find the value(s) of k, correct to one decimal place. (3 marks)

A rival transport company claims that there is a 0.85 probability that each delivery it makes will arrive on time or earlier. Assume that whether each delivery is on time or earlier is independent of other deliveries.

d Assuming that the rival company's claim is true, find the probability that on a day in which the rival company makes eight deliveries, fewer than half of them arrive on time or earlier. Give your answer correct to four decimal places. (3 marks)

e Assuming that the rival company's claim is true, consider a day in which it makes n deliveries.

i Express, in terms of n, the probability that one or more deliveries **will not** arrive on time or earlier. (1 mark)

ii Hence, or otherwise, find the minimum value of n such that there is at least a 0.95 probability that one or more deliveries **will not** arrive on time or earlier. (2 marks)

8 Chapter summary

Continuous random variables

- $P(X = k) = 0$, where k is a discrete outcome.
- The **probability density function** $f(x)$ is a piece-wise-defined function that is used to calculate probabilities.
- Probabilities can be calculated using the area under the graph of the probability density function and are found by integration, $P(a \le x \le b) = \int_a^b f(x)\,dx$, or the area formulas for triangles, rectangles or trapeziums.
- A valid probability density function has $f(x) \ge 0$ for all values of x and $\int_{-\infty}^{\infty} f(x)\,dx = 1$.
- The **cumulative distribution function** $F(x)$ is the integral of $f(x)$ such that

$$F(x) = \begin{cases} 0 & x < a \\ \int_a^x f(t)\,dt & a \le x \le b \\ 1 & x > b \end{cases}$$

- The **expected value** (mean) of X is given by

$$E(X) = \mu = \int_{-\infty}^{\infty} x\,f(x)\,dx$$

- The **median** $x = m$ of X is given by

$$P(X \le m) = \int_{-\infty}^{m} f(x)\,dx = 0.5$$

- The value of the pth **percentile** $(x = k)$ is determined by the integral equation

$$P(X \le k) = \frac{p}{100} \Leftrightarrow \int_{-\infty}^{k} f(x)\,dx = \frac{p}{100}$$

- The **variance** of X is given by

$$\text{Var}(X) = \sigma^2 = \int_{-\infty}^{\infty} (x - \mu)^2 f(x)\,dx$$

or

$$\text{Var}(X) = E(X^2) - E(X)^2 = \int_a^b x^2 f(x)\,dx - \mu^2$$

- The **standard deviation** of X is given by

$$SD(X) = \sqrt{\text{Var}(X)}$$

Linear transformation $Y = aX + b$ of a continuous random variable X

- $E(aX + b) = a\,E(X) + b$
- $\text{Var}(aX + b) = a^2\,\text{Var}(X)$
- $SD(aX + b) = |a|\,SD(X)$

Uniformly distributed continuous random variable

For a **uniformly distributed continuous random variable**, X, defined over $a \le x \le b$:

- the distribution is denoted as $X \sim U[a, b]$
- the probability density function of X is defined as

$$f(x) = \begin{cases} \dfrac{1}{b - a} & a \le x \le b \\ 0 & \text{otherwise} \end{cases}$$

- the cumulative distribution function of X is defined as

$$f(x) = \begin{cases} 0 & x < a \\ \dfrac{x-a}{b-a} & a \le x \le b \\ 1 & x > b \end{cases}$$

- the expected value (mean) and median of X is given by

$$\mathrm{E}(X) = \frac{a+b}{2}$$

- the variance of X is given by

$$\mathrm{Var}(X) = \frac{(b-a)^2}{12}$$

- the standard deviation of X is given by

$$\mathrm{SD}(X) = \frac{b-a}{\sqrt{12}}$$

CHAPTER SUMMARY 8

Triangular continuous random variable

For a **triangular continuous random variable**, X, defined over $a \le x \le b$:

- the maximum value of probability density function of X occurs at the **mode** $x = c$ and has the value

$$f(c) = \frac{2}{b-a}$$

The normal distribution

For a **normally distributed continuous random variable**, X:

- the distribution is denoted as $X \sim N(\mu, \sigma^2)$
- the probability density function is given by the equation $f(x) = \dfrac{1}{\sigma\sqrt{2\pi}} e^{-\frac{1}{2}\left(\frac{x-\mu}{\sigma}\right)^2}$ and is a bell-shaped curve that is symmetrical about its mean μ, which is the same as its median and mode.

The curve satisfies the 68–95–99.7% rule such that

- $\mathrm{P}(\mu - \sigma \le X \le \mu + \sigma) \approx 0.68$
- $\mathrm{P}(\mu - 2\sigma \le X \le \mu + 2\sigma) \approx 0.95$
- $\mathrm{P}(\mu - 3\sigma \le X \le \mu + 3\sigma) \approx 0.997$.

For a **standard normal random variable**, Z:

- the distribution is denoted as $Z \sim N(0, 1)$
- the probability density function is given by the equation

$$f(z) = \frac{1}{\sqrt{2\pi}} e^{-\frac{1}{2}z}$$

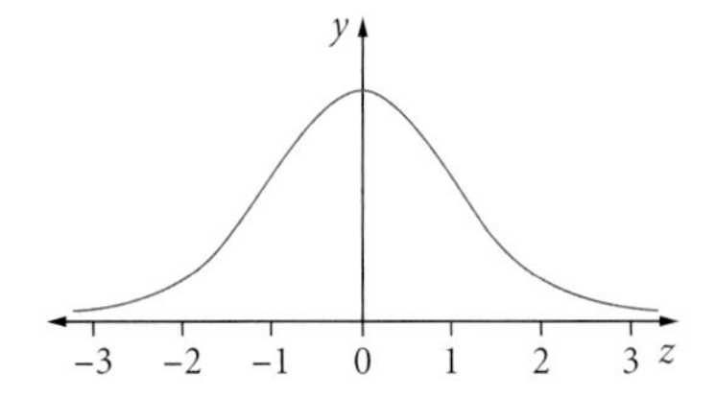

- $\mathrm{P}(Z \le c) = A$, then $\mathrm{P}(Z > c) = 1 - A$, and by symmetry, $\mathrm{P}(Z \ge -c) = A$, so $\mathrm{P}(Z < -c) = 1 - A$

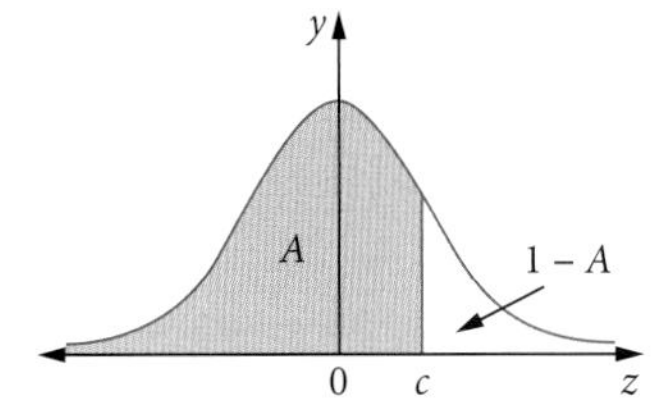

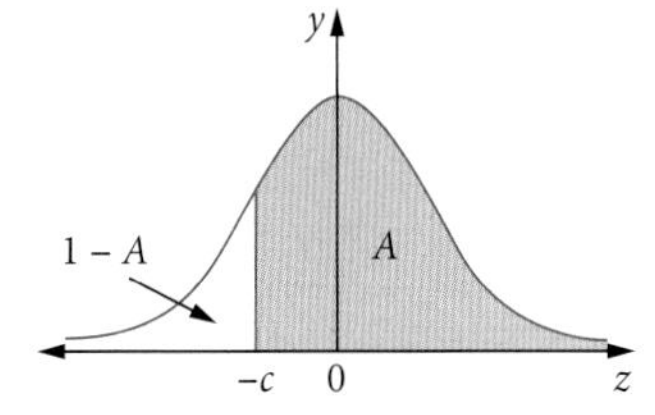

- the values of Z can be obtained from $X \sim N(\mu - \sigma^2)$ using the process of **standardisation**, $Z = \dfrac{X-\mu}{\sigma}$.

Cumulative examination: Calculator-free

Total number of marks: 34 Reading time: 4 minutes Working time: 34 minutes

1 (4 marks) Four identical balls are numbered 1, 2, 3 and 4 and put into a box. A ball is randomly drawn from the box, and not returned to the box. A second ball is then randomly drawn from the box.

a What is the probability that the first ball drawn is numbered 4 and the second ball drawn is numbered 1? (1 mark)

b What is the probability that the sum of the numbers on the two balls is 5? (1 mark)

c Given that the sum of the numbers on the two balls is 5, what is the probability that the second ball drawn is numbered 1? (2 marks)

2 (5 marks) The graph of the distribution of the discrete random variable W is shown below.

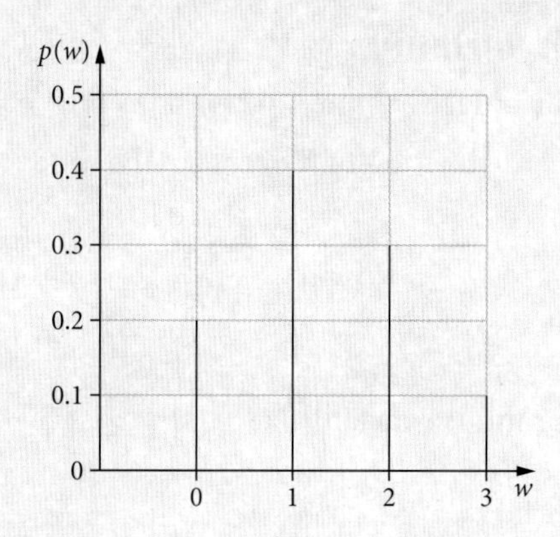

a List the probability distribution. (1 mark)

b Find

i $E(W)$ (2 marks)

ii $Var(W)$. (2 marks)

3 (3 marks) Use the graph of $y = \ln(x)$ below to find the approximate solutions to the equations.

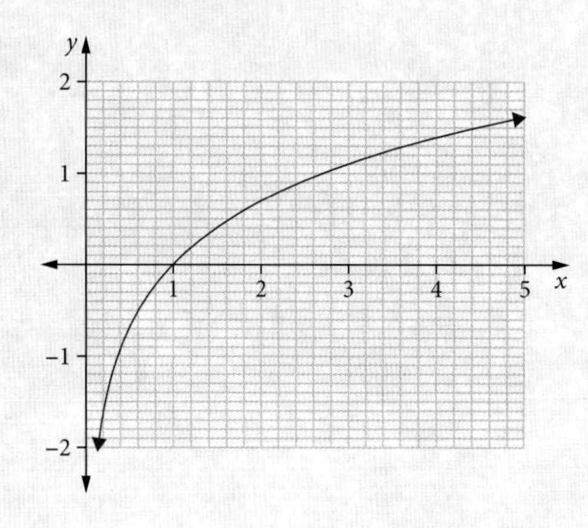

a $\ln(x) = 1.5$ (1 mark)

b $e^{3-2x} = 0.8$ (2 marks)

4 (7 marks) Find

a $\int 2\ln(x^4)\,dx$ (2 marks)

b $\int \frac{6x+15}{x^2+5x-11}\,dx$ (2 marks)

c $\int_1^3 \frac{3}{3x+1}\,dx.$ (3 marks)

5 (5 marks) A continuous random variable X has a probability density function

$$f(x) = \begin{cases} \frac{\pi}{4}\cos\left(\frac{\pi x}{4}\right) & 0 \le x \le 2 \\ 0 & \text{otherwise} \end{cases}$$

a Show that $\frac{d}{dx}\left(x\sin\left(\frac{\pi}{4}x\right)\right) = \frac{\pi x}{4}\cos\left(\frac{\pi x}{4}\right) + \sin\left(\frac{\pi x}{4}\right).$ (2 marks)

b Hence, determine the expected value of X. (3 marks)

6 © SCSA MM2021 Q2 (10 marks) It takes Nahyun between 15 and 40 minutes to get to school each day, depending on traffic conditions. Nahyun leaves home for school at 8:00 am each school day. Let the random variable X be the time, in minutes after 8:00 am, that Nahyun arrives at school. The probability density function of X is shown below.

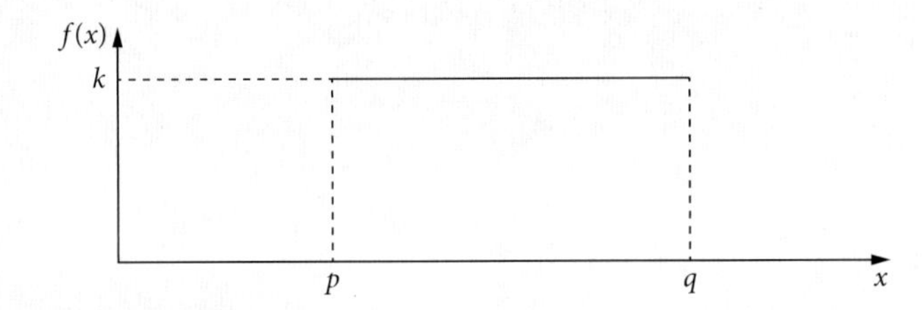

a What is the name of this type of distribution? (1 mark)

b Determine:

i the values of p, q and k (2 marks)

ii the expected value of X (1 mark)

iii the probability that Nahyun arrives at school before 8:25 am. (2 marks)

Nahyun will be late for her first class if she arrives at school after 8:28 am. Otherwise, she will not be late.

c If Nahyun is not late for her first class, what is the probability that she arrives after 8:25 am? (2 marks)

d If Nahyun only wants to be late for her first class at most 4% of the time, what time should she leave home, assuming the 15 to 40 minute travel time remains the same? (2 marks)

Cumulative examination: Calculator-assumed

Total number of marks: 56 Reading time: 6 minutes Working time: 56 minutes

1 (9 marks) A solid block in the shape of a rectangular prism has a base of width x cm. The length of the base is two-and-a-half times the width of the base. The block has a total surface area of 6480 cm^2.

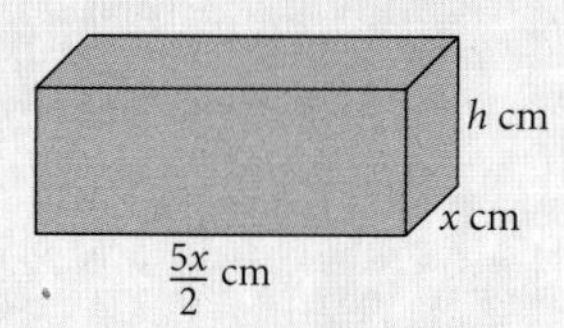

a Show that if the height of the block is h cm, $h = \dfrac{6480 - 5x^2}{7x}$. (2 marks)

b The volume, V cm^3, of the block is given by $V(x) = \dfrac{5x(6480 - 5x^2)}{14}$.

Given that $V(x) > 0$ and $x > 0$, find the possible values of x. (2 marks)

c Find $\dfrac{dV}{dx}$, expressing your answer in the form $\dfrac{dV}{dx} = ax^2 + b$, where a and b are real numbers. (3 marks)

d Find the exact values of x and h if the block is to have maximum volume. (2 marks)

2 © SCSA MM2020 Q15 (9 marks) A chef needs to use an oven to boil 100 mL of water in five minutes for a new experimental recipe. The temperature of the water must reach 100°C in order to boil. The temperature, T, of 100 mL of water t minutes after being placed in an oven set to T_0 can be modelled by the equation below.

$T(t) = T_0 - 175e^{-0.07t}$

In a preliminary experiment, the chef placed a 100 mL bowl of water into an oven that had been heated to $T_0 = 200$°C.

a What is the temperature of the water at the moment it is placed into the oven? (1 mark)

b What is the temperature of the water five minutes after being placed in the oven? (1 mark)

c What change could be made to the temperature at which the oven is set in order to achieve the five-minute boiling requirement? (2 marks)

Assume that T_0 is still 200°C.

d Determine the rate of increase in temperature of the water five minutes after being placed in the oven. Give your answer rounded to two decimal places. (2 marks)

e Explain what happens to the rate of change in the temperature of the water as time increases and how this relates to the temperature of the water. (3 marks)

3 © SCSA MM2020 Q10 (7 marks) Water flows into a bowl at a constant rate. The water level, h, measured in centimetres, increases at a rate given by

$$h'(t) = \frac{4t + 1}{2t^2 + t + 1}$$

where the time t is measured in seconds.

a Determine the rate that the water level is rising when $t = 2$ seconds. (1 mark)

b Explain why $h(t) = \ln(2t^2 + t + 1) + c$. (2 marks)

c Determine the total change in the water level over the first 2 seconds. (1 mark)

The bowl is filled when the water level reaches $\ln(56)$ cm.

d If the bowl is initially empty, determine how long it takes for the bowl to be filled. (3 marks)

4 © SCSA MM2017 Q19 (12 marks) A global financial institution transfers a large aggregate data file every evening from offices around the world to its Hong Kong head office. Once the file is received it must be processed in the company's data warehouse. The time T required to process a file is normally distributed with a mean of 90 minutes and a standard deviation of 15 minutes.

a An evening is selected at random. What is the probability that it takes more than two hours to process the file? (2 marks)

b What is the probability that the process takes more than two hours on two out of five days in a week? (3 marks)

The company is considering outsourcing the processing of the files.

c **i** A quotation for this job from an IT company is given in the table below. Copy and complete the table. (1 mark)

Job duration (minutes)	$T \leq 60$	$60 < T < 120$	$T \geq 120$
Probability			
Cost Y ($)	200	600	1200

ii What is the mean cost? (2 marks)

iii Calculate the standard deviation of the cost. (2 marks)

iv In the following year, the cost (currently $\$Y$) will increase due to inflation and also the introduction of an additional fixed cost, so the new cost $\$N$ is given by: $N = aY + b$. In terms of a and/or b, state the mean cost in the following year and the standard deviation of the cost in the following year. (2 marks)

5 © SCSA MM2018 Q12 (19 marks) The manager of the mail distribution centre in an organisation estimates that the weight, x (kg), of parcels that are posted is normally distributed, with mean 3 kg and standard deviation 1 kg.

a What percentage of parcels weigh more than 3.7 kg? (2 marks)

b Twenty parcels are received for posting. What is the probability that at least half of them weigh more than 3.7 kg? (3 marks)

The cost of postage, ($) y, depends on the weight of a parcel as follows:

- a cost of $5 for parcels 1 kg or less
- an additional variable cost of $1.50 for every kilogram or part thereof above 1 kg to a maximum of 4 kg
- a cost of $12 for parcels above 4 kg.

c Copy and complete the probability distribution table for Y. (4 marks)

x	≤ 1	$1 < x \leq 2$	$2 < x \leq 3$	$3 < x \leq 4$	$x > 4$
y	$5				
$P(Y = y)$					

d Calculate the mean cost of postage per parcel. (2 marks)

e Calculate the standard deviation of the cost of postage per parcel. (3 marks)

f If the cost of postage is increased by 20% and a surcharge of $1 is added for all parcels, what will be the mean and standard deviation of the new cost? (3 marks)

g Show one reason why the given normal distribution is not a good model for the weight of the parcels. (2 marks)

CUMULATIVE EXAMINATION: CALCULATOR-ASSUMED

CHAPTER

INTERVAL ESTIMATES FOR PROPORTIONS

Syllabus coverage

TOPIC 4.3: INTERVAL ESTIMATES FOR PROPORTIONS

Random sampling

4.3.1 examine the concept of a random sample

4.3.2 discuss sources of bias in samples, and procedures to ensure randomness

4.3.3 use graphical displays of simulated data to investigate the variability of random samples from various types of distributions, including uniform, normal and Bernoulli

Sample proportions

4.3.4 examine the concept of the sample proportion $\hat{p}$ as a random variable whose value varies between samples, and the formulas for the mean p and standard deviation $\sqrt{\frac{p(1-p)}{n}}$ of the sample proportion $\hat{p}$

4.3.5 examine the approximate normality of the distribution of $\hat{p}$ for large samples

4.3.6 simulate repeated random sampling, for a variety of values of p and a range of sample sizes, to illustrate the distribution of $\hat{p}$ and the approximate standard normality of $\frac{\hat{p}-p}{\sqrt{\frac{\hat{p}(1-\hat{p})}{n}}}$ where the closeness of the approximation depends on both n and p

Confidence intervals for proportions

4.3.7 examine the concept of an interval estimate for a parameter associated with a random variable

4.3.8 use the approximate confidence interval $\left(\hat{p} - z\sqrt{\left(\frac{\hat{p}(1-\hat{p})}{n}\right)}, \hat{p} + z\sqrt{\left(\frac{\hat{p}(1-\hat{p})}{n}\right)}\right)$ as an interval estimate for p, where z is the appropriate quantile for the standard normal distribution

4.3.9 define the approximate margin of error $E = z\sqrt{\frac{\hat{p}(1-\hat{p})}{n}}$ and understand the trade-off between margin of error and level of confidence

4.3.10 use simulation to illustrate variations in confidence intervals between samples and to show that most, but not all, confidence intervals contain p

Video playlists (4):

9.1 Random sampling

9.2 The sampling distribution of sample proportions

9.3 Confidence intervals for proportions

WACE question analysis Interval estimates for proportions

Worksheets (6):

9.2 Sample proportions • Sample proportion calculations • Sample proportion probabilities

9.3 Sample proportion confidence intervals • Margin of error for standard normal variables • Sample sizes

Nelson MindTap

To access resources above, visit **cengage.com.au/nelsonmindtap**

Video playlist
Random sampling

9.1 Random sampling

The language of random sampling

Suppose the administration of a local high school containing 1500 students wanted to collect data from every single student around a particular statistical variable, X. In some cases, this level of large-scale data collection of an entire **population** may be impractical. To do so, a **census** is needed. When a census is conducted to gain data on an entire population, characteristics of that particular population can be calculated, such as the mean or standard deviation of X. These are called **population parameters**.

However, perhaps to be more practical or time efficient, the school administration may consider only collecting data from a **sample** of 100 students from that school around that particular statistical variable. This is then called a **survey** of a sample group of **sample size** 100. From this sample data, **sample statistics** can be calculated, such as the mean and standard deviation, and then used as **point estimates** of the population mean and standard deviation.

In most situations involving data collection, we cannot take a census of a whole population and so population parameters are often unknown. As a result, we must rely on the use of sample statistics obtained from surveys to make inferences about the larger population from which the data was obtained.

WORKED EXAMPLE 1 Identifying population, parameters, sample size and statistics

The first 20 people leaving a convenience store after 6:30 pm were asked how much they spent. The smallest amount was \$5.40, the largest was \$47.60 and the average amount was \$22.40. Identify the

a population **b** sample size **c** sample statistics.

Steps	Working
a Determine the population (the whole group being investigated).	The population is all customers of that particular convenience store.
b Identify the number of people from whom data was collected.	The sample size is 20.
c Identify any characteristics from this information.	Some possible statistics of this sample are: • the minimum amount spent – \$5.40 • the maximum amount spent – \$47.60 • the range – \$42.20 • the mean amount spent – \$22.40.

When data are collected from sample groups, it must be considered as to whether that data gives useful, accurate and reliable information about the population parameters. Things that need to be considered include

- sample size – is it sufficiently large enough to represent the population?
- randomness and bias – are the data free from biases that could affect the reliability of the data being used to estimate population parameters?

When a sample meets these conditions, it can be called a **fair and representative sample** of the population. For example, the sample data collected in Worked example 1 may not be considered a fair and representative sample and, hence, could be considered a **biased sample** as:

- it is only a small sample of 20 customers of all customers of that store
- the data was only collected once, at a particular time of day (after 6:30 pm) on one particular day.

As a result, when collecting data from samples, procedures to ensure randomness and minimise any sources of **bias** need to be considered so that valid conclusions can be made about the population of interest.

Random sampling procedures and sources of bias

The best way to reduce sampling bias and obtain fair and representative samples is to use a **probability sampling method**. This is a method that involves random selection in which each member of the population or subset of a population has an equally likely chance of being chosen. Commonly used probability sampling methods are listed as follows.

1 **Simple random sampling** – every member of the entire population has an equally likely chance of being selected; for example, through the use of a random number generator to generate n numbers and then those numbers are used to select the sample group.

2 **Systematic sampling** – the population is ordered on the basis of an unbiased characteristic (e.g. alphabetical order based on first name or last name, age, height) and every kth member of the population is chosen to create the sample of size n.

3 **Stratified sampling** – the population is divided into subgroups, called strata, on the basis of common characteristics (e.g. year groups, profession) and then simple random or systematic selections are made from each subgroup proportional to the size of the subgroup.

4 **Cluster sampling** – the population is divided into subgroups, called clusters, on the basis of common characteristics (e.g. location, time) and then simple random or systematic selections are made from each subgroup, but not necessarily in proportion.

Exam hack

If you are asked to describe a method to ensure randomness when sampling, it is not the name of the sampling method that is the important feature but rather the description of the random process.

WORKED EXAMPLE 2 Describing a random sampling process

A company needs to select 100 houses in a particular street of 421 houses to collect data for an employment survey. Describe a sampling procedure that would ensure randomness.

Steps	Working
1 Choose a probability sampling method that ensures each house has an equally likely chance of being selected.	Consider a systematic sampling method. Divide 421 by 100 and round to the nearest integer. $\frac{421}{100} = 4.21 \approx 4$
2 Describe the process.	Use a random number generator to pick a starting house number between 1 and 421; for example, 27. Collect data from every 4th house number, starting from 27 (i.e. 27, 31, 35, 39 … 419) and start the sequence again from 2 until 100 households are surveyed.

Sampling procedures that are not random are called **non-probability sampling methods**, in which each member of the population does not have an equally likely chance of being selected. Commonly used non-probability sampling methods are listed as follows.

1 **Convenience sampling** – the sample is chosen such that the data is conveniently collected from the most accessible data source; for example, surveying family members in the same household about political views.

2 **Quota sampling** – the sample size is pre-determined and no more data is collected once that limit has been reached; for example, the first 10 employees who arrive at work are surveyed about transport methods.

3 **Volunteer sampling** – the sample is collected by asking for volunteers to opt-in to the data collection. An example is a radio show asking for callers to share information on the number of Australian states they have visited.

It is often in these cases that different types of bias will arise and can lead to an under- or over-representation of particular subgroups of a population. Common types of bias are given below.

1 Spatial bias – the sample is non-representative of the population because of the location from which it is collected.
2 Temporal bias – the sample is non-representative of the population because of the time at which it is collected.
3 Self-selection bias – the sample is non-representative of the population because the participants choose to take part voluntarily.
4 Non-response bias – the participants chosen to participate may choose to not give a response.
5 Leading question bias – the data may be collected in a way that encourages a particular response.

Exam hack

Once again, it is not the name of the bias that is the important feature but rather the description of the source of the bias; that is, where has the bias come from and what effect does it have on the data collection process.

WORKED EXAMPLE 3 Discussing sources of bias and ensuring randomness

A legal firm with 2000 employees wants to collect data on employment satisfaction and sends out an all-staff email at 9:00 pm on a Friday night containing a link to an optional survey.

a Identify and explain **two** possible sources of bias with this sampling method.

b Suggest a random sampling procedure that will minimise bias in this data collection.

Steps	Working
a 1 Consider the four main types of bias: spatial, temporal, self-selection, leading question. Identify which of the four are applicable to this situation.	Temporal bias – sending out the survey at 9:00 pm on a Friday night means that there is an increased chance that only employees who are working on the Friday night and over the weekend respond to the survey.
2 Identify the source of bias in the sampling method and give a brief explanation as to why it creates bias.	Self-selection bias – only those who want to respond to the survey will, meaning that there could be an over-representation of those who either are or are not satisfied with their job.
b 1 Choose a probability sampling method that ensures each employee has an equally likely chance of being selected. **2** Describe the process.	List the employees alphabetically by surname and send the email during working hours to every 10th employee in the ordered list to complete a compulsory survey.

Variability of random samples

Suppose the legal firm in Worked example 3 randomly surveyed three different sample groups of employees, each of sample size 200. Each of these samples would have produced different sample statistics due to the randomness in the sampling process. This is referred to as the **variability of random samples**. That is, even though all 600 employees came from the same population of 2000, each set of sample data has the potential to provide very different representations of the population.

The variability of samples can be seen through a process of **simulation**, whereby technology can be used to model the events of a random probability experiment or mimic the data collection process of a sample from a population. However, when simulating sample data, assumptions need to be made about the nature of the **population distribution** (also called the **parent distribution**) from which the samples are being taken. Based on our knowledge of random variables, we can simulate sample data from populations that are uniformly, normally, Bernoulli or binomially distributed.

USING CAS 1 Simulating sample data from a uniform distribution

Simulate two different samples of 100 scores from the continuous uniform random variable, $X \sim U[30, 70]$. Compare the mean and standard deviation of the samples to the mean and standard deviation of X.

ClassPad

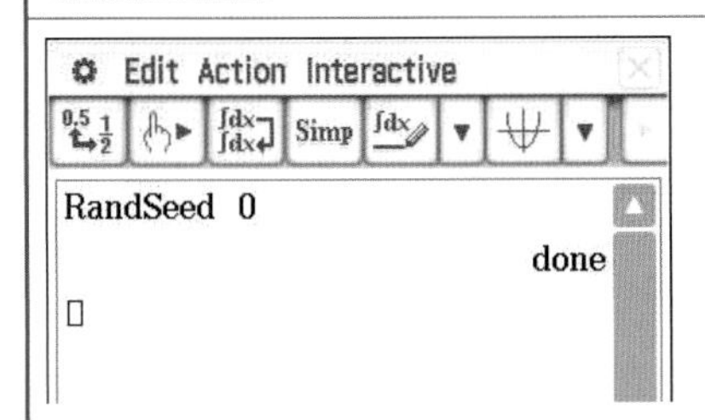

1 Tap **Decimal**.

2 Open the **Keyboard** > **Catalog**.

3 Tap **R** then scroll down to select **RandSeed**.

4 Enter a number from 0 to 9. The default is **0**.

5 Press **EXE**. This sets a new starting point for generating random numbers.

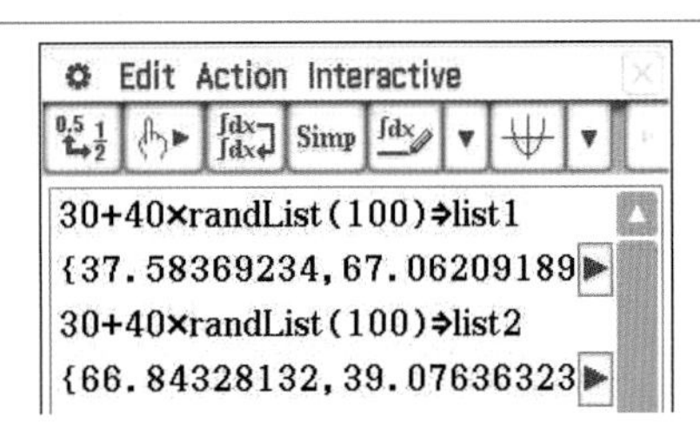

6 Open the **Keyboard** > **Catalog**.

7 Tap **R** then scroll to select **randList**.

8 Use the formula $a + (b - a) \times$ **randList**(m) to generate m values from the distribution $U[a, b]$, as shown above.

9 Store the first sample as **list1**.

10 Repeat for the second sample and store as **list2**.

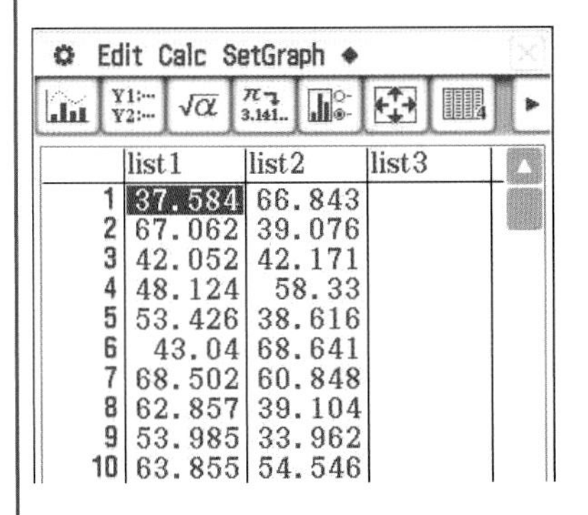

11 Tap **Menu** > **Statistics**.

12 The randomly generated values will appear in **list1** and **list2**.

13 Tap **SetGraph** > **Setting**.

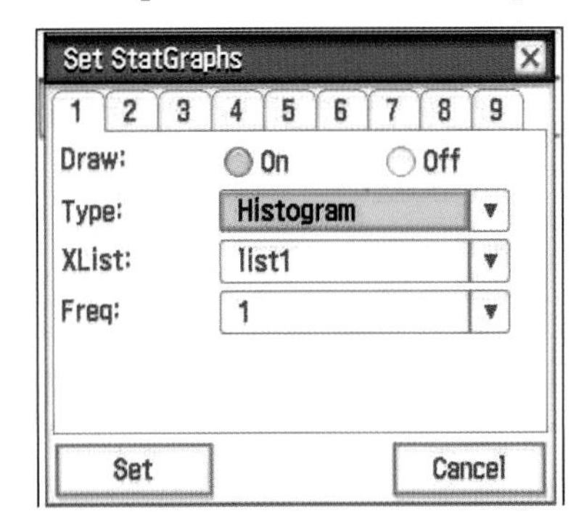

14 Tap on the **1** tab.

15 Change the **Type:** field to **Histogram** and keep the **XList** field as **list1**.

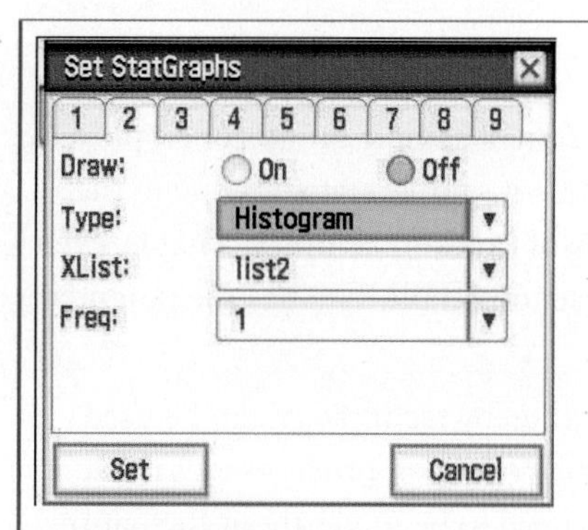

16 Tap on the **2** tab.

17 Change the **Type:** field to **Histogram** and change the **XList** field to **list2**.

18 Tap **Set**.

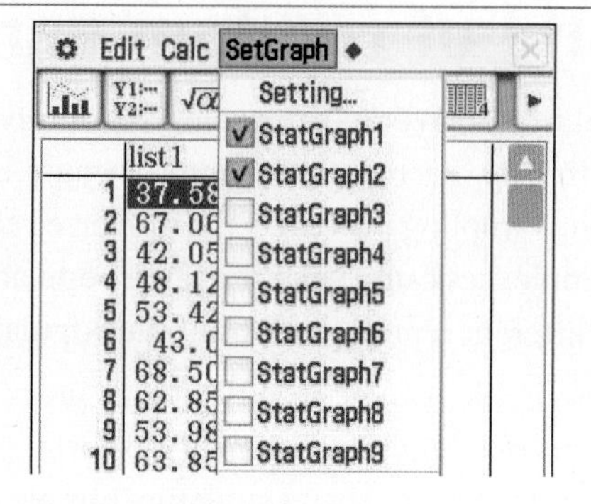

19 Tap **SetGraph**.

20 Tap to select **StatGraph1** and **StatGraph2**.

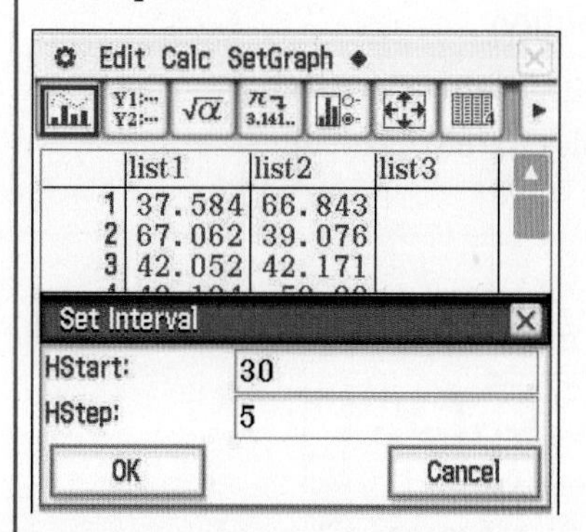

21 Tap **Graph**.

22 Keep the **HStart:** field as **30** and change the **HStep:** field to **5** (optional).

23 Tap **OK**.

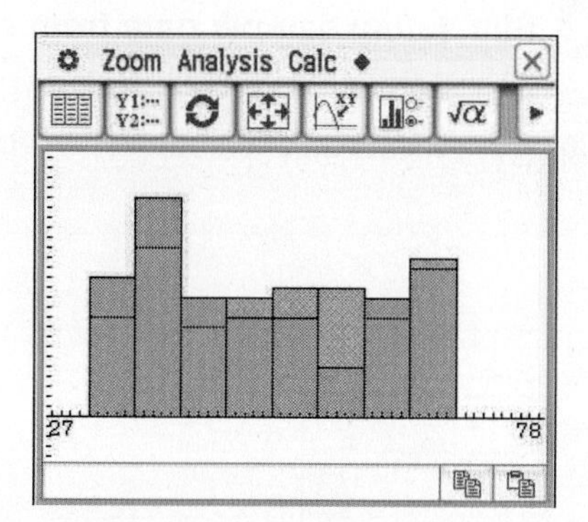

24 The histograms of the data will appear in the lower window (the windows have been swapped for this screen).

25 Tap the upper window to select **Statistics**.

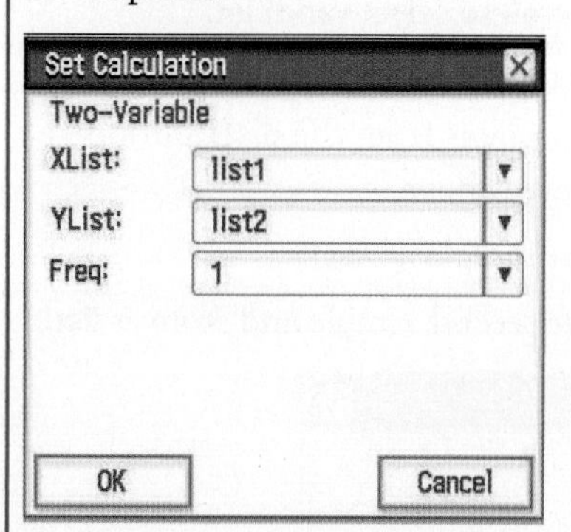

26 Tap **Calc** > **Two-Variable**.

27 Keep the **XList:** field as **list1** and change the **YList:** field to **list2**.

28 Tap **OK**.

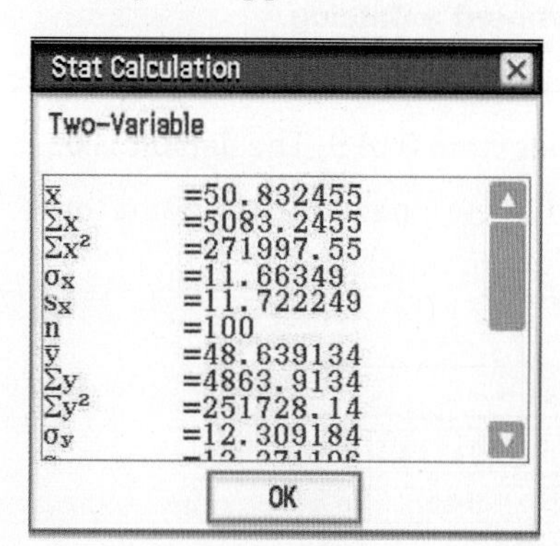

29 The summary statistics will appear in the **Two-Variable** window.

30 Scroll down to view all the statistics.

$E(X) = \dfrac{30 + 70}{2} = 50$. Both simulated means are approximately 50, but vary.

$SD(X) = \dfrac{70 - 30}{\sqrt{12}} = 11.55$. Both simulated standard deviations are close to 11, but vary.

TI-Nspire

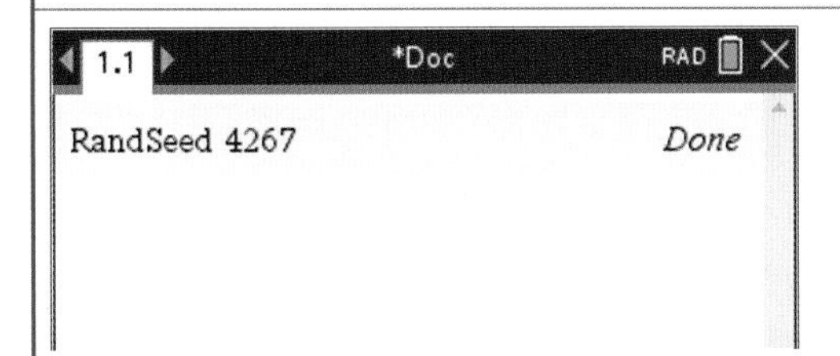

1 Press **catalog** > **R** to jump to the functions starting with the letter R.

2 Scroll down and select **RandSeed**.

3 Enter a 4-digit number and press **enter**. This sets a new starting point for generating random numbers.

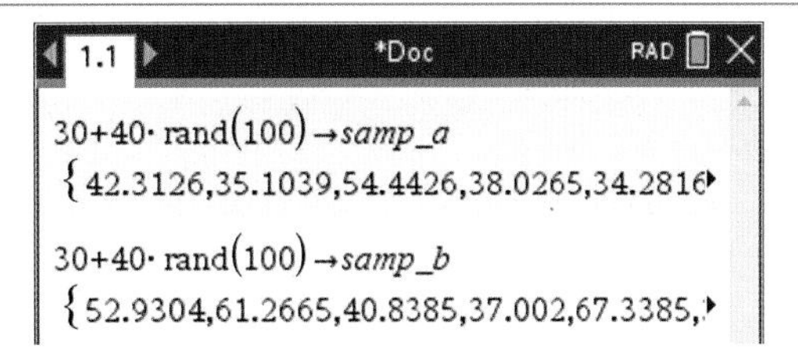

4 Press **catalog** > **rand**.

5 Use the formula $a + (b - a) \times \textbf{randList}(m)$ to generate m values from the distribution $U[a, b]$, as shown above.

6 Store the first sample as **samp_a**.

7 Repeat for the second sample and store as **samp_b**.

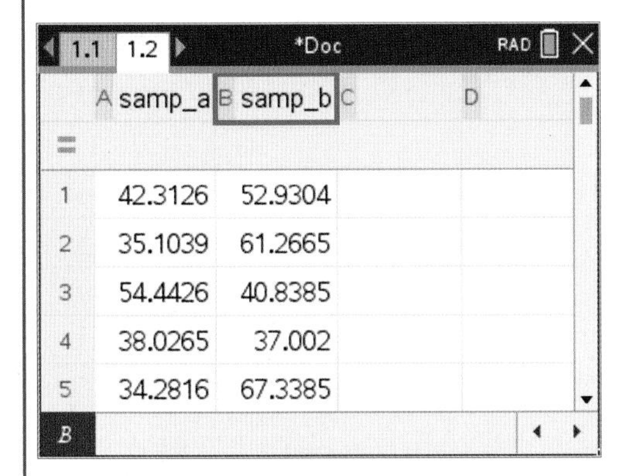

8 Add a **Lists & Spreadsheet** page.

9 Tap in the cell next to the **A**.

10 Press **var** > **Link To:** and select **samp_a**.

11 Tap in the cell next to the **B**.

12 Press **var** > **Link To:** and select **samp_b**.

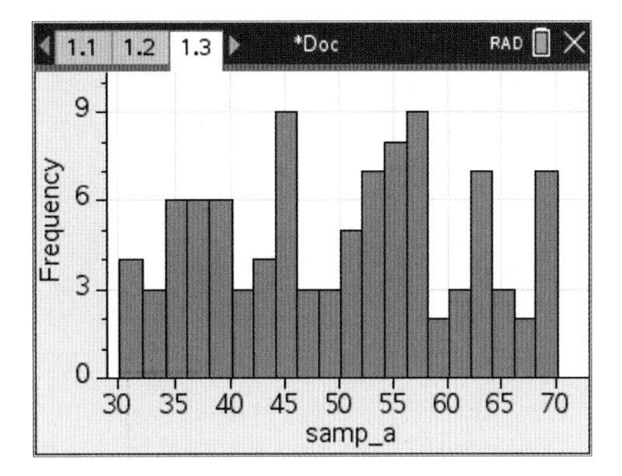

13 Add a **Data & Statistics** page.

14 For the horizontal axis, select **samp_a**.

15 Press **menu** > **Plot Type** > **Histogram**.

16 The sample a data will be displayed as a histogram.

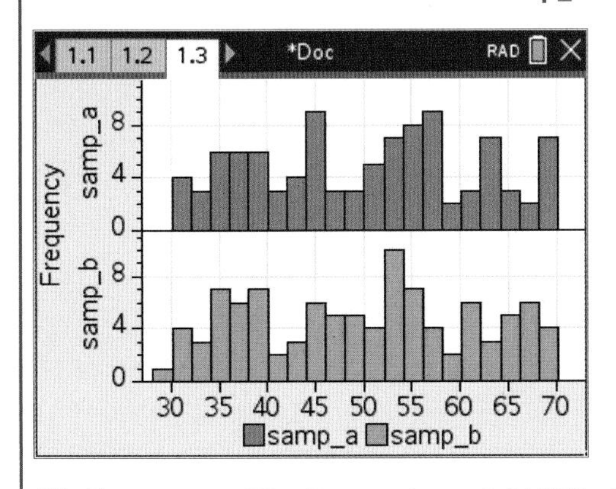

17 Tap **menu** > **Plot Properties** > **Add X Variable**.

18 Select **samp_b**.

19 The data for both samples will be displayed as histograms.

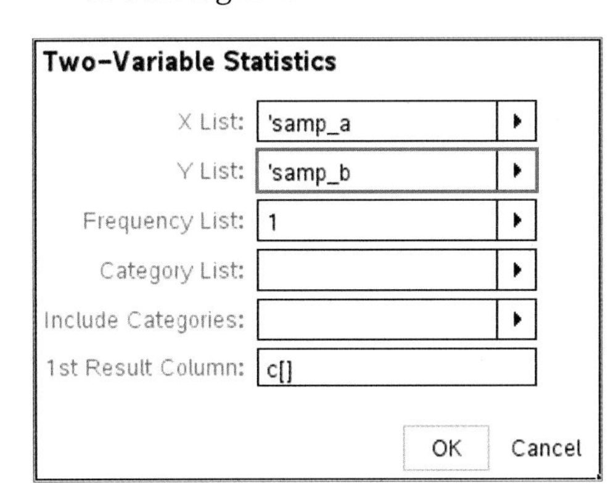

20 Return to the **Lists & Spreadsheet** page.

21 Press **menu** > **Statistics** > **Stat Calculations** > **Two-Variable**.

22 In the **X List:** field, press the right arrow and select **samp_a**.

23 In the **Y List:** field, press the right arrow and select **samp_b**.

24 Press **enter**.

	samp_a	samp_b	C	D
=				=TwoVar(
2	35.1039	61.2665	x̄	50.0956
3	54.4426	40.8385	Σx	5009.56
4	38.0265	37.002	Σx²	263374.
5	34.2816	67.3385	sx := sn-...	11.1992
6	44.8403	30.3555	σx := σn...	11.1431

D2 =50.095642024361

25 The summary statistics will be displayed in columns **C** and **D**.

	samp_a	samp_b	C	D
=				=TwoVar(
8	57.2401	67.2434	ȳ	49.8084
9	63.1764	37.4991	Σy	4980.84
10	40.6772	61.529	Σy²	260839.
11	30.1584	52.1873	sy := sn-...	11.349
12	38.3587	35.6654	σy := σn...	11.2921

D8 =49.808404640982

26 Scroll down to view all the statistics.

$E(X) = \frac{30 + 70}{2} = 50$. Both simulated means are approximately 50, but vary.

$SD(X) = \frac{70 - 30}{\sqrt{12}} = 11.55$. Both simulated standard deviations are close to 11, but vary.

Note that if we were to repeat the above simulation with a greater number of scores, we would *expect* the distribution to become *more uniform*; however, due to the nature of random sampling, we could get a simulation that does not become more uniform. For example, the histogram on the right shows a simulation in which 250 scores were generated.

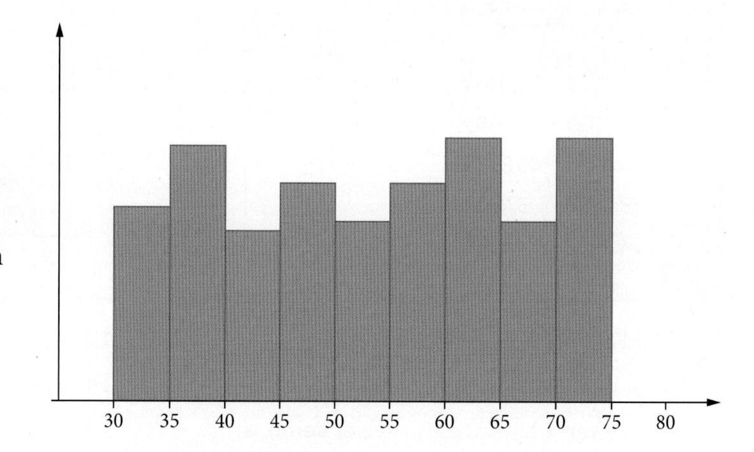

USING CAS 2 Simulating sample data from a normal distribution

Simulate two different samples of 30 scores from the continuous normal random variable, $Y \sim N(25, 5^2)$. Compare the mean and standard deviation of the samples to the mean and standard deviation of Y.

ClassPad

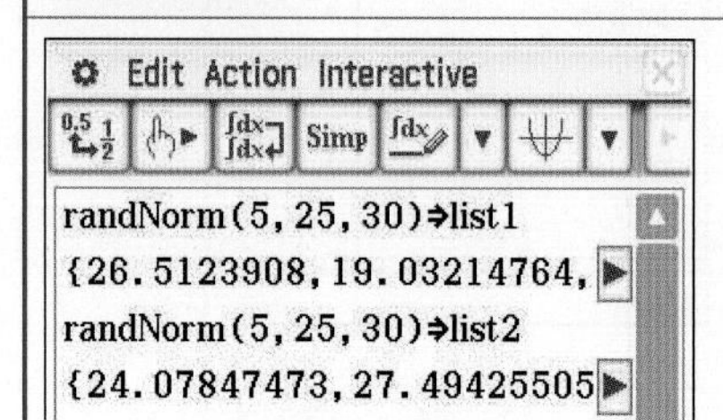

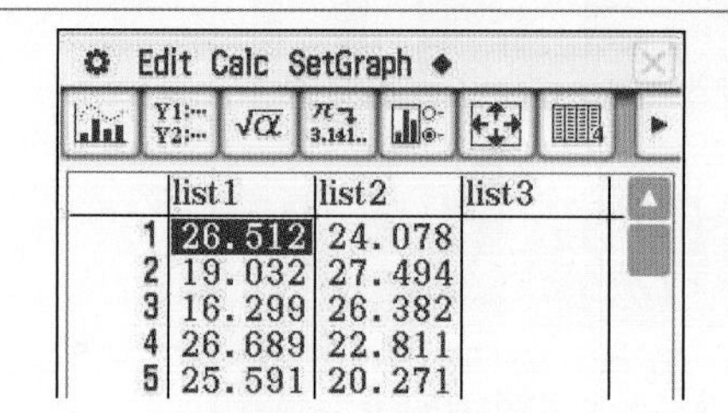

1 Open the **Keyboard** > **Catalog**.
2 Tap **R** then scroll down to select **randNorm**.
3 Use the input **randNorm**(μ, σ, m) to generate m values from the distribution $N(\mu, \sigma^2)$, as shown above.
4 Store the first sample as **list1**.
5 Repeat for the second sample and store as **list2**.
6 Tap **Menu** > **Statistics**.
7 The randomly generated values will appear in **list1** and **list2**.
8 Tap **SetGraph** > **Setting**.

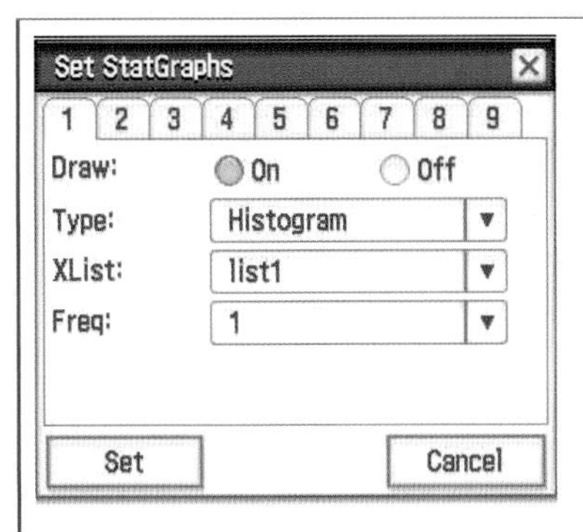

9 Tap tab **1**.

10 Ensure the **Type:** field is **Histogram** and the **XList:** field is **list1**.

11 Tap tab **2**.

12 Ensure the **Type:** field is **Histogram** and the **XList:** field is **list2**.

13 Tap **Set**.

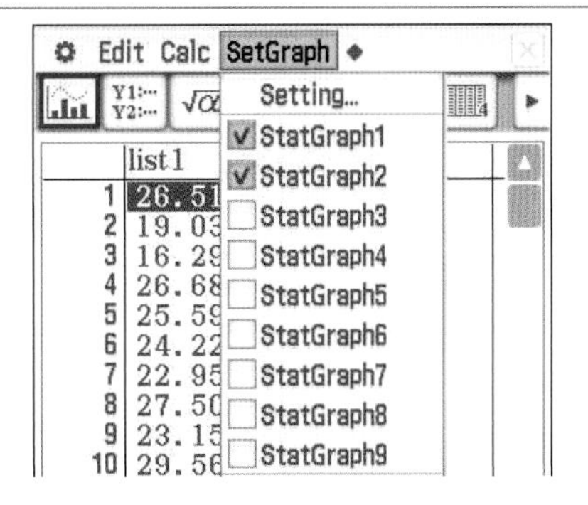

14 Tap **SetGraph**.

15 Ensure **StatGraph1** and **StatGraph2** are selected.

16 Tap **Graph**.

17 In the **Set Interval** dialogue box, keep the defaults settings and tap **OK**.

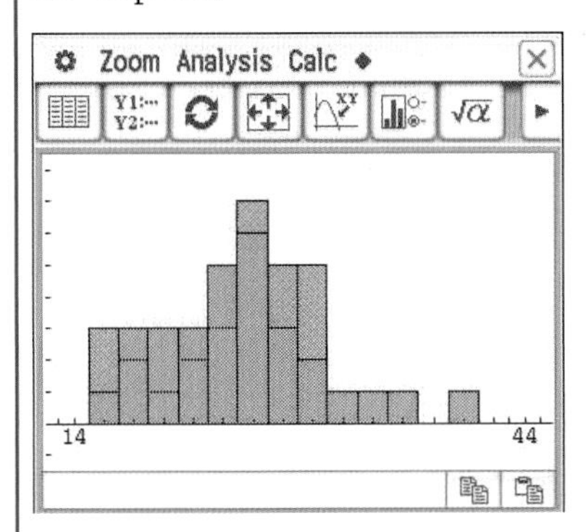

18 The histograms of the data will appear in the lower window (the windows have been swapped for this screen).

19 Tap the upper window to select **Statistics**.

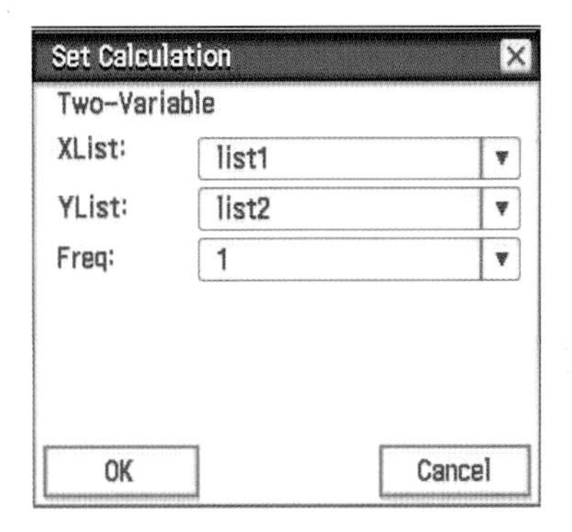

20 Tap **Calc** > **Two-Variable**.

21 Ensure the **XList:** field is **list1** and the **YList:** field is **list2**.

22 Tap **OK**.

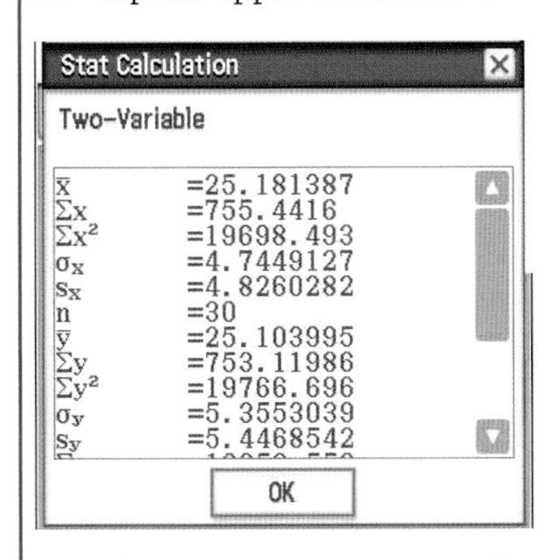

23 The summary statistics will appear in the **Two-Variable** window.

24 Scroll down to view all the statistics.

$E(Y) = 25$. Both simulated means are approximately 25, but vary.

$SD(Y) = 5$. Both simulated standard deviations are close to 5, but vary.

TI-Nspire

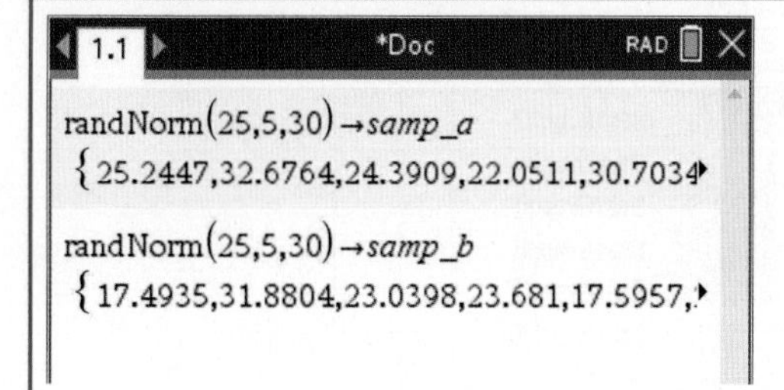

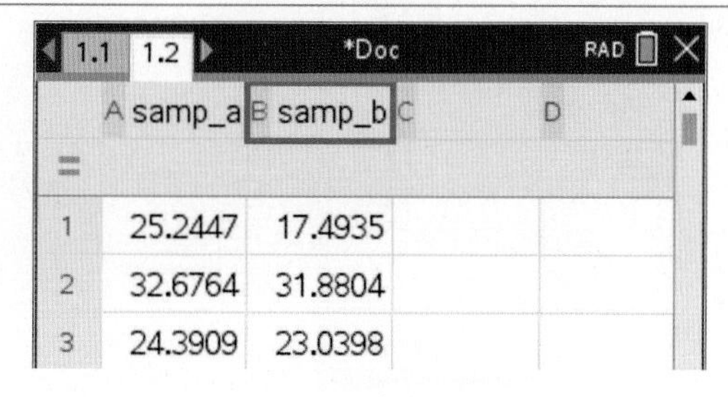

1 Press **catalog** > **randNorm**.

2 Use the input **randNorm**(μ, σ, m) to generate m values from the distribution $N(\mu, \sigma^2)$, as shown above.

3 Store the sample as **samp_a**.

4 Repeat for the second sample and store as **samp_b**.

5 Add a **Lists & Spreadsheet** page.

6 Tap in the cell next to the **A**.

7 Press **var** and select **samp_a**.

8 Tap in the cell next to the **B**.

9 Press **var** and select **samp_b**.

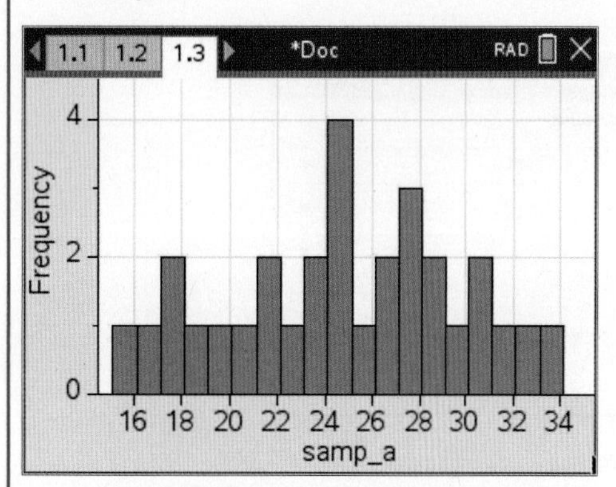

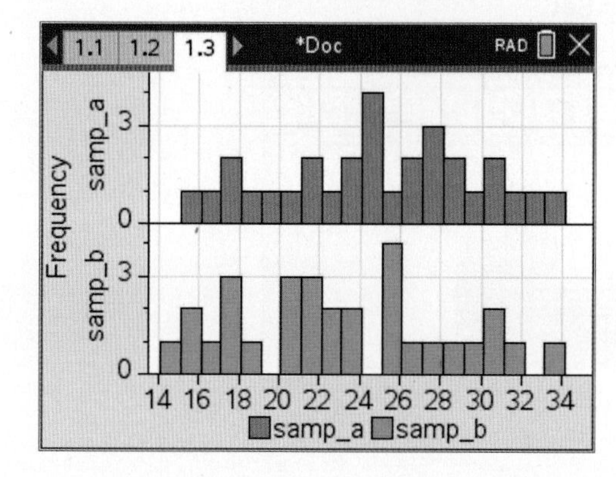

10 Add a **Data & Statistics** page.

11 For the horizontal axis, select **samp_a**.

12 Press **menu** > **Plot Type** > **Histogram**.

13 The sample a data will be displayed as a histogram.

14 Tap **menu** > **Plot Properties** > **Add X Variable**.

15 Select **samp_b**.

16 The data for both samples will be displayed as histograms.

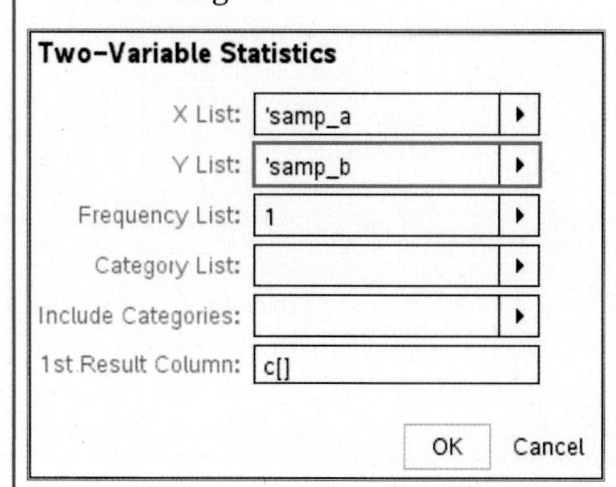

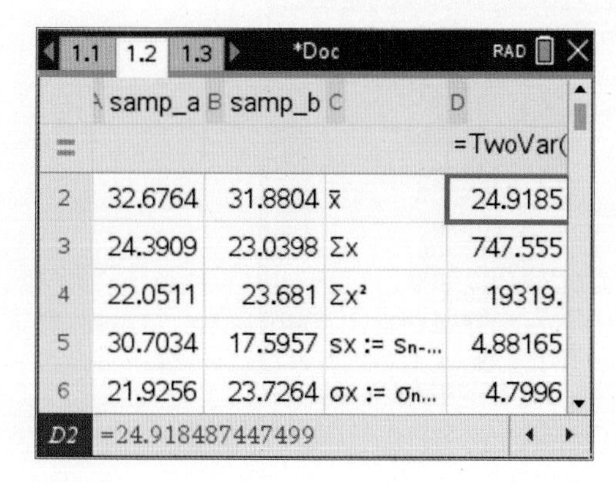

17 Return to the **Lists & Spreadsheet** page.

18 Press **menu** > **Statistics** > **Stat Calculations** > **Two-Variable**.

19 In the **X List:** field, press the right arrow and select **samp_a**.

20 In the **Y List:** field, press the right arrow and select **samp_b**.

21 Press **enter**.

22 The summary statistics will be displayed in columns **C** and **D**.

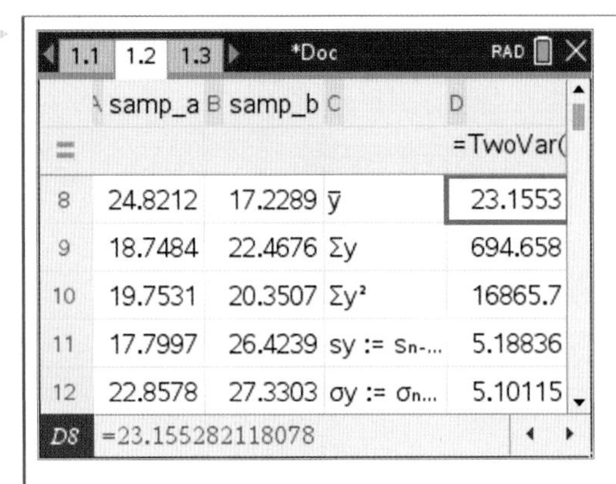

23 Scroll down to view all the statistics.

E(*Y*) = 25. Both simulated means are approximately 25, but vary.

SD(*Y*) = 5. Both simulated standard deviations are close to 5, but vary.

Similarly, if we were to repeat the above simulation with a greater number of scores, we would *expect* the distribution to become *more normal*; however, due to the nature of random sampling, we could get a simulation that does not become more normal. For example, the histogram on the right shows a simulation in which 250 scores were generated.

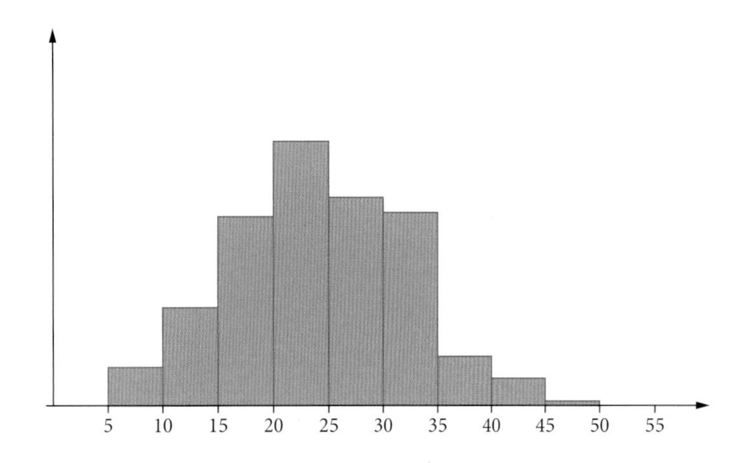

USING CAS 3 Simulating sample data from a Bernoulli distribution

Simulate two different samples of 20 scores from the discrete Bernoulli random variable, $Z \sim \text{Bern}(0.8)$. Compare the mean and standard deviation of the samples to the mean and standard deviation of Z.

ClassPad

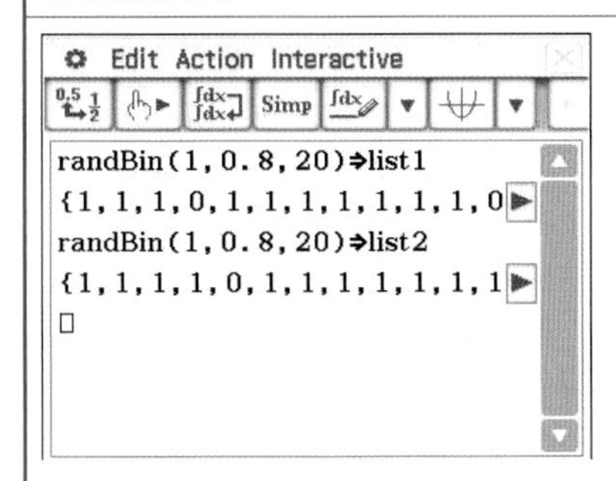

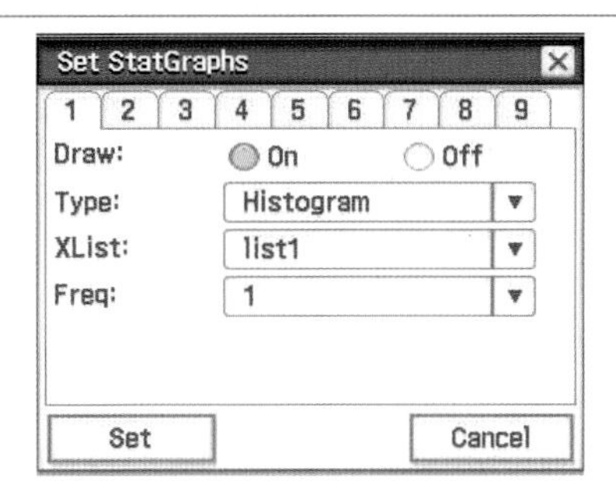

1 Open the **Keyboard** > **Catalog**.

2 Tap **R** then scroll down to select **randBin**.

3 Use the input **randBin**$(1, p, m)$ to generate m values from the distribution $\text{Bin}(1, p)$, as shown above.

4 Store the first sample as **list1**.

5 Repeat for the second sample and store as **list2**.

6 Tap **Menu** > **Statistics**.

7 Tap **SetGraph** > **Setting**.

8 For tab **1**, ensure the **Type:** field is **Histogram** and the **XList**: field is **list1**.

9 For tab **2**, ensure the **Type**: field is **Histogram** and the **XList**: field is **list2**.

10 Tap **Set**.

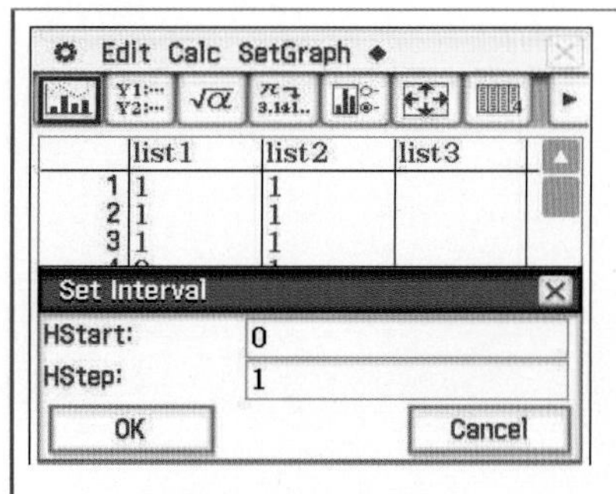

11 Tap **SetGraph** and ensure **StatGraph1** and **StatGraph2** are selected.

12 Tap **Graph**.

13 In the **Set Interval** dialogue box, change the **HStep**: field to **1** and tap **OK**.

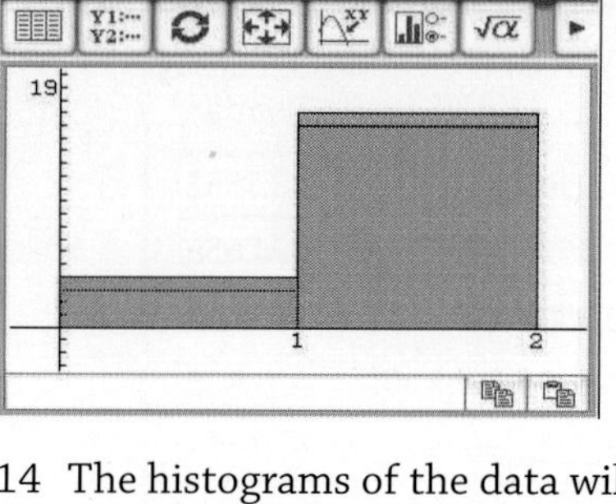

14 The histograms of the data will appear in the lower window (the windows have been swapped for this screen).

15 Tap the upper window to select **Statistics**.

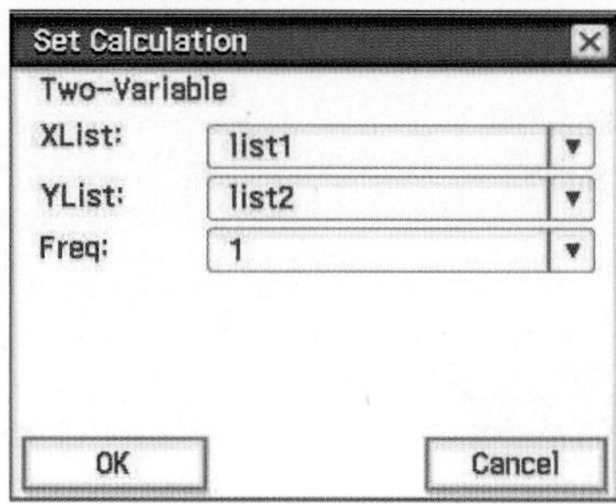

16 Tap **Calc** > **Two-Variable**.

17 Ensure the **XList**: field is **list1** and the **YList**: field is **list2**.

18 Tap **OK**.

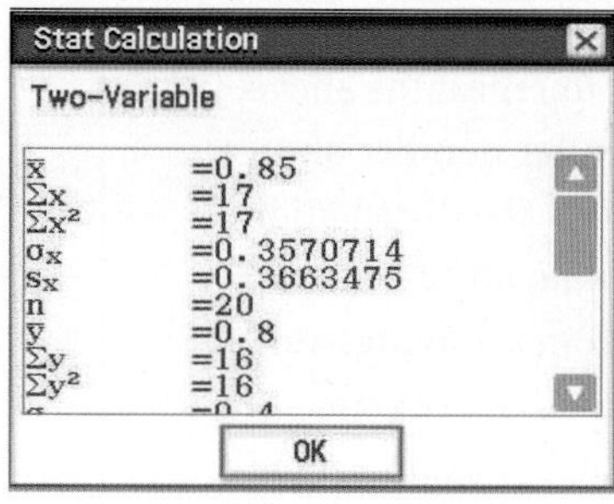

19 The summary statistics will appear in the **Two-Variable** window.

20 Scroll down to view all the statistics.

E(Z) = 0.8. Both simulated means are approximately 0.8, but vary.

SD(Z) = 0.4. Both simulated standard deviations are close to 0.4, but vary.

TI-Nspire

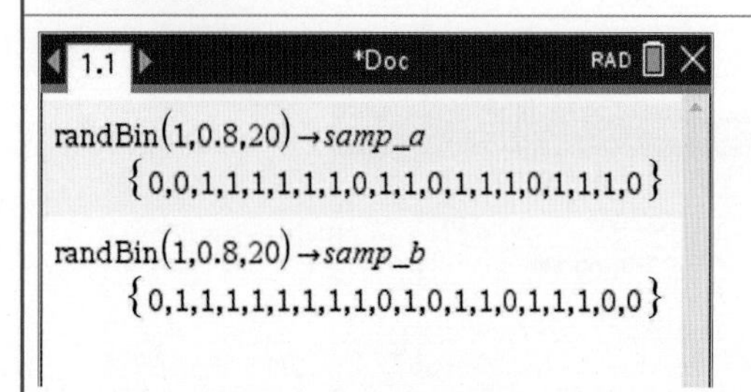

	A samp_a	B samp_b	C	D
=				
1	0	0		
2	0	1		
3	1	1		

1 Press **catalog** > **randNorm**.

2 Use the input **randBin(1, *p*, *m*)** to generate m values from the distribution Bin(1, p), as shown above.

3 Store the first sample as **samp_a**.

4 Repeat for the second sample and store as **samp_b**.

5 Add a **Lists & Spreadsheet** page.

6 Tap in the cell next to the **A**.

7 Press **var** and select **samp_a**.

8 Tap in the cell next to the **B**.

9 Press **var** and select **samp_b**.

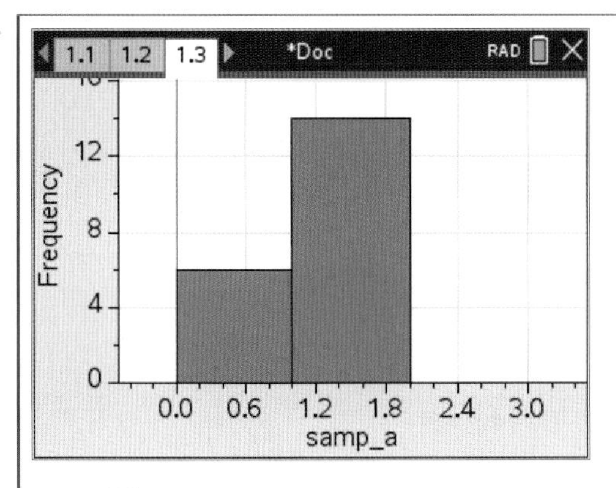

10 Add a **Data & Statistics** page.

11 For the horizontal axis, select **samp_a**.

12 Press **menu** > **Plot Type** > **Histogram**.

13 The sample a data will be displayed as a histogram.

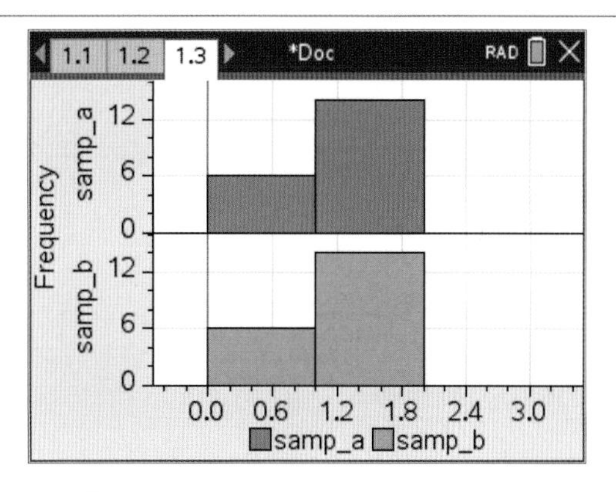

14 Tap **menu** > **Plot Properties** > **Add X Variable**.

15 Select **samp_b**.

16 The data for both samples will be displayed as histograms.

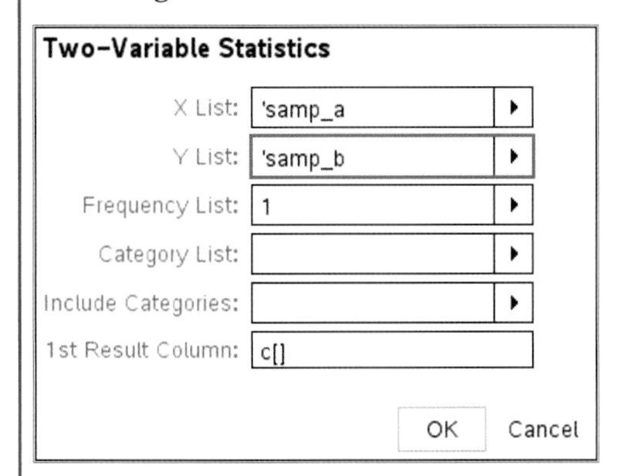

17 Return to the **Lists & Spreadsheet** page.

18 Press **menu** > **Statistics** > **Stat Calculations** > **Two-Variable**.

19 In the **X List**: field, press the right arrow and select **samp_a**.

20 In the **Y List**: field, press the right arrow and select **samp_b**.

21 Press **enter**.

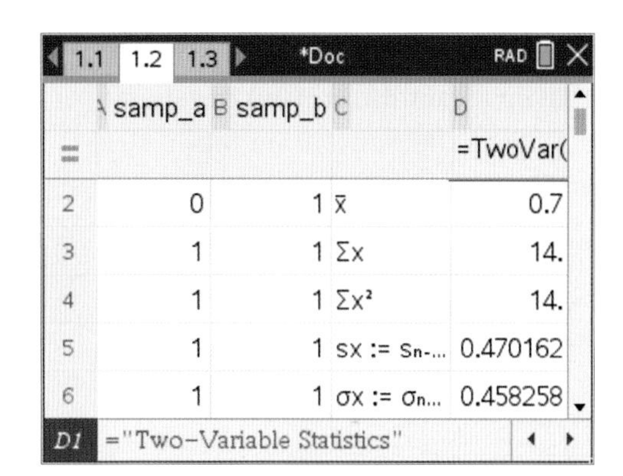

22 The summary statistics will be displayed in columns **C** and **D**.

23 Scroll down to view all the statistics.

$E(Z) = 0.8$. Both simulated means are approximately 0.8, but vary.

$SD(Z) = 0.4$. Both simulated standard deviations are close to 0.4, but vary.

USING CAS 4 Simulating sample data from a binomial distribution

Simulate two different samples of 50 scores from the discrete binomial random variable, $Z \sim \text{Bin}(16, 0.4)$. Compare the mean and standard deviation of the samples to the mean and standard deviation of Z.

ClassPad

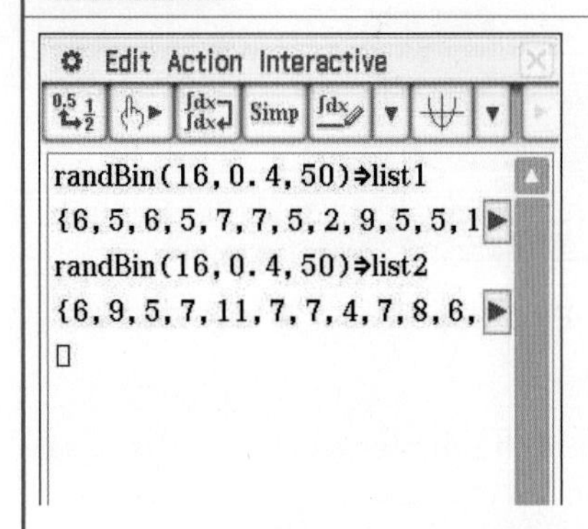

1 Open the **Keyboard** > **Catalog**.

2 Tap **R** then scroll down to select **randBin**.

3 Use the input **randBin(*n*, *p*, *m*)** to generate *m* values from the distribution $\text{Bin}(n, p)$, as shown above.

4 Store the first sample as **list1**.

5 Repeat for the second sample and store as **list2**.

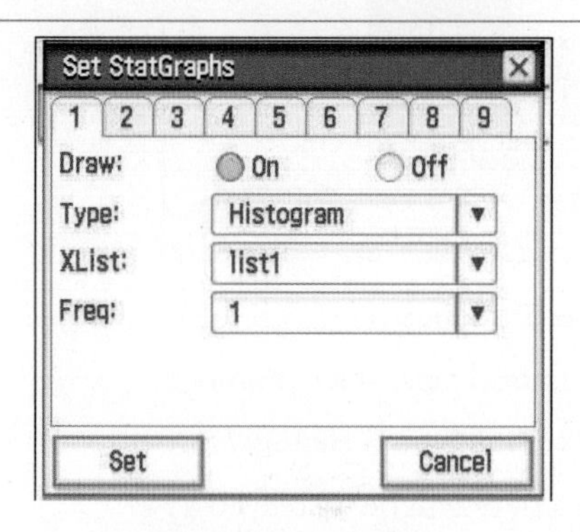

6 Tap **Menu** > **Statistics**.

7 Tap **SetGraph** > **Setting**.

8 For tab **1**, ensure the **Type**: field is **Histogram** and the **XList**: field is **list1**.

9 For tab **2**, ensure the **Type**: field is **Histogram** and the **XList**: field is **list2**.

10 Tap **Set**.

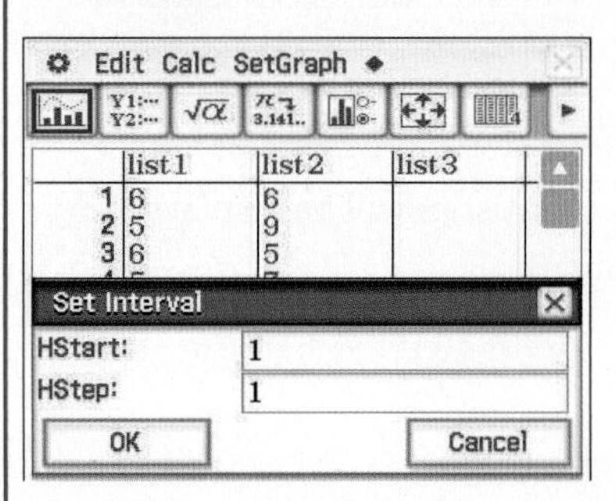

11 Tap **SetGraph** and ensure **StatGraph1** and **StatGraph2** are selected.

12 Tap **Graph**.

13 In the **Set Interval** dialogue box, change the **HStart**: and **HStep**: fields to **1** and tap **OK**.

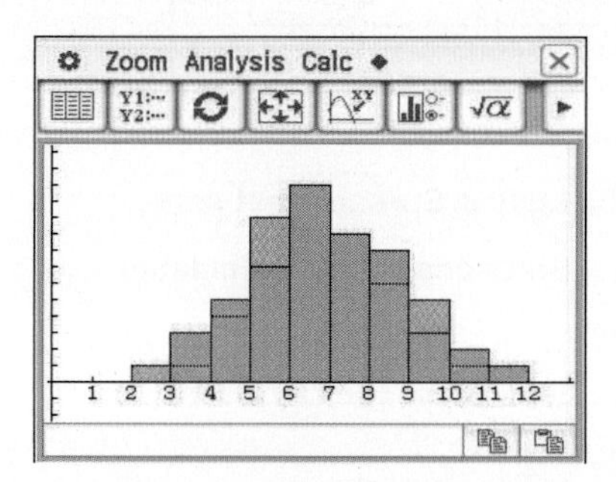

14 The histograms of the data will appear in the lower window (the windows have been swapped for this screen).

15 Tap the upper window to select **Statistics**.

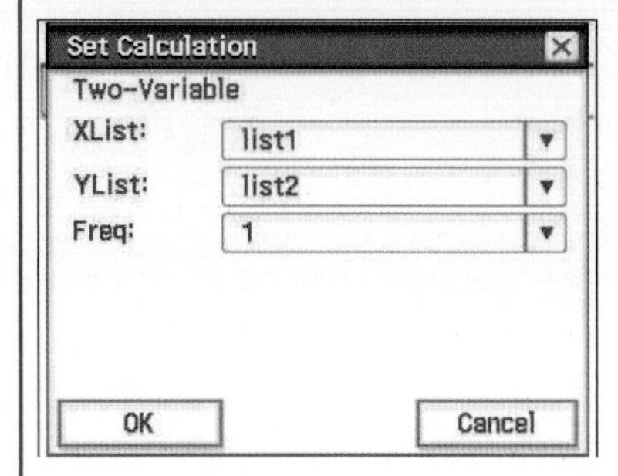

16 Tap **Calc** > **Two-Variable**.

17 Ensure the **XList**: field is **list1** and the **YList**: field is **list2**.

18 Tap **OK**.

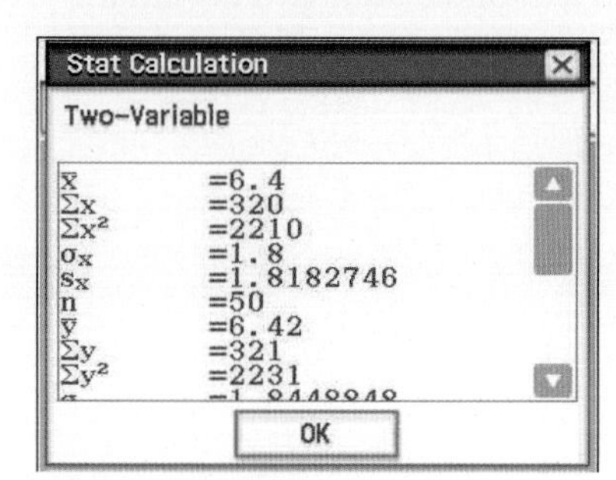

19 The summary statistics will appear in the **Two-Variable** window.

20 Scroll down to view all the statistics.

$\text{E}(Z) = 6.4$. Both simulated means are approximately 6.4, but vary.

$\text{SD}(Z) = 1.96$. Both simulated standard deviations are approximately 2, but vary.

TI-Nspire

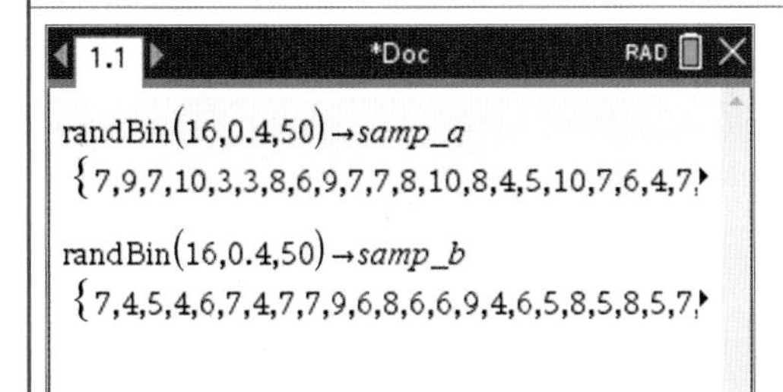

1 Press **catalog** > **randNorm**.

2 Use the input **randBin(1, *p*, *m*)** to generate *m* values from the distribution Bin(1, *p*), as shown above.

3 Store the first sample as **samp_a**.

4 Repeat for the second sample and store as **samp_b**.

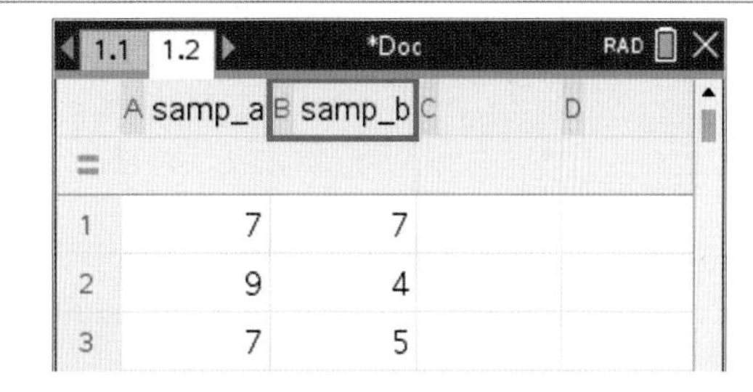

5 Add a **Lists & Spreadsheet** page.

6 Tap in the cell next to the **A**.

7 Press **var** and select **samp_a**.

8 Tap in the cell next to the **B**.

9 Press **var** and select **samp_b**.

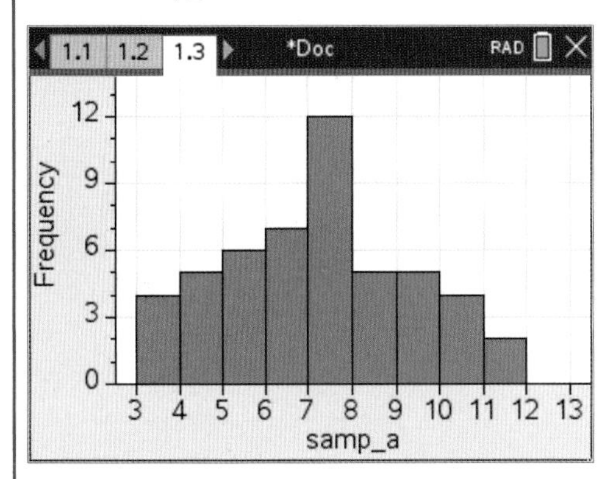

10 Add a **Data & Statistics** page.

11 For the horizontal axis, select **samp_a**.

12 Press **menu** > **Plot Type** > **Histogram**.

13 The sample a data will be displayed as a histogram.

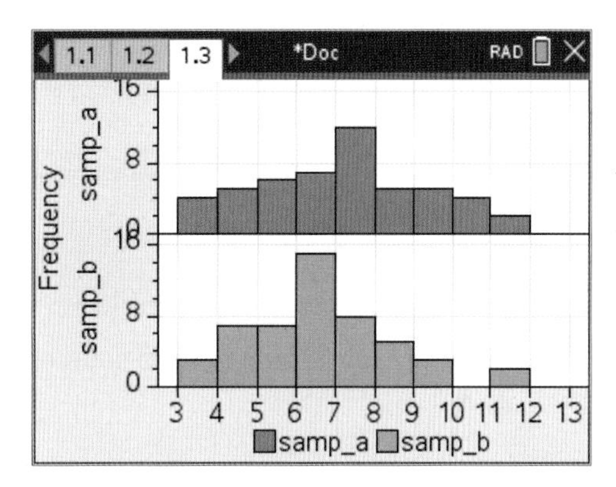

14 Tap **menu** > **Plot Properties** > **Add X Variable**.

15 Select **samp_b**.

16 The data for both samples will be displayed as histograms.

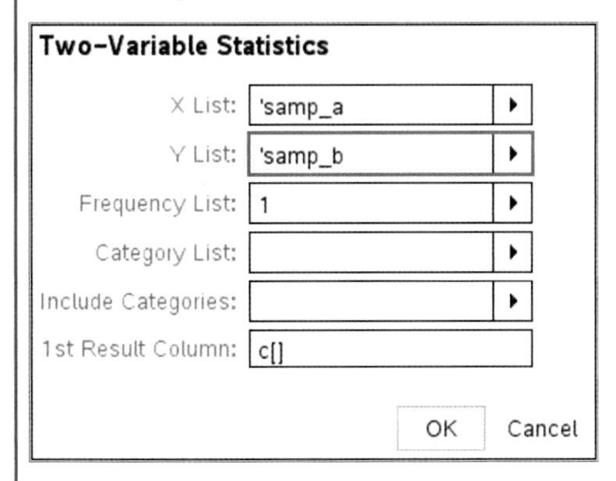

17 Return to the **Lists & Spreadsheet** page.

18 Press **menu** > **Statistics** > **Stat Calculations** > **Two-Variable**.

19 In the **X List**: field, press the right arrow and select **samp_a**.

20 In the **Y List**: field, press the right arrow and select **samp_b**.

21 Press **enter**.

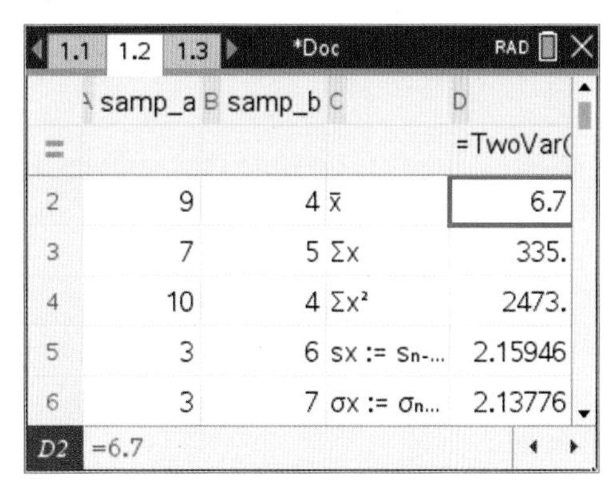

22 The summary statistics will be displayed in columns **C** and **D**.

23 Scroll down to view all the statistics.

E(Z) = 6.4. Both simulated means are approximately 6.4, but vary.

SD(Z) = 1.96. Both simulated standard deviations are approximately 2, but vary.

Variability of samples and as $n \to \infty$

For each sample of size n taken from a parent distribution defined by the random variable X, the sample statistics and shape of the distribution will vary, but will approximate the parameters and shape of the population distribution.

As $n \to \infty$:

- the mean of a sample will generally tend towards $E(X)$, but can still vary
- the shape of the distribution will better represent the shape of the distribution of X.

EXERCISE 9.1 Random sampling

ANSWERS p. 409

Mastery

1 WORKED EXAMPLE 1 The quality control officer at a local pie factory recorded the weights of six meat pies produced. The weights, to the nearest gram, were 110 g, 105 g, 110 g, 98 g, 101 g and 102 g. Identify the

a population **b** sample size **c** sample statistics.

2 WORKED EXAMPLE 2 A hotel wishes to conduct a survey regarding the quality of room service to be rated on a scale from 1 – Very Poor to 5 – Excellent. Describe a sampling procedure that would ensure randomness.

3 WORKED EXAMPLE 3 For each of the following situations

i identify and explain **one** possible source of bias with this sampling method

ii suggest and describe a random sampling procedure that will minimise bias in this data collection.

a The first 30 people that arrive at the Perth Domestic Airport to catch flights are asked the question 'How many times so far this year have you travelled interstate?'.

b On a Sunday afternoon from 2:00 pm to 3:00 pm, a supermarket store worker asks customers in the self-service checkouts the question 'How many bags have you brought with you today?'.

4 Using CAS 1 Simulate two different samples of 80 scores from the continuous uniform random variable, $X \sim U[15, 75]$. Compare the mean and standard deviation of the samples to the mean and standard deviation of X.

5 Using CAS 2 Simulate two different samples of 50 scores from the continuous normal random variable, $Y \sim N(40, 2.5^2)$. Compare the mean and standard deviation of the samples to the mean and standard deviation of Y.

6 Using CAS 3 Simulate two different samples of 30 scores from the discrete Bernoulli random variable, $Z \sim \text{Bern}(0.35)$. Compare the mean and standard deviation of the samples to the mean and standard deviation of Z.

7 Using CAS 4 Simulate two different samples of 40 scores from the discrete binomial random variable, $T \sim \text{Bin}\left(75, \frac{1}{3}\right)$. Compare the mean and standard deviation of the samples to the mean and standard deviation of T.

8 © SCSA MM2017 Q12a (4 marks) The Slate Tablet Company produces a variety of electronic tablets. It wants to gather information on consumers' interest in its tablets. In each of the following cases, comment, giving reasons, whether or not the proposed sampling method introduces bias.

a A Slate Tablet Company representative stood outside an electronics store on a Saturday morning and asked people entering the store 'If you were to purchase an electronic tablet would you choose a Slate Tablet or an inferior brand?' (2 marks)

b Fifteen hundred randomly selected mobile phone numbers were telephoned and people were asked 'Which brand of electronic tablet do you prefer?' (2 marks)

9 © SCSA MM2018 Q17c (2 marks) Tina believes that approximately 60% of the mangoes she produces on her farm are large. She takes a random sample of 500 mangoes from a day's picking. Tina decides to select the mangoes for her sample as they pass along the conveyor belt to be sorted. Describe briefly how Tina should select her sample.

10 © SCSA MM2019 Q13bc (4 marks) The proportion of working adults who miss breakfast on week days is estimated to be 40%. Tom takes a random sample of 400 adults to investigate this theory. He obtained his sample by selecting the first 400 workers he met in a busy mall in Perth city during lunchtime.

a Discuss briefly **two** possible sources of bias in Tom's sample. (2 marks)

Amir suggests that a better sampling scheme is to obtain a random sample of 400 voters and contact them by telephone.

b Outline **one** source of bias in Amir's sampling scheme. (1 mark)

c Which of Tom's or Amir's sampling scheme is better? Provide a reason for your choice. (1 mark)

11 © SCSA MM2020 Q14c (4 marks) A suburban council hires a consultant to estimate the proportion of residents of the suburb who use its library. The consultant decides to select the sample by standing on the roadside outside the library at lunchtime and asking a random sample of the passers-by whether they use the library. Identify and explain **two** possible sources of bias with this sampling scheme.

Calculator-assumed

12 (9 marks) CAS is used to simulate three different samples of varying sample sizes, as seen in the histograms provided.

Sample A

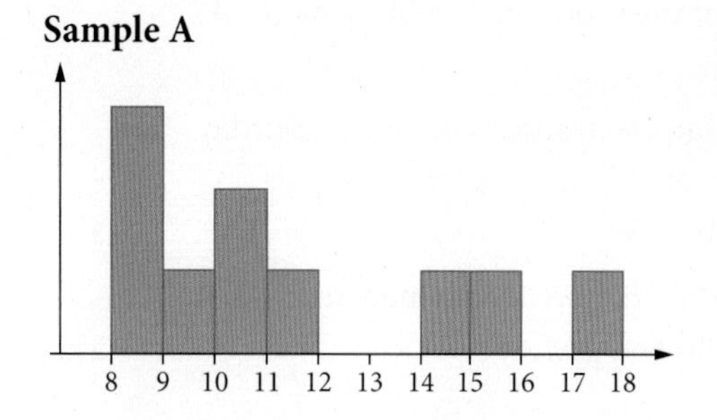

Sample B

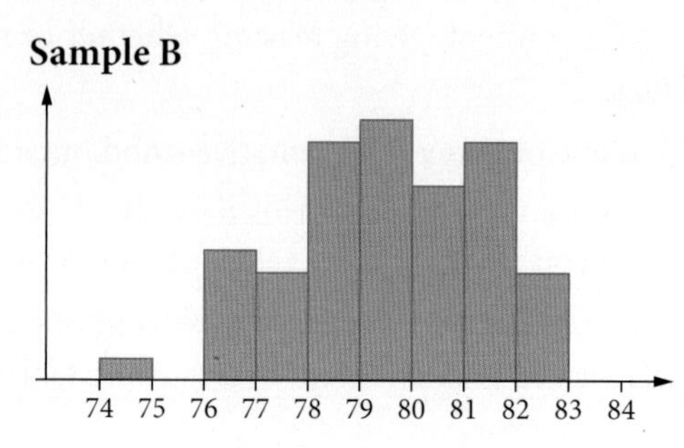

Sample C

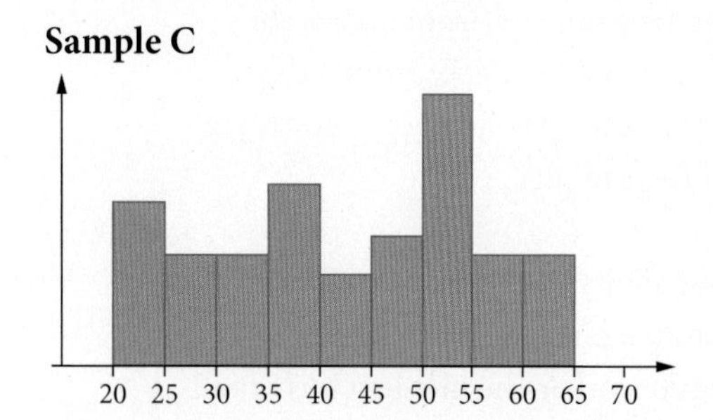

a Explain why it is not likely that these three samples came from the same population. (2 marks)

It is later known that the samples were simulated from a normal random variable X, a uniform random variable Y and a binomial random variable Z.

b **i** Identify the sample that is most likely simulated from the binomial random variable. (1 mark)

ii Justify whether the value of p in $Z \sim \text{Bin}(n, p)$ is less than or greater than 0.5. (2 marks)

c **i** Identify the sample that is most likely simulated from the normal random variable. (1 mark)

ii Estimate the value of $\text{E}(X)$. (1 mark)

d **i** Identify the sample that is most likely simulated from the uniform random variable. (1 mark)

ii Suppose the sample size simulated from this variable doubled. Describe the likely effect on the shape of the distribution. (1 mark)

Video playlist The sampling distribution of sample proportions

9.2 The sampling distribution of sample proportions

Worksheets Sample proportions

Sample proportion calculations

Sample proportion as a random variable, $\hat{p}$

Suppose that a random sample of 500 Australians was taken, and the eye colour of each individual was noted. Now imagine in this sample, it was found that 140 of the 500 had blue eyes. Then it can be said that the **sample proportion** of Australians with blue eyes is $\frac{140}{500}$ or 0.28 or 28%. This sample proportion, denoted as $\hat{p}$, is the relative frequency or experimental probability of a particular 'success condition' out of the sample size n.

Sample proportion as a value

For a sample of size n, the value of the single sample proportion is given by:

$$\hat{p} = \frac{\text{number of observed successes}}{n}$$

Sample proportions can be written as fractions, decimals or percentages.

9780170477536

Now suppose another random sample of 500 Australians found that 108 had blue eyes. Then the sample proportion of Australians with blue eyes for this sample is $\frac{108}{500} = 0.216 = 21.6\%$. As we have previously seen, due to the nature of random sampling, we can expect there to be variability between samples and so we would expect the proportion of Australians with blue eyes to change from sample to sample.

9.2

The purpose of collecting different random samples and examining the sample proportion of Australians with blue eyes is to find a way to estimate the true **population proportion**, p, of Australians with blue eyes because it is impractical and largely impossible to take a census of eye colour of the entire Australian population. As a result, each sample proportion $\hat{p}$ acts as a point estimate for p, which we assume remains constant and does not change like the sample proportions.

WORKED EXAMPLE 4 Calculating and using sample proportions

A school has a population of 1080 students. A random sample of 200 students was taken and it was found that 43 were not born in Australia.

a Calculate the sample proportion of students born overseas.

b Hence, use this sample proportion as a point estimate to estimate the total number of students born overseas.

Steps	Working
a Express the number of students born overseas as a fraction of the sample size.	$\hat{p} = \frac{43}{200} = 0.215 = 21.5\%$
b 1 Use the proportion to estimate for a population of size 1080.	$0.215 \times 1080 = 232.2$
2 Round to the nearest whole.	It is estimated that 232 students from this school population were born overseas.

In each sample of 500 Australians, we can consider each Australian a Bernoulli random variable or a Bernoulli trial, X_i, for $1 \le i \le 500$ where $x_i = 1$ means that the ith Australian surveyed has blue eyes and $x_i = 0$ means that the ith Australian surveyed does not have blue eyes. The distribution can be written as $X_i \sim \text{Bern}(p)$ where p is the probability of success of an Australian having blue eyes. This is the population proportion that is currently unknown. We also know that $\text{E}(X_i) = p$ and $\text{Var}(X_i) = p(1 - p)$.

Now when asking each of the 500 Australians, if we assume that each of the trials, $X_1, X_2 \ldots X_{500}$ are identically and independently distributed Bernoulli trials, then we have a binomially distributed random variable X, which represents the number of Australians with blue eyes such that $X \sim \text{Bin}(n, p)$. We should also remember that $\text{E}(X) = np$ and $\text{Var}(X) = np(1 - p)$.

So, the sample proportion $\hat{p}$ is a random variable that can change from sample to sample, such that the number of observed successes out of a sample of size n is determined by a binomially distributed random variable, X.

Sample proportion as a random variable

Let $X \sim \text{Bin}(n, p)$, where n is the sample size and p is the probability of success (i.e. true population proportion). Then the random variable $\hat{p}$ is defined as

$$\hat{p} = \frac{X}{n}$$

and represents the set of all possible sample proportions that can exist when a sample of size n is taken. It is considered a random variable because its outcomes are the result of a probability experiment.

Imagine we now took lots of different samples of 500 Australians and calculated the sample proportion of Australians with blue eyes for each sample. The distribution of all of the values of $\hat{p}$ is called the **sampling distribution of sample proportions**.

The mean, variance and standard deviation of $\hat{p}$

Given that $\hat{p} = \frac{X}{n}$, we can consider it a binomial random variable scaled by a factor of $\frac{1}{n}$. As a result, we can use the fact that for $X \sim \text{Bin}(n,p)$, $\text{E}(X) = np$ and $\text{Var}(X) = np(1-p)$ to deduce the expected value, variance and standard deviation of $\hat{p}$ as a random variable.

Expected value, variance and standard deviation of $\hat{p}$

Considering $\hat{p} = \frac{1}{n}X$ as a linear change of scale, then:

$$\begin{aligned}\text{E}(\hat{p}) &= \text{E}\left(\frac{1}{n}X\right)\\ &= \frac{1}{n}\text{E}(X)\\ &= \frac{np}{n}\\ &= p\end{aligned}$$

That is, if the population proportion is p, then the expected value (or mean) of all the sample proportions taken from different samples of fixed size n, is p. That is, the mean of the sampling distribution of sample proportions is a good and unbiased estimator of the true population proportion because it is independent of n.

$$\begin{aligned}\text{Var}(\hat{p}) &= \text{Var}\left(\frac{1}{n}X\right)\\ &= \frac{1}{n^2}\text{Var}(X)\\ &= \frac{np(1-p)}{n^2}\\ &= \frac{p(1-p)}{n}\end{aligned}$$

$$\text{SD}(\hat{p}) = \sqrt{\frac{p(1-p)}{n}}$$

As sample size increases, i.e. $n \to \infty$, $\text{Var}(\hat{p}) \to 0$ and $\text{SD}(\hat{p}) \to 0$. That is, there should be very little variation in the different values of $\hat{p}$ taken from different samples of a significantly large, fixed size n.

WORKED EXAMPLE 5 Expected value, variance and standard deviation of $\hat{p}$ from X with a known p

Given that $X \sim \text{Bin}(50, 0.6)$, determine

a $\text{E}(\hat{p})$ **b** $\text{Var}(\hat{p})$ **c** $\text{SD}(\hat{p})$, correct to three decimal places.

Steps	Working
a Use the fact that $\text{E}(\hat{p}) = p$.	$\text{E}(\hat{p}) = 0.6$
b Use the formula $\text{Var}(\hat{p}) = \frac{p(1-p)}{n}$.	$\text{Var}(\hat{p}) = \frac{0.6(0.4)}{50} = 0.0048$
c Use the formula $\text{SD}(\hat{p}) = \sqrt{\text{Var}(\hat{p})}$.	$\text{SD}(\hat{p}) = \sqrt{0.0048} = 0.069$

Let's now revisit the context of Australians with blue eyes. The problem in this example is that although we know the value of $n = 500$, we do not know the true value of p. That is, we do not know the true probability of success of selecting someone from the Australian population with blue eyes. This type of situation whereby p is unknown is most common.

In these situations, we have to use a sample proportion $\hat{p}$ from one specific random sample, hopefully fair and representative, as a point estimate for p. For example, suppose we let the sample proportion obtained from the first sample of 500 Australians, $\hat{p} = 0.28$, be a point estimate for the true value of p. Then we assume that $X \sim \text{Bin}(n, \hat{p})$.

It then follows that the sampling distribution of sample proportions $\hat{p}$ will have:

$$\text{E}(\hat{p}) = \hat{p}$$

$$\text{Var}(\hat{p}) = \frac{\hat{p}(1-\hat{p})}{n}$$

$$\text{SD}(\hat{p}) = \sqrt{\frac{\hat{p}(1-\hat{p})}{n}}$$

So, for our example where $X \sim \text{Bin}(500, 0.28)$, we will have:

$$\text{E}(\hat{p}) = 0.28$$

$$\text{Var}(\hat{p}) = \frac{0.28(0.72)}{500} = 0.000\,4032$$

$$\text{SD}(\hat{p}) = \sqrt{0.000\,4032} = 0.0201$$

WORKED EXAMPLE 6 **Expected value, variance and standard deviation of $\hat{p}$ from X with an unknown p**

From a sample of 20 people, it was found that 8 had colds last winter.

a State the sample proportion of people who had colds last winter, $\hat{p}$.

b Using this value of $\hat{p}$ as a point estimate for the true population proportion of people who had colds last winter, determine

i $\text{E}(\hat{p})$ **ii** $\text{Var}(\hat{p})$ **iii** $\text{SD}(\hat{p})$, correct to three decimal places.

Steps	Working
a Express the number of people who had colds last winter as a proportion of the sample size.	$\hat{p} = \frac{8}{20} = 0.4 = 40\%$
b i Use the fact that $\text{E}(\hat{p}) = \hat{p}$.	$\text{E}(\hat{p}) = 0.4$
ii Use the formula $\text{Var}(\hat{p}) = \frac{\hat{p}(1-\hat{p})}{n}$.	$\text{Var}(\hat{p}) = \frac{0.4(0.6)}{20} = 0.012$
iii Use the formula $\text{SD}(\hat{p}) = \sqrt{\text{Var}(\hat{p})}$.	$\text{SD}(\hat{p}) = \sqrt{0.012} = 0.110$

WORKED EXAMPLE 7 **Using standard deviation of $\hat{p}$ to find an unknown parameter**

In a large population of fish, the true population proportion of angel fish is known to be $\frac{1}{4}$. Let $\hat{p}$ be the random variable that represents the sample proportion of angel fish for samples of fixed size n taken from the population. Find the smallest integer value of n such that the standard deviation of $\hat{p}$ is less than or equal to 0.01.

Steps	Working
1 Establish an inequality involving $\text{SD}(\hat{p})$ using the given values.	$\sqrt{\frac{\frac{1}{4}\left(\frac{3}{4}\right)}{n}} \le 0.01$
2 Solve for n using CAS. (See CAS screens on following page.)	$n \ge 1875$

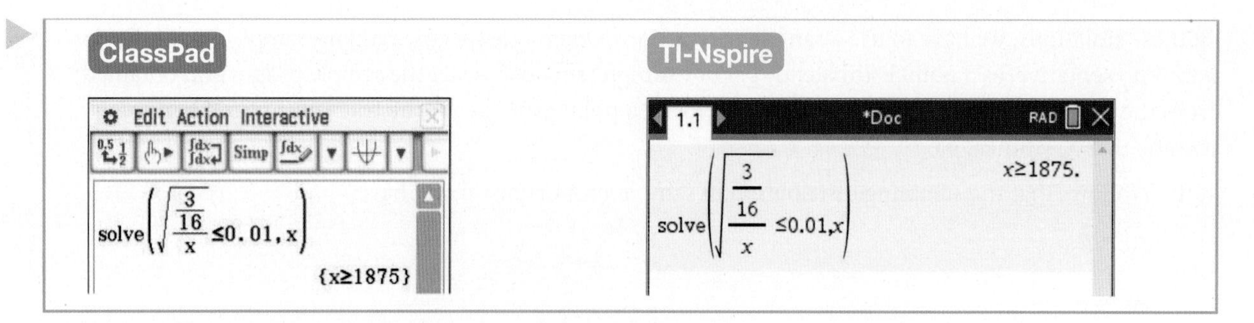

Alone, these calculations involving $E(\hat{p})$, $Var(\hat{p})$ and $SD(\hat{p})$ are just statistics and we cannot do much with them until we know more about the shape of the sampling distribution of sample proportions, $\hat{p}$.

Approximate normality and the central limit theorem

From your work with binomial random variables in Chapter 5, you will recall that the shape of the frequency histogram of a binomial distribution depends on the values of n and p. We can use CAS to simulate binomial distributions with a fixed n value and differing p values to observe the effect.

USING CAS 5 Simulating binomial distributions

Simulate 50 different observations from each of the following binomial random variables, graphing each of the results.

a $X \sim \text{Bin}(10, 0.2)$ **b** $X \sim \text{Bin}(10, 0.5)$ **c** $X \sim \text{Bin}(10, 0.8)$

ClassPad

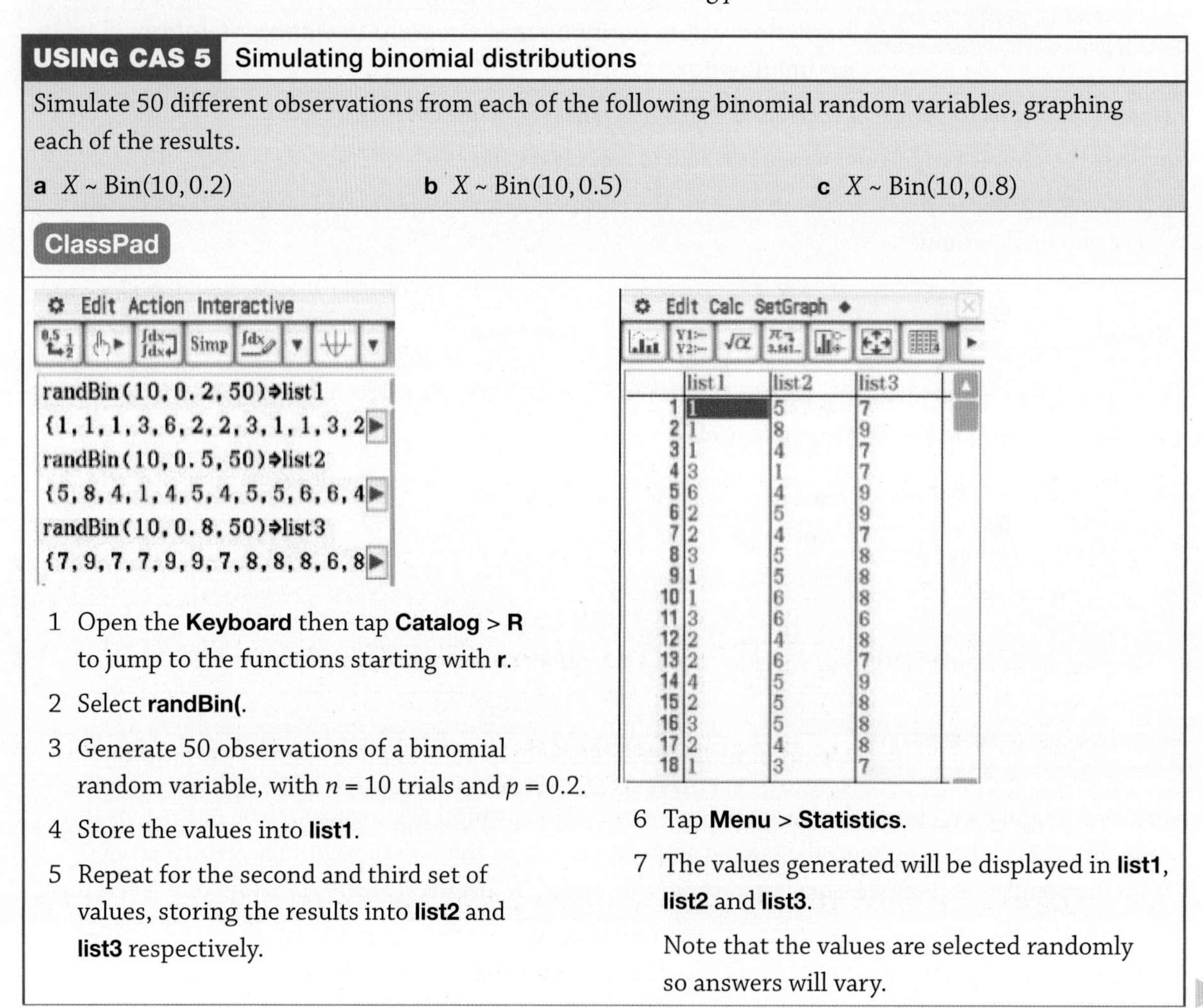

	list1	list2	list3
1	1	5	7
2	1	8	9
3	1	4	7
4	3	1	7
5	6	4	9
6	2	5	9
7	2	4	7
8	3	5	8
9	1	5	8
10	1	6	8
11	3	6	6
12	2	4	8
13	2	6	7
14	4	5	9
15	2	5	8
16	3	5	8
17	2	4	8
18	1	3	7

1 Open the **Keyboard** then tap **Catalog** > **R** to jump to the functions starting with **r**.
2 Select **randBin(**.
3 Generate 50 observations of a binomial random variable, with $n = 10$ trials and $p = 0.2$.
4 Store the values into **list1**.
5 Repeat for the second and third set of values, storing the results into **list2** and **list3** respectively.
6 Tap **Menu** > **Statistics**.
7 The values generated will be displayed in **list1**, **list2** and **list3**.

Note that the values are selected randomly so answers will vary.

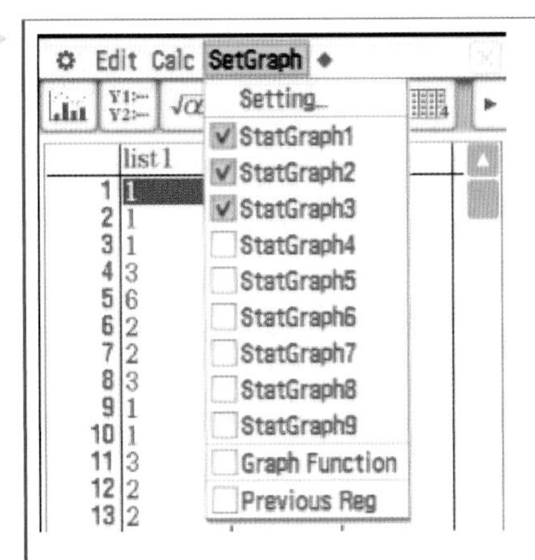

8 Tap **SetGraph**.

9 Tap to select **StatGraph1**, **StatGraph2** and **StatGraph3**, as shown above. Tick or untick these to view individual graphs.

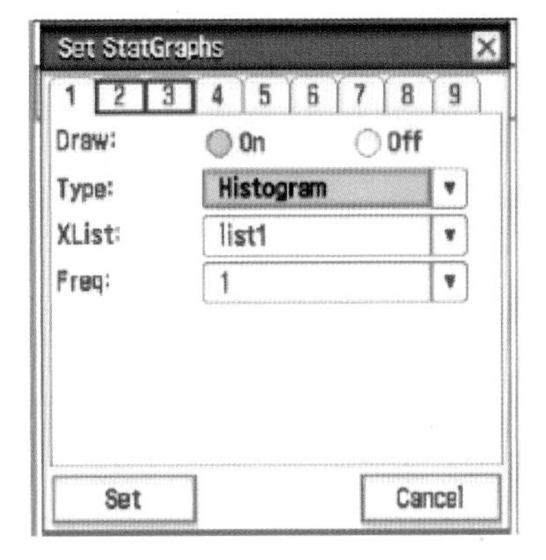

10 Tap **SetGraph** > **Setting**.

11 For the **Type:** field, select **Histogram**.

12 Tap tab **2** at the top of the page.

13 Set the **Type**: field to **Histogram**.

14 Change the **XList**: field to **list2**.

15 Repeat for tab **3** by selecting **Histogram** and changing the **XList**: field to **list3**.

16 Tap **Set**.

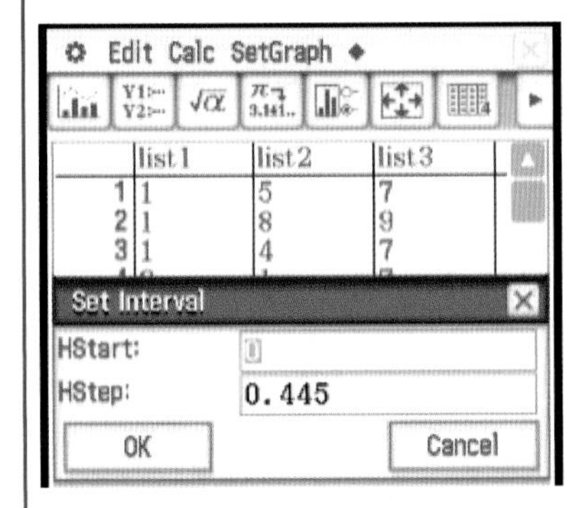

17 Tap **Graph**.

18 When the **Set Interval** dialogue box is displayed, tap **OK** to accept the default settings.

NOTE: You can change the HStart and HStep if you want to view the histograms on a different scale. For example, try a HStep of 1.

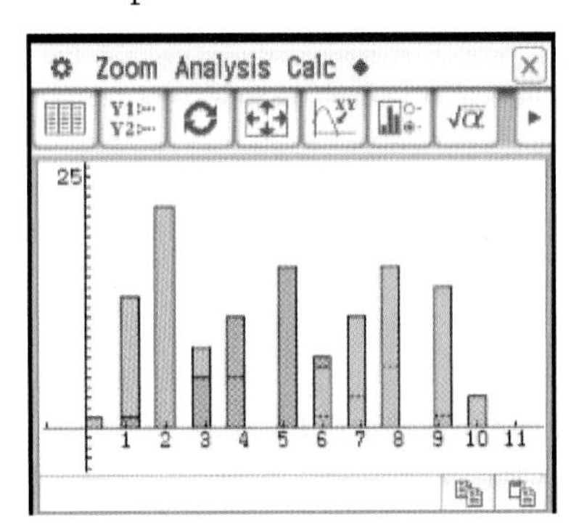

19 Histograms of the three random samples will be displayed.

20 Compare the alignment of the histograms with their respective probabilities.

StatGraph1 – purple

StatGraph2 – orange

StatGraph3 – green

TI-Nspire

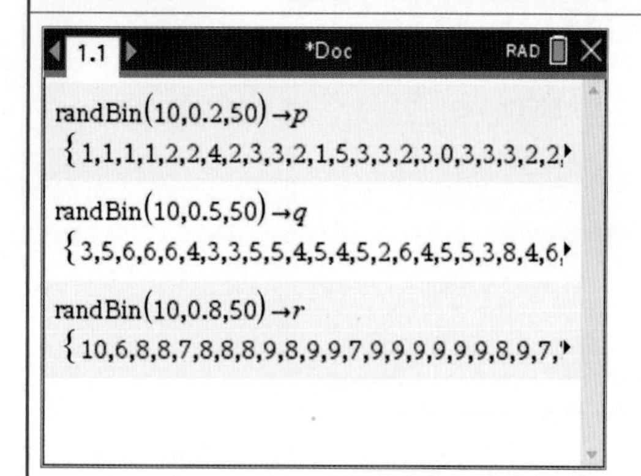

1. Press **menu** > **Probability** > **Random** > **Binomial**.
2. Generate 50 observations of a binomial random variable, with $n = 10$ trials and $p = 0.2$.
3. Press **ctrl** + **var** to store the result in **p**.
4. Repeat for the second and third set of values, storing the results in **q** and **r** respectively.

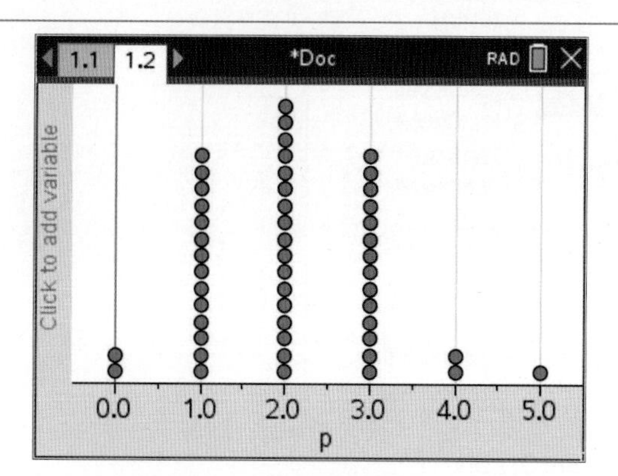

5. Add a **Data & Statistics** page.
6. For the horizontal axis, click to select the variable **p**.
7. A dot plot of the probabilities will be displayed.

   Note that the values are selected randomly so answers will vary.

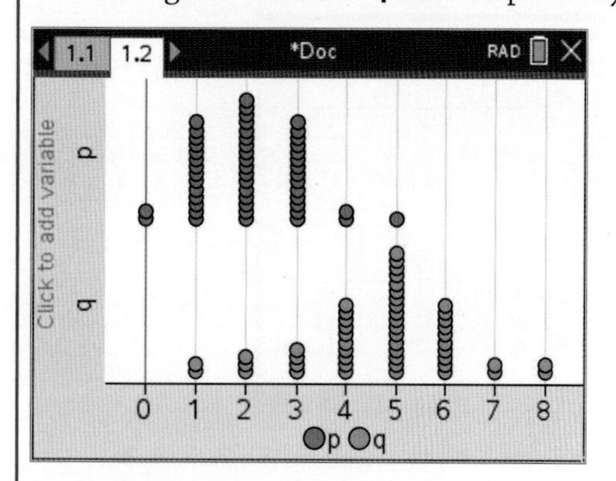

8. Press **menu** > **Plot Properties** > **Add X Variable**.
9. Select the variable **q**.
10. Parallel dot plots for **p** and **q** will be displayed.

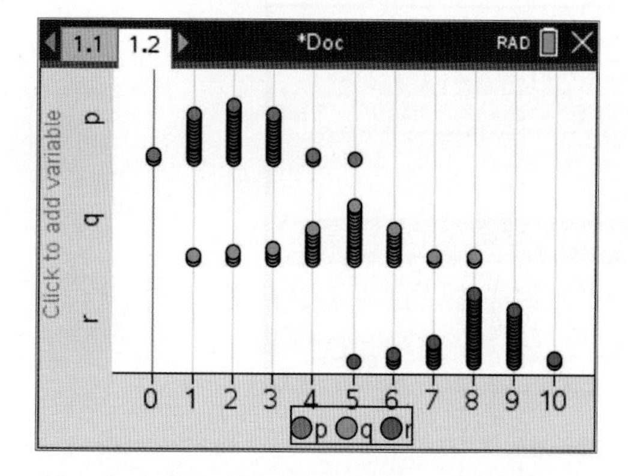

11. Repeat to display the dot plot for the variable **r**.
12. All three dot plots will be displayed.
13. Compare the alignment of the dot plots with their respective probabilities.

From these simulations, we should notice something about the shape of each of the distributions.

- For $X \sim \text{Bin}(10, 0.2)$, the distribution was positively skewed, with an expected value (mean) of $10(0.2) = 2$.
- For $X \sim \text{Bin}(10, 0.5)$, the distribution was approximately symmetrical, with an expected value (mean) of $10(0.5) = 5$.
- For $X \sim \text{Bin}(10, 0.8)$, the distribution was negatively skewed, with an expected value (mean) of $10(0.8) = 8$.

We can also run a simulation exercise for sample proportions by dividing each of the simulated observations from the binomial random variables by the value of n.

USING CAS 6 Simulating the sampling distributions of sample proportions

Simulate 30 different observations from each of the following binomial random variables, and graph the distributions of each of the corresponding distributions of $\hat{p} = \frac{X}{n}$.

a $X \sim \text{Bin}(20, 0.35)$ **b** $X \sim \text{Bin}(20, 0.5)$ **c** $X \sim \text{Bin}(20, 0.75)$

ClassPad

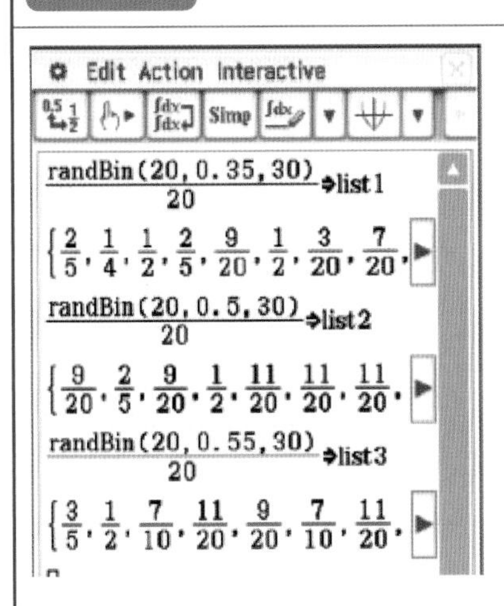

1 Open the **Keyboard** then tap **Catalog** > **R** to jump to the functions starting with r.

2 Select **randBin(**.

3 Generate 50 observations of a binomial random variable, with $n = 10$ trials and $p = 0.35$, and divide the set by 20.

4 Store the values into **list1**.

5 Repeat for the second and third set of values, storing the results into **list2** and **list3** respectively.

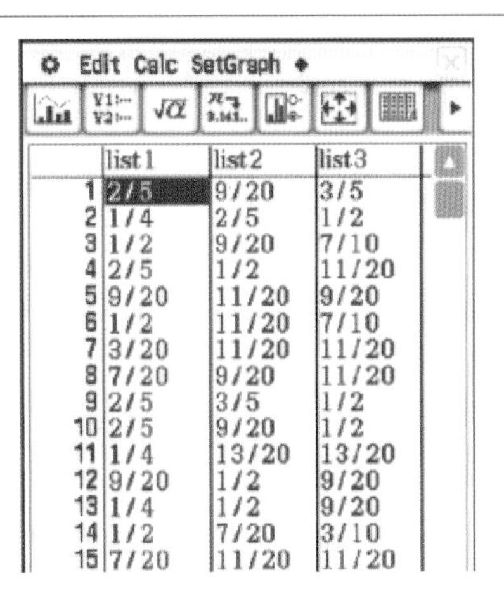

6 Tap **Menu** > **Statistics**.

7 The values generated will be displayed in **list1**, **list2** and **list3**.

Note that the values are selected randomly so answers will vary.

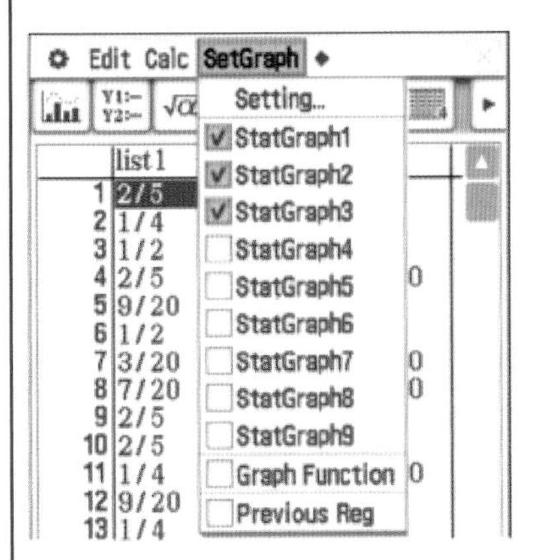

8 Tap **SetGraph**.

9 Tap to select **StatGraph1**, **StatGraph2** and **StatGraph3** as shown above. Tick or untick these to view individual graphs.

10 Tap **SetGraph** > **Setting**.

11 For the **Type:** field, select **Histogram**.

12 Tap tab **2** at the top of the page.

13 Set the **Type:** field to **Histogram**.

14 Change the **XList:** field to **list2**.

15 Repeat for tab **3** by selecting **Histogram** and changing the **XList:** field to **list3**.

16 Tap **Set**.

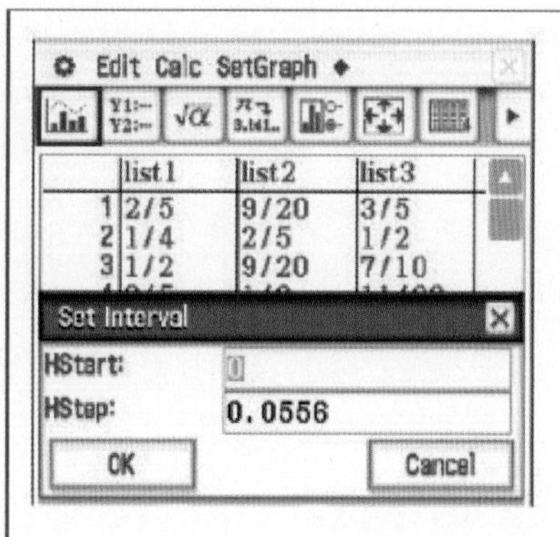

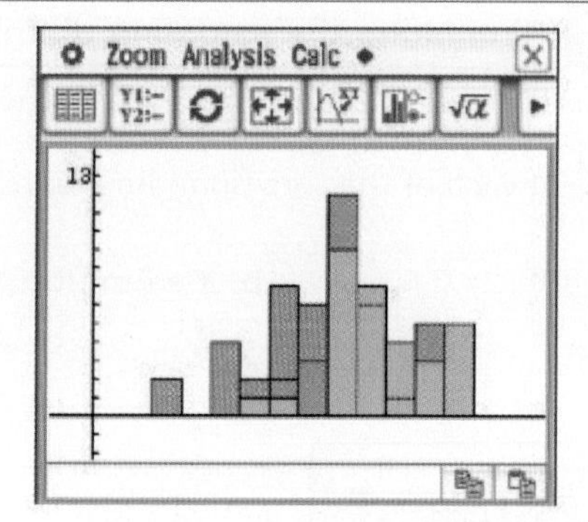

17 Tap **Graph**.

18 When the **Set Interval** dialogue box is displayed, tap **OK** to accept the default settings.

NOTE: You can change the HStart and HStep if you want to view the histograms on a different scale. For example, try a HStep of 1.

19 Histograms of the three random samples will be displayed.

20 Compare the alignment of the histograms with their respective probabilities.

StatGraph1 – purple

StatGraph2 – orange

StatGraph3 – green

TI-Nspire

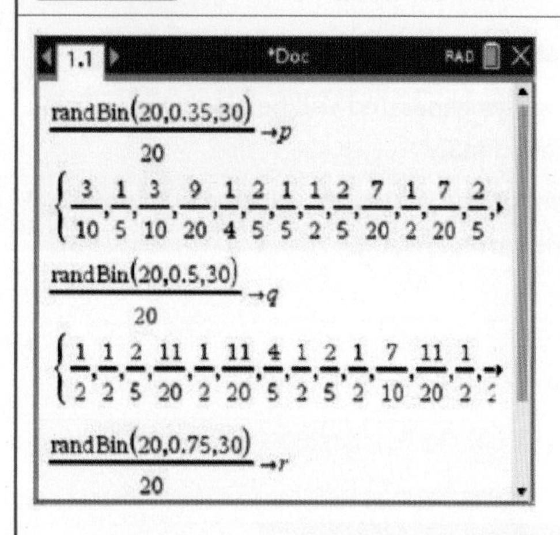

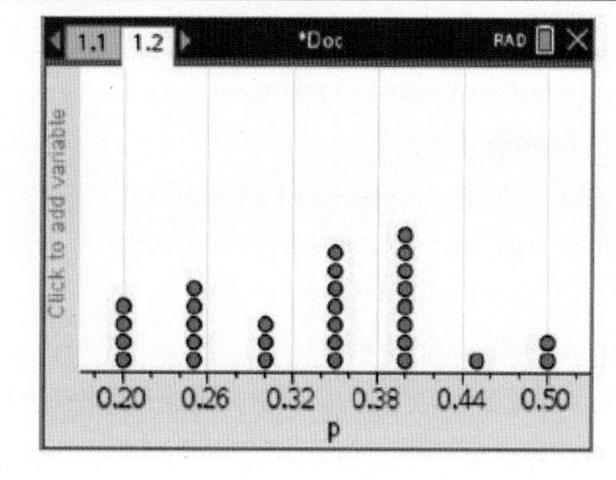

1 Press **menu** > **Probability** > **Random** > **Binomial**.

2 Generate 50 observations of a binomial random variable, with $n = 10$ trials and $p = 0.35$, and divide the set by 20.

3 Press **ctrl** + **var** to store the result in **p**.

4 Repeat for the second and third set of values, storing the results in **q** and **r** respectively.

5 Add a **Data & Statistics** page.

6 For the horizontal axis, click to select the variable **p**.

7 A dot plot of the probabilities will be displayed.

Note that the values are selected randomly so answers will vary.

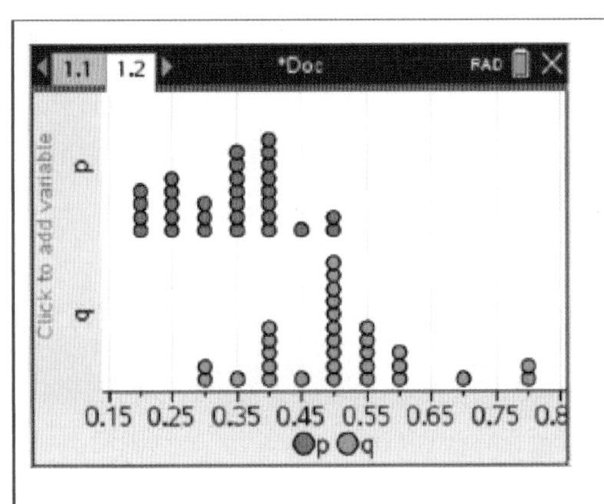

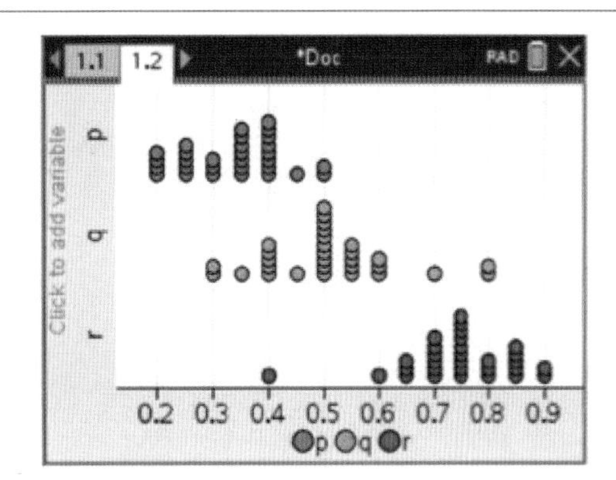

8 Press **menu** > **Plot Properties** > **Add X Variable**.

9 Select the variable **q**.

10 Parallel dot plots for **p** and **q** will be displayed.

11 Repeat to display the dot plot for the variable **r**.

12 All three dot plots will be displayed.

13 Compare the alignment of the dot plots with their respective probabilities.

The shape of each of the corresponding sampling distributions of sample proportions will be the same as the shape of the binomial distribution, but with a rescaled horizontal axis by a factor of $\frac{1}{n}$.

It should be more obvious that these histograms are approximately symmetrical about their expected values of 50, 125 and 200, respectively. Given the apparent symmetry about the mean, we could now consider it appropriate for each of these binomial random variables to be suitably modelled using a normally distributed random variable.

The crucial question is: *when is it appropriate to assume approximate normality*? Let's consider which of the simulations for $n = 10$ gave an approximately symmetrical distribution and hence could be appropriately modelled by a normal distribution.

- For $X \sim \text{Bin}(10, 0.2)$, $np = 2$ and the distribution was positively skewed due to $p = 0.2$. The normal distribution is not appropriate!
- For $X \sim \text{Bin}(10, 0.5)$, $np = 5$ and the distribution was approximately symmetrical due to $p = 0.5$. The normal distribution could be appropriate!
- For $X \sim \text{Bin}(10, 0.8)$, $np = 8$ and the distribution was negatively skewed due to $p = 0.8$. The normal distribution is not appropriate!

Suppose we repeat the above simulation exercises with a much larger sample size, say $n = 250$, but the same values of $p = 0.2$, 0.5 and 0.8. Running this simulation once for these three binomial random variables, we might observe three such histograms.

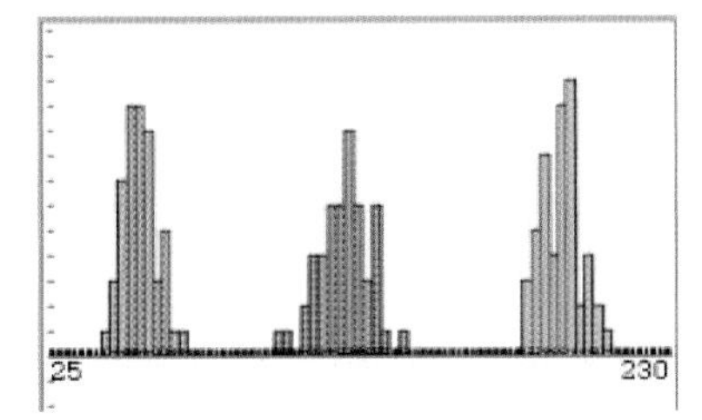

TI-Nspire

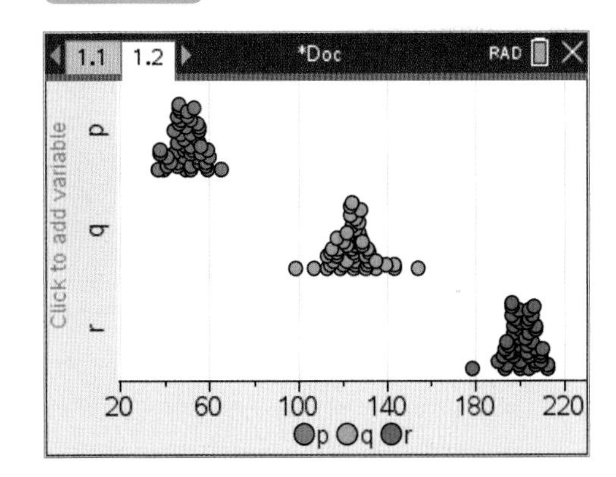

The approximate normality of a binomial distribution with $p \approx 0.5$

For a binomially distributed random variable $X \sim \text{Bin}(n,p)$, if $p \approx 0.5$, and n is sufficiently large, the values of x can be approximated by a normally distributed random variable X_N with parameters $\mu = np$ and $\sigma^2 = np(1-p)$. That is,

$$X_N \sim N(np, np(1-p))$$

Exam hack

In these cases, 'sufficiently large' is often considered as $n \geq 30$, but when p is approximately 0.5, some leniency can be given due to the symmetry of the distribution. Remember that you could always check three standard deviations either side of the p value to ensure that the normal approximation is still appropriate!

Now let's consider the cases where n was significantly larger.

- For $X \sim \text{Bin}(250, 0.2)$, $np = 50$ and the distribution was approximately symmetrical even though $p = 0.2$. The normal distribution could be appropriate!
- For $X \sim \text{Bin}(250, 0.5)$, $np = 125$ and the distribution was approximately symmetrical due to $p = 0.5$. The normal distribution could be appropriate!
- For $X \sim \text{Bin}(250, 0.8)$, $np = 200$ and the distribution was approximately symmetrical even though $p = 0.8$. The normal distribution could be appropriate!

The approximate normality of a binomial distribution with a sufficiently large n

For a binomially distributed random variable $X \sim \text{Bin}(n,p)$, if n is sufficiently large enough to cater for a value of p that deviates from 0.5, the values of x can be approximated by a normally distributed random variable X_N with parameters $\mu = np$ and $\sigma^2 = np(1-p)$. That is,

$$X_N \sim N(np, np(1-p)).$$

Some mathematicians like to put a minimum condition on the size of np and $n(1-p)$ instead of having a 'sufficiently large n'. A commonly used restriction is that both $np \geq 10$ and $n(1-p) \geq 10$. That is, there are at least 10 'successful' and 10 'failed' observations. This may vary in different texts.

As a result, when a binomially distributed random variable X is approximately normal, then it can be said the corresponding sampling distribution of sample proportions taken from X, $\hat{p} = \frac{X}{n}$, is also approximately normal.

The approximate normality of the sampling distribution of sample proportions $\hat{p}$

For a binomially distributed random variable $X \sim \text{Bin}(n,p)$ that can be approximated by a normally distributed random variable X_N with parameters $\mu = np$ and $\sigma^2 = np(1-p)$, then $\hat{p} = \frac{X_N}{n}$ is approximately normal such that:

$$\hat{p} \sim N\left(p, \frac{p(1-p)}{n}\right)$$

When a point estimate is used to estimate p, then

$$\hat{p} \sim N\left(\hat{p}, \frac{\hat{p}(1-\hat{p})}{n}\right)$$

A good way of checking whether n was sufficiently large is by ensuring that all values of $\hat{p}$ within three standard deviations of the mean (i.e. approximately 99.7% of $\hat{p}$ values) lie between $0 \leq \hat{p} \leq 1$, as we cannot have negative sample proportions, or sample proportions larger than 1.

The above result is also known as the **central limit theorem** for the sampling distribution of sample proportions.

WORKED EXAMPLE 8 Describing the sampling distribution of sample proportions

On a Friday afternoon, a random sample of 58 car spaces at Westfield shopping centre carpark were observed and it was found that 24 of the spaces were occupied by a car.

a State the sample proportion of car spaces that were occupied, $\hat{p}$, correct to four decimal places.

b Determine $E(\hat{p})$ and $SD(\hat{p})$, correct to four decimal places.

c Describe the distribution of the random variable $\hat{p}$, justifying your answer.

Steps	Working
a Express the number of occupied car spaces as a proportion of total car spaces observed.	$\hat{p} = \frac{24}{58} = 0.4138$
b **1** Assume the observed sample proportion is a point estimate for the value of p.	Let $\hat{p} = 0.4138$ be a point estimate of p.
2 Use the fact that $E(\hat{p}) = \hat{p}$.	$E(\hat{p}) = \frac{24}{58} = 0.4138$
3 Use the formula $SD(\hat{p}) = \sqrt{\frac{\hat{p}(1-\hat{p})}{n}}$.	$SD(\hat{p}) = \sqrt{\frac{\frac{24}{58}\left(\frac{34}{58}\right)}{58}} = 0.0647$
c **1** Identify the conditions on n and $\hat{p}$ for the distribution to be considered approximately normal. **2** Give the appropriate mathematical calculations supporting your reasons.	Use any combination of the following reasons • $\hat{p} \approx 0.5$ and n is sufficiently large • $np = 24 \geq 10$ and $n(1-p) = 34 \geq 10$ • $E(\hat{p}) - 3SD(\hat{p}) = 0.2198 > 0$ and $E(\hat{p}) + 3SD(\hat{p}) = 0.6078 < 1$.
3 State the distribution (i.e. normal) and its corresponding parameters.	By the central limit theorem, the distribution of $\hat{p}$ is approximately normal such that $\hat{p} \sim N(0.4138, 0.0647^2)$.

Exam hack

When $\hat{p}$ gives an answer that is a non-terminating decimal, unless you have been asked to round it to a certain number of decimal places, leave it in the exact fractional form. Regardless, you should use the fractional form or the full decimal value in your calculator to avoid any accuracy errors.

Probability problems involving $\hat{p}$

Worksheet
Sample proportion probabilities

If it is appropriate to model the sampling distribution of sample proportions by an **approximate normal distribution**, probability calculations can be carried out using the knowledge of normally distributed random variables from Chapter 8.

Note that some texts encourage the use of a **continuity adjustment/correction** to account for the fact that a binomially distributed discrete random variable is being approximated using a normally distributed continuous random variable. In these cases, a default value of 0.5 is added to or subtracted from the discrete integer bounds to 'correct' the error going from a discrete to continuous random variable.

Strictly speaking, this approach is not an expectation of our course, but instead you may be asked to compare the probability obtained when using the approximate normal distribution to the equivalent probability calculation using the binomial distribution.

WORKED EXAMPLE 9 Using an approximate normal distribution

From a random sample of 40 Year 12 students, 19 were found to have heights greater than 180 cm. Let $\hat{p}$ be the sampling distribution of sample proportions of Year 12 students with a height greater than 180 cm.

a Describe the distribution of $\hat{p}$, justifying your answer.

b Hence, determine the probability correct to four decimal places that in a randomly selected sample of 40 Year 12 students

i more than 25% of the students sampled will have a height greater than 180 cm

ii between 45% and 65% of the students sampled will have a height greater than 180 cm.

Steps	Working
a 1 Calculate $\hat{p}$ as a point estimate for p.	Let $\hat{p} = \frac{19}{40} = 0.475$ be a point estimate for p.
2 Identify the conditions on n and $\hat{p}$ for the distribution to be considered approximately normal.	Use either of the following reasons: • $\hat{p} \approx 0.5$ and n is sufficiently large • $np = 19 \geq 10$ and $n(1-p) = 21 \geq 10$.
3 Give the appropriate mathematical calculations supporting your reasons.	$\text{SD}(\hat{p}) = \sqrt{\frac{\frac{19}{40}\left(\frac{21}{40}\right)}{40}} = 0.0790$
4 State the distribution (i.e. normal) and its corresponding parameters.	By the central limit theorem, the distribution of $\hat{p}$ is approximately normal such that $\hat{p} \sim N(0.475, 0.0790^2)$.
b i 1 Interpret the question as a probability statement.	More than 25% means $\hat{p} > 0.25$.
2 Use CAS to calculate the probability correct to four decimal places.	$\text{P}(\hat{p} > 0.25) = 0.9978$

ClassPad

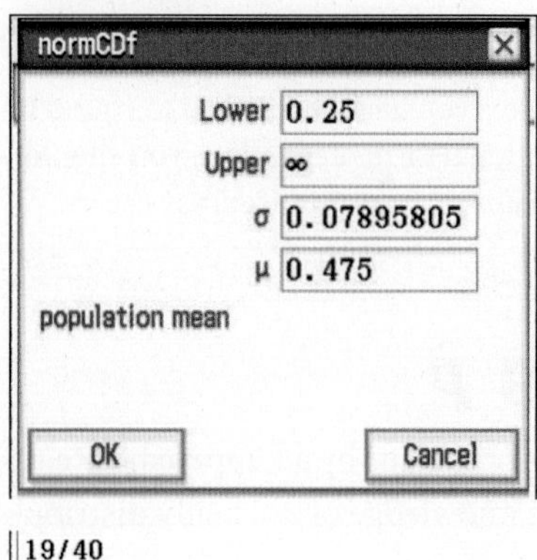

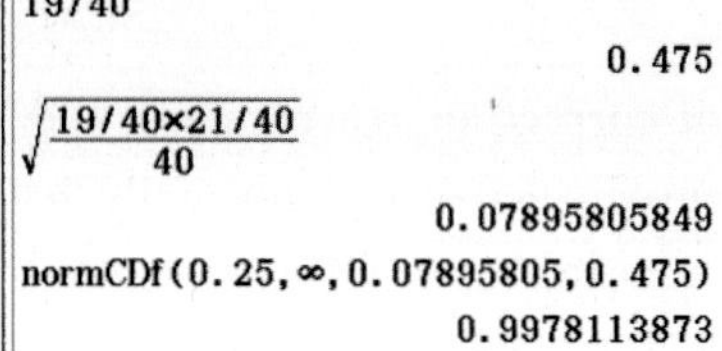

TI-Nspire

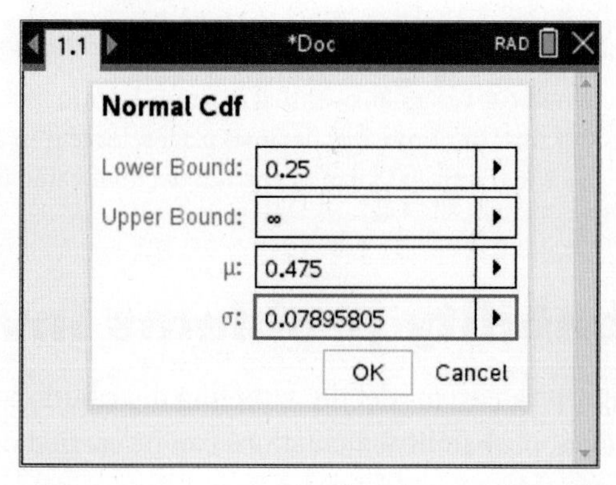

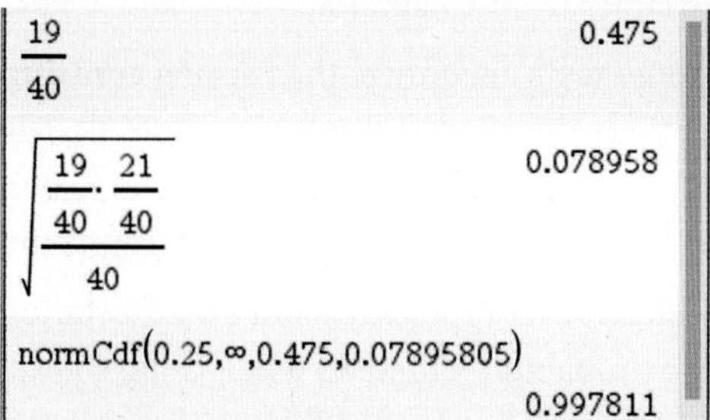

ii 1 Interpret the question as a probability statement.	Between 45% and 65% means $0.45 < \hat{p} < 0.65$.
2 Use CAS to calculate the probability correct to four decimal places.	$P(0.45 < \hat{p} < 0.65) = 0.6109$

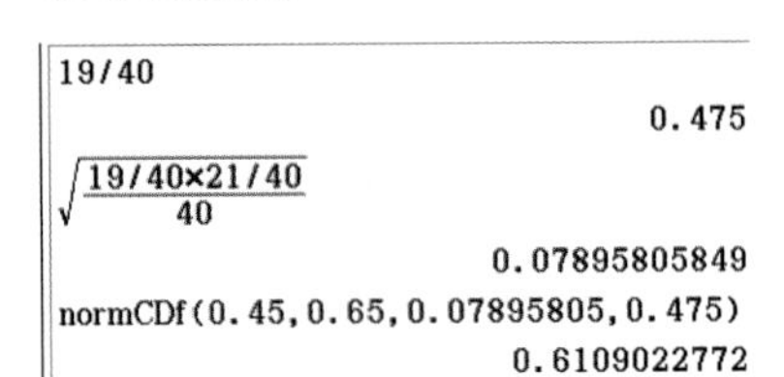

TI-Nspire

$\frac{19}{40}$ 0.475

$\sqrt{\frac{\frac{19}{40}\cdot\frac{21}{40}}{40}}$ 0.078958

normCdf(0.45,0.65,0.475,0.07895805) 0.610902

When it is not appropriate to model the sampling distribution of sample proportions using an approximate normal distribution, calculations for a binomial distribution should be used.

WORKED EXAMPLE 10 Justifying the sampling distribution of sample proportions

In a sample of 24 Year 12 students, 2 were colour blind.

a State the sample proportion of Year 12 students that are colour blind, $\hat{p}$.

b Explain why it is not appropriate for the sampling distribution of sample proportions to be approximated by a normal distribution.

c Hence, estimate the probability that in another sample of 24 Year 12 students, at least 1 student is colour blind.

d Compare the probability in part **c** to the probability if an approximate normal distribution was inappropriately used.

Steps	Working
a Express the number of colour blind Year 12 students as a proportion of the sample size.	$\hat{p} = \frac{2}{24} = 0.08\dot{3}$
b Use the values of n and $\hat{p}$ to justify the skewness of the distribution and conclude it is not approximately normal.	Given a very small value of $\hat{p}$ far from 0.5 and an insufficiently large n, the distribution will be positively skewed (i.e. not symmetrical) and, hence, an approximate normal distribution will not be appropriate.
c 1 Define the binomial random variable and its parameters.	Let X be the number of colour blind Year 12 students in a sample of 24. Then $X \sim \text{Bin}\left(24, \frac{2}{24}\right)$.
2 Express the probability using the binomial pdf and complement rule.	$P(x \geq 1) = 1 - P(X = 0)$ $= 1 - \binom{24}{0}\left(\frac{2}{24}\right)^0\left(\frac{22}{24}\right)^{24}$ $= 0.8761$

3 Use CAS to calculate the probability.

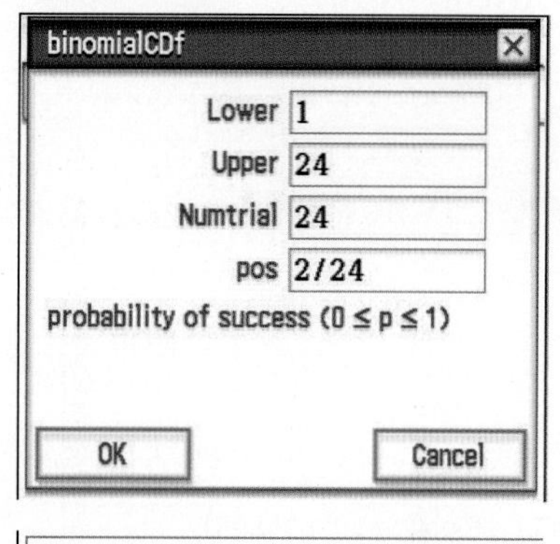

$\text{binomialCDf}\left(1, 24, 24, \frac{2}{24}\right)$

0.8760990779

TI-Nspire

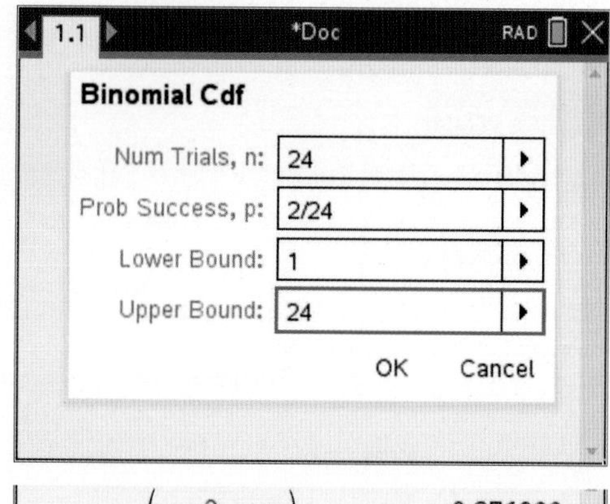

$\text{binomCdf}\left(24, \frac{2}{24}, 1, 24\right)$ 0.876099

d 1 Define the approximate normal random variable and its parameters.

Assume $\hat{p} \sim N\left(\frac{2}{24}, 0.0564^2\right)$.

$P\left(\hat{p} \geq \frac{1}{24}\right) = 0.7699$

2 Use CAS to calculate the probability.

3 Compare the probability from the normal distribution to the probability from the binomial distribution.

The estimate using a non-suitable normal distribution is 0.1062 less than using the binomial distribution.

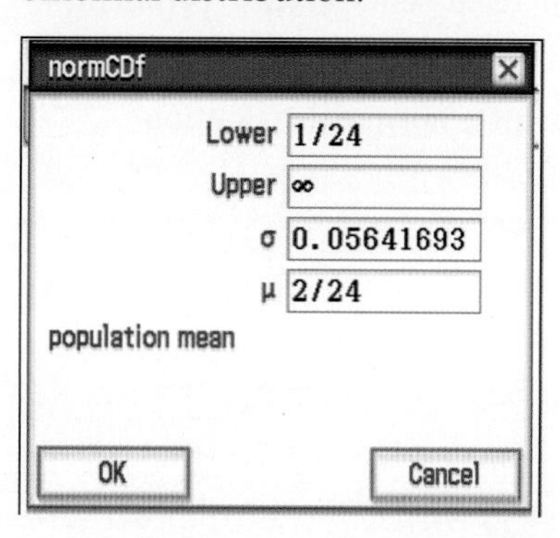

$\text{normCDf}\left(\frac{1}{24}, \infty, 0.05641693, \frac{2}{24}\right)$

0.7699095457

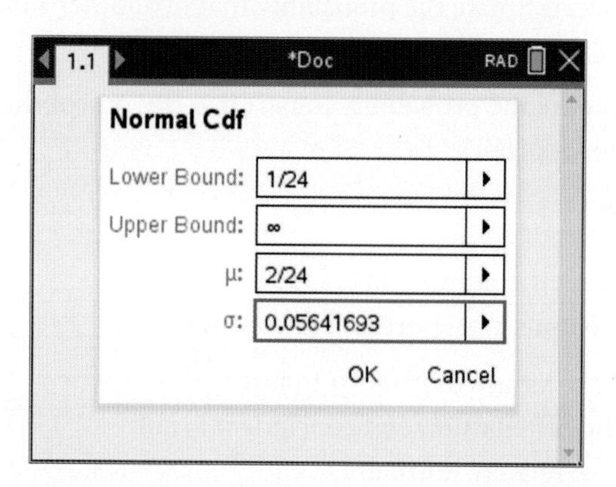

$\text{normCdf}\left(\frac{1}{24}, \infty, \frac{2}{24}, 0.05641693\right)$ 0.76991

The standard normal distribution with $\hat{p}$

In some contexts, once an approximate normal distribution has been used to model the sampling distribution of sample proportions, you may be asked to solve problems involving the use of the **approximate standard normal distribution** for the sampling distribution of sample proportions.

The approximate standard normal distribution

For the approximate normal distribution

$$\hat{p} \sim N\left(p, \frac{p(1-p)}{n}\right)$$

where the value of p is known, then an approximate standard normal distribution $Z \sim N(0, 1)$ can be obtained using the linear transformation

$$Z = \frac{\hat{p} - p}{\sqrt{\frac{p(1-p)}{n}}}$$

When the value of p is unknown and a specific sample proportion $\hat{p}_1$ is used as a point estimate for p, then

$$Z = \frac{\hat{p} - \hat{p}_1}{\sqrt{\frac{\hat{p}_1(1-\hat{p}_1)}{n}}}$$

WORKED EXAMPLE 11 Finding a z-score using p

Repeated random samples of size 140 are taken from a population with $p = 0.48$ to form the sampling distribution of sample proportions, $\hat{p}$.

a Give one reason why it is appropriate to model $\hat{p}$ using an approximate normal distribution.

b Hence, determine the number of standard deviations that a sample proportion of 0.43 is from the true population proportion.

Steps	Working
a Use the value of n or the value of p to justify the approximate normality of $\hat{p}$.	Given that $p = 0.48 \approx 0.5$ with a sufficiently large $n = 140$, the distribution will be fairly symmetrical and so an approximate normal distribution is appropriate.
b **1** State the expected value and standard deviation of $\hat{p}$.	$\text{E}(\hat{p}) = p = 0.48$
2 Use the standard score formula to find z.	$\text{SD}(\hat{p}) = \sqrt{\frac{0.48(0.52)}{140}}$ $z = \frac{0.43 - 0.48}{\sqrt{\frac{0.48(0.52)}{140}}} = -1.18$
3 Interpret the result as a number of standard deviations from the mean.	So, a sample proportion of 0.43 is 1.18 standard deviations below the population proportion p.

WORKED EXAMPLE 12 Solving for unknowns using $\hat{p}$ and Z

For the approximate normal distribution $\hat{p} \sim N(0.39, 0.002\,379)$, determine the

a sample size, n

b value of k such that $P(Z < k) = P(\hat{p} > 0.4)$, correct to four decimal places

c value of $\hat{p}$ that corresponds to the value of k such that $P(Z \geq k) = 0.01$, correct to two decimal places.

Steps	Working
a 1 Identify the value of $\hat{p}$ used as the point estimate of p.	$\hat{p} = 0.39$
2 Use the formula $\text{Var}(\hat{p}) = \frac{\hat{p}(1-\hat{p})}{n}$ to solve for n.	$0.002\,379 = \frac{0.39(0.61)}{n}$ $n = \frac{0.39(0.61)}{0.002\,379}$ $n = 100$
b 1 Draw a normal curve to represent $P(\hat{p} > 0.4)$.	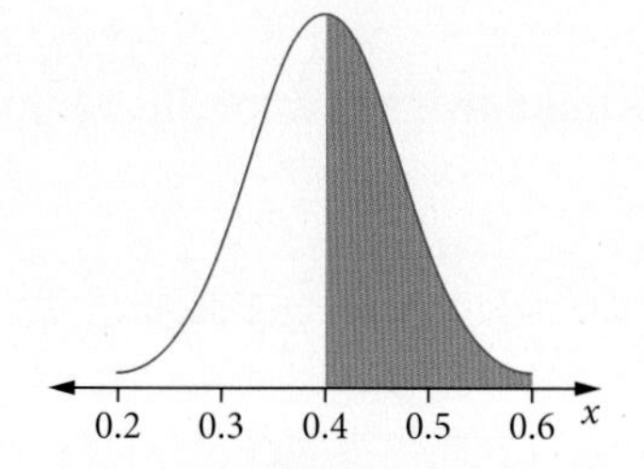
2 Use the symmetry about $\hat{p} = 0.39$ to shade the region with equivalent area, $P(\hat{p} < 0.38)$.	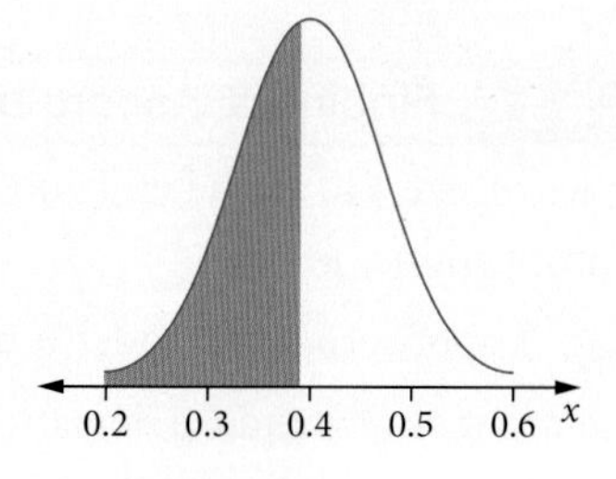
3 Use the standard score formula to find the value of k.	$k = \frac{0.38 - 0.39}{\sqrt{0.002\,379}}$ $= -0.2050$
c 1 Find the z-score, k, corresponding to the 0.99 quantile.	$P(Z \geq k) = 0.01 \Rightarrow k = 2.3263$
2 Use the z-score formula for sample proportions to determine the value of $\hat{p}$.	$z = \frac{\hat{p} - \hat{p}_1}{\sqrt{\frac{\hat{p}_1(1-\hat{p}_1)}{n}}}$ $2.3263 = \frac{\hat{p} - 0.39}{\sqrt{0.002\,379}}$ $\hat{p} = 0.50$

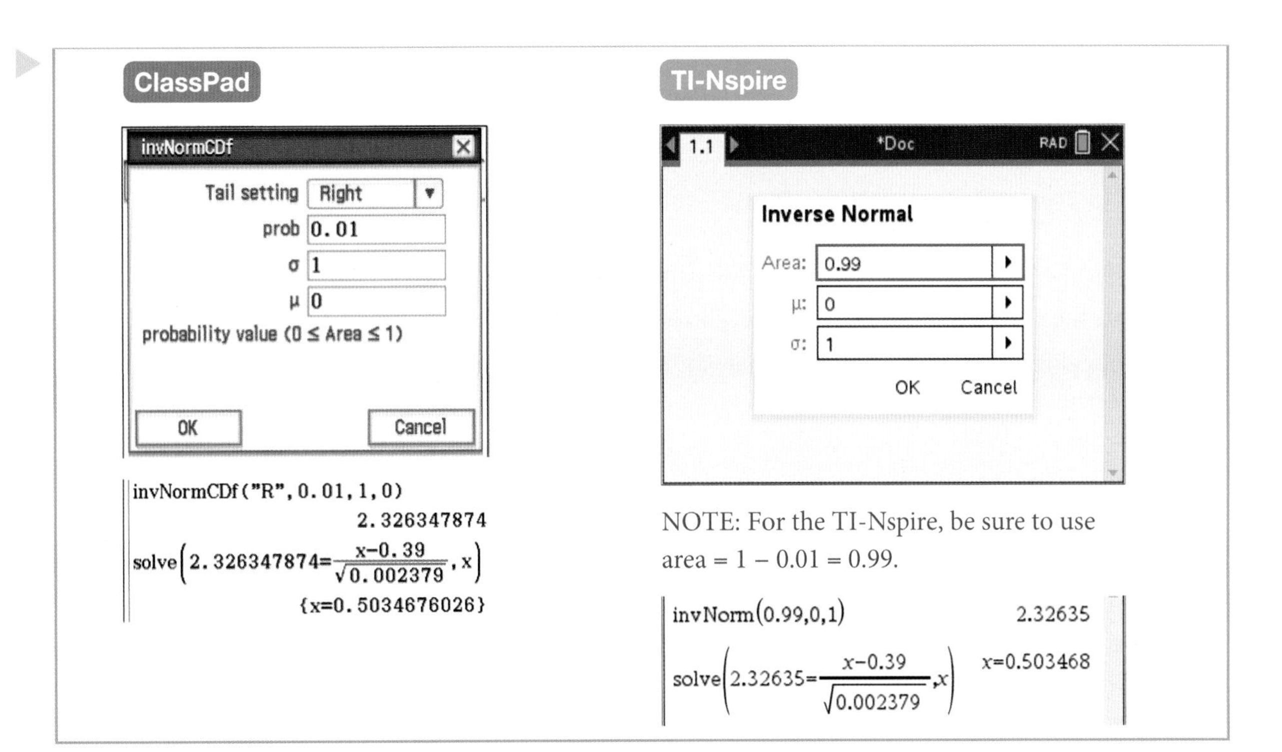

EXERCISE 9.2 The sampling distribution of sample proportions

ANSWERS p. 409

Recap

1 A media officer for a local football team wanted to collect information about how its supporters travelled to and from matches. Ten different groups of 5 team supporters standing close to each other were surveyed and asked how they travelled to the match.

- **a** Identify the population of interest.
- **b** Identify and explain one source of bias in this sampling.

2 The difference between a statistic and a parameter is that

- **A** a statistic is found from a measurement, but a parameter is a known quantity.
- **B** they measure the same quantity, but a parameter is more reliable than a statistic.
- **C** a statistic is a measure of a sample, while a parameter is a measure of a population.
- **D** a statistic is a measure of a population, while a parameter is a measure of a sample.
- **E** a statistic is found by measuring some aspect of a sample, but a parameter is an approximation of the same measure of the population from which the sample is taken.

Mastery

3 WORKED EXAMPLE 4 A school has a population of 980 high-school students. A random sample of 156 students was taken and it was found that 28 had bought food from the school cafeteria that day.

- **a** Calculate the sample proportion of students who had bought food from the cafeteria.
- **b** Hence, use this sample proportion as a point estimate to estimate the total number of students at this school who buy food from the cafeteria.

4 WORKED EXAMPLE 5 Given that $X \sim \text{Bin}(125, 0.8)$, determine

a $E(\hat{p})$

b $\text{Var}(\hat{p})$, correct to four decimal places

c $SD(\hat{p})$, correct to three decimal places.

5 WORKED EXAMPLE 6 From a sample of 200 people, it was found that 9 had red hair.

a State the sample proportion of people who have red hair, $\hat{p}$.

b Using this value of $\hat{p}$ as a point estimate for the true population proportion of people who have red hair, determine

i $E(\hat{p})$

ii $\text{Var}(\hat{p})$, correct to four decimal places

iii $SD(\hat{p})$, correct to three decimal places.

6 WORKED EXAMPLE 7 A box contains 20 000 marbles that are either blue or red. There are more blue marbles than red marbles. Random samples of 100 marbles are taken from the box. Each random sample is obtained by sampling with replacement. Let $\hat{p}$ be the random variable representing the proportion of blue marbles selected. If the standard deviation of $\hat{p}$ is 0.03, determine the number of blue marbles in the box.

7 Using CAS 5 Simulate 50 different observations from each of the following binomial random variables, graphing each of the results and describing the shape of the distributions.

a $X \sim \text{Bin}(10, 0.4)$ b $X \sim \text{Bin}(10, 0.6)$ c $X \sim \text{Bin}(10, 0.9)$

8 Using CAS 6 Simulate 30 different observations from each of the following binomial random variables, and graph the distributions of each of the corresponding distributions of $\hat{p} = \frac{X}{n}$.

a $X \sim \text{Bin}(100, 0.3)$ b $X \sim \text{Bin}(100, 0.55)$ c $X \sim \text{Bin}(100, 0.7)$

9 WORKED EXAMPLE 8 From a sample of 22 students studying Year 12 Mathematics Methods, 10 students were also studying Chemistry.

a State the sample proportion of Year 12 Mathematics Methods students studying Chemistry, $\hat{p}$, correct to four decimal places.

b Determine $E(\hat{p})$ and $SD(\hat{p})$, correct to four decimal places.

c Describe the distribution of the random variable $\hat{p}$, justifying your answer.

10 WORKED EXAMPLE 9 From a random sample of 500 Australian high-school students, 122 were found to be able to speak a second language. Let $\hat{p}$ be the sampling distribution of sample proportions of high-school students who can speak a second language.

a Describe the distribution of $\hat{p}$, justifying your answer.

b Hence, determine the probability, correct to four decimal places, that in a randomly selected sample of 500 Australian high-school students

i more than 25% of the students sampled can speak a second language

ii between 15% and 25% of the students sampled can speak a second language.

11 WORKED EXAMPLE 10 A stove manufacturer checked the 125 stoves leaving the factory on one day for faults. Eight were found to have faults in the paintwork or other problems that would make them non-sellable items.

a State the sample proportion of non-sellable items, $\hat{p}$.

b Explain why it is not appropriate for the sampling distribution of sample proportions to be approximated by a normal distribution.

c Hence, estimate the probability that in another sample of 125 stoves, more than five are considered non-sellable items.

d Compare the probability in part **c** to the probability if an approximate normal distribution was used.

12 WORKED EXAMPLE 11 Repeated random samples of size 120 are taken from a population with $p = 0.41$ to form the sampling distribution of sample proportions, $\hat{p}$.

a Give one reason why it is appropriate to model $\hat{p}$ using an approximate normal distribution.

b Hence, determine the number of standard deviations that a sample proportion of 0.5 is from the true population proportion.

13 WORKED EXAMPLE 12 If repeated random sampling from the same population gives sample proportions which have an approximate normal distribution $\hat{p} \sim N(0.56, 0.003\,08)$, determine the

a sample size, n

b value of k such that $P(-k < Z < k) = P(0.5 < \hat{p} < 0.62)$, correct to four decimal places

c value of $\hat{p}$ that corresponds to the value of k such that $P(Z \leq k) = 0.01$, correct to two decimal places.

Calculator-free

14 (4 marks) In a random sample of 75 Australian households, it was found that 12% of them had more than three bedrooms.

a State the number of households in this sample that had more than three bedrooms. (1 mark)

b Give one reason why it may not be considered appropriate to model the sampling distribution of sample proportions of Australian households with more than three bedrooms using an approximate normal distribution. (1 mark)

c Hence, write an expression that can be used to estimate the probability that exactly two houses in a random sample of 10 Australian households have more than three bedrooms. *Do not evaluate this expression.* (2 marks)

15 (3 marks) For random samples of five Australians, let $\hat{p}$ be the random variable that represents the proportion who live in a capital city. Suppose that the value of p, the true population proportion, is known. If $P(\hat{p} = 0) = \frac{1}{243}$, determine the value p.

16 (7 marks)

a A random variable $\hat{p}$ representing a sample proportion has a standard deviation of 0.08. If $p = 0.2$, determine the value of n. (3 marks)

b A random variable $\hat{p}$ representing a sample proportion has a standard deviation of 0.04. If $n = 100$ and $p < 0.5$, determine the exact value of p. (4 marks)

Calculator-assumed

17 © SCSA MM2016 Q14abd (6 marks) The simulation of a loaded (unfair) five-sided die rolled 60 times is recorded with the following results.

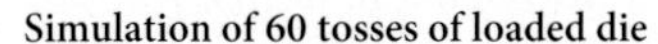

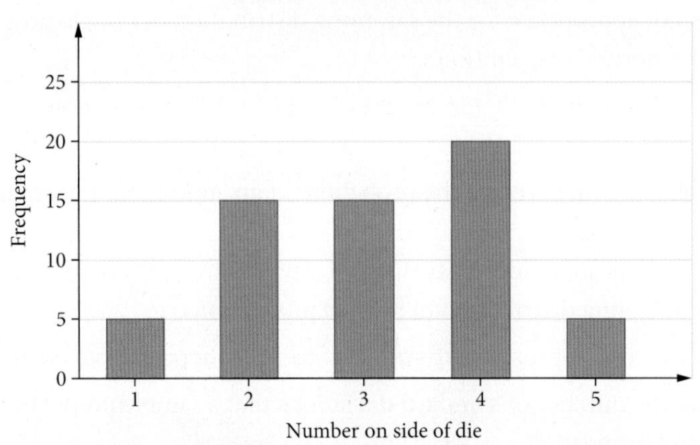

a Calculate the proportion of prime numbers recorded in this simulation. (2 marks)

b Determine the mean and standard deviation for the sample proportion of prime numbers in 60 tosses, using the results above. (2 marks)

c This simulation of 60 rolls of the die is performed another 200 times, with the proportion of prime numbers recorded each time and graphed. Comment briefly on the key features of this graph. (2 marks)

18 © SCSA MM2017 Q18d MODIFIED (6 marks) Alex is a beekeeper and has noticed that some of the bees are very sleepy. She takes a random sample of 320 bees and finds that 15 of them are indeed so-called *lullabees* that fall asleep easily. It turns out that the true proportion of lullabees is 0.02. Now that Alex knows this, she decides to take a new sample.

a Suppose a new sample of 290 bees was taken. Given that the true proportion of lullabees is 0.02 and assuming an approximate normal distribution for $\hat{p}$, what is the probability that the sample proportion in this new sample is at most 0.03? (3 marks)

b Show one mathematical calculation that suggests an approximate normal distribution for $\hat{p}$ may not be appropriate. (1 mark)

c If Alex takes a larger sample, will the above probability increase or decrease? Explain. (2 marks)

19 © SCSA MM2018 Q17abd (6 marks) Tina believes that approximately 60% of the mangoes she produces on her farm are large. She takes a random sample of 500 mangoes from a day's picking.

a Assuming Tina is correct and 60% of the mangoes her farm produces are large, what is the approximate probability distribution of the sample proportion of large mangoes in her sample? (3 marks)

b What is the probability that the sample proportion of large mangoes is less than 0.58? (2 marks)

A random sample of 500 contains 250 large mangoes.

c On the basis of this data, estimate the proportion of large mangoes produced on the farm. (1 mark)

20 © SCSA MM2019 Q13a (4 marks) The proportion of working adults who miss breakfast on weekdays is estimated to be 40%. A study takes a random sample of 400 working adults. For this sample

a what is the (approximate) distribution of the sample proportion of workers who miss breakfast (2 marks)

b what is the probability that the sample proportion of workers who miss breakfast is greater than 44%? (2 marks)

21 © SCSA MM2021 Q10acd MODIFIED (6 marks) A charity organisation has printed 'Lucky 7' scratchie tickets as a fundraiser for use at two special events. The tickets contain two panels. Each ticket has the same numbers as the sample ticket shown below, arranged randomly and hidden within each panel.

A player scratches one section of each panel to reveal a number. The two numbers revealed are then added together. If the total is seven or higher, the player wins a prize.

At the first event, 400 tickets are purchased, and a prize is won on 124 occasions. Let p denote the probability that a prize is won.

a Determine the sample proportion of times that a prize is won at the first event. (1 mark)

It can be shown that the probability p of winning a prize is $\frac{7}{24}$.

b Calculate the mean and standard deviation of the sample proportion of times a prize is won when 400 tickets are purchased. (2 marks)

c At a second event, 400 scratchie tickets are again purchased. If the sample proportion was 0.6 standard deviations from the population proportion, how many prizes were won at the second event? (3 marks)

9.3 Confidence intervals for proportions

Video playlist Confidence intervals for proportions

Worksheets Sample proportion confidence intervals

Margin of error for standard normal variables

Sample sizes

Approximate confidence intervals for p

In Section 9.2, we introduced the idea that a single sample proportion, $\hat{p}$, can act as a point estimate for the true population proportion, p, when p is not known. However, the limitation of a single point estimate does not allow for the possible error in the estimate that arises from the nature of random samples and the variability of samples. As a result, it is common to consider an **interval estimate for p**; that is, a range of possible values that p falls within an interval determined by a sample proportion $\hat{p}$. Because p is unknown in these situations, the true value of $\text{SD}(\hat{p}) = \sqrt{\frac{p(1-p)}{n}}$ is also often unknown and so the **standard error** $\text{SD}(\hat{p}) = \sqrt{\frac{\hat{p}(1-\hat{p})}{n}}$ is used as a point estimate in its place. The standard error is another term used to describe the standard deviation of the random variable $\hat{p}$ when a single point estimate for p is used.

To calculate an interval estimate for p, first we need to ensure that the distribution of the random variable $\hat{p}$ is approximately normal. Once we have an approximate normal distribution for $\hat{p}$, we know that the mean of $\hat{p}$ is $\text{E}(\hat{p}) = \hat{p}$ and the standard error is $\text{SD}(\hat{p}) = \sqrt{\frac{\hat{p}(1-\hat{p})}{n}}$. We can then use a number of standard deviations from the mean, i.e. a z-score, to define a **margin of error**, E, on either side of the point estimate used.

Structure of an interval estimate for p

For an approximately normal sampling distribution of sample proportions formed using $\hat{p}$ as a point estimate for p, an interval estimate for p is of the form

$$\hat{p} - E \le p \le \hat{p} + E$$

where E is a margin of error.

The specific type of interval estimate we use in this course is called a confidence interval, or more specifically an **approximate confidence interval**, because the normal distribution being used is only approximate. The term confidence will be explored in more detail soon, but for now let's look at the construction of an approximate confidence interval.

Each approximate confidence interval has a **confidence level**, $100c\%$, where c represents the proportion of $\hat{p}$ values that we want to account for in our estimate. The value of c is used to the calculate the corresponding z-scores such that $P(-z \le Z \le z) = c$. For example, if we choose a 95% confidence level, then $P(-z \le Z \le z) = 0.95$ gives a value of $z = 1.960$ to three decimal places. Other common confidence levels include

- a 90% confidence level such that $P(-z \le Z \le z) = 0.90$ gives a z-score of $z = 1.645$ to three decimal places
- a 99% confidence level such that $P(-z \le Z \le z) = 0.99$ gives a z-score of $z = 2.576$ to three decimal places.

The corresponding z-score for any confidence level can be found using the appropriate inverse normal calculation.

Once we have the number of standard deviations z for a given level of c, we can define the value of the margin of error.

Margin of error of an approximate $100c\%$ confidence interval

For a confidence level of $100c\%$ with a corresponding z-score, z, the margin of error of an approximate $100c\%$ confidence interval is given by

$$E = z\,\text{SD}(\hat{p}) = z\sqrt{\frac{\hat{p}(1-\hat{p})}{n}}$$

From this we can define the approximate 100c% confidence interval.

Approximate 100c% confidence interval

For a confidence level of 100c% with a corresponding z-score, z, and a margin of error of $z\sqrt{\dfrac{\hat{p}(1-\hat{p})}{n}}$,

an approximate 100c% confidence interval for p is given by

$$\hat{p} - z\sqrt{\frac{\hat{p}(1-\hat{p})}{n}} \le p \le \hat{p} + z\sqrt{\frac{\hat{p}(1-\hat{p})}{n}}$$

90%, 95% and 99% confidence levels are the most commonly used for confidence intervals, with the following standard scores.

Confidence level	90%	95%	99%
Standard score (z-score)	1.645	1.960	2.576

Note that you may also see this written in a different interval notation such as

$$\left[\hat{p} - z\sqrt{\frac{\hat{p}(1-\hat{p})}{n}}, \hat{p} + z\sqrt{\frac{\hat{p}(1-\hat{p})}{n}}\right]$$

or with open bounds

$$\hat{p} - z\sqrt{\frac{\hat{p}(1-\hat{p})}{n}} < p < \hat{p} + z\sqrt{\frac{\hat{p}(1-\hat{p})}{n}}$$

$$\left(\hat{p} - z\sqrt{\frac{\hat{p}(1-\hat{p})}{n}}, \hat{p} + z\sqrt{\frac{\hat{p}(1-\hat{p})}{n}}\right)$$

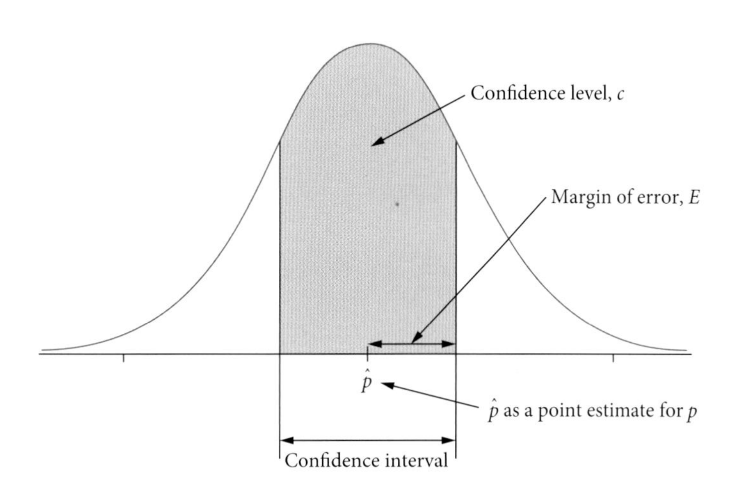

Exam hack

Always show the line of working involving the construction of the confidence interval with the values of $\hat{p}$, z and n substituted.

Width of an approximate 100c% confidence interval

The width, w, of a 100c% confidence interval is twice the margin of error

$$w = 2E = 2z\sqrt{\frac{\hat{p}(1-\hat{p})}{n}}$$

Although we still haven't discussed the meaning and interpretation of the word 'confidence' in the context of interval estimates, let's first construct some confidence intervals.

WORKED EXAMPLE 13 **Expressing an approximate confidence interval without CAS**

A random sample of 40 USBs from a large consignment found that 15 had packaging defects. Let $\hat{p}$ be the random variable for the proportion of USBs with packaging defects in samples of size 40.

a State the distribution of $\hat{p}$. Justify your answer.

b Hence, write expressions for the approximate confidence intervals of p with the following levels of confidence. *Do not evaluate these intervals.*

i 90% confidence level **ii** 95% confidence level **iii** 99% confidence level

Steps	Working
a **1** Calculate $\hat{p}$ as a point estimate for p.	Let $\hat{p} = \frac{15}{40}$ be a point estimate for p.
2 Identify the conditions on n and $\hat{p}$ for the distribution to be considered approximately normal.	Using the fact that $np = 15 \geq 10$ and $n(1-p) = 25 \geq 10$.
3 Give the appropriate mathematical calculations supporting your reasons.	By the central limit theorem, the distribution of $\hat{p}$ is approximately normal such that $\text{E}(\hat{p}) = \frac{15}{40} = \frac{3}{8}$. $\text{SD}(\hat{p}) = \sqrt{\frac{\frac{3}{8}\left(\frac{5}{8}\right)}{40}} = \sqrt{\frac{15}{64 \times 40}} = \frac{\sqrt{15}}{16\sqrt{10}} = \frac{\sqrt{3}}{16\sqrt{2}}$
4 State the distribution (i.e. normal) and its corresponding parameters.	$\hat{p} \sim N\left(\frac{15}{40}, \frac{3}{512}\right)$
b Use the structure of the confidence interval $\hat{p} - z\sqrt{\frac{\hat{p}(1-\hat{p})}{n}} \leq p \leq \hat{p} + z\sqrt{\frac{\hat{p}(1-\hat{p})}{n}}$ with each of the appropriate z-scores.	
i $z = 1.645$	$\frac{15}{40} - 1.645\frac{\sqrt{3}}{16\sqrt{2}} \leq p \leq \frac{15}{40} + 1.645\frac{\sqrt{3}}{16\sqrt{2}}$
ii $z = 1.960$	$\frac{15}{40} - 1.960\frac{\sqrt{3}}{16\sqrt{2}} \leq p \leq \frac{15}{40} + 1.960\frac{\sqrt{3}}{16\sqrt{2}}$
iii $z = 2.576$	$\frac{15}{40} - 2.576\frac{\sqrt{3}}{16\sqrt{2}} \leq p \leq \frac{15}{40} + 2.576\frac{\sqrt{3}}{16\sqrt{2}}$

It is likely that you will have the assistance of CAS to evaluate an approximate confidence interval for p.

USING CAS 7 Constructing an approximate confidence interval for p

Three hundred carrot seeds were moistened and placed in an incubator. When they were checked five days later, 250 were found to have germinated. Let $\hat{p}$ be the random variable representing the germination rate of samples of carrot seeds taken from this population. Assuming the approximate normality of $\hat{p}$, determine an approximate 95% confidence interval for the true germination rate of carrot seeds from this population, correct to three decimal places.

ClassPad

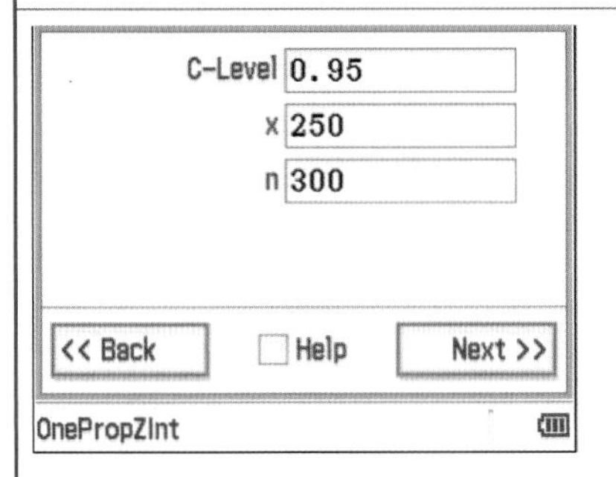

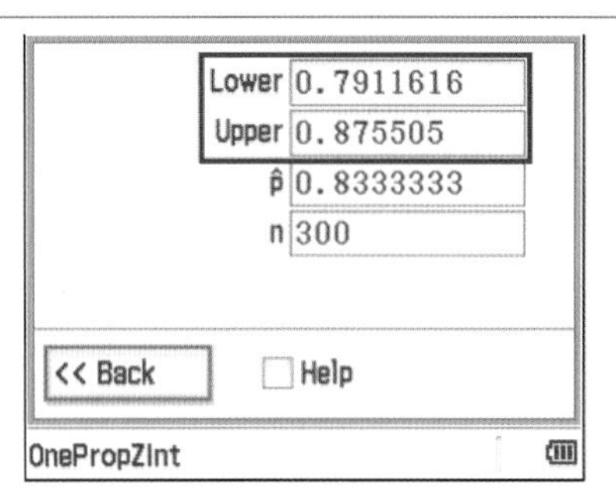

1 Open the **Statistics** application.
2 Tap **Calc** > **Interval**.
3 In the lower window, select **One-Prop Z Int** from the dropdown menu then tap **Next**.
4 Enter the values as shown above then tap **Next**.
5 The values will be displayed in the lower window.
6 The **Lower** and **Upper** values of the confidence interval are highlighted above.

TI-Nspire

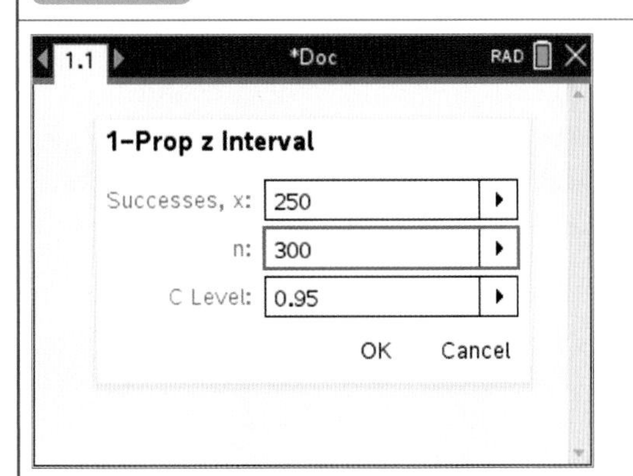

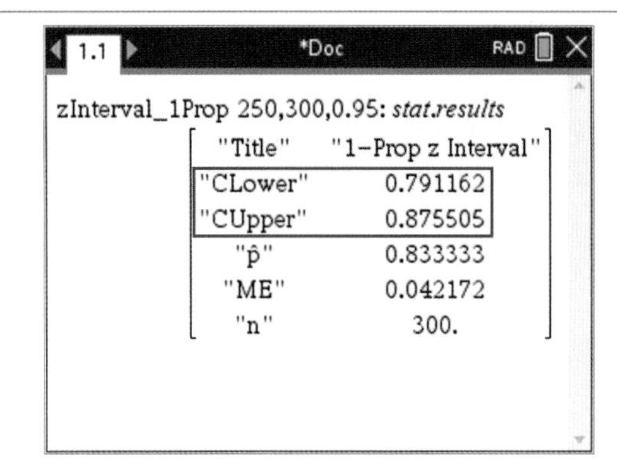

1 Press **menu** > **Statistics** > **Confidence Intervals** > **1-Prop z Interval**.
2 Enter the values as shown above.
3 Press **enter**.
4 The z-interval table will be displayed.
5 The **CLower** and **CUpper** values of the confidence interval are highlighted above.

The approximate 95% confidence interval for the true germination rate of carrot seeds from this population is $0.791 \le p \le 0.876$.

Interpreting confidence intervals and the containment of p

So, here is the big question: *What is the meaning of confidence in the construction of an approximate confidence interval*? The simple answer is that the construction of a single confidence interval serves no purpose other than to act as a single interval estimate for p, but no formal conclusions can be drawn from a single, constructed confidence interval. This is a largely debated concept in statistics, but, for this course, we use the **frequentist interpretation** of a confidence interval.

Frequentist interpretation of confidence intervals

Upon the repeated construction of a large number of approximate $100c\%$ confidence intervals for p, from multiple random samples of sample size n, we can expect (on average) that $100c\%$ of all confidence interval yet to be constructed will contain the true value of p.

The frequentist perspective is about a long-run relative frequency of confidence intervals that are expected to contain the true population proportion.

With this interpretation comes four very important conclusions.

1 Most, but not all, confidence intervals contain p.

2 Because p is unknown, and due to the nature of random sampling, it can never be known for certain whether a confidence interval contains p.

3 Because p is constant, once a confidence interval is constructed, the probability that the given confidence interval contains p is either 0 or 1. It either does not contain p or it does, but we can never know for certain because p is unknown.

4 No single constructed confidence interval is any more or less likely to contain p than any other single constructed confidence interval.

The diagram shows what we could expect to occur if we were to construct 100 different 95% confidence interval from random samples. That is, we would expect (on average) 95 of the 100 to contain the true value of p (shown in blue) and 5 to not contain the true value of p (shown in red). However, this can never be known for certain due to the nature of random sampling but can be shown by simulation.

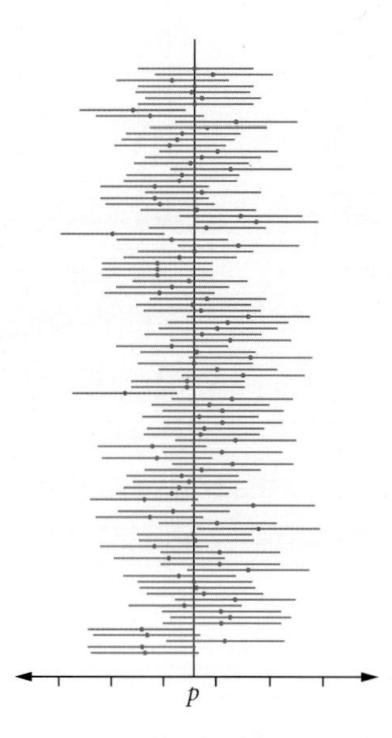

Exam hack

If you are asked 'which of the following confidence intervals are more likely to contain the true value of p?', it is a trick question! The likelihood that a confidence interval contains p cannot be inferred from a single confidence interval.

9780170477536

WORKED EXAMPLE 14 Discussing containment of p

Boxes of a particular brand of breakfast cereal are labelled with a weight of 485 g. In a random sample of 350 boxes, it is found that 160 were underweight. Let $\hat{p}$ be the random variable representing the sample proportion of underweight boxes of breakfast cereal in samples of size 350.

a Assuming the approximate normality of $\hat{p}$, determine an approximate 99% confidence interval for p, correct to three decimal places.

b A second random sample of 350 boxes was taken, the number of underweight boxes was observed and an approximate 99% confidence interval was found to be $(0.372, 0.508)$. Which of the two confidence intervals is more likely to contain the true value of p? Justify your answer.

c If a further 500 random samples of 350 boxes were to be taken and approximate 99% confidence intervals were to be constructed for p, how many of the confidence intervals could be expected to contain the true value of p?

Steps	Working
a 1 State the distribution of $\hat{p}$.	$\hat{p} \sim N\left(\frac{160}{350}, 0.0266^2\right)$
2 Establish the confidence interval, stating the value of z.	For 99% confidence, $z = 2.576$
3 Use CAS to evaluate the confidence interval.	$\frac{160}{350} - 2.576(0.0266) \le p \le \frac{160}{350} + 2.576(0.0266)$ $0.389 \le p \le 0.526$

ClassPad

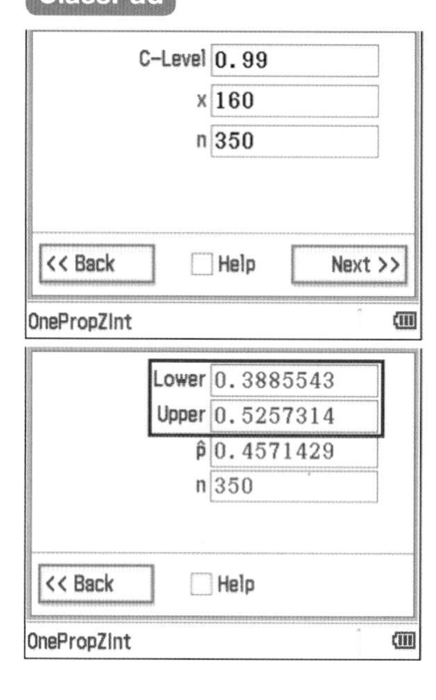

TI-Nspire

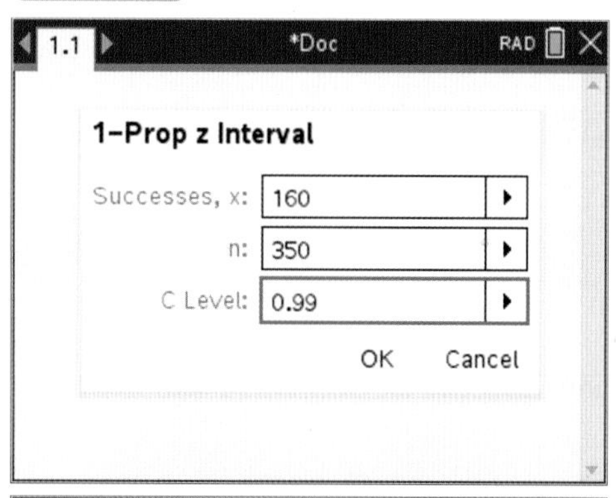

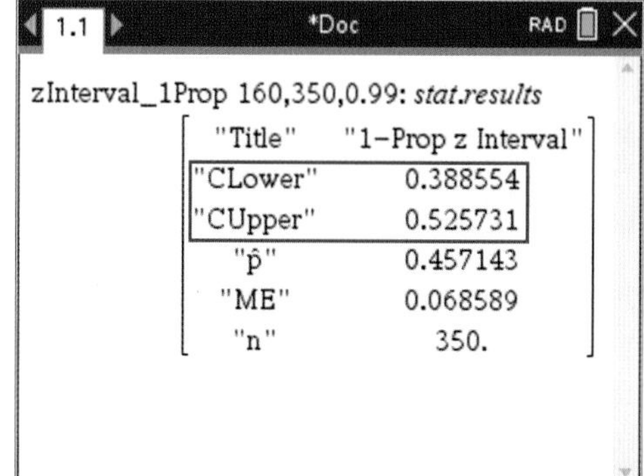

Steps	Working
b Recognise that both confidence intervals have been calculated and, hence, they either do or do not contain p.	Neither one is more likely than the other to contain p, as once observed, the probability that a confidence interval contains p is either 0 or 1. Hence, it cannot be determined.
c 1 Recognise that the 500 confidence intervals are yet to be constructed.	$0.99 \times 500 = 495$
2 Calculate 99% of 500 and estimate the number of confidence intervals expected to contain p.	It can be expected that approximately 495 of the 500 confidence intervals will contain the true value of p.

Another common misconception regarding approximate confidence intervals is that a greater level of confidence means that it is more likely that an observed confidence interval will contain the true value of p. This is not an accurate statement for a single, constructed confidence interval.

Changing confidence levels

If the values of $\hat{p}$ and n remain unchanged, as the confidence level $100c\%$ increases

- the value of z increases, and so
- the value of the margin of error E increases, and so
- the width of the confidence interval increases.

Changing sample size

If the values of $\hat{p}$ and z remain unchanged, as the sample size n increases

- the value of the standard error $\sqrt{\dfrac{\hat{p}(1-\hat{p})}{n}}$ decreases, and so
- the value of the margin of error E decreases, and so
- the width of the confidence interval decreases.

WORKED EXAMPLE 15 Describing changes to confidence intervals

An approximate 90% confidence interval for p for a random sample of size 400 is found to be $(0.25, 0.33)$. State the effect on the width of the confidence interval constructed

a if the confidence level was increased to 98%, but $\hat{p}$ and n remain unchanged

b if the sample size is reduced to 100, but $\hat{p}$ and z remain unchanged.

Steps	Working
a Describe the sequence of effects if confidence level is increased.	If confidence level is increased, the value of z increases, and so the margin of error increases, increasing the width of the confidence interval.
b Describe the sequence of effects if sample size is reduced.	If sample size is reduced, the standard error increases, and so the margin of error increases, increasing the width of the confidence interval.

Using confidence intervals to calculate unknowns

In some cases, as in Worked example 16, we may be given the lower and upper bounds of a confidence interval. When these bounds are given with sufficient information, the values of $\hat{p}$, E, z, n or $\text{SD}(\hat{p})$ can be calculated.

WORKED EXAMPLE 16 Calculating unknowns given a confidence interval

An approximate 95% confidence interval for p using a random sample of size n is found to be $(0.46, 0.52)$. Calculate the

a sample proportion $\hat{p}$ used to construct the approximate confidence interval

b margin of error of the confidence interval

c standard error of $\hat{p}$ to three decimal places

d sample size used.

Steps	Working
a Use the symmetry of the confidence interval about $\hat{p}$ to calculate the sample proportion.	$\hat{p} = \frac{0.46 + 0.52}{2} = 0.49$
b Calculate half of the width of the confidence interval.	$E = \frac{0.52 - 0.46}{2} = 0.03$
c Divide the margin of error by the value of z corresponding to the confidence level.	$E = z\,\text{SD}(\hat{p})$ $\text{SD}(\hat{p}) = \frac{0.03}{1.960}$ $= 0.015$
d **1** Use the formula $\text{SD}(\hat{p}) = \sqrt{\frac{\hat{p}(1-\hat{p})}{n}}$ to solve for n.	$0.015 = \sqrt{\frac{0.49(0.51)}{n}}$ $n = 1066.65$
2 Round to the nearest integer.	$n = 1067$

WORKED EXAMPLE 17 Calculating confidence level given sufficient information

An approximate C% confidence interval for p using a random sample of size 80 is found to be (0.80, 0.94). Find the level of confidence, correct to one decimal place, used to construct the confidence interval.

Steps	Working
1 Find $\hat{p}$ and E.	$\hat{p} = \frac{0.80 + 0.94}{2} = 0.87$ $E = \frac{0.94 - 0.80}{2} = 0.07$
2 Calculate $\text{SD}(\hat{p})$.	$\text{SD}(\hat{p}) = \sqrt{\frac{0.87(0.13)}{80}} = 0.0375\ldots$ $= 0.0376$ to three decimal places
3 Find z using $E = z\,\text{SD}(\hat{p})$.	$z = \frac{E}{\text{SD}(\hat{p})} = \frac{0.07}{0.0375\ldots} = 1.8617\ldots$
4 Use CAS to find $P(-1.8617 \le Z \le 1.8617)$.	$P(-1.8617 \le Z \le 1.8617) = 0.937$
5 State the confidence level as a percentage, rounded to one decimal place.	Therefore, a confidence level of 93.7% was used.

ClassPad

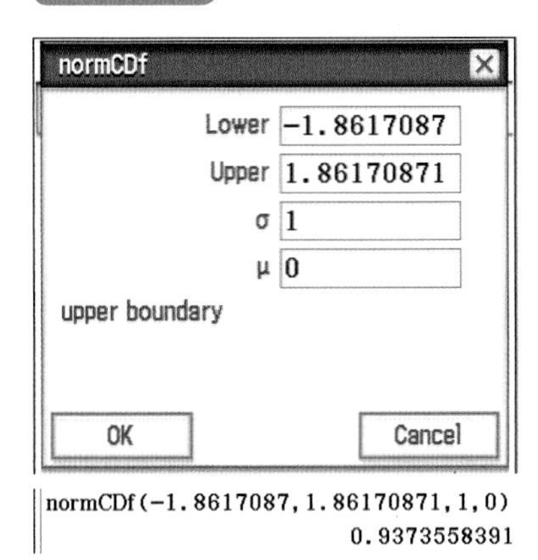

TI-Nspire

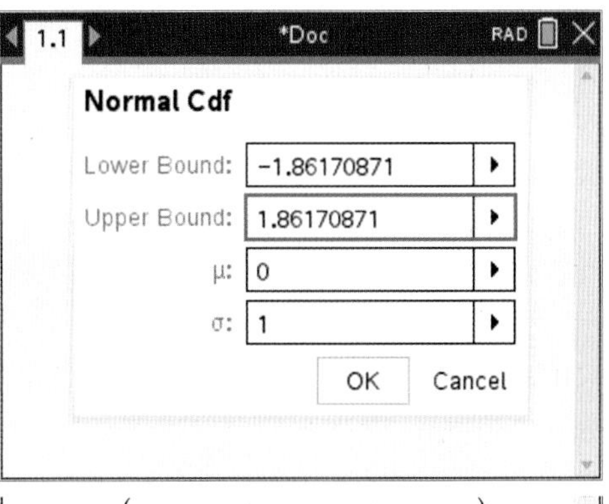

Another typical calculation involving confidence intervals is to determine a minimum sample size required to ensure a certain margin of error is obtained in the construction of an approximate confidence interval. However, the issue with these problems is that the sample proportion is not yet known and so the calculation for n cannot be carried out unless there is a given value of $\hat{p}$. It is common for this value to come from historical data and so we can use it as a point estimate for p.

WORKED EXAMPLE 18 **Determining a minimum sample size given a point estimate for p**

A previous study suggests that about 60% of Year 12 students obtain their driver's licence before they complete Year 12. Determine the minimum sample size that would be needed to obtain an approximate 90% confidence interval for p with a maximum width of 14%.

Steps	Working
1 Calculate E given the width and state the other known values.	$E = \frac{0.14}{2} = 0.07$ $z = 1.645$ $\hat{p} = 0.6$
2 Use the formula $E = z\sqrt{\frac{\hat{p}(1-\hat{p})}{n}}$ to solve for n.	$0.07 = 1.645\sqrt{\frac{0.6(0.4)}{n}}$ $n = 132.54$
3 If n is a decimal value, note that the integer smaller than n will give a width larger than 14%. The integer value larger than n will give a width smaller than 14%. Round up!	The minimum sample size to ensure a maximum width of 0.14 is 133 Year 12 students.

If, in a similar problem, the value of $\hat{p}$ is not known, then we assume the worst case scenario and adopt the value of $\hat{p}$ that gives the maximum margin of error.

It can be shown using calculus techniques that the value of $\hat{p}$ that maximises the function $E = z\sqrt{\frac{\hat{p}(1-\hat{p})}{n}}$ is 0.5.

$$E = z\sqrt{\frac{\hat{p}(1-\hat{p})}{n}}$$

Given that z and n are just constants, then we can simply maximise the function:

$$f(\hat{p}) = \sqrt{\hat{p}(1-\hat{p})} = \sqrt{\hat{p}-\hat{p}^2}$$

$$f'(\hat{p}) = \frac{1}{2}(\hat{p}(1-\hat{p}))^{-\frac{1}{2}} \cdot (1-2\hat{p})$$

$$0 = \frac{(1-2\hat{p})}{2\sqrt{\hat{p}(1-\hat{p})}}$$

$$1 - 2\hat{p} = 0$$

$$\hat{p} = \frac{1}{2}$$

By observing the graph of $f(\hat{p})$ or considering the second derivative, $f''(\hat{p}) = -\frac{\sqrt{\hat{p}(1-\hat{p})}}{4\hat{p}^2(\hat{p}-1)^2}$, we can show that it is indeed a maximum.

$$f''\left(\frac{1}{2}\right) = -2$$

So, f is concave down at $\hat{p} = \frac{1}{2}$ and so gives a maximum turning point.

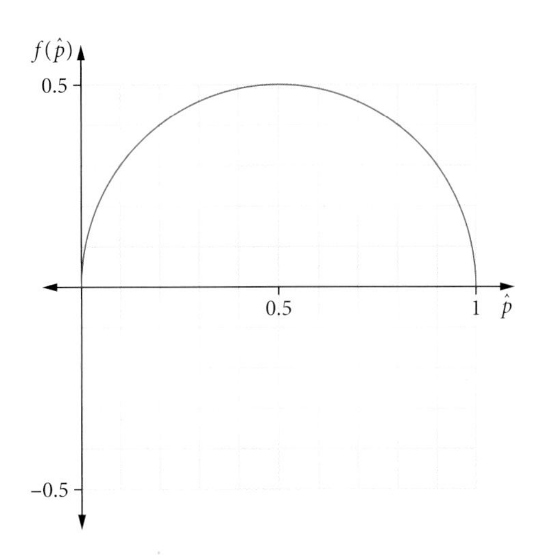

WORKED EXAMPLE 19	Determining a minimum sample size when $\hat{p}$ is unknown
A survey wants to establish an interval estimate for the proportion of Year 12 students from single-parent families to within 3% at a confidence level of 99%. Determine the minimum number of Year 12 students who need to be surveyed.	
Steps	**Working**
1 Assume $\hat{p} = 0.5$ and state the known values of E and z.	$\hat{p} = 0.5$ $E = 0.03$ $z = 2.576$
2 Use the formula $E = z\sqrt{\frac{\hat{p}(1-\hat{p})}{n}}$ to solve for n.	$0.03 = 2.576\sqrt{\frac{0.5^2}{n}}$ $n = 1843.27$
3 If n is a decimal value, round up to nearest integer value.	At least 1844 Year 12 students should be surveyed.

Population claims, historical data and the comparison of samples

Further to the construction of single approximate confidence intervals and solving for unknown values given sufficient information, there is not much more we can do with confidence intervals without the formal topic of hypothesis testing (which is not in the Year 12 Mathematics Methods course). However, there is a common tendency in WACE exam questions to ask you to infer results from confidence intervals in three different situations, the first two of which are explained below.

1 Compare an approximate confidence interval from a sample to a claimed value of p and suggest whether there is sufficient evidence to suggest that the sample appears to come from the same population.

2 Compare an approximate confidence interval from a sample to a claimed value of p or one that arises from historical data and suggest whether the claim or data is supported by the sample.

Comparing a claimed value of p to an approximate confidence interval for p

If the claimed value of p lies within an approximate confidence interval obtained from a random sample, then

- there is insufficient evidence to suggest the sample came from a different population, or
- there is insufficient evidence to suggest that the claimed value of p shouldn't be accepted.

If the claimed value of p lies outside of an approximate confidence interval obtained from a random sample, then

- there is insufficient evidence to suggest the sample came from the same population, or
- there is insufficient evidence to suggest that the claimed value of p can be accepted based on this sample.

Exam hack

Your responses to these questions should always be written in the context of the question.

WORKED EXAMPLE 20 Comparing a confidence interval to a claimed p

A local fashion designer claims that approximately 70% of Perth residents wear dark-coloured clothing to work. Over the course of a week, a random sample of 120 Perth residents going to work was taken and it was observed that 70 of them were wearing dark-coloured clothing. Use an approximate 95% confidence interval to comment on the fashion designer's claim.

Steps	Working
1 Obtain the distribution of $\hat{p}$.	$\hat{p} \sim N\left(\frac{70}{120}, 0.045^2\right)$
2 Construct a 95% confidence interval using CAS.	$\frac{70}{120} - 1.960(0.045) \le p \le \frac{70}{120} + 1.960(0.045)$
3 Observe the location of the claimed p in relation to the confidence interval.	$0.495 \le p \le 0.672$
4 Conclude appropriately in context of the question.	Given that the claimed $p \approx 0.70$ lies outside of the interval estimate $0.495 \le p \le 0.672$, based on this sample, there is insufficient evidence to suggest that the fashion designer's claim can be accepted.

The third case that may arise is the following:

3 Compare approximate confidence intervals from two or more samples after changed conditions and suggest whether the changed conditions have affected the sample proportions.

Comparing two or more samples

If a second sample proportion $\hat{p}_2$ lies within an approximate confidence interval obtained using a first sample proportion $\hat{p}_1$, OR if the two approximate confidence intervals overlap such that one or both sample proportions are contained within the other, then

- there is insufficient evidence to suggest that the samples came from different populations, or that a changed condition had an impact on the population.

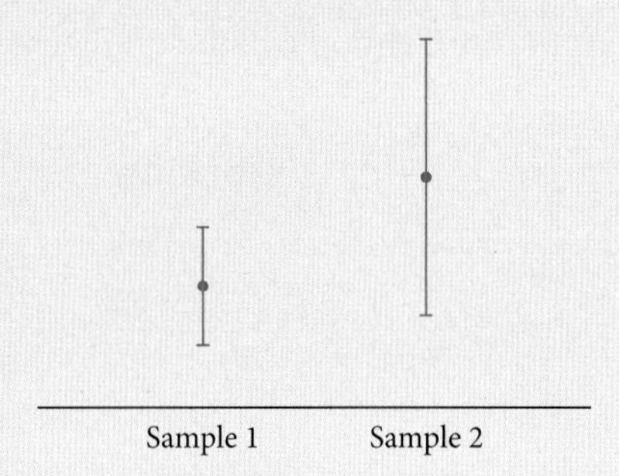

If the approximate confidence interval obtained from $\hat{p}_2$ does not overlap at all with an approximate confidence interval obtained using $\hat{p}_1$, then

- there is sufficient evidence to suggest that the samples may have come from different populations, or that a changed condition may have an impact on the population, but we cannot say for certain.

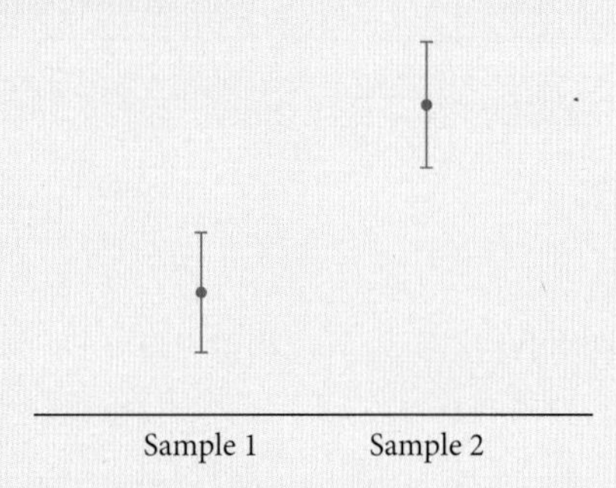

If the approximate confidence interval obtained from $\hat{p}_2$ partially overlaps an approximate confidence interval obtained using $\hat{p}_1$, such that neither value of $\hat{p}$ is contained within the other, then

- there is insufficient evidence to conclude anything definitive about the two samples.

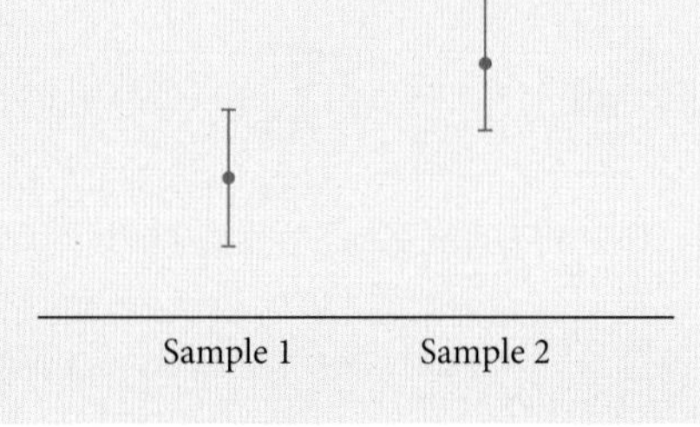

WORKED EXAMPLE 21 Comparing two samples

One Saturday, two weeks before a local election, a random sample of 320 people were asked about their voting intentions. It was found that in a two-party preferential vote between Liberal and Labor, 147 indicated they would vote Labor.

a Construct an approximate 90% confidence interval for the true proportion of Labor voters in the electorate, correct to four decimal places.

In the following week, Labor intensified their advertising campaign and in a poll on the following Saturday, it was found that 71 out of 105 people surveyed said they would vote Labor.

b How likely is it to obtain a sample proportion greater than that found in the second random sample when using the sampling distribution of sample proportions of the first sample?

c Perform the necessary calculations to comment on whether the increased advertising campaign improved Labor's polling results.

Steps	Working
a 1 Obtain the distribution of $\hat{p}$ for the first sample.	$\hat{p} \sim N\left(\frac{147}{320}, 0.027\,86^2\right)$
2 Construct a 90% confidence interval using CAS.	$\frac{147}{320} - 1.645(0.028) \le p \le \frac{147}{320} + 1.645(0.028)$ $0.4136 \le p \le 0.5052$ **Exam hack** If you are rounding values on your page, be sure to use the full decimal values in CAS.
b 1 Interpret the question as a probability statement.	For $\hat{p} \sim N\left(\frac{147}{320}, 0.027\,86^2\right)$
2 Use CAS to calculate the probability.	$P\left(\hat{p} > \frac{71}{105}\right) = 3.548 \times 10^{-15}$
3 Comment on the likelihood.	It is extremely unlikely to obtain a sample proportion greater than $\frac{71}{105}$ in the first sampling distribution, as the probability is close to 0.
c 1 Obtain the distribution of $\hat{p}$ for the second sample.	$\hat{p} \sim N\left(\frac{71}{105}, 0.045\,67^2\right)$
2 Construct a 90% confidence interval using CAS.	$\frac{71}{105} - 1.645(0.046) \le p \le \frac{71}{105} + 1.645(0.046)$ $0.6011 \le p \le 0.7513$
3 Draw a diagram representing the intervals and observe the location of the second confidence interval in relation to the first confidence interval.	0 0.1 0.2 0.3 0.4 0.5 0.6 0.7 0.8 0.9 1 The two confidence intervals do not overlap.
4 Conclude appropriately in context of the question.	There is sufficient evidence to suggest that the increased advertising campaign may have improved Labor's polling result, but we cannot say for certain.

WACE QUESTION ANALYSIS

© SCSA MM2020 Q12 Calculator-assumed (21 marks)

It is estimated that 20% of small businesses fail in the first year. A business advisory group takes a random sample of 500 new businesses that started in January 2018. An analyst employed by the group suggests the use of the binomial distribution is appropriate in this case.

a What is the probability that at most 120 of the businesses fail in the first year? (2 marks)

b What is the approximate distribution of the sample proportion of small businesses that fail by the end of the year in this sample? Justify your answer. (3 marks)

c What is the probability that the sample proportion of businesses that fail by the end of the year is less than 0.18? (2 marks)

d By January 2019, 90 of the 500 new businesses had failed. Calculate a 95% confidence interval for the proportion of new businesses that fail in the first year. (2 marks)

The business advisory group believes that the proportion of new businesses that fail within a year can be reduced by providing financial advice. They took another random sample of 500 businesses that started in January 2019 and provided them with regular financial advice. In this random sample, at the end of the year 80 businesses had failed.

e Calculate the sample proportion and its margin of error at the 95% confidence level. (2 marks)

f Calculate a 95% confidence interval for the proportion of businesses that failed. What do you conclude regarding the value of the financial advice provided to the new businesses? (4 marks)

g If the sample size was reduced, what would be the effect on the confidence interval? Justify your answer. (2 marks)

h State **two** assumptions that the analyst made in recommending the use of the binomial model in this case and discuss whether they are valid. (4 marks)

Video WACE question analysis: Interval estimates for proportions

Reading the question

- Make note of the given values of n, p and $\hat{p}$ throughout the question.
- Highlight when you are told to use a specific distribution, e.g. binomial distribution.
- Highlight the key high-order command words, e.g. *justify, conclude, discuss.*

Thinking about the question

- Make sure you know when to use the normal distribution rather than the binomial distribution.
- Recall the possible situations and conclusions when comparing confidence intervals for two samples after changed circumstances.

Worked solution (✓ = 1 mark)

a Let the random variable X denote the number of new businesses that fail out of the 500.

$X \sim \text{Bin}(500, 0.2)$

$P(X \leq 120) = 0.9877$

defines a binomial random variable and its parameters ✓

calculates the probability ✓

b Given that n is sufficiently large and $np = 100 \geq 10$ and $n(1 - p) = 400 \geq 10$, then $\hat{p}$ is approximately normal, such that $\hat{p} \sim N(0.2, 0.0179^2)$.

uses sample size and p to justify approximate normality ✓

states the mean of the distribution ✓

states the variance of the distribution ✓

c $\mathbf{P(\hat{p} < 0.18) = 0.1318}$ ✓

uses the approximate normal distribution to calculate $\mathbf{P(\hat{p} < 0.18)}$ ✓

d A new sample proportion is defined and so this must be used in the construction of the confidence interval for p.

$\hat{p} = \frac{90}{500} = 0.18$

$z = 1.960$

$0.18 - 1.960\sqrt{\frac{0.18(0.82)}{500}} \le p \le 0.18 + 1.960\sqrt{\frac{0.18(0.82)}{500}}$

$0.1463 \le p \le 0.2137$

shows the construction of the confidence interval ✓

obtains the correct bounds of the confidence interval ✓

e A new sample proportion is defined and so this must be used in the construction of the confidence interval for p.

$\hat{p} = \frac{80}{500} = 0.16$

$\mathbf{z = 1.960}$ ✓

$\mathbf{E = 1.960\sqrt{\frac{0.16(0.84)}{500}} = 0.0321}$ ✓

f $0.16 - 0.0321 \le p \le 0.16 + 0.0321$

$0.1279 \le p \le 0.1921$

Comparing $0.1463 \le p \le 0.2137$ and $0.1279 \le p \le 0.1921$, $\hat{p}_1 \in [0.1279, 0.1921]$and $\hat{p}_2 \in [0.1463, 0.2137]$. Given the overlap of the confidence intervals such that the sample proportions are mutually contained, there isn't sufficient evidence to suggest that the financial advice has reduced the proportional of businesses that fail in the first year.

shows the construction of the second confidence interval ✓

obtains the correct bounds of the second confidence interval ✓

notes the location of the sample proportions in relation to the overlapping confidence intervals ✓

concludes correctly in context of the question ✓

g As n decreases, standard error increases and so margin of error increases and, hence, the width of the confidence interval increases.

describes the effect on the standard error and margin of error ✓

describes the effect on the width of the confidence interval ✓

h The underlying assumptions of a binomial distribution are as follows:

- The success or failure of each business is independent of the success or failure of any other business. This is unlikely to be valid as similar businesses may both fail or both survive, or be affected by the competition within the market of other similar businesses.
- The probability of a business failing in the first year is a fixed, constant value of p. This is unlikely to be valid, as different types of business are expected to have different probabilities of failure, possibly at different times in the year.

states assumption that a binomial distribution involves independent trials in context ✓

explains why this assumption is invalid ✓

states assumption that each trial has a fixed probability of success in context ✓

explains why this assumption is invalid ✓

Exam hack

- When dealing with a large question with multiple marks, check to see whether successive parts of the question rely on earlier answers or whether they can be answered independently.
- Answer to four decimal places to be safe, even if not asked to.
- Always state the distribution and its parameters the first time it is being used.
- Be clear and specific with explanations and justifications and involve the language of the mathematical concepts.

EXERCISE 9.3 Confidence intervals for proportions

ANSWERS p. 410

Recap

1 Which of the following binomially distributed random variables would give the best approximation of a normal distribution?

A $X \sim \text{Bin}(20, 0.8)$ **B** $X \sim \text{Bin}(15, 0.6)$ **C** $X \sim \text{Bin}(40, 0.25)$

D $X \sim \text{Bin}(40, 0.5)$ **E** $X \sim \text{Bin}(90, 0.01)$

2 Samples of bread rolls from bakeries around Western Australia were weighed. The samples all had the same number of rolls. The mean of the sampling distribution of sample proportions of underweight rolls was found to be about 0.09. The standard deviation of the sampling distribution of sample proportions was found to be about 0.03.

The sample size was approximately

A 3 **B** 9 **C** 10 **D** 90 **E** 100

Mastery

3 WORKED EXAMPLE 13 A random sample of 50 T-shirts from a large consignment found that 25 had embroidery defects. Let $\hat{p}$ be the random variable for the proportion of T-shirts with embroidery defects in samples of size 50.

a State the distribution of $\hat{p}$. Justify your answer.

b Hence, write expressions for the approximate confidence intervals of p with the following levels of confidence. *Do not evaluate these intervals.*

i 90% confidence level

ii 95% confidence level

iii 99% confidence level

4 Using CAS 7 Of 140 randomly sampled songs played on a radio station during 'drive time' over several weeks, it was found that 95 of them were less than 3 minutes long. Assuming the approximate normality of $\hat{p}$, determine an approximate 90% confidence interval for the true proportion of songs during 'drive time' on this radio station that are less than 3 minutes long, correct to three decimal places.

5 WORKED EXAMPLE 14 Blocks of a particular brand of chocolate are labelled with a weight of 250 g. In a random sample of 300 blocks, it is found that 110 were underweight. Let $\hat{p}$ be the random variable representing the sample proportion of underweight blocks of chocolate in samples of size 300.

a Assuming the approximate normality of $\hat{p}$, determine an approximate 95% confidence interval for p, correct to three decimal places.

b A second random sample of 300 blocks was taken, the number of underweight blocks was observed and an approximate 95% confidence interval was found to be $(0.27, 0.37)$. Which of the two confidence intervals is more likely to contain the true value of p? Justify your answer.

c If a further 240 random samples of 300 blocks were to be taken and approximate 95% confidence intervals were to be constructed for p, how many of the confidence intervals could be expected to contain the true value of p?

6 WORKED EXAMPLE 15 An approximate 99% confidence interval for p for a random sample of size 200 is found to be $(0.09, 0.22)$. State the effect on the width of the confidence interval constructed

a if the confidence level was decreased to 97%, but $\hat{p}$ and n remain unchanged

b if the sample size is increased to 300, but $\hat{p}$ and z remain unchanged.

7 WORKED EXAMPLE 16 An approximate 90% confidence interval for p using a random sample of size n is found to be $(0.408, 0.558)$. Calculate the

a sample proportion $\hat{p}$ used to construct the approximate confidence interval

b margin of error of the confidence interval

c standard error of $\hat{p}$ to three decimal places

d sample size used.

8 WORKED EXAMPLE 17 An approximate C% confidence interval for p using a random sample of size 25 is found to be $(0.148, 0.652)$. Find the level of confidence, correct to the nearest percentage, used to construct the confidence interval.

9 WORKED EXAMPLE 18 A previous study suggests that about 28% of Year 12 students work a part-time job during their final year of schooling. Determine the minimum sample size that would be needed to obtain an approximate 95% confidence interval for p with a maximum width of 10%.

10 WORKED EXAMPLE 19

a An advertising company wants to conduct a small survey of consumers to establish a baseline for the proportion of consumers who were aware of a particular brand of ice cream before a marketing campaign. Determine the minimum number of consumers that need to be surveyed to obtain an estimate for p accurate to within 3% at a confidence level of 95%.

b An insurance company wants to conduct a small survey of customers to gain insight into the proportion of customers who have fire and theft cover for their car. Determine the minimum number of customers that need to be surveyed to obtain an estimate for p accurate to within 10% at a confidence level of 90%.

11 WORKED EXAMPLE 20

a A local Perth tour guide claims that approximately 40% of tourists who enquire at the Perth CBD Information Centre ask about attractions in the Perth CBD. Over the course of a week, it was recorded that in a random sample of 75 tourists who enquired at the Information Centre, 36 of them enquired about attractions in the Perth CBD. Use an approximate 95% confidence interval to comment on the tour guide's claim.

b Osborne Park piano tuners north of the Swan River estimate that the proportion of pianos requiring new strings when they are tuned is about 0.30. A piano tuner based in Victoria Park south of the Swan River checked their records and found that of 120 pianos, 50 needed new strings. Use an approximate 95% confidence interval to comment on whether piano tuning demands north and south of the Swan River are the same.

12 WORKED EXAMPLE 21 In a survey of 30 randomly selected Western Australian government officials, it was found that 40% of officials were in favour of Western Australia adopting a new state flag.

a Construct an approximate 95% confidence interval for the true proportion of government officials in favour of adopting a new state flag, correct to four decimal places.

In a follow-up survey, a few newly proposed designs for the flag were included and it was found that 12 out of 50 people surveyed said they were in favour of adopting a new state flag.

b How likely is it to obtain a sample proportion less than that found in the second random sample when using the sampling distribution of sample proportions of the first sample?

c A marketing officer claimed that the newly proposed designs were not better than the current state flag. Perform the necessary calculations to comment on the marketing officer's claim.

9.3

13 Jacinta tosses a coin five times. Albin suspects that the coin Jacinta tossed is not actually a fair coin and he tosses it 18 times. Albin observes a total of 12 heads from the 18 tosses. Based on this sample, construct the approximate 90% confidence interval for the probability of observing a head when this coin is tossed. Use the z value of 1.645. *Do not evaluate the bounds of the interval estimate.*

Calculator-free

14 © SCSA MM2017 Q4 (3 marks) Two independent samples of different sizes were taken from a population. The first sample had sample size n_1 and the second sample had sample size n_2. The sample proportions of males in the samples were the same. When 99% confidence intervals were calculated for each sample, it was found that the corresponding margin of error in the second sample was half that of the first sample.

What is the ratio of the two sample sizes, $\frac{n_2}{n_1}$?

15 © SCSA MM2018 Q5 (3 marks) A 95% confidence interval for a population proportion based on a sample size of 200 has width w. What sample size is required to obtain a 95% confidence interval of width $\frac{w}{3}$?

Calculator-assumed

16 (4 marks) A laptop supplier collects a sample of 100 laptops that have been used for six months from a number of different schools and tests their battery life. The laptop supplier wishes to estimate the proportion of such laptops with a battery life of less than three hours. The laptop supplier finds that, in a particular sample of 100 laptops, six of them have a battery life of less than three hours.

a Determine a 95% confidence interval for the supplier's estimate of the proportion of interest. Give values correct to three decimal places. (3 marks)

b Give one reason as to why the confidence interval in part **a** may not be considered reliable. (1 mark)

17 (3 marks) An opinion pollster reported that for a random sample of 574 voters in a town, 436 indicated a preference for retaining the current council. Determine an approximate 90% confidence interval for the proportion of the total voting population with a preference for retaining the current council, correct to three decimal places.

18 © SCSA MM2016 Q14c (3 marks) The simulation of a loaded (unfair) five-sided die rolled 60 times is recorded with the following results.

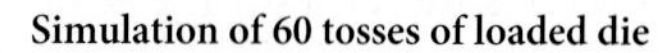

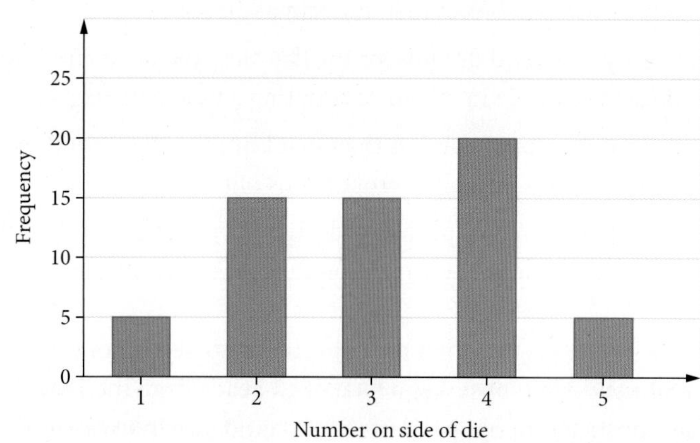

It has been decided to create a confidence interval for the proportion of prime numbers in this simulation. The level of confidence will be chosen from 90% or 95%.

Explain which level of confidence will give the smallest margin of error. State this margin of error.

19 © SCSA MM2017 Q18abc (6 marks) Alex is a beekeeper and has noticed that some of the bees are very sleepy. She takes a random sample of 320 bees and finds that 15 of them are indeed so-called *lullabees* that fall asleep easily.

a Calculate the sample proportion of lullabees. (1 mark)

b Determine a 90% confidence interval for the true proportion of lullabees, rounded to four decimal places. (3 marks)

c What is the margin of error in the above estimate? (2 marks)

20 © SCSA MM2018 Q13 (10 marks) The proportion of caravans on the road being towed by vehicles that have the incorrect towing capacity is p.

a Show, using calculus, that to maximise the margin of error a value of $\hat{p} = 0.5$ should be used. Note: As z and n are constants, the standard error formula can be reduced to $E = \sqrt{\hat{p}(1 - \hat{p})}$. (3 marks)

b A consulting firm wants to determine p within 8% with 99% confidence. How many towing vehicles should be tested at a random check? (3 marks)

c Six months later, the consulting firm carries out a random sampling of towing vehicles. A 99% confidence interval calculated for the proportion of vehicles with incorrect towing capacity is $(0.342, 0.558)$. Determine the number of vehicles in the sample that have an incorrect towing capacity. (4 marks)

21 © SCSA MM2019 Q14 (7 marks)

a What is the minimum sample size required to estimate a population proportion to within 0.01 with 95% confidence? (3 marks)

b Identify **two** factors that affect the width of a confidence interval for a population proportion and describe the effect of each. (4 marks)

22 © SCSA MM2020 Q14ab MODIFIED (6 marks) A suburban council hires a consultant to estimate the proportion of residents of the suburb who use its library.

a The consultant decides to estimate a 95% confidence interval for the proportion to within an error of 0.03. What minimum sample size should be selected? (3 marks)

b If resource limitations dictate that the maximum sample size that can be managed is 500, what is the maximum margin of error in estimating a 99% confidence interval? (3 marks)

9.3

23 © SCSA MM2018 Q17efg (6 marks) Tina believes that approximately 60% of the mangoes she produces on her farm are large. She takes a random sample of 500 mangoes from a day's picking and finds that it contains 250 large mangoes.

a Calculate a 95% confidence interval for the proportion of large mangoes produced on the farm, rounded to four decimal places. (3 marks)

b On the basis of your calculations, how would you respond to Tina's belief that the proportion of large mangoes produced is at least 60%? Justify your response. (2 marks)

c What can Tina do to further test her belief? (1 mark)

24 (7 marks) A company supplies schools with whiteboard pens. As a whiteboard pen ages, its tip may dry to the point where the whiteboard pen becomes defective (unusable). The company has stock that is two years old and, at that age, company historical data suggests that 6% of Grade A whiteboard pens will be defective. A box of 100 Grade A whiteboard pens that is two years old is selected and it is found that 10 of the whiteboard pens are defective.

a Determine an approximate 99% confidence interval for the population proportion from this sample, correct to four decimal places. (3 marks)

b Determine an approximate 90% confidence interval for the population proportion from this sample, correct to four decimal places. (3 marks)

c Using the two confidence intervals constructed, comment on whether it appears that the company's historical data is still relevant for their current stock. (1 mark)

25 (4 marks) Rusty's Robotics manufactures sensor components for robots. Prior company data suggests that approximately 5% of all the sensors manufactured are defective. A random sample of 500 sensors is selected and it is found that 40 sensors in this sample are defective.

a Determine an approximate 95% confidence interval for the proportion of defective sensors, correct to four decimal places. (3 marks)

b Comment on whether it appears that the company's historical data is still relevant for their current manufacturing quality. (1 mark)

26 (7 marks) A local entertainment reviewer claims in a newspaper article that approximately 4% of concerts start more than 15 minutes after the scheduled starting time. For the purposes of customer satisfaction, the owners of the local Mathsland Concert Hall decide to review their operation and study the information from 1000 concerts conducted at their venue, collected as a simple random sample. The sample value for the number of concerts that start more than 15 minutes after the scheduled starting time is found to be 43.

a Describe the sampling distribution of sample proportions, $\hat{p}$, for the proportion of interest. Justify your answer. (3 marks)

b Find an approximate 95% confidence interval for the proportion of concerts that begin more than 15 minutes after the scheduled starting time. Give values correct to three decimal places. (2 marks)

c The owners of the Mathsland Concert Hall claim that the reviewer must have visited their concert hall before writing the review. Comment on the validity of such a claim. (2 marks)

27 © SCSA MM2019 Q8 (7 marks) Big Foods is a large supermarket company. The manager of Big Foods wants to estimate the proportion of households that do the majority of their grocery shopping in their stores. A junior staff member at Big Foods conducted a survey of 250 randomly-selected households and found that 56 did the majority of their grocery shopping at a Big Foods store.

a Calculate the sample proportion of households who did the majority of their grocery shopping at Big Foods. (1 mark)

b Determine the 95% confidence interval for the proportion of households who do the majority of their grocery shopping at Big Foods. Give your answer to four decimal places. (3 marks)

c What is the margin of error of the 95% confidence interval? Give your answer to four decimal places. (1 mark)

An independent research company conducted a large-scale survey of household supermarket preferences and estimated that the true proportion of households that conduct most of their grocery shopping at Big Foods was 0.17 (assume that this is indeed the true proportion).

d With reference to your answer to part **b**, does this result suggest that the junior staff member at Big Foods made a mistake? (2 marks)

28 © SCSA MM2021 Q13defg (8 marks) A carnival game involves five buckets, each containing 5 blue balls and 15 red balls. A player blindly selects a ball from each bucket and wins the game if they select at least 4 blue balls. Let X denote the number of blue balls selected. An observer records the outcome of 100 consecutive games and determines the 90% and 95% confidence intervals for the proportion of wins, p. The confidence intervals are $(0.04, 0.16)$ and $(0.05, 0.15)$.

a Which of these intervals is the 95% confidence interval for p? Justify your answer. (2 marks)

b How many wins were observed out of the 100 games? (2 marks)

c Determine what you would expect to happen to the width of the confidence intervals if 400 games had been observed. (2 marks)

d The true proportion of wins does not lie within either of the above confidence intervals. Does this suggest that a sampling error was made? Justify your answer. (2 marks)

29 © SCSA MM2016 Q10 MODIFIED (11 marks) A survey in Western Australia was conducted on the popularity of a calculator known as Type A. Out of 1450 Year 12 students, the survey found that 986 students used the Type A calculator.

a Determine an approximate 90% confidence interval, to three decimal places, for the proportion of Western Australian Year 12 students who use the Type A calculator, stating any necessary assumptions. (3 marks)

b State the margin of error in this confidence interval. (1 mark)

Another three surveys of Year 12 students were conducted on the use of Type A calculators across Australia.

Survey 2	Survey 3	Survey 4
Type A usage 1772 out of 3221 Year 12 students	Type A usage 1021 out of 1566 Year 12 students	Type A usage 2203 out of 3221 Year 12 students

c Determine approximate 90% confidence intervals for Surveys 2 to 4. (3 marks)

d A data analyst claims that Survey 2 is likely to have been taken outside of Western Australia. Comment on the validity of the analyst's claim. (2 marks)

e Using the sample proportion of Survey 1, determine a sample size that will halve the margin of error for the proportion of Western Australian Year 12 students who use the Type A calculator, with a confidence of 90%. (2 marks)

30 © SCSA MM2016 Q20 MODIFIED (12 marks) A chocolate factory produces chocolates with the machines calibrated such that 80% of chocolates produced are pink. Each box of chocolates contains exactly 30 pieces.

a Identify the probability distribution of X: the number of pink chocolates in a single box. Give the mean and standard deviation of X. (3 marks)

b Determine the probability, to four decimal places, that there are at least 27 pink chocolates in a randomly selected box. (2 marks)

Quality Control collects a sample of 20 boxes of chocolates and finds that 450 chocolates are pink.

c By first stating an appropriate distribution for the sample proportion of pink chocolates, determine an approximate 95% confidence interval for the proportion of pink chocolates in a sample of 20 boxes. (4 marks)

d Quality Control claims that there may be an error with the calibration of the machine. Comment on the validity of the claim. (1 mark)

To check the calibration, Quality Control collects a further three samples and determines an approximate 95% confidence interval each time. It is found that all three contain the value of 0.8.

e Use this finding to account for the results in parts **c** and **d**. (2 marks)

9 Chapter summary

Random sampling

- A **census** collects data from an entire **population** and is used to calculate **population parameters** of a certain characteristic, for example, population mean and standard deviation.
- A **survey** collects data from a **sample** group of a population and is used to calculate **sample statistics** of a certain characteristic, for example, sample mean and standard deviation.
- A sample is **fair and representative** if
  - the sample size is sufficiently large enough to represent the population
  - the data is free from biases that could affect the reliability of it being used to estimate population parameters.
- An unfair and non-representative sample is called a **biased sample**.
- A **probability sampling method** is a data collection process whereby each member of the population has an equally likely chance of being randomly selected. These methods often minimise bias.
- A **non-probability sampling method** is a data collection process whereby each member of the population does not have an equally likely chance of being randomly selected. These methods may introduce different biases.
- Due to the nature of random sampling, there exists **variability in random samples** such that
  - the sample statistics and shape of the distribution will vary, but will approximate the parameters and shape of the population distribution
  - as $n \to \infty$, the mean of a sample will generally tend towards $E(X)$, but can still vary, and the shape of the distribution will better represent the shape of the distribution of X.

The sampling distribution of sample proportions

- For a single sample of size n, the **sample proportion** is $\hat{p} = \dfrac{\text{number of observed successes}}{n}$.
- As a random variable, $\hat{p} = \dfrac{X}{n}$, where $X \sim \text{Bin}(n, p)$ such that n is the sample size and p is the probability of success (i.e. true population proportion).
- The distribution of all possible $\hat{p}$ values has
  - $\text{E}(\hat{p}) = p$
  - $\text{Var}(\hat{p}) = \dfrac{p(1-p)}{n}$
  - $\text{SD}(\hat{p}) = \sqrt{\dfrac{p(1-p)}{n}}$
- As $n \to \infty$, $\text{Var}(\hat{p}) \to 0$ and $\text{SD}(\hat{p}) \to 0$, meaning there is very little variation in the different values of $\hat{p}$ taken from different samples of a significantly large, fixed size n.
- For a binomially distributed random variable $X \sim \text{Bin}(n, p)$,
  - if $p \approx 0.5$, with a sufficiently large n (e.g. $n \geq 30$) or
  - n is sufficiently large such that $np \geq 10$ and $n(1 - \text{p}) \geq 10$,

  then X can be modelled by an **approximate normal distribution** of the form:

  $X_N \sim N(np, np(1-p))$.

- If X is approximately normal, then by the **central limit theorem** $\hat{p} = \frac{X_N}{n}$ is approximately normal such that:

  $$\hat{p} \sim N\left(p, \frac{p(1-p)}{n}\right)$$

- When a **point estimate** is used to estimate p, then

  $$\hat{p} \sim N\left(\hat{p}, \frac{\hat{p}(1-\hat{p})}{n}\right)$$

- An **approximate standard normal distribution**, $Z \sim N(0, 1)$, can be obtained using the linear transformation
  - $Z = \dfrac{\hat{p} - p}{\sqrt{\dfrac{p(1-p)}{n}}}$ when p is known
  - $Z = \dfrac{\hat{p} - \hat{p}_1}{\sqrt{\dfrac{\hat{p}_1(1-\hat{p}_1)}{n}}}$ when the value of p is unknown and a specific sample proportion $\hat{p}_1$ is used

  as a point estimate for p.

Confidence intervals

- An **interval estimate** for p is of the form $\hat{p} - E \leq p \leq \hat{p} + E$, where E is a **margin of error**.
- An **approximate 100c% confidence interval** for p has the margin of error, $E = z\,\text{SD}(\hat{p}) = z\sqrt{\dfrac{\hat{p}(1-\hat{p})}{n}}$,

  where $\text{SD}(\hat{p})$ is called the **standard error**.
- An approximate 100c% confidence interval for p with a corresponding z-score, z, and a margin of error of $E = z\sqrt{\dfrac{\hat{p}(1-\hat{p})}{n}}$, can be given by any of the following notations:

  $$\hat{p} - z\sqrt{\frac{\hat{p}(1-\hat{p})}{n}} \leq p \leq \hat{p} + z\sqrt{\frac{\hat{p}(1-\hat{p})}{n}}$$

  $$\left[\hat{p} - z\sqrt{\frac{\hat{p}(1-\hat{p})}{n}}, \hat{p} + z\sqrt{\frac{\hat{p}(1-\hat{p})}{n}}\right]$$

  $$\hat{p} - z\sqrt{\frac{\hat{p}(1-\hat{p})}{n}} < p < \hat{p} + z\sqrt{\frac{\hat{p}(1-\hat{p})}{n}}$$

  $$\left(\hat{p} - z\sqrt{\frac{\hat{p}(1-\hat{p})}{n}}, \hat{p} + z\sqrt{\frac{\hat{p}(1-\hat{p})}{n}}\right)$$

- 90%, 95% and 99% confidence levels are the most commonly used for confidence intervals, with the following standard scores.

Confidence level	90%	95%	99%
Standard score (z-score)	1.645	1.960	2.576

- The width, w, of a 100c% confidence interval is twice the margin of error

  $$w = 2E = 2z\sqrt{\frac{\hat{p}(1-\hat{p})}{n}}$$

- The frequentist interpretation of confidence intervals says: *Upon the repeated construction of a large number of approximate* 100c% *confidence intervals for p, from multiple random samples of sample size n, we can expect (on average) that* 100c% *of all confidence interval yet to be constructed will contain the true value of p.*
  - Most, but not all, confidence intervals contain p.
  - Because p is unknown and due to the nature of random sampling, it can never be known for certain whether a confidence interval contains p.
  - Because p is constant, once a confidence interval is constructed, the probability that the given confidence interval contains p is either 0 or 1. It either does not contain p or it does, but we can never know for certain because p is unknown.
  - No single constructed confidence interval is any more or less likely to contain p than any other single constructed confidence interval.
- If the values of $\hat{p}$ and n remain unchanged, as the confidence level 100c% increases
  - the value of z increases, and so
  - the value of the margin of error E increases, and so
  - the width of the confidence interval increases.
- If the values of $\hat{p}$ and z remain unchanged, as the sample size n increases
  - the value of the standard error $\sqrt{\frac{\hat{p}(1-\hat{p})}{n}}$ decreases, and so
  - the value of the margin of error E decreases, and so
  - the width of the confidence interval decreases.
- If $\hat{p}$ is unknown, then $\hat{p} = 0.5$ produces the largest margin of error in a confidence interval constructed with a given sample size n.
- Approximate confidence intervals for p constructed from samples can be used to comment on
  - the validity of claimed values of p or a value of p from historical data by observing the location of p with respect to $[\hat{p} - E, \hat{p} + E]$
  - the effect of changed circumstances between samples by observing the location of $[\hat{p}_2 - E, \hat{p}_2 + E]$ in relation to $[\hat{p}_1 - E, \hat{p}_1 + E]$.

Cumulative examination: Calculator-free

Total number of marks: 28 Reading time: 2 minutes Working time: 28 minutes

1 (4 marks) A binomial random variable has mean 20 and variance 4.

a Write two equations in terms of n and p. (2 marks)

b Find the values of n and p. (2 marks)

2 (5 marks)

a Find $\frac{dy}{dx}$ if $y = x^3 \ln(3x)$. (2 marks)

b Hence find $\int x^2 \ln(3x)\,dx$. (3 marks)

3 © SCSA MM2017 Q2 (6 marks) Michelle is a soccer goalkeeper and has built a machine to help her practise. The machine will shoot a soccer ball randomly along the ground at or near a goal that is seven metres wide. The machine is equally likely to shoot the ball so that the centre of the ball crosses the goal line anywhere between point A three metres left of the goal, and point B five metres right of the goal, as shown in the diagram below.

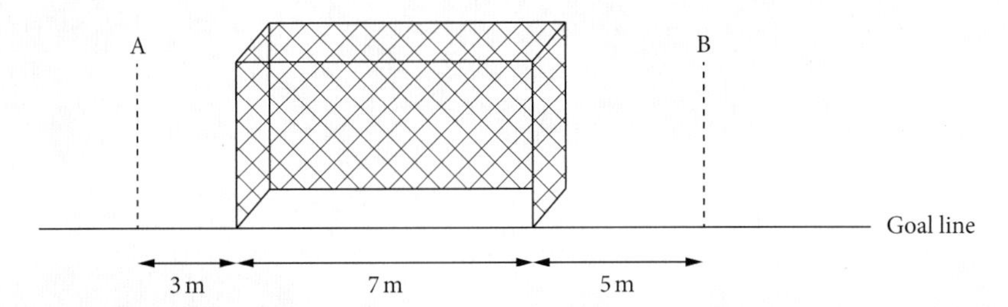

Michelle sets up a trial run without anyone in the goals. Assume the goal posts are of negligible width.

Let the random variable X be the distance the centre of the ball crosses the goal line to the right of point A.

a Copy and complete the graphical representation of the probability density function for the random variable X. (2 marks)

b What is the probability that the machine shoots a ball so that its centre misses the goal to the left? (1 mark)

c What is the probability that the machine shoots a ball so that its centre is inside the goal? (1 mark)

d If the machine shoots a ball so that its centre misses the goal, what is the probability that the ball's centre misses to the right? (2 marks)

4 © SCSA MM2020 Q7 MODIFIED (13 marks) Consider the function $f(x) = e^{2x} - 6e^{x} + 8$.

a Determine the coordinates of the x-intercept(s) of f. You may wish to consider the factorised version of f: $f(x) = (e^{x} - 2)(e^{x} - 4)$. (3 marks)

b Show that there is only one turning point on the graph of f, which is located at $(\ln(3), -1)$. (3 marks)

c Determine the coordinates of the point(s) of inflection of f. (3 marks)

d Sketch the function f, labelling clearly all intercepts, the turning point and point(s) of inflection. Some approximate values of the natural logarithmic function provided in the table below may be helpful. (4 marks)

x	1	2	3	4
$\ln(x)$	0	0.7	1.1	1.4

Cumulative examination: Calculator-assumed

Total number of marks: 66 Reading time: 8 minutes Working time: 66 minutes

1 (3 marks) The volume $V\text{cm}^3$ of water in a vessel is given by $V = \frac{1}{6}\pi x^3$, where $x\text{ cm}$ is the depth of the water in the vessel in cm. By using the increments formula, determine an approximation for the change in depth when the volume of water changes from 200 to 210 cm^3.

2 (7 marks) Let $f(x) = \frac{1}{5}(x-2)^2(5-x)$. The point $P\left(1, \frac{4}{5}\right)$ is on the graph of f, as shown below. The tangent at P cuts the y-axis at S and the x-axis at Q.

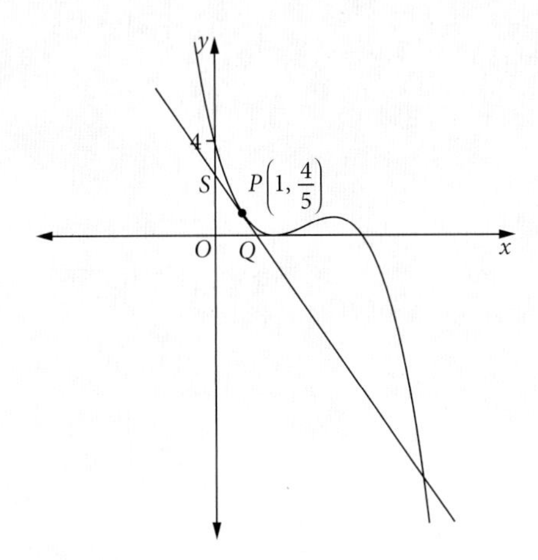

a Write down the derivative $f'(x)$ of $f(x)$. (1 mark)

b **i** Find the equation of the tangent to the graph of f at the point $P\left(1, \frac{4}{5}\right)$. (1 mark)

ii Find the coordinates of points Q and S. (2 marks)

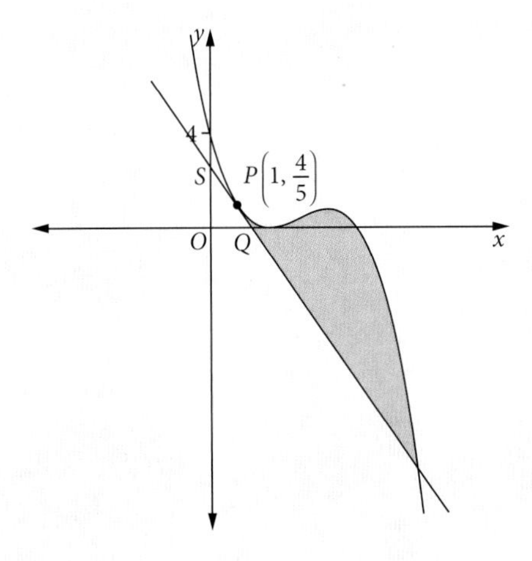

c Find the area of the shaded region in the graph above. (3 marks)

3 (9 marks) Steve, Katerina and Jess are three students who have agreed to take part in a psychology experiment. Each student is to answer several sets of multiple-choice questions. Each set has the same number of questions, n, where n is a number greater than 20. For each question there are four possible options (A, B, C or D), of which only one is correct.

a Steve decides to guess the answer to every question, so that for each question he chooses A, B, C or D at random. Let the random variable X be the number of questions that Steve answers correctly in a particular set.

i What is the probability that Steve will answer the first three questions of this set correctly? (1 mark)

ii Find, to four decimal places, the probability that Steve will answer at least 10 of the first 20 questions of this set correctly. (2 marks)

iii Use the fact that the variance of X is $\frac{75}{16}$ to show that the value of n is 25. (1 mark)

If Katerina answers a question correctly, the probability that she will answer the next question correctly is $\frac{3}{4}$. If she answers a question incorrectly, the probability that she will answer the next question incorrectly is $\frac{2}{3}$.

In a particular set, Katerina answers Question 1 incorrectly.

b Calculate the probability that Katerina will answer questions 3, 4 and 5 correctly. (3 marks)

c The probability that Jess will answer any question correctly, independently of her answer to any other question, is p ($p > 0$). Let the random variable Y be the number of questions that Jess answers correctly in any set of 25.

If $P(Y > 23) = 6P(Y = 25)$, show that the value of p is $\frac{5}{6}$. (2 marks)

4 (9 marks) Toby is learning to speak Spanish before going to South America for 12 months. While completing an online course, the number of words he learns, w, is modelled by the function.

$$w = 100\ln(t + 1) + 150$$

where t is the number of days after he starts his online course.

Toby needs a very basic vocabulary of 600 words for the trip.

a How many Spanish words did Toby know when he started the course? (1 mark)

b How many words did Toby learn in the first day? (2 marks)

c How many Spanish words did Toby know after 5 days? (1 mark)

d How long will it take him to learn the very basic vocabulary? (2 marks)

e Write the equation in the form $t = e^{aw-b} - c$. (3 marks)

5 (14 marks) A train is travelling at a constant speed of w km/h along a straight level track from M towards Q. The train will travel along a section of track $MNPQ$.

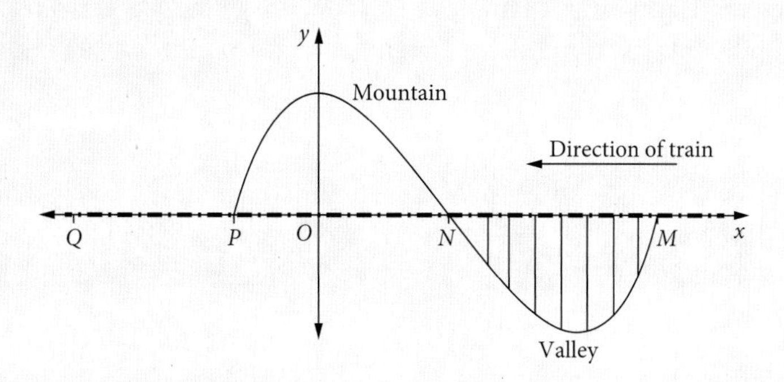

Section MN passes along a bridge over a valley. Section NP passes through a tunnel in a mountain. Section PQ is 6.2 km long.

From M to P, the curve of the valley and the mountain, directly below and above the train track, is modelled by the graph of

$y = \frac{1}{200}(ax^3 + bx^2 + c)$ where a, b and c are real numbers.

All measurements are in kilometres.

a The curve defined from M to P passes through $N(2, 0)$. The gradient of the curve at N is -0.06 and the curve has a turning point at $x = 4$.

i From this information write down three simultaneous equations in a, b and c. (3 marks)

ii Hence show that $a = 1$, $b = -6$ and $c = 16$. (2 marks)

b Find, giving exact values

i the coordinates of M and P (2 marks)

ii the length of the tunnel (1 mark)

iii the maximum depth of the valley below the train track. (1 mark)

The driver sees a large rock on the track at a point Q, 6.2 km from P. The driver puts on the brakes at the instant that the front of the train comes out of the tunnel at P.

From its initial speed of w km/h, the train slows down from point P so that its speed v km/h is given by $v = k\log_e\left[\frac{(d+1)}{7}\right]$ where d km is the distance of the front of the train from P and k is a real constant.

c Find the value of k in terms of w. (1 mark)

d If $v = \frac{120\log_e(2)}{\log_e(7)}$ when $d = 2.5$, find the value of w. (2 marks)

e Find the exact distance from the front of the train to the large rock when the train finally stops. (2 marks)

6 © SCSA MM2017 Q12bcde MODIFIED (9 marks) A common problem with a particular tablet is screen failure. Historically, the manufacturer of Slate Tablets has found that 1% of its tablet screens will fail within three years. A sample of 200 tablets is taken. Let the random variable X denote the number of tablets that have screen failure within three years in the sample of 200.

a State the distribution of X. (2 marks)

b Determine the probability, to four decimal places, that more than four tablets will have screen failure within three years. (2 marks)

In a random sample of 200 Slate Tablets, four of them had screen failure within three years.

c Calculate an approximate 95% confidence interval for the proportion of tablets that have screen failure within three years. Give your answer to four decimal places. (3 marks)

d Comment on whether the company's historical data still appears relevant for current standards of tablet screen quality. (1 mark)

e The company's quality control department wants the proportion of tablets with faulty screens to be no more than 1%. Based on your confidence interval, decide whether the quality control department is meeting its target. Justify your decision. (1 mark)

7 © SCSA MM2021 Q11 (15 marks) A new political party, the Sustainable Energy Party, is planning to have candidates run in the next election. Researchers have collected data that suggests the proportion of voters likely to vote for the party to be 23%.

One year before the next election, random samples of 400 voters were taken in a particular electorate. Let $\hat{p}$ denote the sample proportion of voters who indicated they would vote for the Sustainable Energy Party at the next election.

a State the distribution of $\hat{p}$. (2 marks)

b Calculate the probability that the proportion of voters likely to vote for the Sustainable Energy Party in a sample of 400 is less than 0.20. (2 marks)

One week before the election, researchers believed that the proportion of voters likely to vote for the party in that same electorate had increased. A random sample of 200 voters was taken at this time, and 55 of them indicated they would vote for the Sustainable Energy Party at the next election.

c Based on this sample, estimate the proportion of voters likely to vote for the Sustainable Energy Party in this electorate. (1 mark)

d For a 99% confidence interval, what is the margin of error of the sample proportion of voters likely to vote for the Sustainable Energy Party in this electorate, based on this sample? (2 marks)

e Based on this sample, calculate a 95% confidence interval for the population proportion of voters likely to vote for the Sustainable Energy Party in this electorate. (3 marks)

f Based on the research, did the proportion of voters likely to vote for the Sustainable Energy Party in this electorate increase in the year leading up to the election? Justify your answer. (2 marks)

g The analysis above models the number of voters likely to vote for the Sustainable Energy Party as binomially distributed. State and discuss the validity of any assumptions for the binomial distribution in this context. (3 marks)

Answers

CHAPTER 1

EXERCISE 1.1

1 **a** $-18x^5$ **b** $-x^{-\frac{9}{4}} = -\frac{1}{x^{\frac{9}{4}}}$ **c** $3x^2 + \frac{\sqrt{5}}{x^{\frac{2}{3}}}$

2 **a** $f'(x) = 6x^2 - 2x, f'(-2) = 28$

b $f'(x) = 2x + 1$

$f'(1)$ = undefined (since $f(x)$ is not defined at $x = 1$)

c $f'(x) = 2 + 9x^2$ $f'\left(\frac{1}{8}\right) = 2\frac{9}{64}$

3 $15x^2$ **4** $16\frac{1}{2}$

5 **a** $\frac{\frac{1}{2}}{x^{\frac{1}{2}}} + \frac{\frac{1}{6}}{x^{\frac{5}{6}}} + \frac{\frac{1}{4}}{x^{\frac{3}{4}}}$

b Proof: see worked solutions

6 $\frac{-3(2x+1)}{x^4}$

7 $f'(x) = x^{a-1}$

$4^{a-1} = 16 \Rightarrow a = 3$

8 $3 - 15x^2$

9 $a = -1$ and $b = -3$

10 $-(x+1)^{-3}, -(x-1)^{-3}$

11 **a** $f'(x) = 12x^2 + 5$

b $x^2 \geq 0$ for all x, so $12x^2 + 5 \geq 5$ for all x.

12 $f(x) = a - 2x$

$f'(a) = -a$

Hence $g(x) = -ax + c$.

$g(a) = 0 \Rightarrow -a \times a + c = 0, c = a^2$

13 $x^2 - 1$ **14** $\frac{1-x}{\sqrt{x}}$

EXERCISE 1.2

1 D **2** B

3 $f'(x) = 84x^3 + 84x^2 - 6x - 4$

4 **a** $15x^4 + 4x^3$ **b** $24x + 1$

c $112x - 35$ **d** $7x^6 - 20x^4$

e $24x^5 - 12x^2$ **f** $50x$

g $9x^2 + 2x - 3$ **h** $32x^3 + 30x^2 - 16x - 6$

i $4x^3 + 4x$

5 -18

6 **a** $14x^6 + 44x^3 + 6x^2 + 14x$

b -38

7 14

8 **a** $y = x - 5$ **b** $\left(\frac{7}{4}, \frac{-25}{8}\right)$

9 $y = -3x - 10$ **10** $\frac{81}{4}$

11 $f'(x)g(x) = 2bx(c + dx^2) = 2bcx + 2bdx^3$

$f(x)g'(x) = 2dx(a + bx^2) = 2adx + 2bdx^3$

$2bcx + 2bdx^3 = 2adx + 2bdx^3$

$2bcx = 2adx$

$bc = ad$

12 $a = 5, b = 3$ **13** $-\frac{2}{3}, \frac{3}{4}$

14 **a** $2abx + a^2 + b^2$

b $a = 3, b = 2$ and $a = 2, b = 3$

15 $(2 - \sqrt{7}, 20 + 14\sqrt{7}), (2 + \sqrt{7}, 20 - 14\sqrt{7})$

16 -1.9 **17** $\frac{9}{5}, \frac{162\sqrt{5}}{125}$ **18** $a = \frac{1}{2}$

EXERCISE 1.3

1 E **2** 9 **3** $\frac{-39}{(9x-8)^2}$

4 **a** $\frac{-4}{(2x+3)^2}$ **b** $\frac{1}{(x-5)^2}$ **c** $\frac{2}{(x+1)^2}$

d $\frac{-(2x+1)}{x^2(x+1)^2}$ **e** $\frac{2x^3 + 9x^2 + 3}{(x+3)^2}$ **f** $\frac{x^2 - 1}{x^2}$

g $2x + 1$ **h** $\frac{6x^2 + 30x + 6}{(x^2 - 1)^2}$

5 $\frac{1}{4}$

6 $f'(x) = -\frac{2k}{(x-k)^2}$

$f'(5) = -8 \Rightarrow \frac{2k}{(5-k)^2} = -8$

$4k^2 - 41k + 100 = 0$

$(4k - 25)(k - 4) = 0$

$k = 4$

7 $f'(x) = -\frac{1}{2(x-2)^2}$

8 -5 **9** $y = x - 7, (0, -7), (7, 0)$

10 $a = 5, b = -\frac{1}{2}$

11 $\frac{dy}{dx} = -\frac{1}{(1+x)^2}$

$= -\left(\frac{1}{1+x}\right)^2$

$= -y^2$

12 $\frac{x(x+4)}{(x+2)^2}$ **13** $x = -4.11, x = 0.58$

14 $-\frac{3}{4}$ at $x = 0$ and $-\frac{2}{3}$ at $x = \pm 2$

15 $a = \frac{1}{2}$ **16** $a > -1$

EXERCISE 1.4

1 $\frac{2x^2-6x}{(2x-3)^2}$

2 A

3 a $-\frac{24x^2}{(x^3+1)^5}$ **b** $\frac{x}{\sqrt{x^2-1}}$

4 a $5(x-5)^4$ **b** $16(4x-3)^3$

c $3(6x^2+1)(2x^3+x)^2$ **d** $-24x(8-2x^2)^5$

e $\frac{9}{2}\left(\frac{1}{2}x-6\right)^8$

f $2(3x^2-4x+1)(x^3-2x^2+x+1)$

g $\frac{\sqrt{2}}{\sqrt{2x+3}}$ **h** $\frac{3(1-\sqrt{x})(x-2\sqrt{x})^2}{\sqrt{x}}$

i $\frac{\sqrt{5}}{2\sqrt{x+10}}$ **j** $-\frac{4}{(2x+7)^3}$

k $\frac{1}{2\sqrt{(4-x)^3}}$ **l** $-\frac{15}{2\sqrt{(x-8)^5}}$

5 a $8(2x-1)^3$

b $-6x^2(3-x^3)$

c $28(1+x)(3+4x+2x^2)^6$

d $12(x+3)(x^2+6x)^5$

e $15x^2(1-2x^3)(x^3-x^6+1)^4$

f $(n+1)x^n(x^{n+1}+1)^n$

6 $5(6x-5)(3x^2-5x)^4$

7 $4(x^2-5x)^3(2x-5)$ or $\frac{dy}{dx}=4x^3(x-5)^3(2x-5)$

8 $-\frac{1}{2\sqrt{4-x}}$

9 $3(-9x^2+2x)(-3x^3+x^2-64)^2$

10 $a=1$

11 $\frac{dy}{dx}=\frac{x-a}{\sqrt{1+(x-a)^2}}$

$\sqrt{1+(x-a)^2}$ is positive for all values of x and $x-a$ will be positive if $x>a$.

12 $y=[1-f(x)]^{\frac{1}{2}}$

$y'=\frac{1}{2}[1-f(x)]^{-\frac{1}{2}}(-f'(x))$

$y'=\frac{-f'(x)}{2(1-f(x))^{\frac{1}{2}}}$

13 $f(x)=(x-a)^2g(x)$

$f'(x)=2(x-a)g(x)+g'(x)(x-a)^2$

$f'(x)=(x-a)[2g(x)+(x-a)g'(x)]$

14 $\frac{-4}{27}$ **15** $a=2, b=3$ **16** $a=\frac{9}{4}, b=1$

17 $\frac{dh}{dt}=(2t^3+2t+1)(3t^2+1)$

$t=0.1, \frac{dh}{dt}=1.2\text{ cm/m}$

EXERCISE 1.5

1 A **2** E

3 $2x(5x+1)(2x+1)^2$ **4** $u=3, v=-5$

5 $10(1-p)^8(1-10p)$

6 a, b $\frac{-3}{(2x-1)^2}$

7 $a=3$ **8** Proof: see worked solutions

9 0.5 **10** $a=1, b=3$

11 Proof: see worked solutions

12 a $f'(x)=2(2x-1)(x^2-x+1), g'(x)=3(x+a)^2$

b $f'(0)=-2$

$g'(0)=3a^2$

$f'(0)\times g'(0)=-1 \Rightarrow -2\times 3a^2=-1$ or $6a^2=1$

13 Proof: see worked solutions

14 a $a=1, b=12$ **b** $(3.231, 0.566)$

15 $a=9, b=5$

CUMULATIVE EXAMINATION: CALCULATOR-FREE

1 $35(5x+1)^6$ **2** $30x^2(2x^3+1)^4$

3 a $\frac{2}{(x+2)^2}$ **b** -9

4 $-\frac{1}{\sqrt{1-2x}}$

5 At $(2,4)$ $y'=4$

Substitute $(2,4)$ and $y'=4$ to obtain tangent line $y=4x-4$.

Substitute $x=3$ into tangent line and get $y=4(3)-4=8$.

6 $-\frac{3}{5}(x-4)(x-2)$

7 a $\frac{1}{4}$ **b** 20

CUMULATIVE EXAMINATION: CALCULATOR-ASSUMED

1 $y=41x-31$

2 Proof: see worked solutions

3 $a=4, b=5$

4 $\left(-1,-\frac{10}{3}\right), (-3,-2)$

CHAPTER 2

EXERCISE 2.1

1 0.0063 **2** -10.05 cm^3 **3** -0.475

4 $3\frac{1}{600}$ **5** $-\frac{17}{200}$ **6** 1.6 cm^2

7 -0.002 cm **8** 0.094 cm^3

EXERCISE 2.2

1 0.08 cm

2 $-15.7\,\text{cm}^2$

3 a $\dfrac{-1}{4(x-3)^{\frac{3}{2}}}$ **b** $\dfrac{24}{(x+5)^3}$

c 0 **d** $\dfrac{9}{8x^{\frac{1}{2}}}$

e $84x^5 + 30x^4 + 360x^3 + 108x^2 + 324x + 54$

f 0

4 73

5 Concave down for $x < 0.22$, concave down for $x > 0.22$, point of inflection $(0.22, 0.02)$.

6 The exponential function $f(x) = e^x$ or any function that equals a constant, e.g. $f(x) = 6$.

7 1

8 $a = 1$ and $b = 3$

9 $x = 1$ or $x = 2$

EXERCISE 2.3

1 $\dfrac{2}{27}$

2 approx. $(-0.6, 10.6)$

3 $(2, -4)$, local minimum

4 $(1, 0)$

5 $\left(-1, \dfrac{5}{2}\right)$, local maximum

6 $\left(-\dfrac{5}{4}, \dfrac{2187}{512}\right)$ and $(1, 0)$

7 $(-3, 10)$, local minimum and $\left(1, -\dfrac{2}{3}\right)$, local minimum

8 There is a stationary point of inflection.

9 There is a stationary point of inflection.

10 2

11 a Local minimum at $(0, 0)$ and local maximum at $(2, 4)$.

b

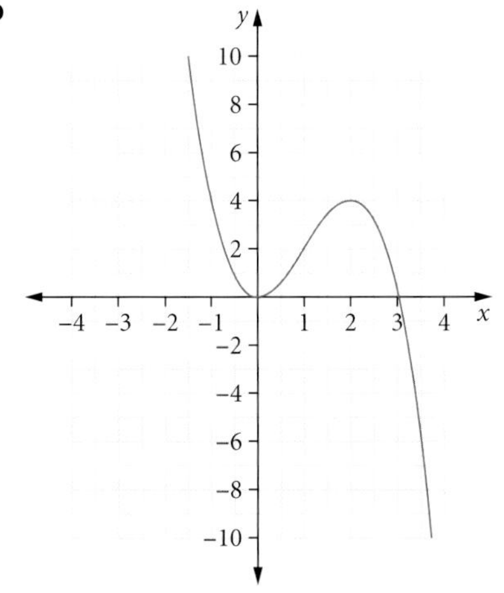

12 a F **b** F **c** T

d T **e** F

13 $a = 3, b = -9$

14 a $\dfrac{dy}{dx} = \dfrac{1}{\sqrt{x}} - 2x + 1$

b Substitute $x = 1$ into $\dfrac{dy}{dx}$ to get

$\dfrac{1}{1} - 2(1) + 1 = 1 - 2 + 1 = 0.$

Thus, $x = 1$ must be a stationary point as $\dfrac{dy}{dx} = 0$.

c $\dfrac{d^2y}{dx^2} = -\dfrac{1}{2x^{\frac{3}{2}}} - 2$, substitute in $x = 1$, $\dfrac{d^2y}{dx^2} < 0$,

so a local maximum.

15 $f'(x) = \dfrac{1}{2}x^{-\frac{1}{2}} + 2x$

$= \dfrac{1}{2\sqrt{x}} + 2x$

Let $f'(x) = 0$ to determine the stationary points.

$\dfrac{1}{2\sqrt{x}} + 2x = 0$

$x^{\frac{3}{2}} = -\dfrac{1}{4}$

As there is no solution for $f'(x) = 0$, there are no stationary points.

16 $f'(x) = 3ax^2 - 2bx + c = 0$

$x = \dfrac{b \pm \sqrt{b^2 - 3ac}}{3a}$

For no solution, $b^2 - 3ac < 0$.

Therefore, $c > \dfrac{b^2}{3a}$.

17 a $12x^2 + 5$

b $x^2 \geq 0$, for all values of x, hence $12x^2 + 5 \geq 5$ for all x.

18 $m = 0, 1$ or 2

EXERCISE 2.4

1 B

2 $(0, -1)$

3

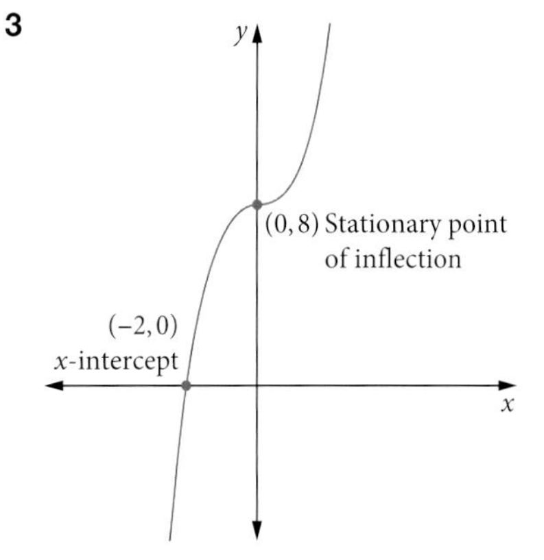

4

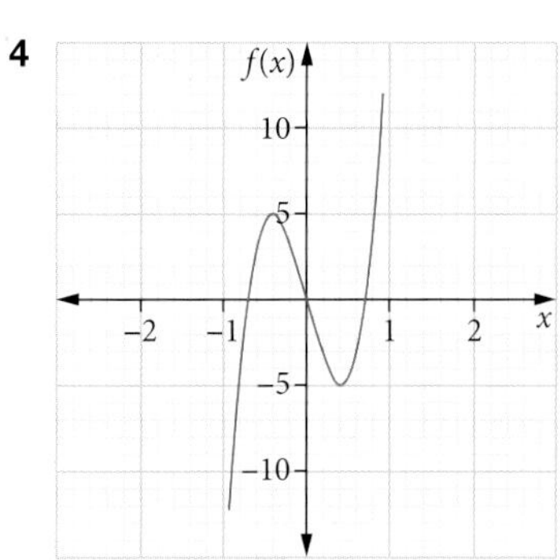

5 a 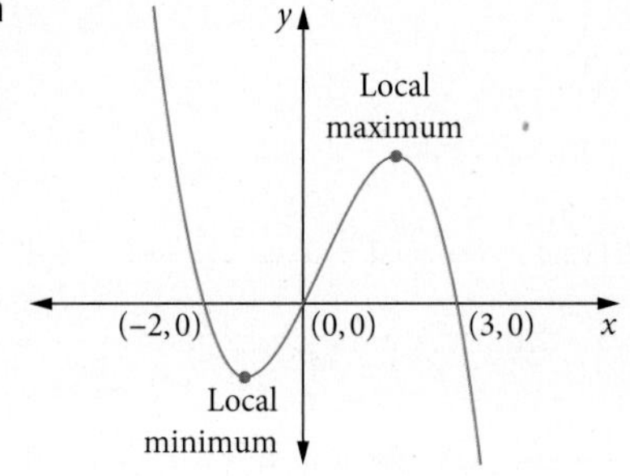

b 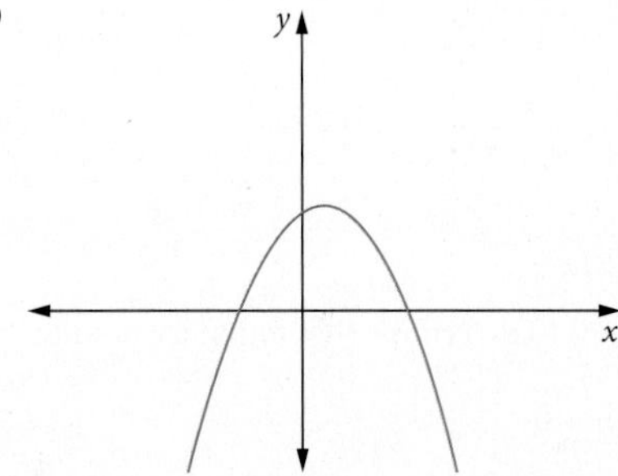

6 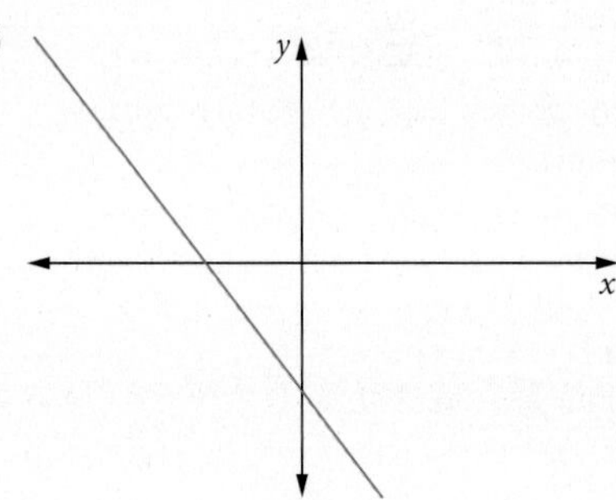

7 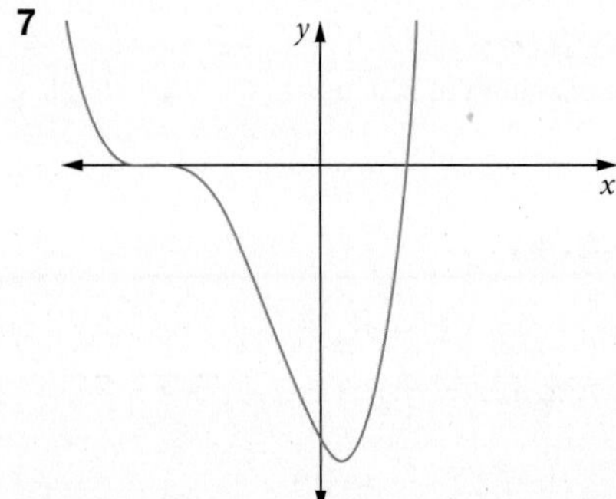

8 a $-3 < x < 3$ **b** $(3, -4)$

EXERCISE 2.5

1 E **2** B

3 a $v(t) = 4t$ **b** $4\,\text{m/s}^2$

4 a $3\,\text{m}$, $-0.5\,\text{m/s}$, $-4\,\text{m/s}^2$

b The ball is moving downwards with a speed of 0.5 m/s.

5 a $v(t) = 6t + 2$ **b** 6

6 a $v = 3t^2 + 12t - 2$, $a = 6t + 12$

b $x(5) = 266\,\text{m}$ **c** $v(5) = 133\,\text{m/s}$

d $a(0) = 12\,\text{m/s}^2$ **e** $a(5) = 42\,\text{m/s}^2$

7 a 4 m/s **b** −2 m/s **c** 14 m/s

d 6 m/s. Part **a** is the average rate of change of displacement = average velocity = $\dfrac{x(4) - x(2)}{4 - 2} = 4\,\text{m/s}$.

e $a = 2\,\text{m/s}^2$ **f** $a = 20\,\text{m/s}^2$

8 a $v(0) = -8\,\text{m/s}$ **b** $a(t) = 4\,\text{m/s}^2$

c $s(5) = 13\,\text{m}$ **d** $t = 2\,\text{s}$

e $s(2) = -5\,\text{m}$

f 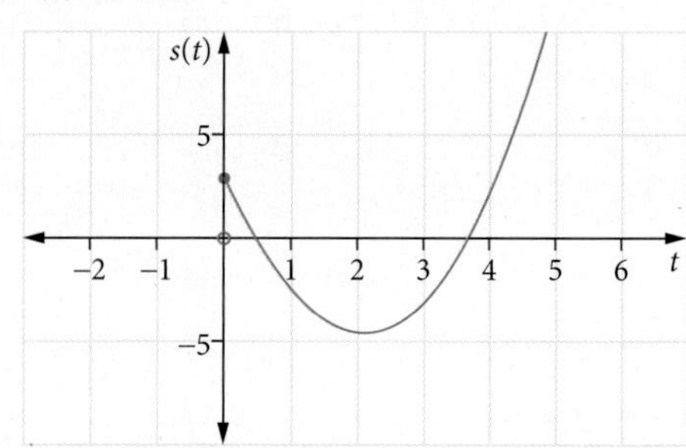

9 a $v(5) = 120\,\text{km/h}$ **b** $a(5) = 20\,\text{km/h}^2$

10 $a(5) = 20\,\text{m/s}^2$ **11** $4\,\text{m/s}^2$

12 a $v(t) = -\dfrac{1}{(2t+5)^2}$

b $a(t) = -\dfrac{4}{(2t+5)^3}$

c $a(t) = -4 \times \dfrac{1}{(2t+5)^2} = -4 \times v(t)$

magnitude of $a(t) = 4 \times$ magnitude of $v(t)$

13 $p = -2$, $q = -12$ and $r = 20$

EXERCISE 2.6

1 −0.55 **2** 1 **3** $1.24\,\text{cm}^3$

4 0.10 units **5** $16\,\text{cm}^3$

6 $P = 8x + 4h$

$\therefore h = \dfrac{P}{4} - 2x$

$V = x^2h$

$V = x^2(\dfrac{P}{4} - 2x)$

$\dfrac{dV}{dx} = \dfrac{Px}{2} - 6x^2$

$\dfrac{Px}{2} - 6x^2 = 0$

$x = 0, x = \dfrac{P}{12}$

Substitute $x = \dfrac{P}{12}$ into $P = 8x + 4h$ to get $h = \dfrac{P}{12}$.

Therefore, the shape must be a cube as $x = h$.

7 a $V = \dfrac{1}{4}xy(P - 4x - 4y)$

b $V = \dfrac{1}{2}x^2(P - 12x)$

$\dfrac{dv}{dx} = 0 \Rightarrow x(P - 18x) = 0 \Rightarrow x = \dfrac{P}{18}$

$V = \dfrac{1}{2}\left(\dfrac{P}{18}\right)^2\left(P - 12 \times \dfrac{P}{18}\right) = \dfrac{P^3}{6 \times 18^2}\,\text{cm}^3$

8 **a** $S = x^2 + y^2 \Rightarrow y = \sqrt{S - x^2}$

product: $p = xy = x\sqrt{S - x^2}$

b $\dfrac{dp}{dx} = \sqrt{S - x^2} - \dfrac{x^2}{\sqrt{S - x^2}}$

$\dfrac{dp}{dx} = 0$

$\sqrt{S - x^2} - \dfrac{x^2}{\sqrt{S - x^2}} = 0$

$x = \dfrac{\sqrt{S}}{\sqrt{2}}$

9 **a** $y = \dfrac{4000\sqrt{3}}{3x^2}$

b $A = 2\left(\dfrac{x^2}{2}.\dfrac{\sqrt{3}}{2}\right) + 3xy$

$= \dfrac{4000\sqrt{3}}{x} + \dfrac{x^2\sqrt{3}}{2}$

c $x = 10\sqrt[3]{4}$ (or $\sqrt[3]{4000}$)

10 1.13 cm

11 $x = 10\,\text{cm}$

12 **a** $S = 10\pi r - 3\pi r^2$ **b** $r = \dfrac{5}{3} = 1\dfrac{2}{3}$ cm

13 **a** $A = 18a - 6a^3$ **b** $A = 12$ when $a = 1$

14 $P(4, 2)$, minimum length $2\sqrt{5}$

15 $3\sqrt{3}$ units^2

16 **a** $G\left(\dfrac{28}{9}, -\dfrac{50}{243}\right)$ **b** $y = -\dfrac{1}{8}x + \dfrac{1}{2}$

c $x = \dfrac{2\left(\sqrt{31} + 7\right)}{9}$ **d** $k = 8m$

e $x = 2^{\frac{2}{3}}$

CUMULATIVE EXAMINATION: CALCULATOR-FREE

1 $y = x + 1$

2 **a** $f'\left(\dfrac{5}{3}\right) = 0,\ f''\left(\dfrac{5}{3}\right) = 10$

b As $f'\left(\dfrac{5}{3}\right) = 0$, this must be a stationary point.

As $f''\left(\dfrac{5}{3}\right) > 0$, it must be a minimum turning point.

3 Let $f(x) = \dfrac{2x^2 + 1}{\sqrt{x}}$

$$f'(x) = \dfrac{x^{\frac{1}{2}}(4x) - \left(2x^2 + 1\right)\left(\frac{1}{2}x^{-\frac{1}{2}}\right)}{x}$$

$$= \dfrac{4x^{\frac{3}{2}} - x^{\frac{3}{2}} - \dfrac{1}{2x^{\frac{1}{2}}}}{x}$$

$$= \dfrac{3x^{\frac{3}{2}} - \dfrac{1}{2x^{\frac{1}{2}}}}{x}$$

$$= \dfrac{6x^2 - 1}{2x^{\frac{3}{2}}}$$

4

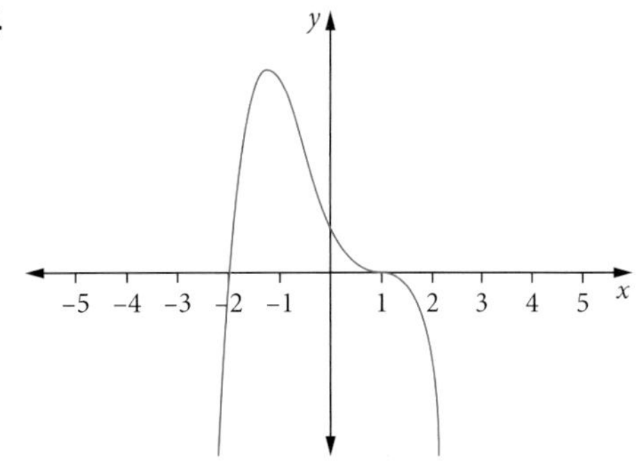

5

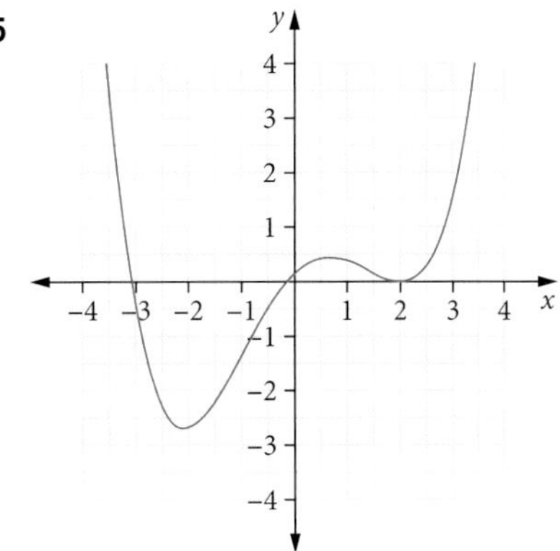

CUMULATIVE EXAMINATION: CALCULATOR-ASSUMED

1 The approximate change in area is 0.87 cm^2.

2 **a** $375 = \pi x^2 h$

$\therefore h = \dfrac{375}{\pi x^2}$

$S = 2\pi x^2 + 2\pi xh$

$= 2\pi x^2 + 2\pi x\left(\dfrac{375}{\pi x^2}\right)$

$= 2\pi x^2 + \dfrac{750}{x}$

b Cans have a radius of 3.9 cm and a height of 7.8 cm to minimise surface area.

3 $a = -18, b = 108$

4 **a** zero **b** 39 m/s

c $t = \dfrac{2}{3}$ or 2 seconds **d** $-8\,\text{m/s}^2$

e 2.37 m

5 **a** Let the height of the tank be h.

$V = xyh = 8 \Rightarrow h = \dfrac{8}{xy}$

$A = xy + 2xh + 2yh$

$= xy + 2x\left(\dfrac{8}{xy}\right) + 2y\left(\dfrac{8}{xy}\right)$

$= xy + \dfrac{16}{x} + \dfrac{16}{y}$

b $\dfrac{dA}{dy} = x - \dfrac{16}{y^2} = 0$ when $y = \dfrac{4}{\sqrt{x}}$

$\therefore A = \dfrac{4x}{\sqrt{x}} + \dfrac{16}{x} + \dfrac{16\sqrt{x}}{4}$

$= 8\sqrt{x} + \dfrac{16}{x}$

$a = 8, b = 16$

ANSWERS

CHAPTER 3

EXERCISE 3.1

1 $y = \frac{x^4}{4} + x^3 - 2x^2 + c$

2 $\frac{1}{16}(4x-1)^4 + c$

3 **a** $\frac{x^3}{3} - \frac{3x^2}{2} + 2x + c$ **b** $\frac{2x^3}{3} - x^2 - 12x + c$

c $\frac{x^2}{2} - 2x + c$ **d** $\frac{2x^{\frac{3}{2}}}{3} + \frac{1}{x} - 3x + c$

e $\frac{(2x-3)^{\frac{3}{2}}}{3} + c$ **f** $\frac{2x^{\frac{7}{2}}}{7} - \frac{4x^{\frac{5}{2}}}{5} + 2x^{\frac{3}{2}} + c$

g $\frac{-1}{2(2x-3)} + c$ **h** $\frac{(3x-4)^{\frac{4}{3}}}{4} + c$

4 $f(x) = x^2 - 3$

5 $y = x^2 + 4x - 11$

6 $f(x) = -3(-2x+4)^{\frac{2}{3}} + 22$

7 $y = x^3 + x^4 - 2x + c$

8 $\frac{-1}{9(3x+4)^3} + c$

9 $\frac{1}{8}(4-2x)^{-4} + c$

10 $f(x) = x^3 - x^2 - 48$

11

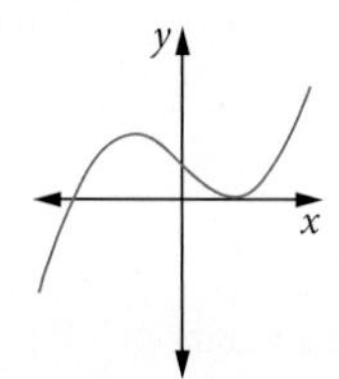

Answer can vary. This is one possible answer. Check with your teacher for other possible answers.

12 **a**

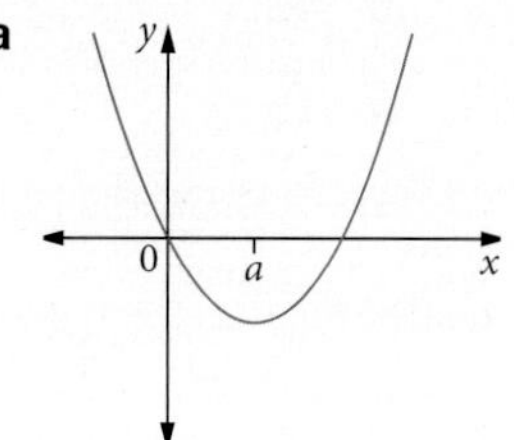

b

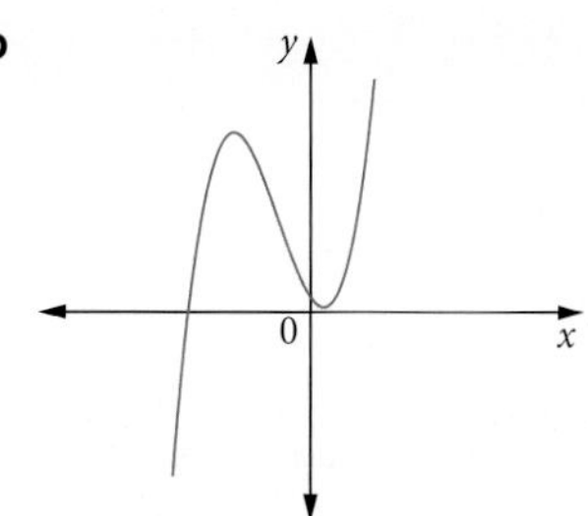

13 $2\sqrt{x^2 - 3x} + c$

14 $f(x) = \frac{2}{3}x^3 - \frac{3}{4}x^{\frac{1}{3}} - \frac{5}{3}$

15 $\frac{-1}{4(2x-1)^2} + c$

16 $f(x) = 2\sqrt{2x-3} - 2$

17 $f(x) = g(x) + 3x + c$

$\therefore f(0) = g(0) + 3(0) + c$

$\therefore 2 = 1 + c \Rightarrow = 1$

$f(x) = g(x) + 3x + 1$

EXERCISE 3.2

1 $y = \frac{x^3}{3} - \frac{3x^2}{2} + c$

2 $f(x) = x^2 - \frac{3}{5}x^{\frac{5}{3}} - \frac{7}{5}$

3 **a** 3.625 **b** 100 **c** 10 **d** 9.33

4 **a** 0.24 units^2 and 0.44 units^2

b 17.45 units^2 and 20.95 units^2

c 1.57 units^2 and 2.17 units^2

5 1.675 units^2

6 $\frac{\pi}{6}\left(1 + \frac{\sqrt{3}}{2} + \frac{1}{2}\right) = \frac{\pi}{12}\left[3 + \sqrt{3}\right]$

7 **a** 17.5 units^2 **b** 15.25 units^2

8 n is larger and h is smaller

9 **a** lower limit $= 20 \times 0.5 + 21 \times 0.5 + 24 \times 0.5$

$= 10 + 10.5 + 12$

$= 32.5$

upper limit $= 21 \times 0.5 + 24 \times 0.5 + 29 \times 0.5$

$= 10.5 + 12 + 14.5$

$= 37$

Therefore, $\int_0^{1.5} f(x)\,dx$ is between these values as this is the area under the curve.

b 75

c By reducing the width of the rectangles and, therefore, using more rectangles to estimate the area, the error in the estimate would be reduced. Another method involves determining the function and using calculus.

10 5.146

11 $f(1) + f(2) + f(3)$ because that gives the heights of the underestimated rectangles.

12 16.25 units^2

EXERCISE 3.3

1 4.75 units^2

2 It is more accurate because there are smaller spaces between rectangles and the curve.

3 **a** 16 **b** 128 **c** 15

d $\frac{49}{3}$ **e** $\frac{1}{3}$ **f** $\frac{1}{4}$

4 **a** $\frac{40}{3}$ **b** $-\frac{40}{3}$

5 **a** $2x^2 + x - 4$ **b** $\frac{3}{x^2 - 1}$

6 -15

7 7

8 a $\frac{4}{3}$ b 2 c 6

d 50 e $\frac{23}{6}$

9 a $\sqrt{x-\pi}$ b $2x^2 - x$

10 $\frac{9x^2}{2} + 2(1-2x^2)$

11 $\frac{23}{3}$

12 a $\int_0^5 x^2 dx$ b $\int_1^7 (x+1)\,dx$

c $\int_{-2}^2 (x^3 - x - 1)\,dx$ d $\int_0^3 (2x+1)\,dx$

13 $\frac{1}{12}$

14 9 15 5 16 4 17 -1

18 $\int_4^8 f(x)\,dx = F(8) - F(4) = F(8) + 6$

$\Rightarrow F(8) = -6 + \int_4^8 f(x)\,dx$

19 25

EXERCISE 3.4

1 $\frac{3}{4}$ 2 4

3 a 7.5 units^2 b 39 units^2 c 74 units^2

4 $24\frac{3}{4}$ units^2 5 $78\frac{1}{12}$ units^2

6 a 9 b 19

7 $\int_{-2}^{-1} f(x)\,dx$ 8 $\int_{-3}^0 f(x)\,dx - \int_0^1 f(x)\,dx$

9 $\frac{4}{3}$ units^2 10 $\frac{4}{15}$ units^2 11 9.5

12 $-\int_{-1}^1 f(x)\,dx + \int_1^4 f(x)\,dx - \int_4^6 f(x)\,dx$

13 $2\sqrt{3}$ 14 40.5 units^2

EXERCISE 3.5

1 D 2 0.5 units^2 3 $\frac{125}{6}$ units^2

4 a $\frac{32}{3}$ units^2 b $\frac{32}{3}$ units^2 c $\frac{125}{24}$ units^2

5 a $\frac{4}{3}$ units^2 b $\frac{500}{3}$ units^2 c 9 units^2

6 a (0,0) and (2,4)

b

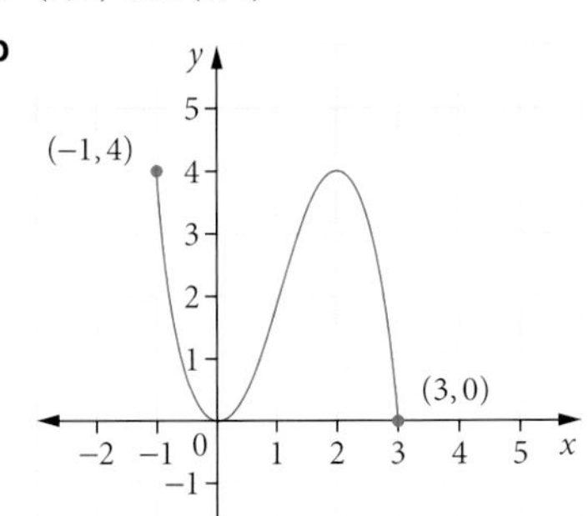

c 6.75 units^2

7 $\int_{-3}^2 [g(x) - f(x)]\,dx$

8 $\int_{-1}^2 f(x)\,dx - \int_{-1}^2 g(x)\,dx$

9 $\int_{-1}^1 [f(x) - g(x)]\,dx + \int_1^4 [g(x) - f(x)]\,dx$

10 4.5 units^2

11 a $\frac{32}{3}$ units^2 b $\frac{49}{2}$ units^2

c $\frac{22}{3}$ units^2 d 36 units^2

12 $\frac{125}{6}$ units^2 13 3.083 units^2

14 $\frac{8}{3}$ units^2 15 $a = 8, m = 4$

EXERCISE 3.6

1 E 2 B 3 $v(t) = 2t^2 + t$

4 a $v(t) = 3t - t^2 + 2$

b $\frac{25}{6}$ m. The particle changes its position by $\frac{25}{6}$ m between $t = 1$ and $t = 2$ seconds.

c Since the particle's velocity is positive between $t = 1$ and $t = 2$, the distance travelled is $\frac{25}{6}$ m.

5 $x(t) = -\frac{t^3}{3} + \frac{t^2}{2} + \frac{2}{3}$ 6 $x(t) = t^4 + t^3$

7 $v(t) = 2t^2 - 8$ 8 $x(t) = \frac{3t^2}{2} - \frac{2t^3}{3}$

9 a $v(t) = 8 + 0.4t$

b $x(t) = 8t + 0.2t^2$

c 100 m

10 81.7 m

11 a $k = 1.5$

b 17 min

c 1125 m below the mountain station

12 a 4 m/s b $\frac{128}{3}$ m

CUMULATIVE EXAMINATION: CALCULATOR-FREE

1 $\sqrt{8} = 2\sqrt{2}$

2 $\frac{5}{2}$

3 $f(x) = \frac{1}{3}x^2 - 5x^2 - 2\sqrt{x} + x + 80$

4 $a = 2, b = 0, c = -6, d = 4$

5 a $46 + 4\pi$ b $20 + 8\pi$ c $\frac{20 + 8\pi}{6} + 18$

6 a $\frac{4}{3}$ units^2

b Both graphs from part **a** have been vertically translated down by 5 units. The shape of both graphs is unchanged. Therefore, the area between them remains unchanged.

CUMULATIVE EXAMINATION: CALCULATOR-ASSUMED

1 −16 **2** $\frac{4}{3}$ units^2

3 a -13 cm/s^2 **b** 12

c $\frac{1}{2}, \frac{5}{3}$ seconds **d** 115.7 cm

4 a i 3 **ii** −2

b 8 units^2

5 a $\left(\frac{2a}{3}, \frac{3}{a}\right)$ **b** $0, \frac{a}{3}, \frac{2a}{3}$ **c** $\frac{1}{8}$ units^2

CHAPTER 4

EXERCISE 4.1

1 a i $N_0 = 200$ **ii** $k = 0.9163$

b 48 828 people

2 a i $B_0 = 100\,000$ **ii** $k = 0.0098$

b 126 389 **c** 71 h

3 a $D_0 = 400$ **b** $k = 0.3567...$ **c** 816.4 mg

4 a 6191 **b** 6439 **c** 4%

5 a $10 = d_0e^m$, $15 = d_0e^{2m}$ **b** $d_0 = 6.667$, $m = 0.405$

6 a $S_0 = 163$ **b** $k = 0.5597$ **b** 76 914

7 a $P_0 = 200$ **b**

$$100 = 200e^{5k}$$
$$\frac{1}{2} = e^{5k}$$
$$2 = \frac{1}{e^{5k}}$$
$$e^{-5k} = 2$$

8 a $k = -0.006\,90 = -6.90 \times 10^{-3}$ **b** 434.42 days

9 a 4 mg/L **b** 3.53 mg/L

10 a $k = 0.0693$ **b** 1.25×10^{19}

c The graph has a horizontal asymptote at $N = 0$, therefore, it will never reach zero where there is no sample left.

EXERCISE 4.2

1 a 50 **b** 1.0986

2 0.0347

3

a	$\lim_{h\to 0}\frac{a^h - 1}{h}$
2.71	0.996 949
2.711	0.997 318
2.712	0.997 686
2.713	0.998 055
2.714	0.998 424
2.715	0.998 792
2.716	0.999 160
2.717	0.999 528
2.718	0.999 896
2.719	1.000 264
2.72	1.000 632

The best approximation is 2.718.

4 a $9e^x$ **b** $e^x + 2x$

c $12e^x(2e^x - 3)^5$ **d** $e^{-3x}(e^{4x} - 2e^{2x} - 3)$

e $2e^{2x-1}$ **f** $\frac{e^{\sqrt{2x+4}}}{\sqrt{2x+4}}$

5 a $xe^x + e^x$ **b** $2xe^x + 5e^x$ **c** $5x^3e^x + 15x^2e^x$

6 $\frac{e^3(e^3 + 6)}{2(e^3 + 1)^{\frac{3}{2}}}$

7 a $\frac{(x - 2)e^x}{x^3}$ **b** $\frac{(6x - 1)e^{6x}}{3x^2}$ **c** $\frac{2(5x - 3)e^{5x}}{5x^4}$

d $\frac{2 - x}{e^x}$ **e** $-\frac{e^x + 2}{e^{2x}}$

8 $\frac{28}{e^4}$ **9** 348.4

10 $80e^6 + e^2$ **11** $x = 2$

12 $y = \frac{\sqrt{e}}{2}(x + 1)$ **13** $y = \frac{e^3}{2}(3x - 4)$

14 a $(4x - 4)e^{2x^2-4x}$ **b** $(16x^2 - 32x + 20)e^{2x^2-4x}$

c $x = 1$ **d** local minimum

15 a 1100 **b** 1246

c 31 swans per month **d** 3.1

16 a 3707 hectares per year

b 2414 hectares per year

c 1058 hectares per year

17 a 200 mg **b** $k = -0.043$

c −6.355 mg/day

18 a $3x^2e^{2x} + 2x^3e^{2x}$ **b** $6e^9$ **c** e^3

19 a Proof: see worked solutions

b $f'(x) = \frac{-(x - 1)(x - 3)}{e^x}$

$f'(1) = 0 = f'(3)$

20 $k = e^a(a - 1)$

21 a 25°C **b** 76.68°C

c 223°C **d** 8.63°C/min

e As time increases, the rate of change in the temperature of the water → 0.

The temperature of the water → the constant value of T_0.

22 a

h	$a = 2.60$	$a = 2.70$	$a = 2.72$	$a = 2.80$
0.1	1.002 65	**1.044 25**	1.052 41	1.084 49
0.001	0.955 97	0.993 75	**1.001 13**	**1.030 15**
0.000 01	0.955 52	**0.993 26**	**1.000 64**	1.029 62

b $a = e \approx 2.71828$

When $a = e$, the table shows that the value of

$\lim_{h\to 0}\left(\frac{a^h - 1}{h}\right)$ is 1.

It follows then from the definition that

$$\frac{d}{dx}(e^x) = e^x \times 1$$
$$= e^x$$

EXERCISE 4.3

1 218.393

2 $3e^6$

3 **a** $-\frac{1}{2}e^{-2x}+c$ **b** $\frac{5}{4}e^{4x}+c$

c $\frac{1}{2}e^{2x+1}+c$ **d** $-\frac{3}{2}e^{-2x}+\frac{1}{4}e^{4x}+c$

e $\frac{1}{3}e^{3x}+e^{-x}+c$ **f** $\frac{1}{6}e^{6x}+\frac{1}{6}e^{-6x}-2x+c$

4 $y=\frac{1}{2}e^{4x}+\frac{3}{2}$

5 $y=\frac{1}{2}e^{2x}+e^{-x}+\frac{19}{2}$

6 $f(x)=-\frac{5}{2}e^{-2x}+\frac{4e^7+5}{2e^6}$

7 **a** e^4-1 **b** $5e^3-5e$ **c** $-e^4+e^2+60$

8 **a** $\frac{e^{\pi}}{6}-\frac{e^{-\pi}}{6}\approx 3.85$ **b** $\frac{e^{3\pi}}{3}+\frac{5}{3}\approx 4132.22$

9 –5306.0

10 $\frac{2e^3}{3}+\frac{1}{3}$ unit^2

11 **a** $2(2x+1)e^{2x}$ **b** $(2x-1)e^{2x}+c$

12 $2\left(2+\frac{1}{e^4}\right)$ units^2

13 **a** $y=-2x+2$ **b** $2e^{\frac{1}{2}}-2$ units^2

14 **a** $xe^{-\frac{x}{5}}$ units^2 **b** $\frac{5}{e}$ at $x=5$ units^2 **c** 1.16 units^2

15 **a** 0.9975 mm^3/mm

b It is the rate of change of the volume with respect to height when the height has reached 0.5 m.

c **i** 5 m/s **ii** 3.488 m^3/s **iii** 8.594 m^3

EXERCISE 4.4

1 $\frac{e^{4x}}{4}-2x-\frac{e^{-4x}}{4}+c$

2 $2e^6-2$

3 $2x\sin(x)+x^2\cos(x)$

4 $\frac{dy}{dx}=-\frac{10\cos(3x)\cos(2x)+15\sin(3x)\sin(2x)}{\sin^2(2x)}$

5 $\frac{dy}{dx}=-(20x^4-1)\sin(4x^5-x)$

6 $\frac{dy}{dx}=-12x^4\sin(x^3)+8x\cos(x^3)$

7 **a** –0.269 **b** 1.047

8 **a** $v=-\frac{3}{2}\sin\left(\frac{t}{2}\right)$ **b** $a=-\frac{3}{4}\cos\left(\frac{t}{2}\right)$

c π s

d $v=-\frac{3}{2}$ m/s, $a=0$ m/s^2 **e** 2π s

9 **a** Proof: see worked solutions

b Proof: see worked solutions **c** (1.9, 2.4)

10 **a** (4.31, 3.78)

b maximum at (5.94, 0.67), minimum at (2.08, –1.74)

11 $h'(x)=\frac{e^{-x}\sin(x)-e^{-x}\cos(x)}{\cos^2(x)}$, $h'(\pi)=e^{-\pi}$

12 **a** $f'(x)=\sin(x)+x\cos(x)$ **b** $f'\left(\frac{\pi}{2}\right)=1$

13 $\frac{\pi\times\sqrt{3}}{6}+\frac{\pi^2}{36}$

14 **a** $P=\left(\frac{\pi}{6},4\right)$ **b** $y=-6x+\pi+4$

15 **a** 0 micrometres/s

b 45 micrometres/s^2

c –15 micrometres/s

16 –4 m/s^2

17 **a**
$$\tan\theta=\frac{y}{12}$$
$$y=12\tan\theta=\frac{12\sin\theta}{\cos\theta}$$
$$\frac{dy}{d\theta}=\frac{12\cos\theta\cos\theta+12\sin\theta\sin\theta}{\cos^2\theta}=\frac{12}{\cos^2\theta}$$

b 265.465 km/min

18
$$y=\frac{\sin(x)}{\cos(x)}$$
$u=\sin(x)$ $v=\cos(x)$

$\frac{du}{dx}=\cos(x)$ $\frac{dv}{dx}=-\sin(x)$
$$\frac{dy}{dx}=\frac{\cos(x)\times\cos(x)+\sin(x)\times\sin(x)}{(\cos(x))^2}$$
$$\frac{dy}{dx}=\frac{\cos^2(x)+\sin^2(x)}{(\cos(x))^2}=\frac{1}{\cos^2(x)}$$

19 **a** $-\frac{1}{2}$

b
$$y-\frac{\sqrt{3}}{2}=-\frac{1}{2}\left(x-\frac{2\pi}{3}\right)$$
$$y=-\frac{1}{2}x+\frac{\pi}{3}+\frac{\sqrt{3}}{2}$$
At $(c,0)$: $0=-\frac{1}{2}c+\frac{\pi}{3}+\frac{\sqrt{3}}{2}$
$$c=\sqrt{3}+\frac{2\pi}{3}$$

EXERCISE 4.5

1 12 m/s^2

2 $\frac{\pi}{2}$

3 **a** $-3\cos(2x)+c$ **b** $\sin\left(\frac{x}{2}\right)+c$

c $-\frac{6}{5}\cos(5x-7)+c$ **d** $\frac{1}{2}$

e $\sqrt{3}-1$ **f** $\frac{1-\sqrt{2}}{3}$

4 $6x\sin(2x)+3\cos(2x)+c$

5 $f(x)=\sin(x)-\frac{1}{3}\cos(3x)+2$

6 8 units^2

7 a $a(t) = \dfrac{7\pi^2}{320}\cos\left(\dfrac{\pi t}{40}\right)$ m/s^2

b $\dfrac{7\pi}{16}$

c $x(t) = 35 - 35\cos\left(\dfrac{\pi t}{40}\right)$

8 2.75 units^2

9 a $\dfrac{1}{2}\sin(2x+1)+c$

b $f(x) = 2\sin(x) + \dfrac{1}{2}\cos(2x) - 1$

10 $\int x\sin(3x)\,dx = \dfrac{\sin(3x)}{9} - \dfrac{x\cos(3x)}{3} + c$

11 $s(t) = -2\sin(t) + 2\cos(3t) - 2t + 2$

12 $s(t) = -\cos(t) + t + 2$

13 a $2\sin(3x) + 6x\cos(3x)$

b
$$\begin{aligned}
\frac{d(2x\sin(3x))}{dx} &= 2\sin(3x) + 6x\cos(3x)\\
\int \frac{d(2x\sin(3x))}{dx}\,dx &= \int (2\sin(3x) + 6x\cos(3x))\,dx\\
2x\sin(3x) + c_1 &= \int 2\sin(3x)\,dx + 6\int x\cos(3x)\,dx\\
\frac{2x\sin(3x) + c_1}{6} &= \frac{-2\cos(3x)}{18} + c_2 + \int x\cos(3x)\,dx\\
\int x\cos(3x)\,dx &= \frac{2x\sin(3x)}{6} + \frac{2\cos(3x)}{18} + c\\
\therefore \int x\cos(3x)\,dx &= \frac{3x\sin(3x) + \cos(3x)}{9} + c
\end{aligned}$$

14 a

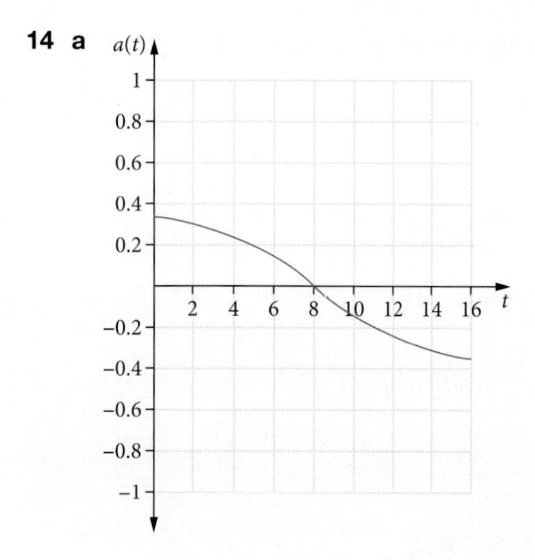

$a(t) = \dfrac{9\pi^2}{256}\cos\left(\dfrac{\pi t}{16}\right)$ m/s^2

b Since the acceleration is positive in the interval $0 < t < 8$, the velocity is increasing in the interval $0 < t < 8$.

Since the acceleration is negative in the interval $8 < t < 16$, the velocity is decreasing in the interval $8 < t < 16$.

c $x(t) = 9 - 9\cos\left(\dfrac{\pi t}{16}\right)$ $\quad x(16) = 18$ m

15 a
$$\begin{aligned}
A(p,q) &= \int_0^q \left(10\sin\left(\frac{x}{15}\right) + p\right)dx\\
&= pq - 150\cos\left(\frac{q}{15}\right) + 150
\end{aligned}$$
$p + q = 500$

$\therefore\ p = 500 - q$
$$\begin{aligned}
A(q) &= q(500-q) - 150\cos\left(\frac{q}{15}\right) + 150\\
&= 500q - 150\cos\left(\frac{q}{15}\right) - q^2 + 150
\end{aligned}$$

b $q \approx 247$, $62\,750$ m^2

16 a 864 cm

b $h'(x) = 4\cos\left(x - \dfrac{3\pi}{2}\right) - 2x + 3\pi$

c $x = 5.74$ m. Hence, the maximum height $h(5.74) = 20.57$ m.

17 $\dfrac{7}{9}$

CUMULATIVE EXAMINATION: CALCULATOR-FREE

1 a gradient = 15 **b** minimum turning point

2 a $30x^2(2x^3+1)^4$ **b** $-2e^{\pi}$

c $\int 3\cos(2x)\,dx = \dfrac{3}{2}\sin(2x) + C$

3 a Use $A = \dfrac{1}{2}bh$ with the following measurements:

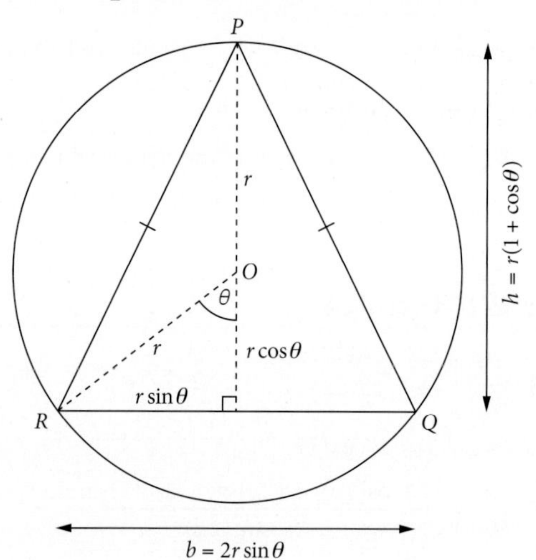

b
$$\begin{aligned}
A &= r^2\sin\theta(1+\cos\theta)\\
\frac{dA}{d\theta} &= r^2\left[\sin\theta(-\sin\theta) + (1+\cos\theta)\cos\theta\right]\\
&= r^2\left[\cos\theta + \cos^2\theta - \sin^2\theta\right]\\
&= r^2\left[\cos\theta + \cos^2\theta - (1-\cos^2\theta)\right]\\
&= r^2\left[2\cos^2\theta + \cos\theta - 1\right]\\
&= r^2(2\cos\theta - 1)(\cos\theta + 1)
\end{aligned}$$

$\dfrac{dA}{d\theta} = 0 \quad \cos\theta = \dfrac{1}{2}, \theta = \dfrac{\pi}{3}, \cos\theta \neq -1, 0 < \theta < \pi$

$$\begin{aligned}
A &= r^2\sin\theta(1+\cos\theta)\\
&= r^2\frac{\sqrt{3}}{2}\left(1+\frac{1}{2}\right)\\
&= \frac{3\sqrt{3}}{4}r^2
\end{aligned}$$

9780170477536

CUMULATIVE EXAMINATION: CALCULATOR-ASSUMED

1 a i $\frac{3}{2}$

ii

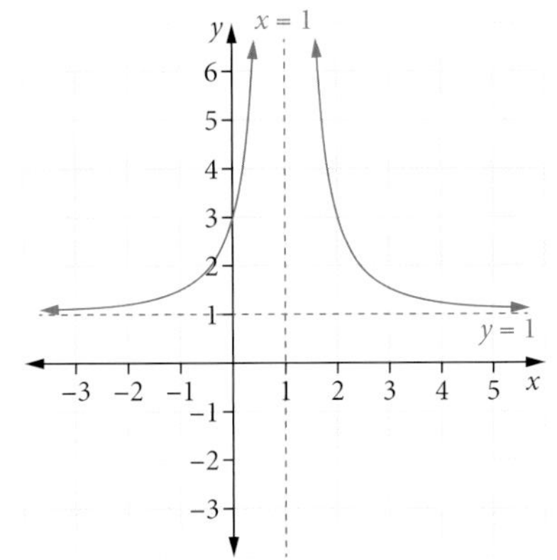

b 2 units^2

2 a 2800

b 176 animals/year

c $t = 14.8$, October 2031

3 a $x(t) = -6\cos\left(\frac{t}{3} + \frac{\pi}{6}\right) + 3\sqrt{3}$

b The drone is 0.27 m (27 cm) due south of the pilot.

c The drone has travelled 12.696 metres.

CHAPTER 5

EXERCISE 5.1

1 $\frac{37}{145}$ **2** $\frac{10}{323}$

3 a 8 6 4 2 / 2 4 6 8

b i $\frac{1}{4}$ **ii** $\frac{1}{4}$

4 0.42 **5** P(A) = 0.15 **6** $\frac{1}{5}$

7 a $\frac{3}{4}$ **b** $\frac{2}{3}$

8 a $\frac{4}{9}$ **b** $\frac{3}{4}$

9 a $\frac{5}{6}$

b i $\frac{19}{32}$ **ii** $\frac{15}{19}$

10 a $\frac{1}{15}$ **b** $\frac{1}{3}$

11 a $\frac{2}{5}$ **b** $1 - \frac{g}{6}$

12

Score	Combinations	Probability
7	3, 4 or 2, 5 or 1, 6	$\frac{2}{8} \times \frac{1}{6} + \frac{2}{8} \times \frac{1}{6} + \frac{4}{8} \times \frac{1}{6} = \frac{8}{48}$
8	3, 5 or 2, 6	$\frac{4}{48}$
9	3, 6	$\frac{2}{48}$

probability of a prize = $\frac{8}{48} + \frac{4}{48} + \frac{2}{48} = \frac{14}{48} = \frac{7}{24}$

13 a $\frac{15}{28}$ **b** $\frac{13}{28}$

14 a 0.49 **b** 0.33

EXERCISE 5.2

1 B

2 a $\frac{4}{49}$ **b** $\frac{43}{49}$

3 a uniform discrete probability distribution

b

x	1	2	3	4	5	6	7	8
P($X = x$)	$\frac{1}{8}$	$\frac{1}{8}$	$\frac{1}{8}$	$\frac{1}{8}$	$\frac{1}{8}$	$\frac{1}{8}$	$\frac{1}{8}$	$\frac{1}{8}$

4 X represents the number of tails.

x	0	1	2
P($X = x$)	$\frac{9}{25}$	$\frac{12}{25}$	$\frac{4}{25}$

5

x	0	1	2
P($X = x$)	$\frac{3}{28}$	$\frac{15}{28}$	$\frac{5}{14}$

6 $p = \frac{3}{5} = 0.6$ **7** $k = \frac{1}{10}$ **8** 0.29

9 Adding all probabilities give $0.6p^2 - p + 0.4 = 0$, giving $(3p - 2)(p - 1) = 0$, so $\frac{2}{3}$ or $p = 1$.

10 $p = \frac{1}{2}$ **11** 0.3

12 a 0.008 **b** $\frac{29}{64}$

13 a

		Roll two			
	Sum of two rolls	1	2	3	4
Roll one	1	1 + 1 = 2	3	4	5
	2	3	4	5	6
	3	4	5	6	7
	4	5	6	7	8

b i

x	2	3	4	5	6	7	8
P($X = x$)	$\frac{1}{16}$	$\frac{2}{16}$	$\frac{3}{16}$	$\frac{4}{16}$	$\frac{3}{16}$	$\frac{2}{16}$	$\frac{1}{16}$

ii $\frac{10}{16} = \frac{5}{8}$

EXERCISE 5.3

1 C **2** B

3 E(X) = 2.85 **4** $a = 0.3$, $b = 0.1$

5 a $E(X) = 2.8$

b $Var(X) = 9 - 2.8^2 = 1.16$

c $SD(X) = 1.077$

6 $Var(X) = 1$

7 a 27 **b** 48 **c** $a = 2, b = 70$

8 $E(X) = 14.30, Var(X) = 26.51, SD(X) = 5.15$

9 1.5

10 a $\frac{28}{15}$ **b** $\frac{8}{15}$

11 a

n	0	1	2	3
$P(N = n)$	$\frac{2}{10}$	$\frac{4}{10}$	$\frac{3}{10}$	$\frac{1}{10}$

b 1.3

12 a 80

b 1936

c $a = \frac{15}{22} \approx 0.682, b \approx \frac{195}{22} = 8.864$

13 a 0.12

b 0.729

c **i** $p^2 + (1 - p)(p - 0.2) = 0.7$ gives $p = 0.75$

ii 1.21

EXERCISE 5.4

1 B **2** E

3 a Bernoulli distribution, $X \sim Bern(0.05)$

b mean = 0.05, variance = 0.0475

4 a 0.216 **b** 0.177 **c** 0.885

5 a 0.046 **b** 0.8122

6 a **i** $6p^5(1 - p)$ **ii** p^6

b $p = \frac{6}{7}$

7 a $n = 150, p = 0.6$ **b** 0.07

8 a $n = 13$ **b** $n = 11$

9 a

y	0	1
$P(Y = y)$	$\frac{1}{5}$	$\frac{4}{5}$

b It is a Bernoulli distribution.

c $\mu = \frac{4}{5}, \sigma = \frac{2}{5}$ **d** $X \sim Bin\left(5, \frac{4}{5}\right)$ **e** $\frac{32}{625}$

10 a $\frac{16}{81}$ **b** $\frac{65}{81}$ **c** $\left(\frac{2}{3}\right)^{24}$

11 a **i** p^3 **ii** $3p^2(1 - p)$

b $p = 0, p = \frac{3}{4}$

12 $\frac{5}{16}$

13 a $p = 0.2$ **b** $E(X) = 0.6$

c $Var(X) = 0.84$ **d** 0.000 45

14 a $X \sim Bin(5, 0.05)$

b 1 The alarms fail independently of each other.

2 The probability that an alarm fails is constant/unchanging/same for all alarms.

c 0.000 03 **d** 0.999 52

15 a 0.9153 **b** 0.086

16 a 0.9015 **b** 0.9311

17 a 0.1147 **b** 23

18 a $X \sim Bin\left(30, \frac{4}{5}\right)$

$u = 24, \sigma = \sqrt{30\frac{4}{5}\left(1 - \frac{4}{5}\right)} = 2.191$

b 0.123

19 18

CUMULATIVE EXAMINATION: CALCULATOR-FREE

1 $\frac{d^2y}{dx^2} = 12x^2 + 12$ **2** $P\left(\frac{8}{5}, \frac{4}{5}\right)$

3 a $\frac{125}{512}$

b

x	$P(X = x)$
0	$\left(\frac{5}{8}\right)^3$ or $\frac{125}{512}$
1	$\binom{3}{1}\left(\frac{5}{8}\right)^2\left(\frac{3}{8}\right)$ or $\frac{225}{512}$
2	$\binom{3}{2}\left(\frac{5}{8}\right)\left(\frac{3}{8}\right)^2$ or $\frac{135}{512}$
3	$\left(\frac{3}{8}\right)^3$ or $\frac{27}{512}$

c $\mu = \frac{9}{8}, \sigma = \frac{45}{64}$

4 a $\frac{1}{10}$ **b** $m = 19n - 20$

5 a $\frac{297}{625}$ **b** $\frac{6^3}{5^4 - 2^4}$

CUMULATIVE EXAMINATION: CALCULATOR-ASSUMED

1 a $P'(t) = 2\sin(3t) + 6t\cos(3t)$\$/year

b $6\sqrt{3} - \frac{\pi}{2}$\$/year^2

c The approximate change in profit is $-\frac{1}{6}$ million dollars $\left(\frac{1}{6}\text{ million dollar loss}\right)$.

2 a

Amount won	20	10	0
Probability	$\frac{6}{36}$	$\frac{10}{36}$	$\frac{20}{36}$

b \$6.11

c Liu Yang is better off in the long term. In the long term, Liu Yang will likely win \$1.11 per game.

d \$7.64

3 a $X \sim \text{Bin}(5, 0.25)$ **b** $\frac{1}{64}$

c If the carnival organisers only charge \$2 per game, then, on average, they will lose approximately 34 c per game.

4 a i 0.310

ii No, the events are not independent. $\text{P}(H|S)$ is not equal to $\text{P}(H)$.

b 0.190 **c** 0.0227

CHAPTER 6

EXERCISE 6.1

1 a $\log_7(49) = 2$ **b** $\log_3(27) = 3$

c $\log_2(16) = 4$ **d** $\log_5(125) = 3$

e $\log_{11}(1) = 0$ **f** $\log_2(1) = 0$

g $\log_5\left(\frac{1}{25}\right) = -2$ **h** $\log_4\left(\frac{1}{16}\right) = -2$

2 a $5^2 = 25$ **b** $4^2 = 16$ **c** $5^3 = 125$ **d** $2^4 = 16$

e $3^1 = 3$ **f** $7^2 = 49$ **g** $2^7 = 128$ **h** $5^0 = 1$

3 a 6 **b** 2 **c** 4 **d** 3

e 3 **f** 0 **g** 1 **h** 5

i 5 **j** 5 **k** 4 **l** −3

4 a 1 **b** 1 **c** $-\frac{1}{3}$ **d** 9

e 3 **f** 0 **g** 1 **h** 2

5 a $\log_4(x^4)$ **b** $\log_7(x^2)$ **c** $\log_6\left(\frac{1}{x}\right)$

d $\log_2\left(\frac{1}{x+2}\right)$ **e** $\log_4(x-1)$ **f** $\log_3(1) = 0$

6 a $\frac{1}{2}$ **b** $\frac{3}{2}$ **c** $\frac{3}{2}$

d $\frac{3}{2}$ **e** $\frac{2}{3}$

EXERCISE 6.2

1 a 5 **b** 3 **c** −4

2 a 5 **b** 4

3 a $\log_3(7) + 5$ **b** $\log_2(12) - 3$

c $\frac{1}{3}\ln(9)$ **d** $\frac{1}{2}(\ln(2) - 3)$

4 a $x = \ln(8)$ **b** $x = \log_5(4)$

c $x = \frac{1}{3}\log_2(3)$ **d** $x = 0, x = \frac{1}{2}\log_3(2)$

5 a $\frac{13}{2}$ **b** $\frac{7}{9}$ **c** $\frac{5}{6}$

d 2.2 **e** 7 **f** 22

6 a 4 **b** 7 **c** $\frac{8}{17}$

7 a 145 **b** $y = \frac{26x}{51}$

8 a $a = 2, b = 10$ **b** $a = 5, b = 21$

9 $x = \frac{\ln(9)}{2}$

10 a 32 **b** $y = -\frac{1}{2}x$

11 a $x = \frac{18}{25}$ **b** $x = \frac{e^2 - 10}{6}$ **c** $x = \frac{10}{3}$

12 a Proof: see worked solutions

b $a = 53, b = \log_3(2)$

EXERCISE 6.3

1 17 **2** $\ln(4)$

3

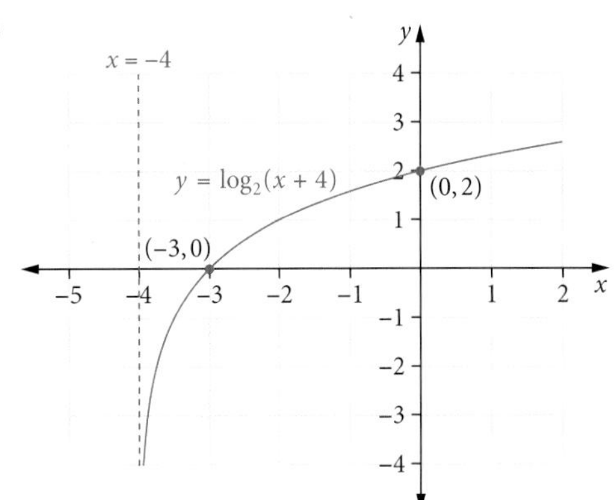

4

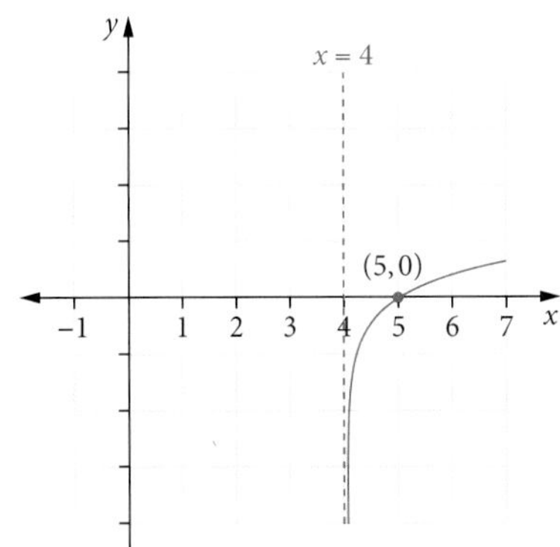

5 Vertical translation 1 unit down and horizontal translation 2 units left.

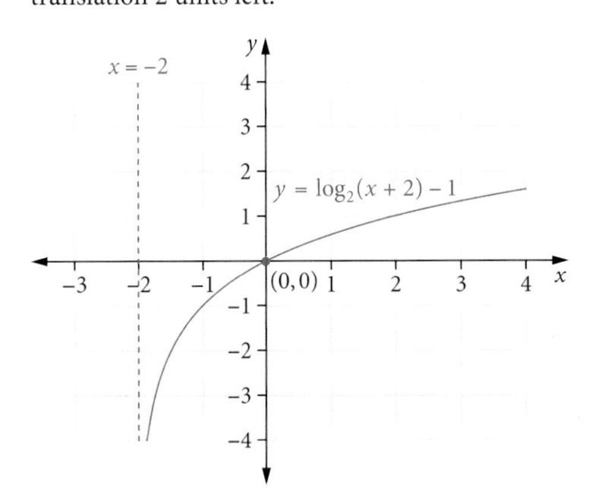

6

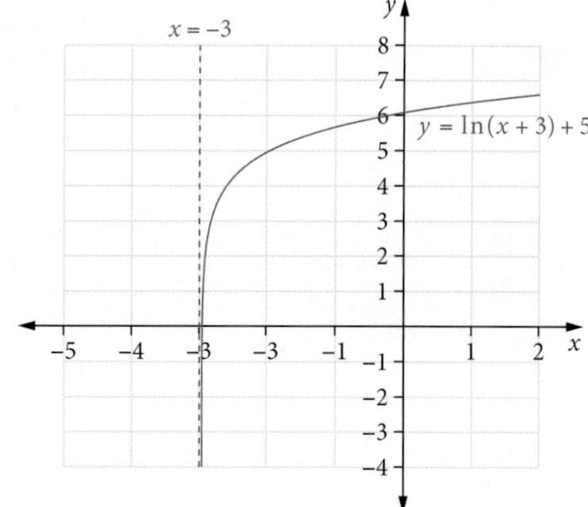

7 $y = 3\ln(x + 4)$

8 $b = 7, c = 1$

9

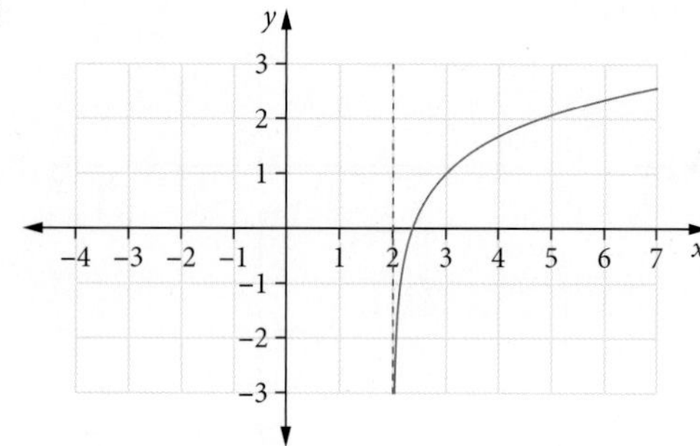

10 **a** x-intercept $(72, 0)$, y-intercept $(0, -2)$, vertical asymptote $x = -9$

b x-intercept $(0, 0)$, y-intercept $(0, 0)$, vertical asymptote $x = -8$

11 $y = \log_5(x - 2) + 8$

12 **a**

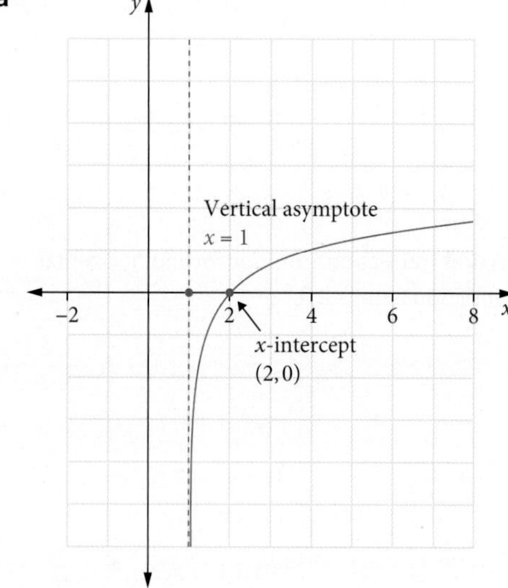

b $m = a + 1$ **c** $(a^{-c} + 1 - b, 0)$

13 $m = -4, p = 5, q = 2$

EXERCISE 6.4

1 $b = 20, c = 10$ 2 $b = 2, c = 8$

3 **a** $n = 3.7$ **b** $n = 2.5$

4 **a** \$507 **b** 58 sheep

c between 14 and 138 sheep

5 **a** 549 **b** 3225

c $t = 2.457$, that is 4 January

6 The amplitude of the waves in Mexico is $10^{1.1}$(or 12.59) times higher than that of the waves in San Francisco Bay.

7 **a** **i** 100 **ii** 1000

b $A = 0.5, B = 2$

c 100 000

d $N = 100(10)^{0.5t}$

8 **a** $p = 4$ **b** $p = 0.1$

9 **a**

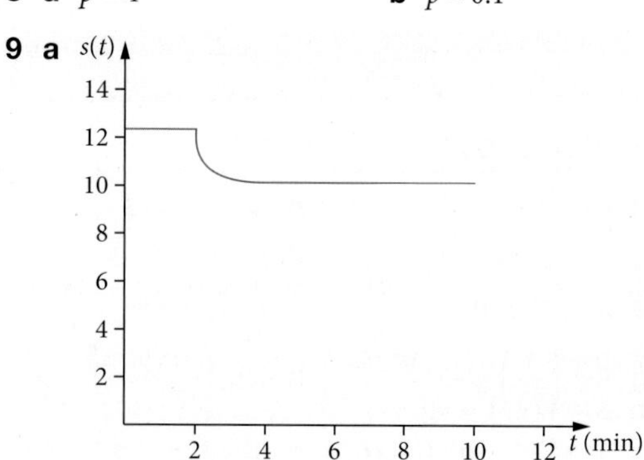

b She runs at 10 km/h when she has run for 2.99 minutes.

c She has zero acceleration for the first 2 minutes of her run and at the instant $t = 6.30$ minutes.

10 $\frac{A_{NZ}}{A_H} = \frac{10^{5.5}}{10^{3.4}} = 10^{2.1}$

11 **a** 97 dB

b $I_0 = \frac{1 \times 10^{-5}}{10^7} = 1 \times 10^{-12}$ watt/m^2

CUMULATIVE EXAMINATION: CALCULATOR-FREE

1 $\frac{dy}{dx} = 1 + x(x^2 - 4)^{-\frac{1}{2}}$

$\frac{d^2y}{dx^2} = (x^2 - 4)^{-\frac{1}{2}} - x^2(x^2 - 4)^{-\frac{3}{2}}$

$\text{LHS} = (x^2 - 4)\left((x^2 - 4)^{-\frac{1}{2}} - x^2(x^2 - 4)^{-\frac{3}{2}}\right)$
$+ x\left(1 + x(x^2 - 4)^{-\frac{1}{2}}\right) - \left(x + (x^2 - 4)^{\frac{1}{2}}\right) = 0 = \text{RHS}$

2

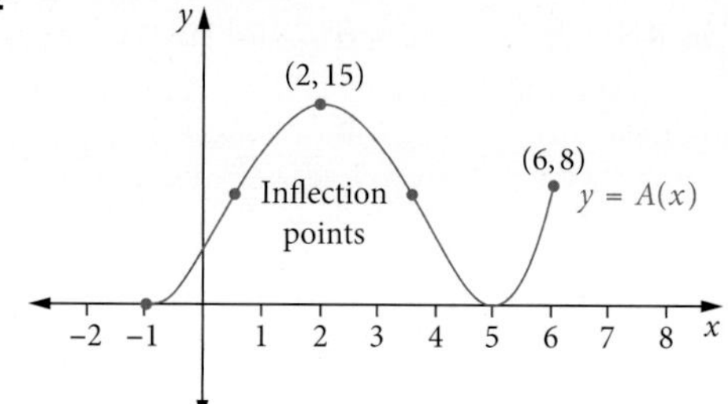

3 **a** $\frac{\ln(11) - 3}{2}$ **b** $\ln(3)$

c $\frac{18}{3}$ **d** $\frac{e^3 - 2}{4}$

4 $a = -6, b = 4$

CUMULATIVE EXAMINATION: CALCULATOR-ASSUMED

1 a $(0,1)$ **b** $c = 1$

c

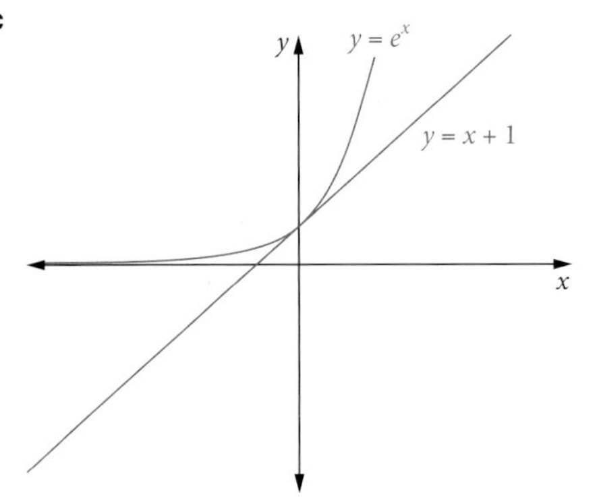

d $1 - \frac{(\ln(2))^2}{2} - \ln(2)$ **e** area $= -\int_{\ln 2}^{1} \ln(x)\,dx$

2 a 0.486 **b** 0.017 **c** 0.312

3 a $\$10\,475.71 \approx \$10\,476$

b

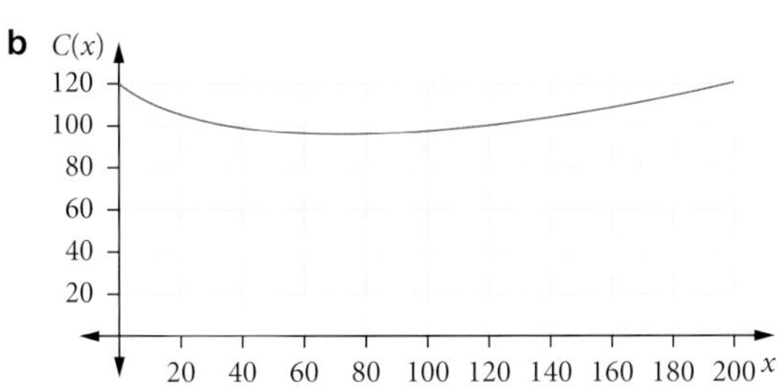

c The minimum is at $x = 74.205$.

$C(74) = 95.4307$ i.e. \$9543.07

$C(75) = 95.4320$ i.e. \$9543.20

The company should manufacture 74 components.

4 a Since $P(0) = 4$, we require $\ln(a) = 1$, giving $a = e$.

b Profit will be approximately \$4 087 000.

c Maximum profit is approximately \$4 436 000.

This occurs when $x \approx 1.79$, so during the second week.

d The model predicts during the 6th week.

e Profit should exceed \$5 million during the 8th week after the changes.

CHAPTER 7

EXERCISE 7.1

1 a $\frac{8}{8x-5}$

b $\frac{6x+6}{3x^2+6x} = \frac{2x+2}{x^2+2x}$

c $\frac{12x^3+24}{x^4+8x}$

2 a $\frac{1}{x}$ **b** $\frac{3}{x}$ **c** $\frac{8}{4x-3}$

d $\frac{1}{4(x-4)}$ **e** $\frac{6}{2x+1}$

3 a $(2x-2)\ln(x) + x - 2$

b $3x^2 + 9x^2\ln(x)$

c $\frac{-\ln(x)+1}{x^2}$

4 $\frac{x^2 - 3x^2\ln(2x)}{x^6} = \frac{1 - 3\ln(2x)}{x^4}$

5 a $\frac{-25}{(5x+4)^2}$ **b** $\frac{-64}{(4x+1)^2}$

6 a $\frac{4}{4x-3}$ **b** $\frac{-16}{(4x-3)^2}$

7 $\frac{2\cos(2)}{e}$

8 a Proof: see worked solutions **b** $-\frac{5}{24}$

9 $x + 2x\ln(x)$

10 $1 + \ln(x)$

11 a $\frac{1 - 2\log_e(x)}{x^3}$ **b** 1

12 0.8

13 $x^2 + 3x^2\ln(2x)$

EXERCISE 7.2

1 a $\frac{dy}{dx} = \frac{-2}{5-2x}$

b $\frac{dy}{dx} = \frac{3x^2+2x}{x^3+x^2} = \frac{3x+2}{x^2+x}$

2 $\frac{-1}{(x+6)^2}$

3 a $(4, \ln 16)$

b $f''(x) = \frac{-2x^2+16x-64}{(8x-x^2)^2}$, $f''(4) = -\frac{1}{8}$, local maximum

4 a $N'(t) = \frac{400(6-t)}{(-t^2+12t+13)}$

b local maximum at $t = 6$ weeks

c 778 frogs

5 $y = \frac{x}{e^2} + 2$

6 $y = \frac{1}{3}x - 1 + \ln(3)$ $\frac{-72}{5(2t+1)^2}$

7 $y = 3x - 9$

8 $\ln(3.003) \approx 1.0996$

9 a $v(t) = \frac{16}{2t+1}$ **b** 1.5 hours **c** $-\frac{32}{9}$ km/h^2

10 a The company will make a loss for a selling price between \$1.50 and \$2.00. The profit then increases to approximately \$2.25 per item sold for a selling price of approximately \$3.25, and then decreases steadily to a value of less than \$1 per item sold for a selling price of \$10.

b $x = 2e^{\frac{1}{2}}$

11 $1 + \ln(2x)$ **12** $\ln(2) + 0.01 = 0.703$

13 a $x(1 + 2\ln x)$

b $\frac{dy}{dx} = x(1 + 2\ln x)$

$\frac{dy}{dx} = 0, \ln x = -\frac{1}{2}, x \neq 0$

Only one point where derivative is zero hence only one stationary point.

14 a $\frac{36}{3(2t + 1)}$ b $-\frac{72}{5(2t + 1)^2}$

c $-\frac{72}{125}$ km/h² d 1 h 24 min

EXERCISE 7.3

1 $e^x \ln(x) + \frac{e^x}{x}$ 2 $\frac{4}{3}$

3 a $2\ln(x) + c$ b $\frac{6}{5}\ln(x) + c$ c $\frac{1}{3}\ln(x) + c$

4 a $\ln(x^2 + 11x - 15) + c$

b $5\ln(x^3 - 13) + c$

c $2\ln(3x^3 + 4x^2 + 1) + c$

5 a $\ln(x - 6) - \ln(x - 5) + c$

b $\frac{1}{4}\ln(2x - 5) - \frac{1}{4}\ln(2x + 5) + c$

6 a $2x\ln(2x) + x$ b $\frac{1}{2}x^2\ln(2x) - \frac{1}{4}x^2 + c$

7 a $3 + \log_e(x^3)$ b $x\log_e(x^3) - 3x + c$

8 a $\frac{1}{5}\ln(5x + 3) + c$ b $\frac{3}{2}\ln(2x - 5) + c$

9 a $\ln(5)$ b $3\ln(2)$ c $\frac{1}{3}\ln\left(\frac{19}{16}\right)$

d $\frac{1}{3}\ln\left(\frac{7}{4}\right)$ e $2\ln(2)$

10 $y = \frac{7}{3}\ln(3x - 5) + 7$

11 $y = 9\ln(x - 3) - 4x - 11$

12 $m = \frac{5e^7 + 1}{3}$ 13 $k = \frac{9}{2}$

14 a $\log_e(3x) + 1$ b $\log_e(12)$

15 a $b = \frac{9}{2}$ b $p = \frac{1}{2}$

16 a $1 + \ln(x)$

b $\frac{d}{dx}(x\ln(x)) = 1 + \ln(x)$

$\int \frac{d}{dx}(x\ln(x))\,dx = \int (1 + \ln(x))\,dx$

$x\ln(x) = x + \int \ln(x)\,dx + c$

$\int \ln(x)\,dx = x\ln(x) - x + c$

17 a Proof: see worked solutions

b $f(x) = 6\ln(x - 1) + x - 1$

c 7 d $y = 7x - 13$

EXERCISE 7.4

1 $m = 3$ 2 $f(x) = \ln(x + 3) + 12$

3 $\frac{1}{3}\ln(2)$ units² 4 12 units²

5 a $b = \ln(7)$ b $7\ln(7) - 6$ units²

6 a $a = 2$ b $e^2 - 1$

7 a $2x\log_e(x) + x$ b $\frac{9}{2}\ln(3) - 2$ units²

8 a i $(\ln 2, 2)$

ii $2\ln(2) - 1$ units²

b $k = \ln 3$

9 a 1 unit² b $2\ln(2) - 1$ units²

c $a\ln(a) - a + 1$ units²

10 a 549 b 3225

c $t = 2.457$, that is 3 January

11 a $a = 1.949$ b 168 days

c 0.012 s/day

d It would take him 1314 days, or more than $3\frac{1}{2}$ years. This is probably impossible to maintain.

12 a $a = 1.820$

b 15 weeks

c At $t = 4$, $\frac{dN}{dt} = 0.063$ skateboards per day.

d At $t = 10$, $\frac{dN}{dt} = 0.165$ skateboards per day.

13 a

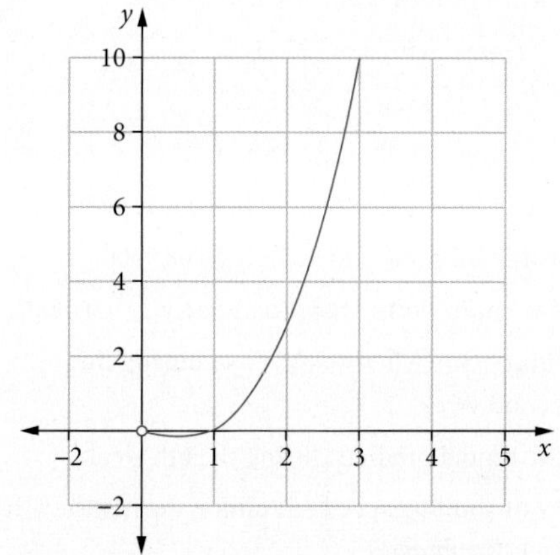

b

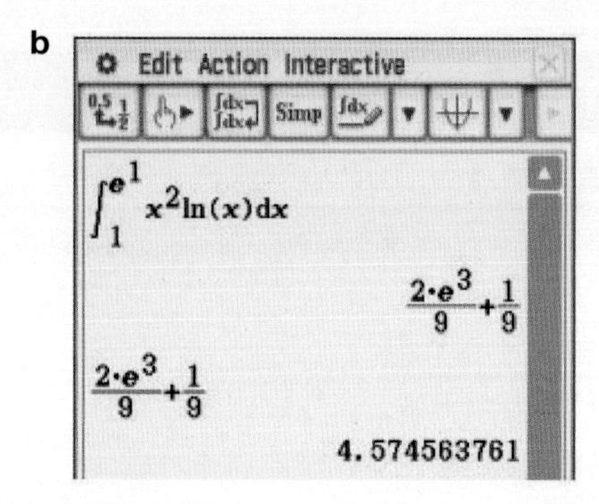

14 a i $-\frac{7}{a}$ ii $x = \sqrt{a}$

b i 7 ii $b = e^{-1}$

c i $\frac{7(a^2 - 1)}{2a}$ ii $a = \sqrt{2} + 1$

iii The area under the curve is less than the area of the trapezium.

Hence $\int_1^a f(x)\,dx < 7$.

From part **b i**

$\int_1^e f(x)\,dx = 7$ but $\int_1^a f(x)\,dx < 7$, so $a < e$.

d $m = e^{\frac{5}{14}}, n = e^{\frac{1}{14}}$

CUMULATIVE EXAMINATION: CALCULATOR-FREE

1 a $d_0e^0 = 2$ and $d_0e^{2m} = 10$ **b** $d_0 = 2, m = \frac{1}{2}\ln(5)$

2 a $p = \frac{1}{3}$

b

x	0	1	2	3
$P(X = x)$	$\frac{1}{27}$	$\frac{2}{9}$	$\frac{4}{9}$	$\frac{8}{27}$

c $\frac{26}{27}$ **d** $\frac{6}{13}$

3 $\left(\ln\left(\frac{3}{2}\right), 0\right)$ and $(\ln(2), 0)$ **4** $\frac{1}{2}\ln(6)$

5 a i Two distinct cases in which the upper bound is twice the lower bound. Shade under the curve from $x = 1$ to $x = 2$, and then from $x = 2$ to $x = 4$.

Other possibilities would be $x = 1.5$ to $x = 3$, $x = 2.5$ to $x = 5$ or $x = 3$ to $x = 6$.

ii $b = 3a$

b i Using the rectangles that estimate $y = \frac{1}{x}$ on the left side of each interval gives

$\int_2^3 \frac{1}{x}\,dx < \frac{1}{2} \times \frac{1}{2} + \frac{2}{5} \times \frac{1}{2} = \frac{9}{20}$.

This is an overestimate of the integral as the top of the rectangles lie above the graph.

Using the rectangles that estimate $y = \frac{1}{x}$ on the right side of each interval gives

$\int_2^3 \frac{1}{x}\,dx > \frac{2}{5} \times \frac{1}{2} + \frac{1}{3} \times \frac{1}{2} = \frac{11}{30}$.

This is an underestimate of the integral as the top of the rectangles lie below the graph.

Hence,

$\frac{11}{30} < \int_2^3 \frac{1}{x}\,dx < \frac{9}{20}$.

ii $\int_2^3 \frac{1}{x}\,dx = [\ln(x)]_2^3 = \ln(3) - \ln(2) = \ln(1.5)$

Hence

$\frac{11}{30} < \ln(1.5) < \frac{9}{20}$.

6 $f(x) = \frac{x}{2} - \frac{1}{2}\log_e(2x - 2) - 1 + \frac{1}{2}\log_e(2)$

7 a $2xe^{kx} + kx^2e^{kx} = xe^{kx}(kx + 2)$

b $k = 1$

c $\int_0^2 x^2e^{kx} + \frac{2xe^{kx}}{k}\,dx$

d $k = \ln(2)$

CUMULATIVE EXAMINATION: CALCULATOR-ASSUMED

1 a $x = \frac{1}{a}$ and $x = \frac{ab+1}{4a}$ **b** $a = 0$

c $a = \frac{3}{b}$ **d** 2 **e** $p = 4$

2 −1

3 a $f'(x) = x^3 + 4x^3\ln(4x)$ **b** $\frac{1}{4}x^4\ln(4x) - \frac{1}{16}x^4 + c$

c $\frac{1}{2}\ln(2) - \frac{225}{4096}$ **d** 28 m

4 a $(0, 1)$ **b** $c = 1$

c

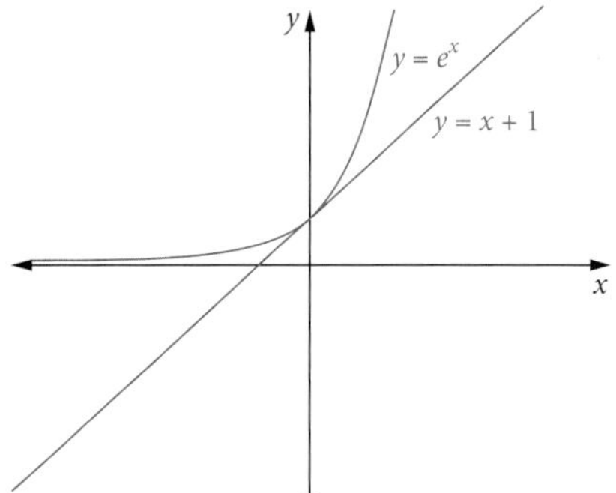

d area $= 1 - \frac{(\ln(2))^2}{2} - \ln(2)$

e area $= -\int_{\ln 2}^1 \ln(x)\,dx$

CHAPTER 8

EXERCISE 8.1

1 a $n(20 \le t < 25) = 3, n(35 \le t < 40) = 6$

b i $\frac{2}{28}$ **ii** $\frac{2}{8}$ **iii** $\frac{20}{28}$ **iv** $\frac{20}{28}$

v $\frac{3}{5}(3) + \frac{3}{5}(2) = \frac{15}{5} = 3$, so $\frac{3}{28}$.

2 $k = \frac{6}{125}$

3 a 0 **b** $\frac{4}{9}$ **c** $\frac{1}{3}$ **d** $\frac{3}{4}$

4 a $\frac{15}{16}$ **b** $\frac{1}{\sqrt{e}} - \frac{1}{\sqrt[5]{e^4}}$ **c** $\frac{1}{3}$

5 $a = 5$ (reject −5 as $a > 0$)

6 a 0.68 **b** 0.36 **c** 0.64

7 a $F(x) = \begin{cases} 0 & x < 0 \\ \frac{x^2}{9} & 0 \le x \le 3 \\ 1 & x > 3 \end{cases}$ **b** $F\left(\frac{5}{2}\right) = \frac{25}{36}$

8 **a** $\frac{3}{8}$

b $f(x) = \begin{cases} \frac{1}{4} - \frac{x}{16} & 0 \le x < 4 \\ \frac{x}{16} - \frac{1}{4} & 4 \le x \le 8 \\ 0 & \text{otherwise} \end{cases}$

c

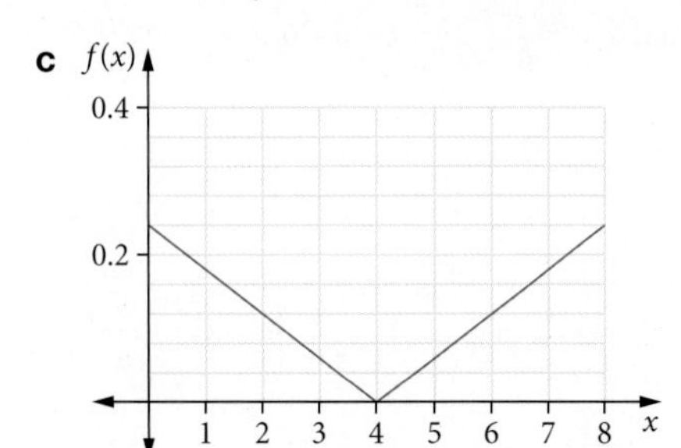

d $\int_0^4 \frac{1}{4} - \frac{x}{16}\,dx + \int_4^x \frac{t}{16} - \frac{1}{4}\,dt$

$= \frac{1}{2} + \left[\frac{t^2}{32} - \frac{t}{4}\right]_4^x$

$= \frac{1}{2} + \frac{x^2}{32} - \frac{x}{4} - \frac{1}{2} + 1$

$= \frac{x^2}{32} - \frac{x}{4} + 1$

9 **a** 0.1 **b** 0.26 **c** 0.008

10 **a** 0.12 **b** $\frac{1}{28}$ **c** 10% **d** $k = \frac{27}{560}$

11 **a** $\int_0^4 \frac{x+1}{k}\,dx = \frac{1}{5}\int_0^4 x + 1\,dx$

$= \frac{1}{k}\left[\frac{x^2}{2} + x\right]_0^4$

$= \frac{1}{k}[8 + 4]$

$\frac{12}{k} = 1$

$k = 12$

b $k = 3$ (reject -5 as $k > 0$) **c** $\frac{7}{15}$

12 **a** $\lim_{k\to\infty}\left(\int_a^k \frac{a}{x^2}\,dx\right) = \lim_{k\to\infty}\left(\left[-\frac{a}{x}\right]_a^k\right) = \lim_{k\to\infty}\left(-\frac{a}{k} + 1\right)$

As $k \to \infty$, $-\frac{a}{k} + 1 \to 1$. All values of $f(x)$ are positive and sum of all probabilities equals 1. Therefore, a valid pdf.

b $1 - P(X \le 2a) = 1 - \int_a^{2a} \frac{a}{x^2}\,dx = \frac{1}{2}$

13 $\frac{7}{12}$ or 0.5833 (4 d.p.)

14 **a** $a = -\frac{3}{4}$ **b** 0.5248 (4 d.p.)

15 **a**

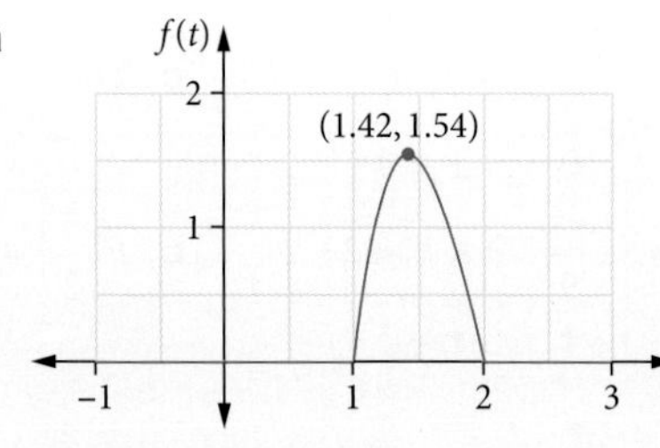

b 0.191

c $k = 1.77\,\text{h}$, $k \approx 106$ minutes

16 **a**

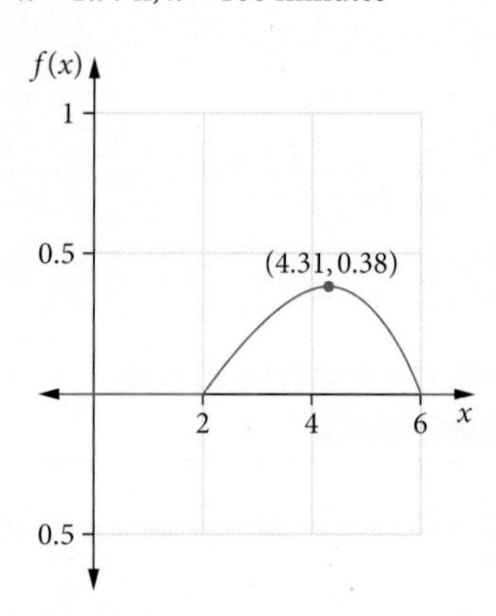

b 0.1211

17 **a** 14.66

b $\int_k^{k+1} \cos(t) + 1dt = [\sin(t) + t]_k^{k+1}$

$= \sin(k+1) + k + 1 - \sin(k) - k$

$= \sin(k+1) - \sin(k) + 1$

$1 = \sin(k+1) - \sin(k) + 1$

$\sin(k+1) = \sin(k),\ 0 < k < 2$

$k = 1.070796327\ldots$

$= \frac{\pi - 1}{2}$

18 **a** $\int_0^4 \frac{y}{16}\,dy + \lim_{k\to\infty}\left(\int_4^k 0.25e^{-0.5(y-4)}dy\right)$

$= \left[\frac{y^2}{32}\right]_0^4 + \lim_{k\to\infty}\left(\left[-0.5e^{-0.5(y-4)}\right]_4^k\right)$

$= 0.5 + \lim_{k\to\infty}\left(-0.5e^{-0.5(k-4)} + 0.5e^0\right)$

$= 0.5 + \lim_{k\to\infty}\left(-0.5e^{-0.5(k-4)} + 0.5e^0\right)$

$= 0.5 + 0 + 0.5$

$= 1$

All values of $f(y)$ are positive and the sum of all probabilities equals 1. Therefore, it is a valid pdf.

b 0.4155

EXERCISE 8.2

1 D **2** E **3** 6 **4** $\frac{14}{9}$

5 **a** $\int_0^1 k\sin(\pi x)\,dx = \left[-\frac{k}{\pi}\cos(\pi x)\right]_0^1$

$= \left(-\frac{k}{\pi}\cos(\pi) + \frac{k}{\pi}\cos(0)\right)$

$= \frac{k}{\pi} + \frac{k}{\pi}$

$= \frac{2k}{\pi}$

$\frac{2k}{\pi} = 1$

$k = \frac{\pi}{2}$

b $\frac{1}{2}$

6 $m = \sqrt{13}$ (reject $-\sqrt{13}$ as $m > 1$) or 3.61 (2 d.p.)

7 $k = \frac{2\sqrt{30}}{5}$ (reject $-\frac{2\sqrt{30}}{5}$ as $k > 0$) or 2.19 (2 d.p.)

8 $\text{Var}(X) = \frac{1}{18} = 0.0\dot{5}$

$\text{SD}(X) = \frac{\sqrt{2}}{6} = 0.2357$

9 a $\text{E}(Y) = \frac{2}{3}(3) - 1 = 1$

b $\text{Var}(Y) = \frac{4}{9}(9) = 4$

c $\text{SD}(Y) = 2$

10 a $\text{Var}(X) = \text{E}(X^2) - \text{E}(X)^2$

$\text{E}(X^2) = \text{Var}(X) + \text{E}(X)^2$

$= 5 + 2^2$

$= 9$

b 8

c $\frac{1}{5}$

11 a $\int_1^2 \frac{k}{x^2}\,dx = \left[-\frac{k}{x}\right]_1^2$

$= -\frac{k}{2} + \frac{k}{1}$

$= -\frac{k}{2} + \frac{2k}{2}$

$= \frac{k}{2}$

$\frac{k}{2} = 1$

$k = 2$

b $2\ln(2)$

12 $5\ln(2)$

13 a $\int_0^5 ax(5 - x)\,dx = a\int_0^5 5x - x^2\,dx$

$= a\left[\frac{5x^2}{2} - \frac{x^3}{3}\right]_0^5$

$= a\left(\frac{125}{2} - \frac{125}{3}\right)$

$= a\left(\frac{125}{6}\right)$

$\frac{125a}{6} = 1$

$a = \frac{6}{125}$

b Given that the pdf is a parabola with roots at $x = 0$ and $x = 5$, the line of symmetry has the equation $x = \frac{5}{2}$. So, $f(x)$ is symmetrical about $x = \frac{5}{2}$ over $0 \le x \le 5$ and, hence, $\text{E}(X) = \frac{5}{2}$.

c $\text{Var}(X) = \text{E}(X^2) - \text{E}(X)^2$

$= \frac{6}{125}\int_0^5 5x^3 - x^4\,dx - \left(\frac{5}{2}\right)^2$

$= \frac{6}{125}\left[\frac{5x^4}{4} - \frac{x^5}{5}\right]_0^5 - \frac{25}{4}$

$= \frac{6}{125}\left(\frac{5^5}{4} - \frac{5^5}{5}\right) - \frac{25}{4}$

$= \frac{6}{125}\left(\frac{5^5}{20}\right) - \frac{25}{4}$

$= \frac{6(25)}{20} - \frac{25}{4}$

$= \frac{6(25)}{20} - \frac{5(25)}{20}$

$= \frac{25}{20}$

$= \frac{5}{4}$

$\text{SD}(X) = \frac{\sqrt{5}}{2}$

14 $m = 1.2285$

15 a $\text{E}(X) = \frac{4}{9} \approx 27$ minutes

b $\text{SD}(X) = \sqrt{\frac{13}{162}} = 0.2833 \approx 17$ minutes

16 a $\lim_{k\to\infty}\left(\int_0^k \frac{1}{8}e^{-\frac{x}{8}}\,dx\right) = \lim_{k\to\infty}\left(\left[-e^{-\frac{x}{8}}\right]_0^k\right)$

$= \lim_{k\to\infty}\left(-e^{-\frac{k}{8}} + 1\right)$

$= 1$

All values of $f(x)$ are positive and the sum of all probabilities equals 1. Therefore, it is a valid pdf.

b i $\text{E}(X) = 8$ **ii** $\text{Var}(X) = 64$

c $k = 2.301$

17 a 1

b $k = 2.5101$ (reject $k = 0.0501$ as $k > 1$)

c $\text{E}(X) = 2.0973$

d $\text{Var}(X) = 0.1760$

e $\text{SD}(Y) = 2\text{SD}(X) = 0.8392$

18 a $\text{E}(X) = 3.0458$

b $P(X > 4) = 0.2617$

$0.2617 \times 200 \approx 52$

19 a $\text{E}(W) = 306$ g

b $P(W < 306) = 0.5248$

c $\text{SD}(X) = 8$ g

d $P(W > 314) = 0.1792$

EXERCISE 8.3

1 E **2** A

3 a 0 **b** $\frac{5}{6}$ **c** $\frac{2}{5}$

ANSWERS

4 a $b - 7 = 13$

$b = 20$

b $f(x) = \begin{cases} \dfrac{1}{13} & 7 \le x \le 20 \\ 0 & \text{otherwise} \end{cases}$ **c** $\dfrac{1}{2}$

5 a $f(x) = \begin{cases} \dfrac{1}{50} & -2 \le x \le 48 \\ 0 & \text{otherwise} \end{cases}$ **b** 23

c $\dfrac{625}{3} = 208.\dot{3}$ **d** $\dfrac{25\sqrt{3}}{3} = 14.43$

6 a

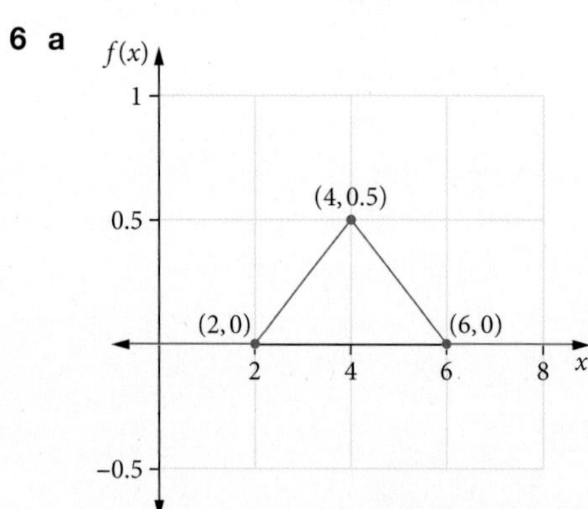

It is a symmetrical triangular distribution about $x = 4$.

b **i** $\dfrac{1}{8}$ **ii** $\dfrac{1}{4}$ **iii** 4 **iv** $\dfrac{2}{3}$

7 a $\int_0^{\frac{1}{2}} 2x\,dx + \int_{\frac{1}{2}}^{2} -\dfrac{2}{3}x + \dfrac{4}{3}\,dx$

$= \left[x^2\right]_0^{\frac{1}{2}} + \left[-\dfrac{x^2}{3} + \dfrac{4x}{3}\right]_{\frac{1}{2}}^{2}$

$= \dfrac{1}{4} + \left(-\dfrac{4}{3} + \dfrac{8}{3} + \dfrac{1}{12} - \dfrac{2}{3}\right)$

$= \dfrac{3 + 16 + 1 - 8}{12}$

$= 1$

All values of $f(x)$ are positive and the sum of all probabilities equals 1. Therefore, it is a valid pdf.

b $\int_0^{\frac{1}{2}} 2x\,dx + \int_{\frac{1}{2}}^{k} -\dfrac{2}{3}x + \dfrac{4}{3}\,dx = 0.7$

$\int_{\frac{1}{2}}^{k} -\dfrac{2}{3}x + \dfrac{4}{3}\,dx = 0.45$

$k = 1.05$ (reject $k = 2.95$ as $k < 2$)

c $\int_0^{\frac{1}{2}} 2x^2\,dx + \int_{\frac{1}{2}}^{2} -\dfrac{2}{3}x^2 + \dfrac{4}{3}x\,dx$

$= \left[\dfrac{2x^3}{3}\right]_0^{\frac{1}{2}} + \left[-\dfrac{2x^3}{9} + \dfrac{2x^2}{3}\right]_{\frac{1}{2}}^{2}$

$= \dfrac{1}{12} + \left(-\dfrac{16}{9} + \dfrac{8}{3} + \dfrac{1}{36} - \dfrac{1}{6}\right)$

$= \dfrac{3 - 64 + 96 + 1 - 6}{36}$

$= \dfrac{30}{36}$

$= \dfrac{5}{6}$

8 a $\text{E}(L) = 248$ mm

$\text{SD}(L) = 1.73$ mm

b $\text{P}(L > 250) = \dfrac{1}{6}$

c Let X be the number of pipes with length greater than 250 mm.

$X \sim \text{Bin}\left(60, \dfrac{1}{6}\right)$

$\text{E}(X) = 10$

d $\text{P}(X \ge 10) = 0.5536$

9 a $f(t) = \begin{cases} \dfrac{t}{3} & 0 \le t < 1.5 \\ -\dfrac{t}{5} + \dfrac{4}{5} & 1.5 \le t \le 4 \\ 0 & \text{otherwise} \end{cases}$

b $\text{P}(T < 1) = \dfrac{1}{6}$

c $\text{P}(1 < T < 3) = \dfrac{11}{15}$

10 a

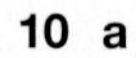

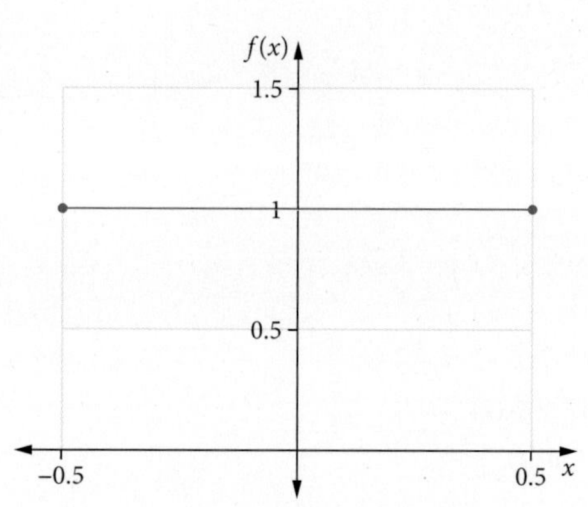

It is a continuous uniform distribution.

b $\text{P}(X \ge 0.35) = 0.15$

c $\text{P}(X < -0.35 \mid X < 0) = 0.3$

d $\text{P}(X^2 < 0.09) = \text{P}(-0.3 < X < 0.3) = 0.6$

e $\text{Var}(X) = \dfrac{1}{12}$

11 a

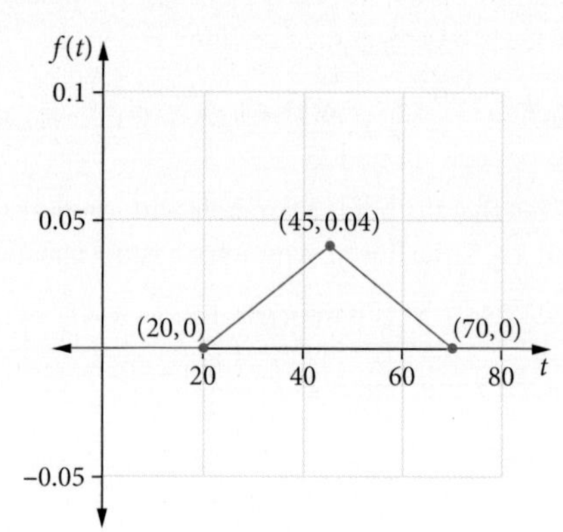

b It is a symmetrical triangular distribution about $t = 45$.

c **i** 0.8

ii 0.0244

iii $k = 39.3649$ (reject $k = 0.6351$ as $k > 20$)

12 a

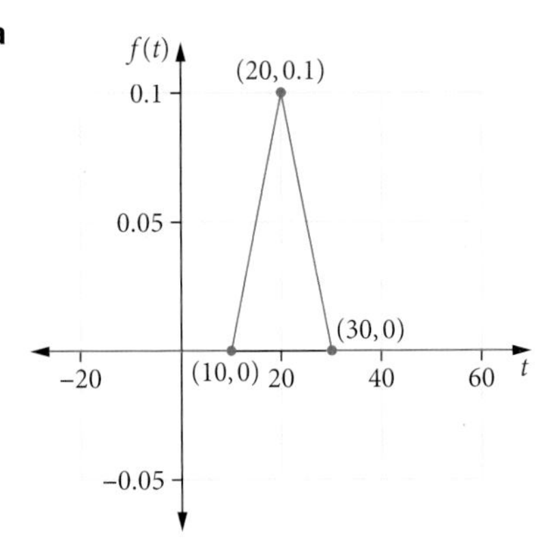

b It is a symmetrical triangular distribution about $t = 20$.

c **i** $\frac{7}{8}$ **ii** $\frac{1}{7}$

13 a $\frac{3}{8}$ **b** $\mu = 251, \sigma = 2.3094$

c Let X be the number of bottles containing less than 250 mL.

$X \sim \text{Bin}\left(15, \frac{3}{8}\right)$

$\text{P}(X = 5) = 0.2025$

14 a $\text{E}(W) = 249\,\text{g}, \text{SD}(W) = 2.3094$

b $\text{P}(W > 250) = \frac{3}{8}$

c Let X be the number of packets that weigh more than 250 g.

$X \sim \text{Bin}\left(100, \frac{3}{8}\right)$

$\text{E}(X) = 37.5$

approx. 38 packets

d $\text{P}(X \geq 25) = 0.9971$

15 a Using area under the curve, based on average height $-a \leq x \leq b$: $(a + b)\left(\frac{3}{4}\right) = 1 \Rightarrow a + b = \frac{4}{3}$.

Using area under the curve, calculating separately for the two parts of the function:

$\frac{1}{2}(a)(2a) + \frac{1}{2}(2a + b)(b) = 1 \Rightarrow a^2 + ab + \frac{b^2}{2} = 1$

b $a = \frac{\sqrt{2}}{3}, b = \frac{4 - \sqrt{2}}{3}$

c $1 - \text{P}(X \leq 0) = 1 - \left(\frac{\sqrt{2}}{3}\right)^2 = \frac{7}{9}$

EXERCISE 8.4

1 B **2** B

3 a 0.95 **b** 0.16 **c** 0.0015

4 a Three standard deviations give 2100 kg and $1500 - 2100 = -600$ gives a negative mass.
Not appropriate, as a mass of 0 kg is 2.14 standard deviations below the mean.

b The number of followers is a discrete variable.
Not appropriate, as not continuous.

5 a 0.8186 **b** 0.64

6 a **i** $z = -3$ **ii** $z = 2$ **iii** $z = 3.5$

b $1 - 0.975 = 0.025$

7 a 0.0056 **b** 0.2525 **c** 0.6107

8 a $c = 2166.86$ **b** $c = 2245.87$

9 $z = -0.5999, \mu = 170 - 50(-0.5999) = 199.99 \approx 200$

10 a 0.9332 **b** 0.3694 **c** 0.3959

d Let X be the number of eggs weighing more than 69 g being classified as Jumbo.

$X \sim \text{Bin}(80, 0.3959)$

$\text{P}(X = 40) = 0.0151$

11 a 0.5 **b** $b = -\frac{2}{3}$

12 a $b = -2$ **b** $\frac{0.34}{0.5} = 0.68$

13 a 0.16 **b** 0.34 **c** $\frac{0.16}{0.5} = 0.32$

14 $q - \frac{1}{2}$

15 a Her statement is valid as 142 cm is three standard deviations below the mean and 184 cm is three standard deviations above the mean. Therefore, approximately 99.7% of women will have heights in that range, i.e. almost all.

b 0.16%

c 149 cm

16 a **i** 2 **ii** C, as it has the largest spread.

b **i**

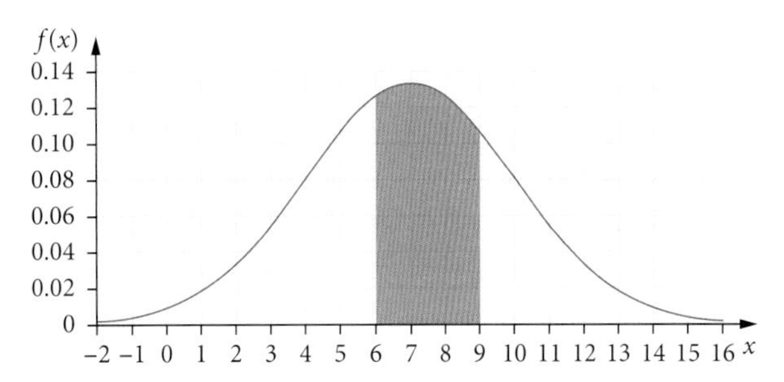

ii No. The total area below the probability density function is 1, and the region shaded above is less than half of that area (i.e. area is less than 0.5). Hence, it corresponds to a probability that is less than 0.5.

c Not normal: a continuous random variable has $\text{P}(Y \geq 2) = \text{P}(Y > 2)$. Since a normally distributed random variable is continuous, it follows that Y is not a normally distributed random variable. It could be binomial, as $\text{P}(Y \geq 2) = \text{P}(Y > 2)$ for a discrete random variable, as long as $\text{P}(Y = 2) \neq 0$. Since the binomial distribution is discrete, it follows that Y could be a binomially distributed random variable.

17 a 5.3 g **b** 0.6530

18 a $\mu = 81.71$ minutes, $\sigma = 16.24$ minutes

b 0.3554

19 a 0.0808 (or 8.08%)

b 0.9832 (or 98.32%)

c 76.45 g

20 a 0.0062

b Let X be the number of pizzas that are delivered free.

$X \sim \text{Bin}(50, 0.0062)$

$P(X > 3) = 0.0003$

c 31.2 minutes **d** 1.6 minutes

21 a 34.51%

b

Classification	Weight W (grams)	P(W)
Small	$W \le 110$	0.1418
Medium	$110 < W \le 155$	0.5131
Large	$155 < W \le 210$	0.3310
Extra large	$W > 210$	$0.3451 - 0.3310 = 0.0141$

c 0.2912

d Let Y be the number of small carrots in a bag.

$Y \sim \text{Bin}(12, 0.1418)$

$P(Y \le 2) = 0.7637$

22 a $a = 1.0134 \approx 1$ minute **b** 0.5467

c $k = -1.5$ or $k = -2.5$ **d** 0.0029

e i $1 - 0.85^n$ **ii** $n = 19$

CUMULATIVE EXAMINATION: CALCULATOR-FREE

1 a $\frac{1}{12}$ **b** $\frac{1}{3}$ **c** $\frac{1}{4}$

2 a

w	0	1	2	3
P($W = w$)	0.2	0.4	0.3	0.1

b i $E(W) = 1.3$ **ii** $\text{Var}(W) = 0.81$

3 a $x = 4.5$ **b** $x = 1.6$

4 a $\frac{8}{x} + c$ **b** $3\ln(x^2 + 5x - 11) + c$

c $\ln(5) - \ln(2)$

5 a $x\frac{\pi}{4}\cos\left(\frac{\pi}{4}x\right) + (1)\sin\left(\frac{\pi}{4}x\right)$

$= \frac{\pi x}{4}\cos\left(\frac{\pi x}{4}\right) + \sin\left(\frac{\pi x}{4}\right)$

b $E(X) = \int_0^2 x\frac{\pi}{4}\cos\left(\frac{\pi x}{4}\right)dx$

$= \int_0^2 \frac{d}{dx}\left(x\sin\left(\frac{\pi}{4}x\right)\right) - \sin\left(\frac{\pi x}{4}\right)dx$

$= \left[x\sin\left(\frac{\pi}{4}x\right) + \frac{4}{\pi}\cos\left(\frac{\pi}{4}x\right)\right]_0^2$

$= 2\sin\left(\frac{\pi}{2}\right) + \frac{4}{\pi}\cos\left(\frac{\pi}{2}\right) - 0 - \frac{4}{\pi}$

$= 2 - \frac{4}{\pi}$

6 a It is a continuous uniform distribution.

b i $p = 15, q = 40, k = \frac{1}{25}$

ii 27.5 minutes

iii $\frac{2}{5}$

c $\frac{3}{13}$ **d** 7:49 am

CUMULATIVE EXAMINATION: CALCULATOR-ASSUMED

1 a $SA = 2\left(\frac{5x^2}{2} + hx + \frac{5xh}{2}\right)$

$6480 = 5x^2 + 7xh$

$7xh = 6480 - 5x^2$

$h = \frac{6480 - 5x^2}{7x}$

b $0 < x < 36$

c $\frac{dV}{dx} = -\frac{75}{14}x^2 + \frac{16\,200}{7}$

d $x = 12\sqrt{3}, h = \frac{120\sqrt{3}}{7}$

2 a 25°C **b** 76.68°C

c 223°C **d** 8.63°C/min

e As time increases, the rate of change in the temperature of the water → 0.
The temperature of the water → the constant value of T_0.

3 a $\frac{9}{11}$ cm/s

b $h'(t)$ is of the form $\frac{f'(x)}{f(x)}$ (the numerator is the derivative of the denominator), so the function $h(t)$ is the natural logarithm of the denominator.

Also, $+ c$ needs to be included in the function, as any constant could be included here.

c $\Delta h = \int_0^2 \frac{4t + 1}{2t^2 + t + 1}dt$

$= \ln(11)$ cm (2.398)

d 5 seconds

4 a 0.0228

b Let X be the number of days that the process takes more than 2 hours.

$X \sim \text{Bin}(5, 0.0228)$

$P(X = 2) = 0.004\,83$

c i

Job duration T (minutes)	$T \le 60$	$60 < T < 120$	$T \ge 120$
Probability	0.0228	0.9545	0.0228
Cost Y (\$)	200	600	1200

ii \$604.55 **iii** \$108.67

iv $\mu = 604.55a + b, \sigma = 108.67a$

5 a 24.2%

b Let M be the number of parcels that weigh more than 3.7 kg.

$M \sim \text{Bin}(20, 0.241\,96)$

$P(M \ge 10) = 0.010\,95$

c

x	≤ 1	$1 < x \le 2$	$2 < x \le 3$	$3 < x \le 4$	$x > 4$
y	\$5	\$6.50	\$8	\$9.50	\$12
P($Y = y$)	0.022 75 (accept 0.021 40)	0.135 91	0.341 34	0.341 34	0.158 66

d \$8.87

e 1.7471

f $\mu = 11.65, \sigma = 2.10$

g Three standard deviations below the mean is a weight of 0. $P(X < 0) > 0$, there is a non-zero probability (small) that the weight will be negative.

CHAPTER 9

EXERCISE 9.1

1 a all pies produced at the local pie factory

b 6

c mode = 110 g, median = 103.5 g, mean = 104.3 g, range = 12 g

2 Survey every 10th customer who check-out. Other answers are possible.

3 a i Spatial bias: people are being asked about flying at the airport and only one airport in the country is surveyed.

ii Use a random number generator to ask randomly selected people upon their check-in/bag drop at various airports around the country. Other answers are possible.

b i Temporal bias: customers are only asked on one day for a one-hour window of time.

ii Ask every 15th customer in line at the self-service check-out across a larger time frame, or across multiple days, or across multiple stores. Other answers are possible.

4 Answers will vary depending on simulation. $E(X) = 45$, $SD(X) = 17.32$

5 Answers will vary depending on simulation. $E(Y) = 40$, $SD(Y) = 2.5$

6 Answers will vary depending on simulation. $E(Z) = 0.35$, $SD(Z) = 0.48$

7 Answers will vary depending on simulation. $E(T) = 25$, $SD(T) = 4.08$

8 a The method is biased due to the people being asked a leading question; the specific time and location used for the survey.

b In this case the question is not biased; however, only mobile phone users were selected, causing bias. Also, many of these people may just hang up, causing non-response bias.

9 Tina could use a random number generator and pick the sample using the numbers she obtains. Other answers are possible.

10 a Any two of the following three reasons:

1 Spatial: only one location, so only those present in that mall will be sampled from.

2 Temporal: lunchtime, so only those present at lunchtime will be sampled from.

3 Selection scheme: quota sampling means that the first 400 workers only are in the sample, so this is not a random sample from all workers.

b Either of the following two reasons:

1 Only those with listed telephone numbers will be selected.

2 Non-response bias: Not everyone will answer their phone when called.

c Amir's, as it samples randomly from the population of workers.

11 1 Temporal: the sample taken is at a fixed time, so only people around at that time will be sampled.

2 Spatial: the location is fixed, so only people at that location will be sampled or not everyone from the suburb will pass by that area, so this is not a random sample of the residents.

12 a The shapes of the distributions are seemingly different, and the ranges of values that the variable takes are not consistent in the samples (e.g. Sample A has scores between 8 and 18, Sample B has scores between 74 and 83 and Sample C has scores between 20 and 65).

b i Sample A

ii $p < 0.5$, as the distribution is positively skewed.

c i Sample B

ii approximately 80

d i Sample C

ii The distribution of the sample should become more uniform; that is, there could be less variation in the heights of the columns and a 'flatter' distribution.

EXERCISE 9.2

1 a supporters of a local football team

b Any of the following reasons:

1 Spatial: only supporters at the game were asked.

2 Spatial: groups of people standing together were asked, meaning they could have travelled together using the same mode of transport.

3 Temporal: data is only collected from one match.

2 C

3 a $\frac{28}{156} = 0.1795 = 17.95\%$

b approx. 176 students

4 a 0.8 **b** 0.0013 **c** 0.036

5 a $\frac{9}{200} = 0.045 = 4.5\%$

b i 0.045 **ii** 0.0002 **iii** 0.015

6 18 000

7 Answers will vary depending on simulation

a The distribution should be slightly positively skewed.

b The distribution should be slightly negatively skewed.

c The distribution should be very negatively skewed.

8 Answers will vary depending on simulation

a The distribution should be positively skewed.

b The distribution should be approximately symmetrical.

c The distribution should be negatively skewed.

9 a 0.4545

b $E(\hat{p}) = 0.4545$, $SD(\hat{p}) = 0.1062$

c Given that $\hat{p} \approx 0.5$, the distribution is approximately symmetrical and, since there are at least 10 each of positive and negative observations, the distribution is approximately normal with $\hat{p} \sim N(0.4545, 0.1062^2)$ (4 d.p.).

10 a Given that n is sufficiently large and that $np = 122 \geq 10$ and $n(1-p) = 378 \geq 10$, $\hat{p}$ is approximately normal with $\hat{p} \sim N(0.244, 0.0192^2)$ (4 d.p.).

b i $P(\hat{p} > 0.25) = 0.3774$

ii $P(0.15 < \hat{p} < 0.25) = 0.6226$

11 a $\frac{8}{125} = 0.064 = 6.4\%$

b Despite the sufficiently large sample size, $np < 10$, and so the distribution may not be appropriately symmetrical for the normal model. Additionally, $3SD(\hat{p}) = 0.0657$ and so, $0.064 - 3SD(\hat{p}) < 0$ meaning a normal distribution is not appropriate.

c Let X be the number of non-sellable items.

$X \sim \text{Bin}\left(125, \frac{8}{125}\right)$

$P(X > 5) = 0.8178$

d $\hat{p} \sim N(0.064, 0.0219^2)$

$P\left(\hat{p} > \frac{5}{125}\right) = 0.8635$

The normal approximation gives a probability that is 0.05 greater than the discrete binomial calculation.

12 a Given that n is sufficiently large and $np = 49.2 \geq 10$ and $n(1-p) = 70.8 \geq 10$, it is appropriate to model $\hat{p}$ using an approximate normal distribution.

b $\hat{p} \sim N(0.41, 0.0449^2)$, $z = 2.00$
i.e. approx. 2 standard deviations above the mean

13 a $n = 80$ **b** $k = 1.0811$ **c** $\hat{p} = 0.43$

14 a 9

b Given that $np = 9 < 10$, the distribution may not be appropriately symmetrical for an approximate normal distribution.

c $\binom{10}{2}(0.12)^2(0.88)^8$

15 If $\hat{p} = 0$, then $X = 0$.

$\binom{5}{0}p^0(1-p)^5 = \frac{1}{243} \Rightarrow 1 - p = \frac{1}{3} \Rightarrow p = \frac{2}{3}$

16 a $n = 25$ **b** $p = \frac{1}{5}$

17 a $\frac{35}{60} = 0.58\dot{3} = 58.\dot{3}\%$

b $E(\hat{p}) = 0.58$, $SD(\hat{p}) = 0.0636$

c Given that $p \approx 0.5$ and $n = 60$ is sufficiently large, the distribution of $\hat{p}$ should be approximately normal with a mean of 0.58. The distribution may be slightly negatively skewed.

18 a $\hat{p} \sim N(0.02, 0.0082^2)$, $P(\hat{p} \leq 0.03) = 0.888$

b $0.02 - 3SD(\hat{p}) = 0.02 - 3(0.0082) = -0.0047 < 0$ or $np = 5.8 < 10$

c Increase; as $n \to \infty$, $SD(\hat{p}) \to 0$ and so the distribution will become narrower, giving a greater area under the curve for $\hat{p} \leq 0.03$.

19 a Given that n is sufficiently large and $p \approx 0.5$, then $\hat{p}$ is approximately normal such that $\hat{p} \sim N(0.6, 0.0219^2)$.

b $P(\hat{p} < 0.58) = 0.1807$

c $\hat{p} = \frac{250}{500} = 0.5$; that is, half of the mangoes.

20 a Given that n is sufficiently large and $p \approx 0.5$, then $\hat{p}$ is approximately normal such that $\hat{p} \sim N(0.4, 0.0245^2)$.

b $P(\hat{p} > 0.44) = 0.0512$

21 a $\hat{p} = \frac{124}{400} = 0.31 = 31\%$

b Using $p = \frac{7}{24}$, $E(\hat{p}) = \frac{7}{24} = 0.2917$, $SD(\hat{p}) = 0.0227$

c $\hat{p} = 0.2917 \pm 0.6(0.0227)$, $400\hat{p} = 111$ or 122

EXERCISE 9.3

1 D **2** D

3 a Given that $\hat{p} = 0.5$, the distribution is approximately symmetrical, and n is sufficiently large ($n \geq 30$), so $\hat{p}$ is approximately normal such that $\hat{p} \sim N(0.5, 0.0707^2)$.

b i $0.5 - 1.645\sqrt{\frac{0.5^2}{50}} \leq p \leq 0.5 + 1.645\sqrt{\frac{0.5^2}{50}}$

ii $0.5 - 1.960\sqrt{\frac{0.5^2}{50}} \leq p \leq 0.5 + 1.960\sqrt{\frac{0.5^2}{50}}$

iii $0.5 - 2.576\sqrt{\frac{0.5^2}{50}} \leq p \leq 0.5 + 2.576\sqrt{\frac{0.5^2}{50}}$

4 $[0.614, 0.743]$

5 a $[0.312, 0.421]$

b Neither one is more likely than the other to contain p, as the probability that a confidence interval contains p is either 0 or 1. Hence, it cannot be determined.

c approx. 228 confidence intervals

6 a If confidence level decreases, z decreases, and so margin of error decreases and, hence, width of the confidence interval decreases.

b If sample size increases, standard error decreases, and so margin of error decreases and, hence, width of the confidence interval decreases.

7 a $\hat{p} = 0.483$ **b** $E = 0.075$

c $SD(\hat{p}) = 0.046$ **d** $n \approx 120$

8 $z = 2.572$, $P(-2.572 < z < 2.572) = 0.9899 \approx 99\%$

9 $n \geq 310$

10 a assume worst case scenario and $\hat{p} = 0.5$ for maximum width; $n \geq 1068$

b assume worst case scenario and $\hat{p} = 0.5$ for maximum width; $n \geq 68$

11 a $[0.3669, 0.5931]$. Given that $0.4 \in [0.3669, 0.5931]$, there is insufficient evidence to suggest that the claimed value of p shouldn't be accepted. It could be 40%, but it cannot be known for certain.

b $[0.3285, 0.5049]$. Given that $0.30 \notin [0.3285, 0.5049]$, there is insufficient evidence to suggest that the piano tuning demands north and south of the Swan River are the same.

12 a $[0.2247, 0.5753]$

b $\hat{p} \sim N(0.4, 0.0894^2)$. $P(\hat{p} < 0.24) = 0.0368$. There is a 3.68% chance of randomly selecting a sample in which less than 24% of the people surveyed were in favour of adopting a new state flag.

c A 95% confidence interval using $\hat{p} = 0.24$ and $n = 50$ gives $[0.1216, 0.3584]$. Given that the two confidence intervals overlap with $0.24 \in [0.2247, 0.5753]$, there is insufficient evidence to suggest that the claim is true and that the newly proposed designs weren't better than the current one.

13 $\frac{2}{3} - 1.645\sqrt{\frac{\left(\frac{2}{3}\right)\left(\frac{1}{3}\right)}{18}} \leq p \leq \frac{2}{3} + 1.645\sqrt{\frac{\left(\frac{2}{3}\right)\left(\frac{1}{3}\right)}{18}}$

14 4

15 1800

16 a $[0.013, 0.107]$

b The distribution of $\hat{p}$ might not be approximately normal, as $\hat{p} = 0.06$ and $np = 6 < 10$, meaning that the distribution might be positively skewed and, hence, a normal distribution might not be appropriate, or

$0.06 - 3\text{SD}(\hat{p}) = 0.06 - 3(0.0237) < 0$ and so, a normal distribution is not appropriate.

17 $[0.730, 0.789]$

18 As level of confidence increases, z increases and so margin of error increases. Therefore, smallest margin of error will be obtained from the 90% confidence interval. $E = 0.10468$

19 a $\hat{p} = \frac{15}{320} = 0.0469 = 4.69\%$

b $[0.0274, 0.0663]$

c 0.0194

20 a

$$E = \sqrt{\hat{p}(1 - \hat{p})}$$

$$\frac{dE}{dp} = \frac{(1 - 2\hat{p})}{2\sqrt{\hat{p}(1 - \hat{p})}}$$

$$0 = 1 - 2\hat{p}$$

$$\hat{p} = \frac{1}{2}$$

$\frac{d^2E}{d\hat{p}^2}, \hat{p} = \frac{1}{2} = -2 \Rightarrow$ concave down and so maximum.

b $n \geq 260$

c approx. 63

21 a 9604

b Any two of the following:

Sample size: increasing the sample size, decreasing the width.

Sample proportion: as the sample proportion moves away from 0.5, the width decreases.

Confidence level: as confidence level increases, the width increases.

22 a 9604

b 0.058; that is, within 5.8%

23 a $[0.4562, 0.5438]$

b Given that $0.6 \notin [0.4562, 0.5438]$, then it is unlikely that Tina is correct.

c Take another random sample and obtain another approximate 95% confidence interval for p.

24 a $[0.0227, 0.1773]$

b $[0.0507, 0.1493]$

c Given that 0.06 lies in both confidence intervals, then there is insufficient evidence to suggest that the company's historical data is no longer relevant.

25 a $[0.0562, 0.1038]$

b Given that 0.05 does not lie in the confidence interval, then there is insufficient evidence to suggest that the company's historical data is still relevant; it may be outdated, but it cannot be known for certain.

26 a Given that n is sufficiently large and $np = 43 \geq 19$ and $n(1 - p) = 957 \geq 10$, then $\hat{p}$ is approximately normal such that $\hat{p} \sim N(0.043, 0.0064^2)$.

b $[0.030, 0.056]$

c Even though 0.04 lies within the confidence interval $[0.030, 0.056]$, it cannot be inferred that the reviewer visited that specific hall.

27 a $\hat{p} = \frac{56}{250} = 0.224$

b $[0.1723, 0.2757]$

c $E = 0.0517$

d No, even though 0.17 does not lie in the confidence interval $[0.1723, 0.2757]$, it cannot be inferred that the junior staff member made a mistake. Due to the nature of random sampling and the construction of 95% confidence interval, it is expected that 95% of the 95% confidence intervals are expected to contain the true proportion and 5% will not.

28 a The 95% confidence interval will have a larger width. $(0.04, 0.16)$ has a width of 0.12 and $(0.05, 0.15)$ has a width of 0.10. Therefore, $(0.04, 0.16)$ is the 95% confidence interval.

b $\hat{p} = 0.1$ meaning that 10 wins were observed out of 100 games.

c $\sqrt{\frac{p(1 - p)}{n}} \rightarrow \sqrt{\frac{p(1 - p)}{4n}} \Rightarrow \sqrt{\frac{p(1 - p)}{n}} \rightarrow \frac{1}{2}\sqrt{\frac{p(1 - p)}{n}}$

Assuming that p remained approximately 0.1, it would be expected that the width of the confidence interval will reduce by a factor of 2 (i.e. halved).

d No mistake has necessarily been made, as not all confidence intervals are expected to contain the true value of p. 90% of 90% confidence intervals and 95% of 95% confidence intervals are expected to contain the true value of p.

29 **a** $[0.660, 0.700]$

It is assumed that $\hat{p}$ is approximately normally distributed such that $\hat{p} \sim N(0.68, 0.0123^2)$.

b 0.0201

c Survey 2: $[0.536, 0.565]$, Survey 3: $[0.632, 0.672]$, Survey 4: $[0.670, 0.697]$

d Given that there is no overlap between the confidence intervals for Survey 1 and Survey 2, there is sufficient evidence to suggest that the samples may have come from different populations, i.e. that Survey 2 did not come from Western Australia, but we cannot say for certain.

e $n = 5800$

30 **a** $X \sim \text{Bin}(30, 0.8)$, $E(X) = 24$ and $SD(X) = 2.191$

b $P(X \geq 27) = 0.1227$

c Given that n is sufficiently large and $np = 450 \geq 10$ and $n(1 - p) = 150 \geq 10$, then $\hat{p}$ is approximately normally distributed such that $\hat{p} \sim N(0.75, 0.0177^2)$. Therefore, $[0.7154, 0.7846]$.

d Given that $0.8 \notin [0.7154, 0.7846]$, then there is insufficient evidence to suggest that there isn't an error with the machine's calibration and it may be off, but it cannot be known for certain.

e Not all confidence intervals are expected to contain the true value of p and it is expected that 95% of confidence intervals to be constructed will contain p. It can be due to the nature of random sampling that 0.8 did not lie in the first confidence interval.

CUMULATIVE EXAMINATION: CALCULATOR-FREE

1 **a** $np = 20$, $np(1 - p) = 4$ **b** $n = 25$, $p = \frac{4}{5}$

2 **a** $3x^2 \ln(3x) + x^2$ **b** $\frac{1}{3}x^3 \ln(3x) - \frac{1}{9}x^3 + c$

3 **a**

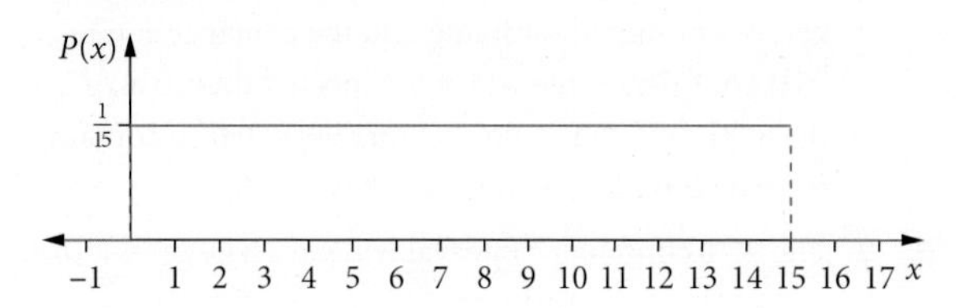

b $\frac{3}{15}$ **c** $\frac{7}{15}$ **d** $\frac{5}{8}$

4 **a** $(\ln(2), 0)$ $(\ln(4), 0)$

b $f'(x) = 2e^{2x} - 6e^x$

$f'(x) = 2e^x(e^x - 3) = 0$, stationary point $(\ln(3), -1)$

c $\left(\ln\left(\frac{3}{2}\right), \frac{5}{4}\right)$

d

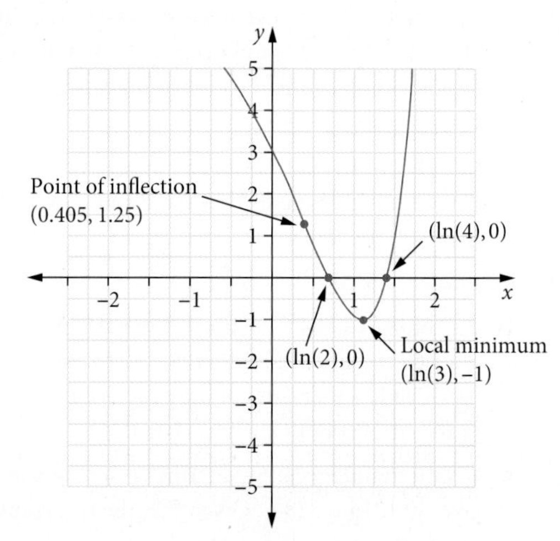

CUMULATIVE EXAMINATION: CALCULATOR-ASSUMED

1 0.121 cm

2 **a** $f'(x) = -0.6(x - 4)(x - 2)$

b **i** $y = -1.8x + 2.6$ **ii** $Q\left(\frac{13}{9}, 0\right), S\left(0, \frac{13}{5}\right)$

c 21.6 units^2

3 **a** **i** X represents the number of questions Steve answers correctly.

$X \sim \text{Bin}(3, 0.25)$

$\Pr(X = 3) = \frac{1}{64}$

ii $X \sim \text{Bin}(20, 0.25)$

$P(X \geq 10) = 0.0139$

iii Proof: see worked solutions

b $\frac{17}{64}$

c $X \sim \text{Bin}(25, p)$

$P(Y > 23) = 6P(Y = 25)$

$P(Y = 24) + P(Y = 25) = 6P(Y = 25)$

$P(Y = 24) - 5P(Y = 25) = 0$

${}^{25}C_{24}\, p^{24}(1 - p)^1 - 5 \times {}^{25}C_{25}\, p^{25}(1 - p)^0 = 0$

$25p^{24}(1 - p) - 5p^{25} = 0$

$25p^{24} - 25p^{25} - 5p^{25} = 0$

$25p^{24} - 30p^{25} = 0$

$5p^{24}(5 - 6p) = 0$

$p = 0, p = \frac{5}{6}$

As $p > 0$, $p = \frac{5}{6}$.

4 **a** 150 words **b** 69 words **c** 329 words

d 89 days **e** $t = e^{0.01w - 1.5} - 1$

9780170477536

5 **a** **i** $0 = \frac{1}{200}(8a + 4b + c)$

$-0.06 = \frac{1}{200}(12a + 4b)$

$0 = \frac{1}{200}(48a + 8b)$

ii Proof: see worked solutions

b **i** $M(2 + 2\sqrt{3}, 0)$, $P(2 - 2\sqrt{3}, 0)$

ii $2\sqrt{3}$ km

iii 80 m

c $k = -\frac{w}{\log_e(7)}$ **d** $w = 120$ **e** 0.2 km

6 **a** $X \sim \text{Bin}(200, 0.01)$

b 0.0517

c $[0.0006, 0.0394]$

d Given that $0.01 \in [0.0006, 0.0394]$, there is insufficient evidence to suggest that the historical data isn't relevant to current standards.

e Given that there are values of $\hat{p}$ within $[0.0006, 0.0394]$ that exceed 0.01, the department cannot conclude definitively that the targets are being met.

7 **a** $\hat{p} \sim N(0.23, 0.0210^2)$

b $(\hat{p} < 0.2) = 0.07697$

c $\frac{55}{200} = 0.275 = 27.5\%$

d $E = 0.08133$

e $[0.2131, 0.3369]$

f Given that $0.23 \in [0.2131, 0.3369]$, then there is insufficient evidence to suggest that the proportion of voters likely to vote for the Sustainable Energy Party in this electorate has increased.

g 1 Voters either vote for the party or not (success or failure).

2 The voters likely to vote for the Sustainable Energy Party are independent of each other. This is a reasonable assumption.

3 The probability of a voter likely to vote for the Sustainable Energy Party is the same for all voters. This is most likely not valid, as the probability may depend on other factors such as the age of the voter, occupation, socio-economic status, and/or employment status.

Glossary and index

68–95–99.7% rule The rule describing the percentages of area under the normal distribution curve one, two and three standard deviations either side of the mean. (p. 290)

acceleration The rate of change of velocity of a moving object, represented by the function $a = \frac{dv}{dt}$, where v is the velocity ('signed speed'). Conversely, $v = \int a(t)\,dt$. (p. 39)

addition rule for probability
$P(A \cup B) = P(A) + P(B) - P(A \cap B)$ (p. 149)

anti-derivative (or **integral** or **primitive**) The opposite of the derivative. The anti-derivative of $f(x)$ is a function $F(x)$ whose derivative is $f(x)$: $F'(x) = f(x)$. (p. 57)

anti-differentiation (or **integration**) The opposite of differentiation; that is, the process of finding the anti-derivative or integral. (p. 57)

(approximate) confidence interval for *p* An interval estimate for the population proportion of the form $\hat{p} - z\sqrt{\frac{\hat{p}(1-\hat{p})}{n}} \le p \le \hat{p} + z\sqrt{\frac{\hat{p}(1-\hat{p})}{n}}$ using an approximate normal distribution for $\hat{p}$. (p. 356)

approximate normal distribution For $\hat{p} = \frac{X}{n}$ where the distribution of $\hat{p}$ is approximately symmetrical based on n and p, then $\hat{p} \sim N\left(p, \frac{p(1-p)}{n}\right)$. When the transformation $Z = \frac{\hat{p} - p}{\sqrt{\frac{p(1-p)}{n}}}$ is applied, Z has an approximate standard normal distribution, $Z \sim N(0, 1)$. (p. 345)

array A rectangular grid that can be used to show the outcomes of a two-step experiment. (p. 148)

Bernoulli distribution A distribution that has two possible outcomes, $x = 1$ and $x = 0$. The outcome $x = 1$, described as success, has probability p and the outcome $x = 0$, described as failure, has probability $1 - p$. (p. 169)

Bernoulli trial (or **binomial trial**) A random experiment with two possible outcomes: 'success' or 'failure'. (p. 169)

bias A feature of a sampling method that makes it an unfair and non-representative sample of the population. This can include spatial (location), temporal (time), self-selection (volunteer), non-response or leading question bias. (p. 318)

binomial distribution A special discrete probability distribution of the results of a series of Bernoulli trials. The Bernoulli distribution is a special case of the binomial distribution where the number of trials (n) is 1. (p. 170)

census The process used to collect data from an entire population. (p. 318)

central limit theorem For a sufficiently large n and a suitable p, $X \sim \text{Bin}(n, p)$ can be modelled by an approximately normal distribution $X_N \sim N(np, np(1-p))$. (p. 344)

chain rule A formula for finding the derivative of a composite function: If $y = f(u)$ and $u = g(x)$, then $\frac{dy}{dx} = \frac{dy}{du} \times \frac{du}{dx}$. (p. 13)

composite function A function of a function, when the results of one function are substituted into another function. (p. 13)

composition of functions If $y = f(x)$ and $w = g(x)$, then z, the composition of functions y and w, is $f(g(x))$. (p. 13)

compound event Two or more simple events being considered together in probability, such as a bus running late and it being a rainy day. (p. 146)

concavity The shape of the curve of a graph in an upward or downward direction. (p. 26)

conditional probability The probability that an event occurs given that another event also occurs; for example, the probability that a soccer team will win a match given that it is a rainy day. $P(A \mid B) = \frac{P(A \cap B)}{P(B)}$, where $P(A \mid B)$ means 'the probability of A occurring, given B occurs'. (p. 150)

confidence level The percentage of a large number of yet to be constructed confidence intervals from random samples of size n that can be expected to contain the true value of the population parameter. (p. 356)

constant A value that is fixed. A number without a variable. (p. 57)

constant of integration A constant value ('+ *c*') added to the anti-derivative of a function. (p. 57)

continuous random variable A random variable that can take the value of any real number over a given continuous interval. (p. 153)

cumulative distribution function The anti-derivative of a probability density function defined as $F(x) = P(X \le x) = \int_a^x f(t)\,dt$. This function describes the cumulative probability distribution of a continuous random variable X. (p. 256)

cumulative probability distribution *See* **cumulative distribution function**.

cumulative relative frequency The sum of the relative frequencies in a frequency table. (p. 249)

decreasing function When a function has a negative gradient $\left(\frac{dy}{dx} < 0\right)$ and its graph is pointing in a downward direction from left to right. (p. 29)

definite integral An integral of the form $\int_a^b f(x)\,dx$, whose value is the (signed) area under a curve (shaded A in the diagram below) and is read 'the integral of $f(x)$ between a and b with respect to x'. (p. 69)

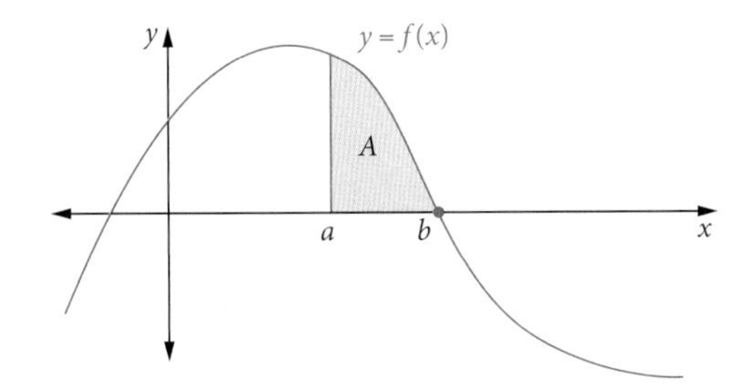

derivative For the function $y = f(x)$, the derivative of y, $\frac{dy}{dx}$ or $f'(x)$, is the function of the instantaneous rate of change of $f(x)$, also called the **gradient function**. (p. 3)

differentiation The process of finding the derivative of a function. (p. 3)

discrete random variable A random variable that can only take a countable number of values; for example, the number of pets in a household. (p. 153)

displacement The 'signed distance' from the origin of a moving object, represented by the function $x(t)$, where displacement is a function of time. (p. 39)

expected value, E(X) The mean of a probability distribution. For discrete probability distributions, $E(X) = \Sigma x \cdot p(x)$. (p. 159)

experimental probability *See* **relative frequency**.

exponent Another name for power. For example, in 2^5, the exponent is 5. (p. 189)

exponential decay A decrease that happens quickly at first, then slows, according to the function $y = be^{kt}$, where t is time and k is a negative constant. (p. 101)

exponential function A function of the form $y = a^x$ where the variable is in the exponent (power). (p. 99)

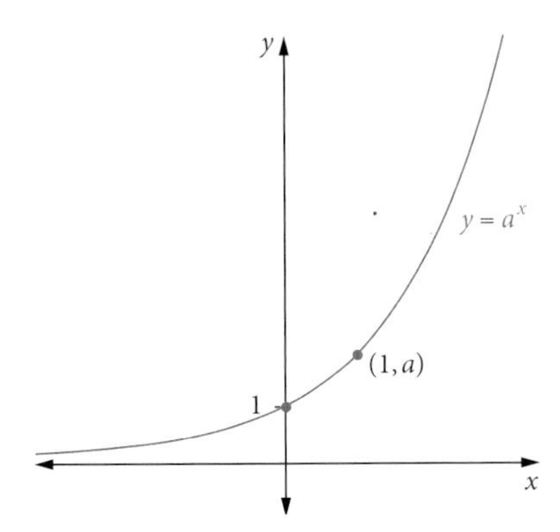

exponential growth An increase that happens slowly at first, then more quickly, according to the function $y = be^{kt}$, where t is time and k is a positive constant. (p. 100)

fair and representative sample A sample with minimal bias and is representative of the population from which it is taken. (p. 318)

frequency histogram A display that shows the frequencies of the outcomes of a continuous numerical data set, whereby the variable being measured is on the horizontal axis and the frequency is on the vertical axis. When the vertical axis measures relative frequency, it is called a relative frequency histogram. (p. 247)

grid A rectangular grid that can be used to show the outcomes of compound events. (p. 148)

increasing function When a function has a positive gradient $\left(\frac{dy}{dx} > 0\right)$ and its graph is pointing up from left to right. (p. 29)

increments formula If δx is small, we can say $\frac{dy}{dx} \approx \frac{\delta y}{\delta x}$, rearranging gives $\delta y \approx \frac{dy}{dx} \times \delta x$. (p. 24)

indefinite integral An integral of the form $\int f(x)\,dx$, which is an anti-derivative function and is read 'the integral of $f(x)$ with respect to x'. (p. 69)

independent events Events whose outcomes (and probabilities) do not depend on each other; for example, the number rolled on a second die does not depend on the number rolled on the first die. For two independent events A and B, $P(A \cap B) = P(A) \times P(B)$. (p. 148)

instantaneous rate of change The rate of change at a point (x_1, y_1) on the graph of a function $f(x)$, given by the gradient of the tangent to the graph at that point, which is the derivative of the function at that point, $f'(x_1)$. (p. 3)

integral (or **anti-derivative** or **primitive**) The opposite of the derivative. The anti-derivative of $f(x)$ is a function $F(x)$ whose derivative is $f(x)$: $F'(x) = f(x)$. (p. 57)

See also **definite integral** and **indefinite integral**.

integrand The function being integrated; the $f(x)$ in $\int f(x)\,dx$. (p. 57)

integration (or **anti-differentiation**) The limiting sum of rectangles under a curve; the opposite of differentiation, the process of finding the anti-derivative or integral. (p. 57)

interval estimate for p A range of possible values that p may fall within determined by a single sample proportion $\hat{p}$. (p. 355)

inverse normal distribution For a normal random variable X, the inverse normal distribution is used to solve quantile problems of the form $P(X \le k) = p$ where k is unknown. (p. 299)

laws of logarithms The rules involving the logarithm of xy, $\frac{x}{y}$, and x^n, related to the index laws. (p. 192)

limit The value of the function $f(x)$ at $x = a$ as x gets infinitely closer to a. (p. 69)

logarithm The exponent (power) to which a fixed number, the base, must be raised to give a number or variable. For example, $2^5 = 32$ so $\log_2(32) = 5$. (p. 189)

logarithmic equation An equation involving $y = \log_a(x)$. (p. 194)

logarithmic function A function of the form $y = \log_a(x)$. (p. 197)

logarithmic scale A scale in which we multiply by the same number to move from one scale mark to the next. In the example below, the log scale involves multiplying by 10 each time.

×10 ×10 ×10

1 10 100 1000 10 000 100 000

This is called a 'log base 10' or '$\log_{10}$' scale. (p. 205)

margin of error (of a confidence interval) The difference between the value of the sample proportion $\hat{p}$ and the upper and lower bounds of a confidence interval constructed using that sample proportion, defined as $E = z\sqrt{\frac{\hat{p}(1-\hat{p})}{n}}$. (p. 356)

maximum rate of increase or decrease (of a function) Where a function is increasing or decreasing most rapidly; where the gradient of its graph is steepest in the positive and negative directions. (p. 126)

mean The expected value or mean of a probability distribution is a measure of centre. For a discrete probability distribution $E(X) = \Sigma x \cdot p(x)$. (p. 159)

measure of centre A central or typical value for a probability distribution. The three measures of centre are mean, mode and median. They provide a measure of an expected or likely outcome for the distribution. (p. 159)

median The middle value of a numerical data set when in ascending or descending order. For a continuous random variable X, the median m is given by $P(X \le m) = 0.5$. (p. 159)

mode The value of a numerical data set with the highest relative frequency. For a continuous random variable X with a probability density function $f(x)$, the mode $x = c$ gives the maximum value of $f(x)$. (p. 159)

natural exponential function A function of the form $y = e^x$ which is used to describe exponential growth and decay. (p. 100)

natural logarithmic function A function of the form $y = \log_e(x) = \ln(x)$ such that $e^y = x$, where e is the exponential constant approximately equal to 2.718. (p. 212)

non-probability sampling method A sampling procedure through which each member of the population does not have an equally likely chance of being chosen. These include convenience sampling, quota sampling and volunteer sampling. (p. 319)

normal distribution curve The curve defined by the probability density function for a normally distributed random variable $X \sim N(\mu, \sigma^2), f(x) = \frac{1}{\sigma\sqrt{\pi}} e^{-\frac{1}{2}\left(\frac{x-\mu}{\sigma}\right)^2}$. (p. 291)

normally distributed random variable A continuous random variable with a symmetrical, bell-shaped distribution about its mean, median and mode whereby the oblique points of inflection of the curve occur at values that are one standard deviation, σ, either side of the mean. $X \sim N(\mu, \sigma^2)$. (p. 290)

parent distribution *See* **population distribution**.

percentile A value of x for which a certain percentage of scores, p%, fall below that value. That is, $P(X \le k) = \frac{p}{100}$ means that $x = k$ is the pth percentile. (p. 267)

point estimate A sample statistic that is used in place of an unknown population parameter. (p. 318)

point of inflection A non-stationary point of inflection is where a function changes concavity. (p. 26)

population The entire group to be studied, from which data could be collected. (p. 318)

population distribution The distribution of the population from which the sample data is being taken or simulated. (p. 321)

population parameter A characteristic or feature of the population. (p. 318)

population proportion p The fraction, decimal or percentage of the entire population that satisfy a particular success condition. (p. 335)

primitive *See* **anti-derivative**.

probability density function The function $f(x)$ that describes the probability distribution of a continuous random variable X. (p. 250)

probability distribution A function that gives the probabilities of all the possible values of a random variable. The probability distribution for a discrete random variable is a **probability mass function**. (p. 153)

probability sampling method A sampling procedure through which each member of the population has an equally likely chance of being chosen. These include simple random sampling, systematic sampling, stratified sampling and cluster sampling. (p. 319)

product rule A formula for finding the derivative of the product of two functions: If $f(x) = u(x) \times v(x)$, $\frac{d}{dx}(uv) = uv' + vu'$. (p. 8)

9780170477536

quantile A value of x for which a certain percentage of scores, $100p\%$, fall below that value. That is, $P(X \le k) = p$ means that $x = k$ is the p quantile. It is the decimal form of a percentile. (p. 299)

quotient rule A formula for finding the derivative of the ratio of two functions: If $f(x) = \frac{u(x)}{v(x)}$, $\frac{d}{dx}\left(\frac{u}{v}\right) = \frac{vu' - uv'}{v^2}$. (p. 11)

random variable A set of numerical quantities, X, whose value is determined by the outcome of a random chance experiment. Random variables can be discrete or continuous. (p. 153)

See also **discrete random variable**.

reciprocal function A function in the form of a fraction where the variable x, appears only in the denominator. (p. 228)

relative frequency The frequency of an outcome as a proportion of the total number of observations. (p. 247)

relative frequency histogram *See* **frequency histogram**.

sample A smaller group taken from the population, from which data is collected. (p. 318)

sample proportion $\hat{p}$ The fraction, decimal or percentage of the sample group that satisfies a particular success condition. (p. 334)

sample size n The number of members in the sample group. (p. 318)

sample statistic A characteristic or feature of the sample group. (p. 318)

sampling distribution of sample proportions The distribution of all possible $\hat{p}$ values when $\hat{p}$ is considered a random variable of the form $\frac{X}{n}$. (p. 335)

second derivative The rate of change of the first derivative function. In other words, the derivative of the derivative is called the second derivative. The second derivative is written as y'' or $f''(x)$ or $\frac{d^2y}{dx^2}$. (p. 25)

simulation The situation when physical or digital technology is used to model the events of a random probability experiment or mimic the data collection process of a sample from a population. (p. 321)

standard deviation A measure of spread of a distribution, equal to the square root of the variance. $\sigma = SD(X) = \sqrt{Var(X)}$ (p. 161)

standard error The standard deviation of the sampling distribution of sample proportions,

$SD(\hat{p}) = \sqrt{\frac{\hat{p}(1-\hat{p})}{n}}$. (p. 355)

standard normal distribution curve The curve defined by the probability density function for a standard normal random variable $Z \sim N(0,1), f(z) = \frac{1}{\sqrt{\pi}}e^{-\frac{1}{2}z^2}$. (p. 294)

standard normal random variable A normally distributed random variable with $\mu = 0$ and $\sigma = 1$, represented by $Z \sim N(0,1)$. (p. 293)

standardisation The process in which a normal random variable $X \sim N(\mu, \sigma^2)$ is transformed to the standard normal random variable $Z \sim N(0,1)$ using the linear transformation $Z = \frac{X-\mu}{\sigma}$. Each value of z corresponding to a value of x is called its standard score. (p. 294)

stationary point Where $f'(x) = 0$ for the function $y = f(x)$. A point on a graph where the gradient equals 0. The graph is flat at the stationary point, neither increasing nor decreasing: $\frac{dy}{dx} = f'(x) = 0$. A stationary point is either a turning point (local maximum or minimum point) or a stationary point of inflection. (p. 3)

stationary point of inflection A stationary point on a graph where the concavity changes but the sign of the gradient stays the same on both sides of the point. If $f''(x) = 0$, further investigation into the nature of the stationary point is required. (p. 29)

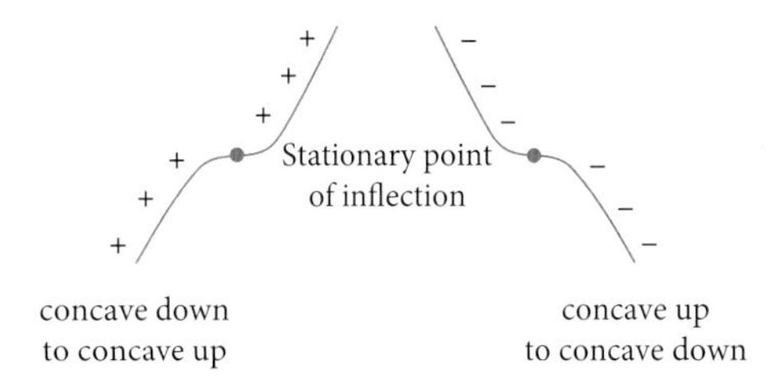

survey The process used to collect data from a sample group. (p. 318)

tangent The line which touches a curve at a point. (p. 3)

tree diagram A diagram of branches for listing all of the possible outcomes of a multi-step chance experiment. (p. 146)

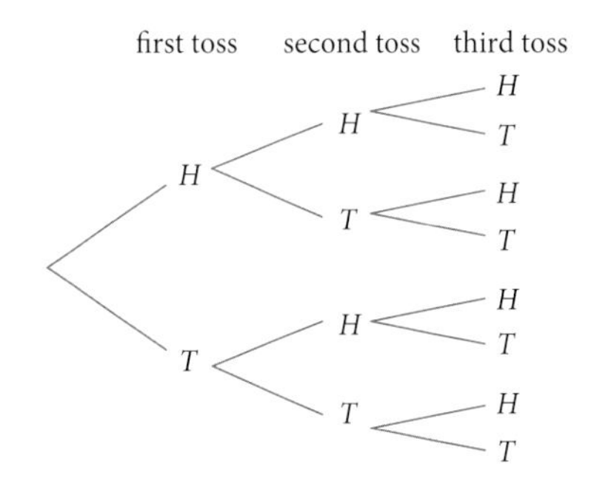

triangular continuous random variable A continuous random variable X defined over the interval $a \le x < b$ with a triangular-shaped graph, which can be symmetrical or asymmetrical. (p. 280)

turning point A stationary point on a graph where the sign of the gradient changes on either side. If it changes from negative (decreasing) to positive (increasing), it is a local minimum point. If it changes from positive (increasing) to negative (decreasing), it is a local maximum point. If $f''(x) > 0$, the turning point is a minimum. If $f''(x) < 0$, the turning point is a maximum. (p. 29)

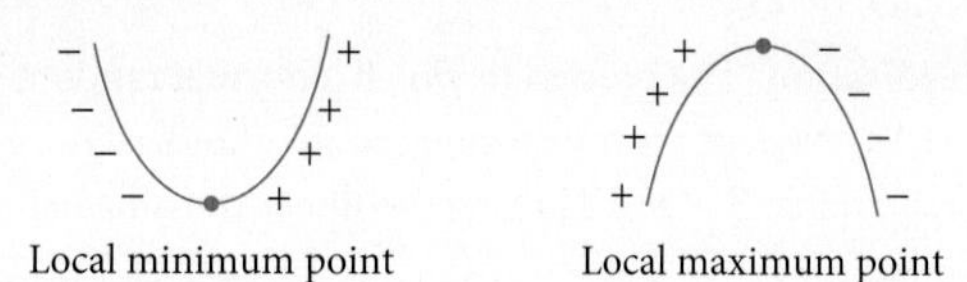

uniform continuous random variable A continuous random variable X defined over the interval $a \le x < b$ such that each value of x has an equally likely chance of occurring. It has a symmetrical, rectangular-shaped graph. (p. 275)

variability of random samples The natural differences that occur between the statistics and shapes of the distributions of different samples, due to the nature of random sampling. (p. 321)

variance A measure of the spread of a distribution, equal to the square of the standard deviation. $\sigma^2 = \text{Var}(X) = \text{E}(X^2) - \mu^2$, where $\mu = \text{E}(X)$ and $\text{E}(X^2) = \Sigma x^2 \cdot p(x)$ for a discrete probability distribution. (p. 161)

velocity The 'signed speed' of a moving object, represented by the function $v = \frac{dx}{dt}$; it is the rate of change of the displacement, x. Conversely, $x = \int v(t)\,dt$. (p. 39)

z-**score** *See* **standardisation**.